COLLINS

BRITISH
WILDLIFE

COLLINS

BRITISH
WILDLIFE

THE DEFINITIVE GUIDE TO BRITAIN'S PLANTS AND ANIMALS

Collins

This book was first published under the title *What is That?* by The Reader's Digest Association Limited, London in association with Collins, an imprint of HarperCollins Publishers.

Collins is an imprint of
HarperCollins Publishers
77–85 Fulham Palace Road
London W6 8JB

www.collins.co.uk

Published in 2008 by HarperCollins Publishers Ltd

10 9 8 7 6 5 4 3 2 1

A catalogue record for this book is available from the British Library

ISBN-13 978-0-00-726353-0

Design and typeset by D & N Publishing, Hungerford, Berkshire
Printed and bound in China

Credits
Publisher: Myles Archibald
Original concept: Paul Sterry, Myles Archibald and David Price-Goodfellow
Picture research and image manipulation: Paul Sterry
Colour reproduction: Nature Photographers Ltd
Production: Keeley Everitt
Administration: Deborah Wadsworth
Cover design: Emma Jern
Edited and designed for HarperCollins by D & N Publishing:
    Project managers: David and Namrita Price-Goodfellow
    Editor: Susi Bailey
    Proof-reader: Michael Jones

Author's notes

As a child, my interest in natural history was first aroused by the wealth of insect and other invertebrate life I came across. Over the years this fascination has broadened to embrace all spheres of British wildlife, and the species included in this book in part reflect my interest in the subject in all its forms. Throughout my years of apprenticeship as a budding natural historian I was forever asking those who knew more than me 'What is that?' So it is fitting that after more than four decades of study I should return the favour for others by writing a book that answers many of the countless questions I have put to my elders and betters over the years.

Choice of Species

The choice of species for inclusion had to be made carefully. It was the intention from the outset to create an overall list that covered almost every commonly encountered, widespread or distinctive species of plant and animal likely to be observed on a day out anywhere in Britain or Ireland. Particular emphasis has been given to groups that arouse special interest among British naturalists, including mammals, birds, fish, butterflies, dragonflies and flowering plants, especially orchids. Throughout the book a number of more unusual but particularly striking species can be found, the reason for their inclusion being the degree of interest their discovery always arouses.

The book has been organised so that trees and flowering plants are covered first, followed by lower plants, fungi, mammals, birds, reptiles and amphibians, and then terrestrial invertebrates. Freshwater life – both plants and animals – are dealt with separately at the end of the book, as is seashore life.

Author's acknowledgement

The author would like to thank the following people for their contributions to the creation of this book: Shane O'Dwyer for his inspirational design; David and Namrita Price-Goodfellow for the tireless work that enabled a concept to become a reality; Andrew Cleave for his guidance in matters photographic and natural historical; and Myles Archibald for his publishing stamina and boundless optimism.

Contents

Non-Flowering plants 172–83

Fungi 184-97

■ Reptiles and Amphibians 326-31

■ Terrestrial Invertebrates 332-417

Freshwater Wildlife 418-41

Seashore and Marine Wildlife 442-65

Introduction

Britain and Ireland are comparatively small and overcrowded islands but they contain an amazing array of plants and animals that thrive alongside the human population. Visit almost any stretch of countryside and you will find something of interest, and many terrestrial habitats, as well as freshwater and marine environments, are positive havens for wildlife. *Collins British Wildlife* covers the flora and fauna of the whole of mainland England, Wales, Scotland and Ireland including the Shetlands, Orkneys, Hebrides, Isle of Man, Isles of Scilly and the Channel Islands. The marine distribution of cetaceans (whales and dolphins), seabirds, and marine fish and invertebrates is also included.

Whether you are a keen amateur naturalist, or someone with a passing interest in wildlife who wants to identify a plant or animal you have seen on a country walk, the coverage of this book will satisfy your needs. The plants and animals described include almost every commonly encountered, widespread or distinctive species likely to be observed on a day out anywhere in Britain and Ireland. Groups of special interest to British naturalists, including mammals, birds, fish, butterflies, dragonflies and flowering plants have been given a particular emphasis. A number of more unusual but particularly striking species are also included, as well as some rare and exciting birds.

How the book is structured

Identification pages

Almost every plant or animal included in the book is described and illustrated in full, sometimes with several pictures. At the start of each species description the most common and current English name is given. This is followed by the scientific name of the animal or plant (*see* box opposite). Where relevant, the subspecies may also be given. Size, location, preferred habitat and key characteristics for identification are also included, as are maps showing the species' distribution in Britain and Ireland.

Introductory pages

Each main section starts with an introduction to the natural history and biology of the plant or animal group in question. It describes the way the group is classified, giving general information about preferred

Habitat pages

Throughout the book are features devoted to all the major habitat types in Britain and Ireland. They explain in simple ecological terms the history, character and species associated with these areas. Following several of these sections are feature pages describing the 'hotspots' – particularly outstanding examples of these habitats.

Common name
Scientific name
Map
oom photos show more detail
Species photo

Classifying plants and animals

To better understand the natural world, biologists assess similarities and differences between living things and divide and subdivide the living world accordingly. This hierarchical system of classification is used by all naturalists and has been used to structure this book.

The most fundamental and profound division places living things into **kingdoms**: plants, fungi and animals are the ones covered in this book. Within each kingdom, the main subdivisions are called **phyla** (singular phylum). Each phylum contains subdivisions called **classes** and within these are groups called **orders**. Each order contains a number of families and a family will comprise a number of **genera** (singular genus). Each genus may embrace several different species and each **species** is given a scientific name made up of two parts, a binomial name.

Among members of the same genus, the first word in each species' binomial scientific name is the same while the second name is unique among members of that particular genus.

Consequently, each species of plant, fungus or animal will have a unique scientific name; this is recognised in all languages and countries, regardless of any colloquial name a species might have.

Taking the Hornet *Vespa crabro* as an example, here is how it is classified:

KINGDOM:	Animalia (animals)
PHYLUM:	Arthropoda (arthropods)
CLASS:	Insecta (insects)
ORDER:	Hymenoptera (wasps, bees and ants)
FAMILY:	Vespidae (social wasps)
GENUS:	*Vespa*
SPECIES:	*crabro*

Hornet
Vespa crabro

How to use the maps

The maps show the distribution and occurrence of each species in the region. The colours used to reflect species' ranges vary from group to group, but within each colour a gradation of intensity reflects the local abundance: intense colour indicates that a species is abundant, while white denotes that the species is absent from a given area in all cases. How the keys to each group work is explained below.

1 Trees
■ Denotes native species.
■ Shows introduced, alien species.

2 Other plants
■ Shows the species' range for both native and introduced plants.

3 Birds
The different colours represent the current ranges of birds at different times of a normal year; the ranges of some species, particularly certain winter visitors and passage migrants, change from year to year.

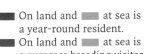
■ On land and ■ at sea is a year-round resident.
■ On land and ■ at sea is a summer breeding visitor.
■ On land and ■ at sea is a winter visitor.
■ On land and ■ at sea is a likely occurrence as a passage migrant.

4 Mammals, reptiles & amphibians
The maps show the current ranges of these animals in the region. The ranges of some marine species change from year to year.

Terrestrial animals (a)
■ Areas where quite common and widespread.
■ Areas where scarce but still widespread.

Marine mammals occurring year-round (b)
■ Areas where observed regularly.
■ Areas where scarce.
□ Areas where rare.
■ Areas where absent.

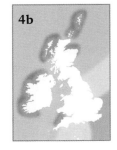

Marine mammals occurring seasonally
■ Areas where observed regularly.
■ Areas where absent.

5 Fish, insects and invertebrates
■ Areas where they are found.

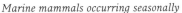

Please bear in mind that for all the maps, given their size, small and isolated populations will not necessarily be featured. Furthermore, the ranges of some species change from year to year; this is particularly true of certain winter visitors and passage migrants.

TREES

Trees are such a familiar part of the landscape that we often take them for granted. But take a closer look at the trees themselves and the woodlands they form and you will discover varied and complex organisms and habitats that support an abundance of wildlife.

Scots Pine.

Tree or shrub?

How do you tell the difference between a tree and a shrub? A tree is often defined as having a single main stem 5m or more tall with a branching crown above this; shrubs typically have numerous stems arising at ground level and normally do not reach the height of a tree. But just to confuse matters, individuals of the same species may become trees or form shrubs, depending on the circumstances in which they grow or on their management. When cut regularly, Hazel forms a multi-stemmed shrub, but in the absence of cutting it may become a medium-sized tree on a single stem.

Pedunculate Oak.

Bark

Bark is the outer skin of a tree's trunk and protects the growing layers of the cells below. It is derived from cambium cells and grows to accommodate the increasing girth of the tree. Depending on the species and its environment, bark may be thin and papery, smooth and shiny, or thick and deeply furrowed.

Sessile Oak bark.

Wood

Trees produce woody tissue that conducts materials around the plant, and leads to the production of permanent shoots that do not die back in winter. Just below the surface of the trunk is a layer called the cambium, which constantly produces new cells. The cells that grow on the inside of the cambium become woody tissue called xylem, which conducts water from the roots to the shoots, buds and leaves; a new layer of this tissue is laid down each year, and it eventually forms the bulk of the trunk and branches. Cells that grow outside the cambium layer form conductive tissue, known as phloem, which carries sugars from the leaves down to the roots.

bark   phloem   cambium   xylem   annual rings

Cross section through Pedunculate Oak trunk.

Leaves

Leaves are thin layers of tissue with cells that trap sunlight energy (in the green pigment chlorophyll), which they use to convert water and carbon dioxide into a simple sugar, glucose. This reaction, called photosynthesis, is arguably the most important chemical reaction in the world, as it is the basis of all food production.

Leaves arrange themselves to absorb the maximum amount of sunlight: spreading canopies or trees growing taller than their neighbours, are ways trees use to maximise the light-gathering power of their leaves. Evergreen trees do not lose all their leaves at the end of every growing season; most leaves stay on the tree during winter. In contrast, deciduous trees generally shed all their leaves at the end of the growing season, before the onset of winter (*see* pp.30–1 for how to identify leaves).

Backlit Hazel leaf, trapping sunlight energy.

Ageing trees

Trees grow vigorously in spring, forming large conductive cells in the trunk that allow the sap to rise. As the season advances, newly produced cells become smaller, with thicker walls for support and a more dense appearance. In winter, cell production slows and then ceases; in spring large cells are produced again. The new growth of large cells immediately adjacent to the thinner layer of dense cells gives the appearance of a ring. On a cut stump you can count the rings and therefore discover the age of the tree.

Close up of growth rings on a Pedunculate Oak.

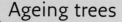

Roots

A mature tree has a network of powerful roots, side-branches and fine root hairs that spread out from the base of the trunk. The main roots are woody and strong, and they help to anchor the tree. Their branches end in fine root hairs only a few cells long. It is these, with their permeable cell walls, that keep the tree alive by supplying it with water and nutrients.

Beech roots.

Reproduction

Conifer flowers lack petals. The male flowers are short-lived, falling off after they have released pollen, but the female flowers, often covered with brightly coloured scales, persist after pollination (fertilisation) and develop into cones containing the seeds. They rely on the wind for pollination and also for seed dispersal. A few close relatives of the conifers, such as yews, produce fleshy fruits instead of cones.

Most flowers of broadleaved trees contain both male and female parts. Some flowers use insects to aid pollination and their flowers often have colourful petals and may also be scented; they usually open in the summer when insects are most active. Wind-pollinated flowers open early in the year before the leaves appear.

The fruits of trees and shrubs vary greatly in appearance. They range from tiny papery seeds with wings (for wind dispersal), through nuts and berries, to large succulent fruits in a variety of shapes and colours. Edible fruits ensure seed dispersal thanks to assistance from animals, which eat the flesh of the fruits but do not digest the seeds.

TYPES OF FLOWERS

Hazel is wind-pollinated: the male catkins shed clouds of pollen (*left*) that are blown onto the tiny, separate female flowers (*above*). Hawthorn blossom (*above right*) is pollinated by insects.

TYPES OF FRUITS

Sycamore winged seeds

Scots Pine; seeds lie between the scales

Guelder-rose berries

Crab Apple fruit

Silver Birch winged seed

The life cycle of an oak

Many acorns are collected and buried by Jays. Those that escape being dug up again will germinate the following spring. A Pedunculate Oak's life begins when an acorn germinates.

The first pair of leaves to appear after germination is quite unlike the true leaves of the tree. They do contain chlorophyll and supply the tiny seedling with its first food made from sunlight energy. The next leaves to appear confirm the young seedling's identity as a Pedunculate Oak.

Once true leaves appear, rapid growth of the seedling begins and continues for many years to come. In the case of a Pedunculate Oak, it could grow up to half a metre each year, both in terms of height and crown size.

The maximum size is reached when the Pedunculate Oak is around 100 years old and, all being well, the tree will continue to produce acorns for at least the next 200 years, with the crown filling out and the trunk expanding.

The end of the oak's life is often heralded by the appearance of fungal fruiting bodies on the trunk. There is something life-affirming about the process of decay in a fallen giant. In the absence of man's intervention, nothing goes to waste and all the nutrients from the tree will enrich the soil, thanks to insect attack and fungal decay, thus creating ideal conditions for the next generation of trees.

Woodlands

Symbols of strength and endurance, trees are justifiably fascinating in their own right. However, their real significance, in ecological terms, lies in the woodland habitat that, collectively, they form. Little wonder then that woods and forests are dear to the hearts of all naturalists.

Native woodlands are home to a select band of tree-loving butterflies. Without doubt, the most majestic of these is the **Purple Emperor** (male is shown here; *see also* p.339), but others include the White Admiral, Silver-washed Fritillary and Speckled Wood. Other invertebrates are also abundant.

Many woodland fungi have intimate associations with tree roots, while others, such as the **Burgundydrop Bonnet** (*see also* p.190) are found on rotting wood and are important in the recycling of nutrients in the woodland ecosystem. Autumn is the time for toadstools – they are fungal fruiting bodies.

The **Great Spotted Woodpecker** (*see also* p.287) is a bold woodland resident while companion species such as Tree-creepers and tits are more unobtrusive. But when breeding begins, even retiring woodland songsters are emboldened, their songs competing with those of warblers and other summer migrants.

Although most wooded areas in Britain harbour some native species and ancient trees, almost no woodland in our region can be described as truly virgin and untouched. For millennia man has interfered with the forested landscape, cutting down trees for fuel and building materials, and in order to create agricultural land. Some types of woodland are more modified than others and a few even exist entirely as a result of human actions. However, when it comes to woodland comprising native species, the history of use does not necessarily detract from its wildlife importance; on the contrary, it often enhances it. Of greater significance than the age of the trees in a given area is a continuity of woodland cover. Many woodlands have had continual tree cover since man first settled the land and the diversity of associated wildlife reflects this venerable ancestry. The fact that ecologists refer to most British and Irish woodland as semi-natural is in no way derogatory.

Even as solitary individuals, native tree species will harbour a good array of wildlife. However, it is when they grow alongside and among other trees, and form woodland, that their true ecological potential becomes apparent. In wildlife terms, the whole - woodland - is greater than the sum of the parts - the trees.

Tree leaves, flowers and fruits are often adorned with oddly shaped growths called galls. Some are caused by fungi, Witches' Broom being an example. But most of those on oaks are caused by tiny gall wasps, including the **Knopper Gall** (*see also* p.398), which deforms both the acorn and its cup.

The **Hazel Dormouse** (*see also* p.201) is our most iconic woodland mammal. Favouring sympathetically managed Hazel and Oak, its fate is now determined by the way we look after our woodlands. It shares its chosen habitats with Yellow-necked Mice, Badgers and a range of bat species.

The carpets of woodland flowers that appear each spring in well-managed lowland sites in Britain are a delight and the envy of botanists the world over – there are few places on the planet where Bluebells, **Wood Anemones** (*see also* p.51), Lesser Celandine and Wood Sorrel can be seen in such profusion.

Woodland habitats & hotspots

In most areas of native woodland typically one or two tree species dominate, either because environmental conditions favour them or because man has encouraged them. Although these types of woodland do not qualify as habitats in the strict sense, some of them are easily recognised and, more significantly, are associated with a classic set of woodland wildlife.

Ash

Huge coppiced Ash stool, more than 600 years old. INSET BELOW: Bluebell

Ash is widespread in Britain, often growing alongside oaks and other deciduous species. However, in some circumstances it comes to dominate if soil conditions suit it – it favours basic soils, and although it can tolerate damp ground it will also grow on limestone pavements. Ash is often encouraged and managed for its timber; some coppiced stools are huge and are hundreds of years old. Hazel coppice is often grown as an understorey beneath its larger cousin. Bluebells, Lesser Celandine, Dog's Mercury and Wood Anemones are common on the woodland floor in spring, with star attractions being Herb-Paris, Goldilocks Buttercup and Early-purple Orchid. In upland limestone-pavement settings, Ash is often accompanied by Bird Cherry and Rowan.

Hotspots *see map:*
1 **Yorkshire Dales National Park** –
   www.yorkshiredales.org.uk.

Pedunculate Oak & Hazel

Typical Pedunculate Oak woodland in Hampshire. INSET: Purple Hairstreak

Pedunculate Oak is common in central and southern England and dominates many woodlands either because soil conditions favour it or because man has encouraged it. There is often an understorey of Hazel, typically planted and managed on a regular basis by coppicing. In spring, carpets of Bluebells, Wood Anemones and Wood Sorrel adorn the woodland floor, intermingling with patches of Dog's Mercury, Early-purple Orchids and Wild Daffodils. Invertebrate life is amazingly rich, the leaves, flowers and wood feeding beetles and moth larvae in abundance. In turn, this community supports a rich diversity of birdlife, along with Hazel Dormice, which depend upon on oak flowers and insects in spring and summer, and on Hazel nuts in autumn.

Hotspots *see map:*
2 **Crab Wood, Hampshire** - www.hants.gov.uk.

Sessile Oak

Hanging Oak woodland, Exmoor, Devon.

Sessile Oak predominates in upland and western parts of Britain, where rainfall is highest. Often the best remaining tracts of Sessile Oak woodland are on steep slopes, the so-called 'hanging oak woodlands'. Many Sessile Oak woods are grazed by sheep and cattle, with the consequence that the shrub layer is considerably reduced. Bluebells and other woodland-floor flowers are usually common, but so are ferns, mosses and liverworts, partly due to reduced competition from more palatable (to grazing animals) flowering plants, but also because of the high humidity. Epiphytic mosses and liverworts often festoon trunks and branches, well off the ground, and Sessile Oak woodlands are the classic domain of nesting Pied Flycatchers and Redstarts.

Hotspots *see map:*
3 **Exmoor National Park** –
   www.exmoor-nationalpark.gov.uk.
4 **Brecon Beacons National Park** –
   www.breconbeacons.org.

Beech

Beech wood in the New Forest.
INSET: Blue Tit

The canopy of a mature Beech wood is so dense that few other trees can grow beneath. Beech tolerates soil types ranging from fairly acid to calcareous; Beech woodlands that cloak the slopes of chalk downs have the fitting name of 'hangers'. Low light levels in summer mean that only shade-tolerant flowers can survive including Sanicle, Woodruff and Bird's-nest Orchid. Beech woods are stunning in autumn due to changes in leaf colour and numerous fungi that adorn the woodland floor.

Hotspots *see map:*
5 **Selborne Hill, Hampshire** –
  www.nationaltrust.org.uk.
6 **Burnham Beeches, Buckinghamshire** –
  www.cityoflondon.gov.uk.

Birch

Birches are often dismissed as being scrub trees and it is certainly true that Silver Birch is invasive on Heathland and cleared woodland on acid soils. However, in its favour it supports large numbers of invertebrates, particularly moths and other insects. In autumn, the leaves turn brilliant yellow and an amazing array of fungi occur around the same time. Downy Birch often grows alongside Silver Birch but replaces it in many western, northern and upland areas. Overall, it supports similar wildlife to its cousin but, in the north, hardy highland species often replace their southern counterparts.

Hotspots *see map:*
7 **New Forest, Hampshire** – www.thenewforest.co.uk.
8 **Birks of Aberfeldy, Perthshire** –
  www.scottish-towns.co.uk/perthshire/aberfeldy.

Birch woodland in the New Forest, Hampshire. INSET: Birch Knight

Carr woodland

As the margins of fens become colonised by vegetation, this leads ultimately to the formation of wet woodland known as carr. Trees tolerant of waterlogged ground – Alder and various willow species – dominate. Carr woodland is hard to explore and, unless the area in question happens to be a boardwalked nature reserve, wet feet are inevitable. However, such areas are usually botanically rewarding. Insect life abounds too, particularly in the form of midges and mosquitoes.

Carr woodland with coppiced Alder.

Hotspots *see map:*
9 **Wicken Fen, Cambridgeshire** – www.wicken.org.uk.

Caledonian

Despite widespread clearance, remnants of ancient Scots Pine forests can still be found in the Scottish Highlands. These Caledonian Pine forests, as they are called, harbour wildlife aplenty, in contrast to conifer plantations elsewhere in Britain. Crested Tits are found only in these forests and the Scottish Crossbill occurs nowhere else in the world. Pine Martens and Red Squirrels also occur in good numbers. Native Caledonian Pine forests are typically open and airy, with a lush ground cover of Bilberry and mosses; floral highlights include wintergreens, Twinflower and Creeping Lady's-tresses.

Caledonian Pine forest in the Scottish Highlands.
INSET: Creeping Lady's-tresses

Hotspots *see map:*
10 **Black Wood of Rannoch** –
   www.jncc.gov.uk/protectedsites.
11 **Cairngorm Forest area** –
   www.jncc.gov.uk/protectedsites.

👓 Find out more about...
- Woodland birds spread through pages **256–321**
- Fungi on **pp.186–97**
- Mosses and liverworts on **pp.178–81**
- Ferns on **pp.174–7**
- Butterflies on **pp.336–51**

Conifers, Juniper & Yew

Conifers bear their seeds in cones; most are evergreen and many have a conical outline in maturity. Their narrow leaves are called needles and related trees share this feature.

Douglas Fir

Pseudotsuga menziesii Pinaceae
Height to 60m

Tall, slender, conical evergreen. **Bark** Greyish green, often blistered. **Branches** In whorls. **Needles** To 3.5cm long, grooved above, with 2 white bands below. **Reproductive parts** Male flowers small and yellow. Female flowers resemble tiny pinkish shaving-brushes. Both sexes grow at tips of twigs. **Status** Native of W North America. Widely planted in Britain for timber; thrives in Scotland.

Norway Spruce

Picea abies Pinaceae
Height to 44m

Narrowly conical tree; the archetypal Christmas tree. **Bark** Brownish, scaly and resinous. **Branches** Almost level. **Needles** 4-angled on short pegs. **Reproductive parts** Male cones small, yellowish and clustered near tips of shoots. Female cones, to 18cm long, are pendulous. **Status** Native of European mountains. Widely planted in Britain as Christmas trees and in shelter-belts.

Sitka Spruce

Picea sitchensis Pinaceae
Height to 52m

Conical evergreen with spire-like crown and buttressed trunk. **Bark** Greyish brown, scaly. **Branches** Ascending with pendent side-shoots. **Needles** To 3cm long, keeled, bright green above with 2 pale blue bands below.
Reproductive parts Female cones yellowish at first, becoming cylindrical and shiny pale brown. **Status** Native of W North America. Planted in Britain for its lightweight, strong timber.

Other larches

Two other larches are also widely grown as forestry trees: **Japanese Larch** *Larix kaempferi*, which has turned-out tips to its cone scales, so they look like woody rosebuds; and **Hybrid Larch** *L. × marschlinsii*, which has cones with reflexed scales and projecting bracts.

Common Larch

Larix decidua Pinaceae
Height to 35m

Deciduous conical conifer. Foliage turns golden before needles fall in autumn. **Bark** Greyish brown, fissured with age. **Branches** Mostly horizontal. **Needles** To 3cm long, in bunches of up to 40. **Reproductive parts** Male flowers are yellow cones. Female cones are red in spring, maturing brown and woody. **Status** Native of central Europe, planted in Britain for timber and ornament.

Scots Pine

Pinus sylvestris Pinaceae
Height to 36m

Conical evergreen that becomes flat-topped with age. **Bark** Grey-brown and scaly low down, red or orange higher up. **Branches** Irregular. **Needles** Paired, grey-green, to 7cm long. Male flowers are yellow, at tips of previous year's shoots. Female flowers grow at tips of new shoots; crimson at first, ripening to brown cones. **Status** Native to parts of Scotland, where it is also planted for timber; now naturalised more widely there and throughout Britain.

Austrian Pine

Pinus nigra ssp. *nigra*
Pinaceae Height to 30m

Broadly conical with a narrow crown. **Bark** Greyish brown, becoming darker and rough in older trees. **Needles** Paired, to 15cm long; stiff with finely toothed margins. **Reproductive parts** Mature cones, to 8cm long, have keeled, spined scales. **Status** Native of central Europe. Widely planted in Britain for shelter or ornament and sometimes naturalised.

Western Hemlock-spruce

Tsuga heterophylla Pinaceae
Height to 45m.

Dense conical evergreen with drooping leading shoot. **Needles** Dark glossy green above, 2 pale bands below, in 2 flattened rows; either 6mm or 2cm long. **Reproductive parts** Male flowers reddish, yellowing with pollen. Female cones pendent, to 3cm long. **Status** Native of W North America. Planted in Britain for timber. **Similar species Eastern Hemlock-spruce** *T. canadensis* has more tapering needles and a further, twisted row of needles along middle of shoots.

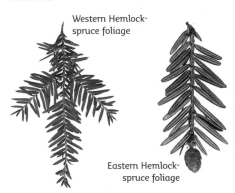

Western Hemlock-spruce foliage

Eastern Hemlock-spruce foliage

Monterey Cypress

The **Monterey Cypress** *Cupressus macrocarpa* is planted in W Britain and Ireland. It is much more common, and grows larger, here than in its native Monterey, California. It is highly tolerant of salt spray and sea winds and thrives near our coasts.

Lawson's Cypress

Chamaecyparis lawsoniana Cupressaceae
Height to 40m

Dense conical evergreen. Trunk often forked. **Bark** Cracks into vertical greyish plates. **Branches** Numerous. **Leaves** Scale-like, to 2mm long, flattened along shoot, in opposite pairs; parsley-scented. **Reproductive parts** Male flowers are cones, to 4mm long. Female cones, to 8mm across, have 4 pairs of scales. **Status** Native of W USA, widely planted in Britain. Numerous cultivars exist with different leaf colours.

Leyland Cypress

× *Cupressocyparis leylandii*
Cupressaceae Height to 35m

Evergreen hybrid, densely foliaged, fast-growing and hardy. **Bark** Reddish brown with vertical ridges. **Branches** Numerous, dense, almost vertical. **Leaves** Scale-like, to 2mm long. **Reproductive parts** Male and female cones are small and seldom seen. **Status** Man-made hybrid, widespread in parks and gardens and often reviled due to its fast growth and the way it blocks out light. 'Haggerston Grey' and 'Leighton Green' are the common cultivars that are most often seen.

Common Juniper

Juniperus communis Cupressaceae
Height to 6m

Aromatic evergreen shrub or small tree. **Bark** Reddish brown, peeling. **Branches** With 3-angled twigs. **Leaves** Needle-like, to 2cm long, in whorls of 3. Foliage is gin- or apple-scented. **Reproductive parts** Male cones small and yellow. Female cones to 9mm long and green, ripening through blue-green to black in 2nd year. **Status** Native of chalk downland in S England and moors and limestone crags in N Britain.

Common Yew

Taxus baccata Taxaceae
Height to 25m

Dense-foliaged broadly conical conifer. **Bark** Reddish, peeling. **Branches** Level or ascending. **Leaves** Flat, needle-like, to 4cm long, dark glossy green with 2 pale yellowish bands below. **Reproductive parts** Male and female flowers on separate trees. Males comprise yellowish anthers. Females are greenish; fruits surrounded by bright red fleshy aril. **Status** Native to Britain; often planted in churchyards.

Did you know?

Britain's finest Yew forest is found at Kingley Vale National Nature Reserve, Sussex. Many of the trees here are ancient, over 1,000 years old, and are among the oldest living things in Britain. Find out more at www.naturalengland.co.uk

Broadleaved trees: willows to birches

Willows, birches and allies are often regarded as invasive scrub species. But in fact these colonisers are incredibly important in ecological terms, harbouring significant invertebrate communities.

Scarce low-growing willows

A number of such willows are found in northern and upland Britain; these are easy to overlook but are of great interest. The commonest is **Dwarf Willow** *Salix herbacea* (**1**), which has round shiny leaves to 2cm long. Scarcer species include **Downy Willow** *S. lapponum* (**2**), whose leaves are downy on both sides; **Mountain Willow** *S. arbuscula* (**3**), whose leaves are shiny above but downy grey below; and **Net-leaved Willow** *S. reticulata* (**4**) with dark green leaves and prominent netted veins. The diminutive **Dwarf Birch** *Betula nana* (**5**) also grows in upland habitats and has tiny rounded leaves just 6–8mm across.

 1 2 3 4 5

Bay Willow

Salix pentandra Salicaceae
Height to 18m

Domed open-crowned tree. **Shoots** Olive-green and glossy. **Leaves** Glossy green, bluish below. **Reproductive parts** Yellow upright male catkins appear with leaves. Female catkins dull yellowish green, pendulous. **Status** Common native of moors and upland woods.

Crack-willow

Salix fragilis Salicaceae
Height to 25m

Large domed tree with a thick trunk. **Bark** Grey-brown with criss-crossed ridges. **Branches** Grow from the base. Shoots Reddish, brightest in spring. **Leaves** Long, glossy, with toothed margins. **Reproductive parts** Male catkins yellow and pendulous; female catkins green and pendulous, on separate trees. **Status** Common in damp habitats.

White Willow

Salix alba Salicaceae
Height to 25m

Broadly columnar tree. **Bark** Dark grey. **Shoots** Yellowish grey, downy at first. **Leaves** Blue-grey. **Reproductive parts** Male catkins small and ovoid. Female catkins longer and green. **Status** Common in damp lowland habitats. **Similar species** **Cricket-bat Willow** *S. alba* var. *caerulea* is the source of timber for cricket bats. **Purple Willow** *S. purpurea* has leaves that are broadest towards the tip. It is a rather local species associated with riverside habitats in lowland areas.

female catkins

Osier

Salix viminalis Salicaceae
Height to 6m

Spreading shrub or small tree. **Shoots** Straight, flexible twigs, shiny with age, often cropped for weaving. **Leaves** Narrow, to 15cm long, woolly below. **Reproductive parts** Male and female catkins, to 3cm long, on separate trees. **Status** Common in wet habitats. **Similar species** **Weeping Willow** *S.* × *sepulcralis* has pendulous branches and leaves to 16cm long. As well as being popular for its ornamental value, it is a classic waterside tree, its trailing foliage often dipping into the water.

Grey Willow

Salix cinerea Salicaceae
Height to 6m

leaf underside

Variable shrub. **Shoots** Downy. **Leaves** Oblong; glossy above but often develop rusty hairs below by autumn. **Reproductive parts** Male catkins ovoid and yellow. Female catkins similar but greener, producing plumed seeds. **Status** Common in damp habitats. **Similar species** **Eared Willow** *S. aurita* has broader leaves with wavy margins, twisted tip and leafy 'ears' at base. Favours moors and heaths.

Goat Willow

Salix caprea Salicaceae
Height to 12m

male catkins

Dense, shrubby tree. **Shoots** Stiff, smooth when bark is peeled. **Leaves** Oval, to 12cm long; twisted point at tip; woolly below. **Reproductive parts** Male and female catkins on separate trees before leaves; to 2.5cm long with silky hairs before opening. Male catkins become yellow with pollen. **Status** Common. Of immense importance to wildlife, especially moth caterpillars.

Creeping Willow

Salix repens Salicaceae Height to 1.5m

Low-growing and creeping shrub. **Shoots** Sometimes downy and usually reddish brown. **Leaves** Ovate, usually untoothed, to 4cm long; hairless above when mature, with silky hairs below. **Reproductive parts** Catkins. **Status** Locally common on moors, heaths and coastal dune slacks.

catkins

White Poplar

Populus alba Salicaceae Height to 20m

Elegant open tree. **Bark** Pale with diamond-shaped scars. **Shoots** Covered with white felt in spring. **Leaves** Lobed, grey-green above, dense white felt below. **Reproductive parts** Male catkins ovoid, white and fluffy; female catkins slender and greenish. **Status** Long-established introduction. **Similar species Grey Poplar** *P. × canescens*, a hybrid between White Poplar and Aspen. Leaves rather rounded and evenly toothed; greyish felt below (*see* p.30). Widespread and common.

Aspen

Populus tremula Salicaceae Height to 18m

Slender tree with a rounded crown. Suckers readily. **Bark** Ridged and fissured with age. **Leaves** Rounded to oval, with shallow marginal teeth, palest below. They rustle in the slightest breeze and turn golden in autumn. **Reproductive parts** Reddish male catkins and greenish female catkins on different trees. **Status** Common on poor, damp soils.

Aspen leaf

Black-poplar leaf

Black-poplar

Populus nigra ssp. *betulifolia* Salicaceae Height to 32m

Spreading tree with a domed crown. **Bark** Blackish, gnarled and burred. **Shoots** Smooth, golden brown when young. **Leaves** Triangular, long-stalked; finely toothed margin. **Reproductive parts** Male catkins pendulous and reddish, female catkins greenish. **Status** Scarce native of heavy, damp soils. It attracts considerable conservation interest.

Common Walnut

Juglans regia Juglandaceae Height to 30m

Spreading domed tree. **Bark** Brown at first, grey and fissured with age. **Branches** Lowest ones spreading; twisted twigs bear purple-brown buds. **Leaves** Compound, with 7–9 leaflets, to 15cm long; thick and leathery. **Reproductive parts** Male catkins yellow, to 15cm long; female flowers small, greenish. Fruits green, rounded, encasing Walnut 'nut'. **Status** Grown since Roman times.

fruits

Silver Birch

Betula pendula Betulaceae Height to 26m

Acquires weeping habit with age. **Bark** Thick, fissured at base, forming rectangular plates; smooth silvery white higher up trunk. **Branches** Ascending; twigs and shoots pendulous. **Leaves** To 7cm long, triangular, toothed; turn yellow in autumn. **Reproductive parts** Male catkins terminal, yellow, pendulous. Female catkins greenish, in leaf axils; seeds are winged. **Status** Common, especially on heaths.

Other poplars

A wide range of alien and hybrid poplars are grown in Britain, either commercially for their timber or as ornamental park specimens. Commonly encountered trees include **Hybrid Black-poplar** *Populus × canadensis* (**1**), **Western Balsam-poplar** *P. trichocarpa* (**2**), hybrid **Balsam-spire** *P.* 'Balsam Spire' (**3**) and **Balm-of-Gilead** *P. × jackii* (**4**).

Downy Birch

Betula pubescens Betulaceae Height to 25m

Elegant tree. **Bark** Reddish in young trees; thick and grey with age, does not break into rectangular plates at base. **Branches** Denser and more untidy than Silver Birch, mostly erect, never pendulous. Twigs with downy white hairs in spring. **Leaves** Rounded at base, evenly toothed; hairy petiole. **Reproductive parts** Catkins similar to those of Silver Birch but seeds have smaller wings. **Status** Commonest in W and N.

Broadleaved trees: Alder to Sweet Chestnut

Oak, Beech and Hornbeam are the most important timber trees in Britain, although species such as Hazel and Sweet Chestnut are also widely used for their wood.

Common Alder

Alnus glutinosa Betulaceae
Height to 25m

Spreading, often multi-stemmed tree. **Bark** Brownish, fissured into squarish plates. **Branches** Ascending in young trees. Young twigs sticky. Buds 7mm long, on 3mm-long stalks. **Leaves** Stalked, to 10cm long, rounded with notched apex. **Reproductive parts** Purplish male catkins, in bunches of 2–3, appear first in winter. Female catkins cone-like, reddish at first, ripening green by summer. **Status** Common beside water.

Female catkins

Hazel

Corylus avellana Betulaceae
Height to 6m

Multi-stemmed shrub, or short tree. **Bark** Smooth, shiny, peeling into papery strips. **Branches** Upright to spreading. Twigs have stiff hairs, buds oval and smooth. **Leaves** Rounded, to 10cm long, hairy above; heart-shaped base and pointed tip. Margins double-toothed. **Reproductive parts** Male catkins to 8cm long, pendulous and yellow. Female flowers red and tiny; produce hard-shelled nuts. **Status** Common, often coppiced.

Hazel nuts

Hornbeam

Carpinus betulus Betulaceae
Height to 30m

Robust tree; trunk gnarled and twisted. **Bark** Silvery grey and fissured. **Branches** Ascending and twisted; twigs greyish brown, hairy. **Leaves** Oval, pointed with rounded base, short stalk, double-toothed margin; 15 pairs of veins. **Reproductive parts** Male catkins, to 5cm long, yellowish green with red scales. Fruits are clusters of winged nutlets and are a favoured food of the Hawfinch. **Status** Locally common, and most frequent in southern parts of England. Grows as a native tree but sometimes also planted and coppiced for the production of its hard, durable timber.

fruits

Beech

Fagus sylvatica Fagaceae
Height to 40m

Imposing deciduous tree with domed crown. **Bark** Smooth and grey. **Branches** Ascending. Buds reddish, to 2cm long, smooth and pointed. **Leaves** To 10cm long, oval, pointed, with wavy margin. **Reproductive parts** Male flowers pendent, clustered. Female flowers paired with brownish bracts. **Fruits** Shiny 3-sided nuts, to 1.8cm long, enclosed in a prickly case. **Status** Common native in S England; widely planted elsewhere.

fruit

Pedunculate Oak

Quercus robur Fagaceae
Height to 36m

Spreading deciduous tree with dense crown. **Bark** Grey, thick and fissured with age. **Branches** Dead branches emerge from canopy of ancient trees. Buds hairless. **Leaves** Deeply lobed with 2 auricles at base; on very short stalks (5mm or less). **Reproductive parts** Flowers are catkins. Fruits are acorns, in groups of 1–3, with long stalks and scaly cups. **Status** Widespread; prefers heavier clay soils to Sessile Oak.

acorns

Sessile Oak

Quercus petraea Fagaceae
Height to 40m

Sturdy domed deciduous tree. **Bark** Grey-brown, fissured. **Branches** Rather straight and radiating. Buds have long white hairs. **Leaves** Lobed, dark green with hairs below on veins; on yellow stalks, 1–2.5cm long, and lacking basal auricles. **Reproductive parts** Flowers are catkins. Acorns egg-shaped, stalkless; sit directly on twig in small clusters. **Status** Common in W and hilly areas of the country on poor soils.

The four seasons

Deciduous trees, such as the Pedunculate Oak, are highly sensitive to seasonal change, and this is reflected in their distinct form at different times of the year.

In SPRING, sap begins to rise up the trunk and fresh green leaves appear, followed by flowers.

In SUMMER, leaves mature, fruits form and the foliage loses the fresh colour it had in spring.

In AUTUMN, nutrients are withdrawn from the leaves, their colour changes and eventually they fall.

In WINTER, the tree is leafless and seemingly lifeless, although in fact it is poised for the arrival of spring.

Sweet Chestnut

Castanea sativa Fagaceae
Height to 35m

Deciduous tree with fine trunk. **Bark** Silvery and smooth at first, spirally fissured and grooved with age. **Branches** Lowest branches spreading, upper ones ascending. **Leaves** Glossy, to 25cm long, lance-shaped and toothed. **Reproductive parts** Male catkins creamy and pendulous. Female flowers green and erect, at base of male catkins; spiny green fruits contain 3 brown nuts. **Status** Native of mainland Europe, planted in Britain since Roman times.

fruit

Horse-chestnut

Aesculus hippocastanum Hippocastanaceae
Height to 25m

Deciduous tree with domed crown. **Bark** Greyish brown, flaking. **Branches** Snap easily. Winter buds shiny brown, sticky, with shield-shaped leaf scars. **Leaves** Long-stalked, palmate, with up to 7 leaflets, each 25cm long. **Reproductive parts** In multi-branched flowerheads, to 30cm tall, comprising 40+ 5-petalled pink-spotted white flowers. Fruits spiny-cased and rounded, containing round seed (conker). **Status** Native of Balkans, long established in Britain.

fruit

conker

Broadleaved trees: elms to Ash

The following trees are seldom found as single-species groups, or stands, but they are nevertheless important components in woods and hedgerows in many parts of Britain and Ireland.

Wych Elm

Ulmus glabra Ulmaceae
Height to 40m

Spreading tree. **Bark** Cracked and ridged with age. **Branches** Main ones spreading. Young twigs have stiff hairs. **Leaves** Oval, to 18cm long, with tapering tip. Unequal base extends beyond petiole. **Reproductive parts** Fruits papery, to 2cm long. **Status** Widespread but much reduced by Dutch elm disease; large trees seldom seen nowadays.

fruits

English Elm

Ulmus procera Ulmaceae
Height to 36m

High-domed lofty tree. **Bark** Dark, grooved with squarish plates. **Branches** Main ones ascending. Twigs reddish, hairy. **Leaves** Rough, rounded to oval; unequal base does not reach beyond petiole. **Reproductive parts** Fruits papery, to 1.5cm long, short-stalked. **Status** Fairly common but declining.

fruits and leaves

Smooth-leaved Elm

Ulmus minor Ulmaceae
Height to 32m

Domed spreading tree. **Bark** Grey-brown, scaly and ridged. **Branches** Ascending, with pendulous shoots. **Leaves** Hornbeam-like, to 15cm long, oval, with short petiole. **Reproductive parts** Papery fruits. **Status** Locally common in S but reduced by Dutch elm disease (*see* 'Did you know?' above). Cornish, Jersey and Plot's elms are locally common clones of this species.

London Plane

Platanus × hispanica Platanaceae
Height to 44m

Deciduous tree with tall trunk and spreading crown. **Bark** Grey-brown, flaking in patches. **Branches** Tangled and twisted. **Leaves** To 24cm long, 5-lobed, palmate. **Reproductive parts** Flowers rounded, in clusters. Greenish spherical fruits have spiky hairs. **Status** Widely planted hybrid in towns and cities.

Sycamore

Acer pseudoplatanus Aceraceae
Height to 35m

fruits

Vigorous spreading deciduous tree. **Bark** Greyish, fissured and flaking. **Branches** Thick, with grey-green twigs and reddish buds. **Leaves** To 15cm long, with 5 toothed lobes. **Reproductive parts** Flowers in pendulous yellow clusters, to 12cm long. Paired wings of fruits spread acutely, and curve in slightly towards tip. **Status** Introduced, widely planted and naturalised.

Norway Maple

Acer platanoides Aceraceae
Height to 30m

Spreading deciduous tree. **Bark** Smooth, grey, ridged. **Branches** Less crowded than those of Sycamore; twigs green, often tinged red. **Leaves** To 15cm long with 5–7 toothed and sharply pointed lobes. **Reproductive parts** Greenish flowers, in erect clusters of 30–40. Paired wings of fruits spread almost horizontally. **Status** Introduced, widely planted and naturalised.

fruit

Field Maple
Acer campestre Aceraceae
Height to 26m

Deciduous tree with rounded crown and twisted trunk. **Bark** Grey-brown, fissured, corky. **Branches** Much-divided and dense. Shoots hairy, sometimes winged. **Leaves** To 12cm long, 3-lobed; turn yellow in autumn. **Reproductive parts** Yellowish flowers in erect clusters. Fruits reddish, winged, in 4s. **Status** Common native of woods and hedgerows, especially on calcareous soils.

fruit

Ash
Fraxinus excelsior Oleaceae
Height to 40m

Deciduous tree with open crown. **Bark** Grey, fissured with age. **Branches** Ascending; grey twigs flattened at nodes, and with conical black buds. **Leaves** Pinnate, to 35cm long, with 7–13 lance-shaped toothed leaflets. **Reproductive parts** Flowers small, purple, clustered. Fruits are single-winged 'keys', in bunches. **Status** Common native; prefers calcareous or base-rich soils.

fruits

Holly
Ilex aquifolium Aquifoliaceae
Height to 15m

Distinctive evergreen. **Bark** Silver-grey, fissured with age. **Branches** Sweep downwards but tips turn up. **Leaves** To 12cm long, leathery, variably wavy with spiny margins. **Reproductive parts** White flowers, 6mm across, 4-petalled, clustered in leaf axils; males and females grow on different trees. Fruits are red berries. **Status** Common native in woods and hedgerows.

leaves and fruit

Rowan
Sorbus aucuparia Rosaceae
Height to 20m

Open deciduous tree. **Bark** Silvery grey, smooth. **Branches** Ascending, with purple-tinged twigs and hairy buds. **Leaves** Pinnate, with 5–8 pairs of ovate toothed leaflets, each to 6cm long. **Reproductive parts** Flowers to 1cm across with 5 white petals; in dense heads. Fruits rounded, scarlet, in clusters. **Status** Locally common native; also widely planted.

leaves and fruit

Common Whitebeam
Sorbus aria Rosaceae
Height to 25m

Deciduous tree or spreading shrub. **Bark** Smooth and grey. **Branches** Spreading; twigs brown above, green below. Buds ovoid, green, tipped with hairs. **Leaves** Oval, to 12cm long, toothed, very hairy below. **Reproductive parts** Flowers white, clustered. Fruits ovoid, to 1.5cm long, red. **Status** Native in S, mainly on chalky soils; also widely planted in towns.

fruits

flowers and leaves

Rare *Sorbus* species

Several rare and local *Sorbus* species are found in Britain, and can be identified by their leaf shape and fruit structure. Among the most distinctive are **French Hales** *S. devoniensis*, a hedgerow tree that grows in Devon; **Bristol Service** *S. bristolensis*, found in the Avon Gorge; and **Cliff Whitebeam** *S. rupicola*, which favours upland limestone areas.

Wild Service-tree
Sorbus torminalis Rosaceae
Height to 25m

Spreading deciduous tree. **Bark** Fissured into squarish plates. **Branches** Straight; twigs shiny, buds green and rounded. **Leaves** To 10cm long with 3–5 pairs of pointed lobes; toothed margin. **Reproductive parts** Flowers white, to 1.5cm across, clustered. Fruits rounded, to 1.8cm across, brown. **Status** Scarce native of ancient woods on heavy soils.

fruit

Fruiting trees & shrubs

A select band of British trees and shrubs bear luscious-looking fruits, some of which are palatable to people as well as wildlife.

Wild Pear

Pyrus pyraster Rosaceae
Height to 15m

Deciduous spreading shrub or small tree. **Bark** Grey-brown, breaking into square plates. **Branches** Spreading and spiny; twigs smooth and greyish brown. **Leaves** To 7cm long, elliptical with toothed margin. **Reproductive parts** Flowers white, 5-petalled, long-stalked; produced in quantity. Fruits rounded, hard, to 3.5cm across, yellowish brown. **Status** Local native.

fruit

Other pears

In addition to Wild Pear, **Common Pear** *Pyrus communis* is often naturalised in hedgerows and sometimes marks the site of former dwellings. Its fruits are considerably larger, and more pear-shaped, than those of its wild cousin. **Plymouth Pear** *P. cordata* is a rare West Country native, and has small rounded fruits.

Common Pear   Plymouth Pear

Wild Crab

Malus sylvestris Rosaceae
Height to 10m

Slender deciduous tree. **Bark** Deep brown, cracking into oblong plates. **Branches** Often spiny; even shoots can be thorny. **Leaves** To 11cm long, oval and toothed. **Reproductive parts** Flowers 5-petalled, to 4cm across, white, sometimes pink-tinged. Fruits to 4cm across, rounded, yellowish green, hard and sour. **Status** Locally common native of hedges and woods.

flowers

fruit

Wild Cherry

Prunus avium Rosaceae
Height to 30m

Deciduous tree with domed crown. **Bark** Reddish brown, shiny, with circular lines; peels horizontally into papery strips. **Branches** Spreading, with reddish twigs. **Leaves** To 15cm long, ovate, toothed. **Reproductive parts** Flowers white, 5-petalled, in clusters of 2–6. Fruits to 2cm long, rounded, ripening dark purple, sometimes yellowish. **Status** Widespread native.

fruit   flowers

Bird Cherry

Prunus padus Rosaceae
Height to 17m

Deciduous tree, domed with age. **Bark** Smooth, dark grey-brown; unpleasant smell if rubbed. **Branches** Mostly ascending, with downy twigs. **Leaves** Elliptical, to 10cm long, toothed, tapering at tip. **Reproductive parts** Flowers white, 5-petalled, in 15cm-long spikes. Fruits to 8mm long, black. **Status** Local, mainly on limestone in N; also widely planted.

fruits

Cherry Laurel

Prunus laurocerasus Rosaceae
Height to 8m

Evergreen shrub or small tree. **Bark** Dark grey-brown, pitted with lenticels. **Branches** Dense, with pale green twigs. **Leaves** Leathery, to 20cm long and oblong. **Reproductive parts** Flowers white, fragrant, in erect spikes to 13cm long. Fruits rounded, green, turning red, ripening blackish purple. **Status** Introduced, widely planted and sometimes naturalised.

leaves and fruit

Blackthorn

Prunus spinosa Rosaceae
Height to 6m

Densely branched shrub. **Bark** Blackish brown. **Branches** Spreading, with spiny twigs. **Leaves** Ovate, toothed, to 4.5cm long. **Reproductive parts** Flowers white, 5-petalled, to 17mm across; produced prolifically (Feb-Mar). Fruits (sloes) to 1.5cm long, ovoid, blue-black with a bloom. **Status** Common. **Similar species Bullace** *P. domestica* ssp. *institia* is spineless, with larger fruits (see p.31).

fruit

flowers

Cherry Plum

Prunus cerasifera Rosaceae
Height to 8m

Deciduous bushy tree. **Bark** Dark brown, pitted with white lenticels. **Branches** Spiny, with glossy green twigs. **Leaves** To 7cm long, ovate, toothed; green, but red in some cultivars. **Reproductive parts** Flowers white, stalked; pink in some cultivars. Fruits, to 3.5cm long, rounded, red or yellow. **Status** Introduced, widely planted and often naturalised.

fruit   flowers

Common Hawthorn

Crataegus monogyna Rosaceae
Height to 15m

Spreading deciduous tree or shrub. **Bark** Fissured with vertical grooves. **Branches** Densely packed, with sharp spines. **Leaves** To 4.5cm long, deeply lobed, with teeth near apex. **Reproductive parts** Flowers white, 15mm across, in flat-topped clusters of 10–18 (May). Fruits (haws) rounded and red. **Status** Common native of hedgerows and scrub, especially on chalk.

fruit

Midland Hawthorn

Crataegus laevigata Rosaceae
Height to 10m

Dense deciduous shrub. **Bark** Grey-brown, cracking into plates. **Branches** Have few, small spines. **Leaves** To 6cm long, with shallow lobes. **Reproductive parts** Flowers white, to 2.4cm across, in lax clusters (May). Fruits red, rounded, to 1cm long. **Status** Widespread but local; less frequent than Common Hawthorn as a native but also widely planted.

flowers

Spindle

Euonymus europaeus Celastraceae
Height to 6m

Twiggy deciduous tree. **Bark** Smooth, grey, fissured and pink-tinged with age. **Branches** Numerous; young green twigs are angular. **Leaves** Ovate, to 10cm long, toothed. **Reproductive parts** Flowers yellowish, 4-petalled, in clusters. Fruits are pink capsules, 1.5cm across with 4 chambers. **Status** Local native of hedgerows and copses, especially on lime-rich soils.

fruit

flowers

Did you know?

Fruits are the structures in which trees and shrubs, and indeed flowering plants generally, produce their seeds. The appearance of these fruits is as varied as the trees themselves, but in all cases the aim is to aid dispersal of the seeds – the idea is that birds and mammals eat the luscious flesh and disperse, but do not digest, the seeds.

Box

Buxus sempervirens Buxaceae
Height to 6m

Dense spreading evergreen shrub. **Bark** Smooth, grey, breaking into squares with age. **Branches** Numerous; young twigs green, angular and hairy. **Leaves** Ovate, to 2.5cm long. **Reproductive parts** Flowers small, green; males have yellow anthers. Fruit is a greenish capsule, 8mm long, with 3 spreading spines. **Status** Local native of chalk slopes; also widely planted.

LEFT: flowers and leaves
ABOVE: fruits

Buckthorn

Rhamnus cathartica
Rhamnaceae
Height to 10m

Spreading deciduous shrub or small tree. **Bark** Dark orange-brown, fissured with age. **Branches** Have slender, slightly spiny shoots.

Buckthorn fruits

Leaves Ovate to rounded, to 6cm long, finely toothed; veins converge towards leaf tip. **Reproductive parts** Flowers fragrant, with 4 green petals. Fruit is black, shiny and 8mm across. **Status** Local native, mainly on chalk.

Alder Buckthorn

Frangula alnus Rhamnaceae
Height to 5m

Sprawling tree. **Bark** Smooth, grey with vertical furrows. **Branches** Spreading, twigs with fine hairs. **Leaves** Ovate, to 7cm long, with entire margins. **Reproductive parts** Flowers greenish, 5-petalled, 3mm across, in small axillary clusters. Fruits berry-like, 1cm across, green ripening yellow, red then black. **Status** Locally common on damp acid soils.

flowers   leaves and fruits

Limes & hedgerow shrubs

Formerly much more common and widespread, native lime species are now used as indicators of ancient woodland. Species such as Guelder-rose and Privet are important scrub and hedgerow shrubs.

Small-leaved Lime

Tilia cordata Tiliaceae
Height to 32m

fruits

Deciduous tree; dense crown, untidy with age. **Bark** Smooth, grey; darkens and flakes with age. **Branches** Ascending; twigs reddish above, olive below. **Leaves** To 9cm long, rounded with heart-shaped base; vein axils hairy below. **Reproductive parts** Flowers 5-petalled, pale with a green bract; project in all directions. Fruit round, hard, 6mm across. **Status** Local native species.

leaves and fruits

Lime

Tilia × europaea Tiliaceae
Height to 46m

fruit

Hybrid between Small-leaved and Large-leaved limes; suckers freely. **Bark** Grey-brown, ridged. **Branches** Ascending and arching; twigs green. **Leaves** To 10cm long, ovate with heart-shaped base; hairs in vein axils below. **Reproductive parts** Flowers yellowish, 5-petalled, clustered on a greenish bract. Fruit hard and rounded. **Status** Widely planted.

Large-leaved Lime

Tilia platyphyllos Tiliaceae
Height to 40m

fruits

Narrow deciduous tree. **Bark** Dark grey, fissured with age. **Branches** Ascending, with pendent tips. Twigs are reddish green. **Leaves** To 9cm long, broadly ovate with irregular heart-shaped base. **Reproductive parts** Flowers yellowish white, clustered on a pale downy bract. Fruit woody, to 1.8cm long, rounded or pear-shaped, ridged, pendent. **Status** Favours lime-rich soils. Local and native.

leaves and fruits

Dogwood

Cornus sanguinea Cornaceae
Height to 4m

Shrub or small tree. **Bark** Grey, smooth. **Branches** Twigs are distinctive dark red in winter. **Leaves** Opposite, oval, with entire margins and 3–4 pairs of prominent veins. **Reproductive parts** Flowers small, white, in large terminal clusters. Fruits are blackish rounded berries, borne in clusters. **Status** Common native on calcareous soils; also widely planted.

leaves and flower

Sea-buckthorn

Hippophae rhamnoides Elaeagnaceae
Height to 11m

Multi-stemmed shrub or small tree. **Bark** Fissured, peeling. **Branches** Thorny twigs have silvery scales. **Leaves** To 6cm long, with silvery scales. **Reproductive parts** Flowers, to 3mm across, lack petals. Male and female flowers on different trees. Fruits are orange berries, up to 8mm long. **Status** Native in coastal E England; planted elsewhere for ornament or to stabilise dunes.

leaves and fruits

Strawberry Tree

flowers

fruit

The **Strawberry Tree** *Arbutus unedo* is popular in gardens. Despite the exotic appearance of its urn-shaped flowers and fruits that resemble miniature strawberries, it is native to a small area of SW Ireland, where it is relatively common. Its main range is SW Europe and the Mediterranean.

Wild Privet

Ligustrum vulgare Oleaceae
Height to 5m

fruit

Branched semi-evergreen shrub. **Bark** Reddish brown with distinct gashes. **Branches** Dense, much divided; twigs downy. **Leaves** Shiny, untoothed, oval, opposite. **Reproductive parts** Flowers creamy white, 4–5mm across, fragrant and 4-petalled; in terminal spikes. Fruits are shiny, black and clustered. **Status** Native on calcareous soils but also widely planted.

Elder

Sambucus nigra Caprifoliaceae
Height to 10m

Untidy deciduous shrub or small tree. **Bark** Grey-brown, furrowed, corky and lichen-covered with age. **Branches** Spreading, twisted, with white central pith. **Leaves** Opposite, compound, with 5–7 pairs of ovate, toothed leaflets, each to 12cm long. **Reproductive parts** Flowers white, with sickly sweet scent, in flat-topped clusters. Fruits are rounded shiny black berries, in pendulous heads. **Status** Common.

LEFT: fruits
BELOW: leaves and flowers

Wayfaring-tree

Viburnum lantana Caprifoliaceae
Height to 6m

Small spreading deciduous tree. **Bark** Brown. **Branches** Have rounded greyish hairy twigs. **Leaves** Opposite, to 14cm long, ovate and toothed. **Reproductive parts** Flowers white, 5-petalled, to 8mm across, in rounded heads about 10cm across. Fruits are oval berries about 8mm long, ripening red to black. **Status** Native, favouring calcareous soils; also planted.

RIGHT: flowers
BELOW: leaves and fruits

Butterfly-bush

Buddleja davidii Buddlejaceae
Height to 4m

Dense perennial shrub. **Bark** Grey-brown. **Branches** Dense and arching. **Leaves** Long, narrow, darker above than below. **Reproductive parts** Flowers pinkish purple, 4-lobed, 2–4mm across, in long spikes; extremely attractive to butterflies. **Status** Naturalised garden escape, which can be extremely invasive and needs to be controlled.

Guelder-rose

Viburnum opulus Caprifoliaceae
Height to 4m

Spreading deciduous tree. **Bark** Reddish brown. **Branches** Sinuous; twigs smooth, angular and greyish. **Leaves** Opposite, to 8cm long, with 3–5 irregularly toothed lobes. **Reproductive parts** Flowers white, in flat heads with showy outer flowers and smaller inner ones. Fruit is a rounded translucent red berry, in clusters. **Status** Favours calcareous soils.

fruit

flowers

Barberry

Berberis vulgaris Berberidaceae
Height to 2m

Small deciduous shrub. **Bark** Brown. **Branches** Have grooved twigs and 3-forked prickles. **Leaves** Oval and sharp-toothed; in tufts from axils of prickles. **Reproductive parts** Flowers are small, yellow, in hanging clusters. Fruits are ovoid reddish berries. **Status** Scarce native, mainly on calcareous soils; also planted and sometimes naturalised.

flowers   fruit

Did you know?

Rhododendrons were favourites with Victorian horticultural collectors. One in particular, *Rhododendron ponticum*, does particularly well in Britain and, in the context of a garden, it may not look out of place. However, in areas with its preferred acid soils, it soon spreads to the countryside and causes real problems because it is so invasive that it swamps most native vegetation. Consequently, conservationists go to great lengths to try to eradicate it.

Identifying the leaves of broadleaved trees & shrubs

The leaves of broadleaved shrubs and trees are among these plants' most conspicuous and distinctive features, and they grow in a huge variety of shapes and sizes. Although even on an individual tree the leaves can be variable, they are still the best feature to use for identification. Below are leaf examples of some of the most widely encountered hedgerow and woodland species.

Silver Birch

Downy Birch

Grey Willow

White Willow

Osier

White Poplar

Aspen

Hybrid Black-poplar

Goat Willow

Crack-willow

Black-poplar

Sweet Chestnut

Grey Poplar

Smooth-leaved Elm

Beech

Hornbeam

Sessile Oak

English Elm

Alder

Wych Elm

Pedunculate Oak

Wild Pear

Whitebeam

Hazel

Wild Service Tree

Rowan

Wild Crab

Midland Hawthorn

Common Hawthorn

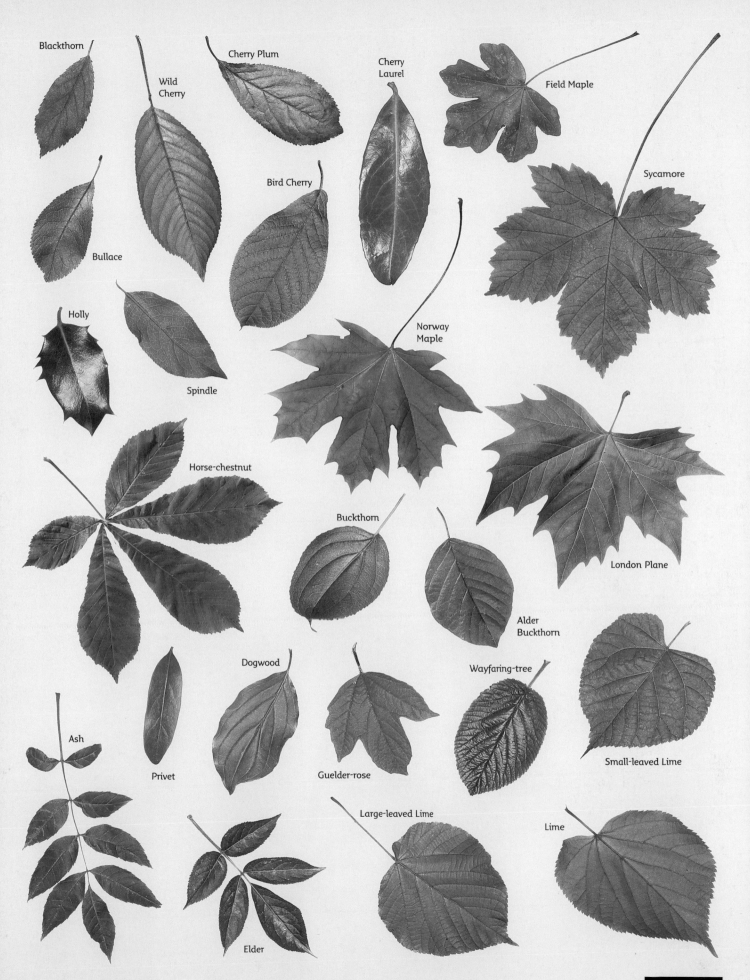

Blackthorn

Wild Cherry

Cherry Plum

Cherry Laurel

Field Maple

Bullace

Bird Cherry

Sycamore

Holly

Spindle

Norway Maple

Horse-chestnut

Buckthorn

London Plane

Alder Buckthorn

Dogwood

Ash

Privet

Guelder-rose

Wayfaring-tree

Small-leaved Lime

Large-leaved Lime

Lime

Elder

31

FLOWERING PLANTS

Despite their beauty, plants do not produce flowers simply to delight the human eye. The flower's role is strictly functional: it is the plant's sex organ and its job is to produce sex cells and ensure the maximum chance of successful fertilisation. The ways in which this is achieved varies considerably throughout the plant kingdom, and differences in flower structure and appearance are vital tools when identifying plants.

Flower structure

In a few species of wildflowers, male and female sex cells are borne in separate flowers, or even on different plants. In most cases they are found together within the same flower. Male sex cells are contained within pollen, tiny grains produced by structures called anthers and borne on slender stems called filaments. Collectively, anthers and filaments are referred to as stamens. The female part of most flowers comprises the ovary, containing the female sex cells, above which is the stigma (which receives the pollen), carried on a stem called the style.

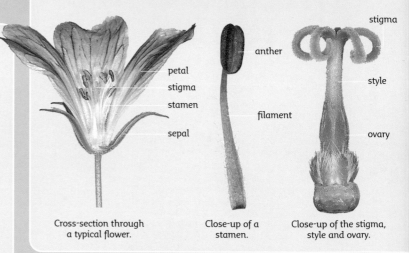

stigma
anther
petal
stigma
stamen
sepal
filament
style
ovary

Cross-section through a typical flower.

Close-up of a stamen.

Close-up of the stigma, style and ovary.

Pollination

In order for a new generation of plants to be created, a flower's male sex cells (contained in the pollen) must fertilise female cells of the same species. This process is called pollination. Some plants can pollinate themselves but most are designed to ensure that cross-pollination takes place, pollen being transferred to other plants of the same species.

Some plants, such as grasses and catkin-bearing shrubs, use the wind to carry their pollen to others of the same species; vast quantities of pollen are needed to achieve a successful outcome with such a random process. Most other species have a more targeted approach and use animals – insects in almost all cases – to transfer pollen. In exchange for a meal of nectar, insects inadvertently carry pollen on their bodies to the next flowers they visit; with luck, a neighbouring plant of the same species will be visited by the insect while it still has pollen on its body.

Bumblebees, and bees generally, are the classic insect pollinators. They visit flowers in search of nectar and unwittingly acquire a dusting of pollen, which is slightly sticky, on their hairy bodies; this is then carried to subsequent flowers they visit.

Fruits and seeds

Fruits follow flowers and are the structures in which a plant's seeds develop and are protected. In many cases, the shape and structure of fruits, and often the seeds themselves, are designed to assist dispersal away from the plant. A number of ingenious methods have evolved to facilitate this process: some seeds are carried by the wind; others attach themselves to the fur of animals; some even float away on water. The study of fruits and seeds is fascinating and can be a valuable aid to correct identification.

RIGHT: Cabbage family members produce fruits known as pods, whose size and shape varies according to species. Those of Wild Candytuft, seen here, are particularly attractive.

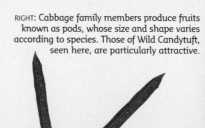

LEFT: Like many other members of the pea family, the fruits of the Bird's-foot Trefoil are elongated pods.

RIGHT: Dandelion seeds have a tuft of hairs that assists wind dispersal. While it remains intact, the collection of seeds and hairs is often referred to as a 'clock'.

Floral variety

Take a close look at the structure of flowers found in the British countryside and you cannot fail to be amazed at the variety and sheer beauty you encounter. Initially, you may be bewildered by the variety but, in time, family similarities will become apparent, usually in terms of the number of petals on show, not necessarily their colour. An appreciation of the broad categories into which flowers can be placed can aid recognition by narrowing down the number of alternatives.

The five-petalled Corncockle.

1 Flowers with three petals

Three-petalled flowers are an unusual arrangement among British wildflowers. Water-plantains and allies provide the main examples (see p.423). In strict botanical terms, members of the iris and orchid families also have three petals, but because these are hard to distinguish from the flower's three colourful sepals, they are categorised below as having six petal-like structures.

Common Water-plantain

2 Flowers with four petals

Examples include cabbage family members (pp.58–63); speedwells (pp.57 and 120–3); bedstraws (pp.108–9); willowherbs (pp.92–3); and poppies and allies (pp.54 and 63).

Sea Rocket

Marsh Speedwell

Spear-leaved Willowherb

Rough Poppy

3 Flowers with five petals

Examples include buttercups (pp.48–51); water-crowfoots (pp.48–9); rock-roses (p.93); St John's-worts (pp.88–9); mallows (p.89); primroses (pp.100–1); loosestrifes and pimpernels (pp.92 and 101); roses and allies (pp.67–71); saxifrages and stonecrops (pp.65–7); sea-lavenders (pp.106–7); Wood-sorrel and allies (p.83); crane's-bills (pp.84–5); mulleins (p.120); violets (pp.90–1); pinks, chickweeds and stitchworts (pp.40–7); forget-me-nots (pp.112–13); pea family members (pp.76–83); sundews (p.64); and centauries (pp.106–7).

Goldilocks Buttercup

Common Rock-rose

Heath Dog-violet

Marsh Stitchwort

4 Flowers with six petal-like structures

Strictly speaking, these flowers comprise three petals and three sepals, but because the latter are often as colourful as the petals, it is often difficult to separate the two. Therefore, for simplicity's sake they are placed together here. Examples include orchids (pp.73 and 150–5); lily family members (pp.142–8); and irises (pp.148–9).

Bee Orchid

Martagon Lily

Spring Squill

Yellow Iris

5 Trumpet- or bell-shaped flowers

Trumpet- and bell-shaped flowers are relatively unusual among British wildflowers. Examples include gentians (p.107); and bindweeds (pp.110–11); heathers and allies (pp.104–5); and bellflowers (pp.129–30).

Marsh Gentian

Sea Bindweed

Bell Heather

Rampion Bellflower

6 Tubular or fused flowers

Examples include eyebright and allies (p.123); fumitories (p.63); comfreys and allies (pp.111–12); figworts (p.121); toadflaxes (pp.120–1); mint family members (pp.114–19); louseworts and cow-wheats (p.124); and broomrapes (p.125).

Common Ramping-fumitory

Common Figwort

Common Toadflax

Common Broomrape

7 Many-flowered and clustered flower heads

Examples include carrot family members (pp.94–9); daisy family members (pp.130–41); and scabiouses (pp.128–9).

Upright Hedge-parsley

Creeping Thistle

Perennial Sow-thistle

Devil's-bit Scabious

Hop, nettles & bistorts

A few of the species described here are among the most successful and widespread of all British plants, although they are not particularly colourful or eye-catching. A number of them have intriguing survival adaptations, ranging from stinging hairs that ward off grazing animals to parasitic lifestyles in which they are reliant on a particular host.

Common Nettle

Urtica dioica Urticaceae
Height to 1m

The familiar stinging nettle. **Flowers** Pendulous catkins; borne on separate sex plants (Jun-Oct). **Fruits** Resemble flowers superficially. **Leaves** Oval, pointed-tipped, toothed and borne in opposite pairs; 8cm long and longer than stalks. **Status** Widespread and common throughout, doing best on nitrogen-enriched and disturbed soils.

Common Nettle leaf

Hop leaf

Bog-myrtle

Myrica gale Myricaceae
Height to 1m

Woody, brown-stemmed and fragrant shrub. **Flowers** Orange, ovoid male catkins and pendulous brown female catkins; borne on separate plants (Apr). **Fruits** Brownish nuts. **Leaves** Oval, grey-green and smell of resin when crushed. **Status** Widespread but local; and absent from much of central England. Sometimes locally dominant and a characteristic plant of boggy habitats, invariably on acid soils.

flowers

Bastard-toadflax

Thesium humifusum Santalaceae
Prostrate

Low-growing plant with sparse branches and a woody base; semi-parasitic on roots of herbaceous plants. **Flowers** Cup-shaped and fused, white inside, yellowish green outside; 4 or 5 pointed lobes create a star-like appearance (Jun-Aug). **Fruits** Greenish and ovoid. **Leaves** 5-15mm long, oval and yellowish green. **Status** Extremely local and habitat-specific, being confined to short turf on chalk grassland.

Hop

Humulus lupulus Cannabaceae
Height to 6m

Twining, hairy climber. **Flowers** Clustered and greenish yellow (male) or green and hop-like (female) (Jun-Aug). **Fruits** Familiar hops that ripen brown in autumn. **Leaves** Divided into 3-5 coarse-toothed lobes. **Status** Widespread but locally common only in the S. Grows on a range of soils and often a relict of cultivation, being found in hedgerows.

fruits

Parasites

Most plants make their own food using photosynthesis and gain water and minerals from the soil via their roots. However, a number of specialists obtain part or all of their nutrition from a host plant. **Mistletoe** *Viscum album* (*right*) is one of the most conspicuous of the semi-parasites, forming large, spherical clumps (up to 1m across) among the branches of trees, mainly apple (often in cultivation), lime and poplars; it is most conspicuous during the winter months when the leaves of host trees have fallen. It has evenly forked branches, oval yellowish leaves in opposite pairs, and white, sticky fruits (*far right*) that are consumed eagerly by birds.

Pellitory-of-the-wall

Parietaria judaica Urticaceae
Height to 7cm

Spreading, downy perennial with reddish stems. **Flowers** Appear in clusters at leaf bases (Jun–Oct). **Fruits** Clustered at leaf bases. **Leaves** Oval, up to 5cm long and long-stalked. **Status** Widespread in England, Wales and Ireland, but most common in coastal areas and in the W. Colonises walls, roadsides and rocky ground.

Common Bistort

Persicaria bistorta Polygonaceae
Height to 60cm

Attractive perennial that forms patches in suitable locations. **Flowers** Pink and borne in dense 30–40mm-long terminal spikes (Jun–Aug). **Fruits** Nut-like. **Leaves** Oval or arrow-shaped, the lower ones stalked, the upper ones almost stalkless. **Status** Locally common in the N but rare in the S. Favours damp, undisturbed meadows.

Water-pepper

Persicaria hydropiper
Polygonaceae Height to 70cm

Upright, branched annual. **Flowers** Pale pink and borne in long spikes that droop at the tip (Jul–Sep). **Fruits** Small and nut-like. **Leaves** Narrow and oval, with a peppery taste when chewed. **Status** Widespread and common, except in the N. Characteristic of damp, bare ground such as winter-wet ruts, and shallow water.

Pale Persicaria

Persicaria lapathifolia
Polygonaceae Height to 60cm

Upright or sprawling annual. Similar to Redshank but the stems are usually greenish and hairy. **Flowers** Greenish white and borne in terminal spikes (Jun–Oct). **Fruits** Nut-like. **Leaves** Narrow and oval. **Status** Widespread and generally common throughout. Found on disturbed ground and at arable field margins.

Knotgrass

Polygonum aviculare
Polygonaceae Height to 1m

Much-branched annual. **Flowers** Pale pink and arise in leaf axils (Jun–Oct). **Fruits** Nut-like and enclosed by the withering flower. **Leaves** Oval, leathery and alternate with a silvery basal sheath; the main stem leaves are larger than those on the side branches. **Status** Widespread and common throughout much of Britain and Ireland on bare ground.

Amphibious Bistort

Persicaria amphibia
Polygonaceae Height to 40cm

Freshwater perennial. The aquatic form has floating stems. **Flowers** Pink; borne in cylindrical spikes (Jun–Sep). **Fruits** Nut-like. **Leaves** Narrow; aquatic forms are hairless, truncate at the base and long-stalked; terrestrial forms are downy, rounded at the base and short-stalked. **Status** Locally common in ponds and water margins.

Redshank

Persicaria maculosa Polygonaceae
Height to 60cm

Upright or sprawling hairless annual with much-branched reddish stems. **Flowers** Pink and borne in terminal spikes (Jun–Oct). **Fruits** Nut-like. **Leaves** Narrow and oval, and usually show a dark central mark; this is most obvious in mature leaves. **Status** Widespread and common throughout the entire region. Found on disturbed ground and at arable field margins, and often associated with damp ground.

Bindweeds, sorrels & docks

Sometimes dismissed and overlooked as 'weeds', most of the plants described here are highly successful colonisers of bare and disturbed ground. They have benefited greatly from man's impact on the environment, and their presence, and often abundance, is not always welcome.

Japanese Knotweed

Fallopia japonica Polygonaceae
Height to 2m

Fast-growing and invasive perennial. **Flowers** Whitish and borne in loose, pendulous spikes that arise from leaf bases (Aug–Oct). **Fruits** Papery. **Leaves** Large, triangular and borne on red zigzag stems. **Status** Alien, but now a widespread garden escape. Quick to colonise roadsides and other wayside places; hard to eradicate.

Black-bindweed

Fallopia convolvulus Polygonaceae
Height to 1m

Extremely common clockwise-twining annual that both trails on the ground and climbs among wayside plants. **Flowers** Greenish and rather dock-like; they are borne in loose spikes that arise from leaf axils (Jul–Oct). **Fruits** Nut-like and blackish. **Leaves** Arrow-shaped and borne on angular stems. **Status** Widespread and common on cultivated and waste ground.

Sheep's Sorrel

Rumex acetosella Polygonaceae
Height to 25cm

Short upright perennial. **Flowers** Greenish and borne in loose, slender spikes (May–Aug). **Fruits** Nut-like. **Leaves** Arrow-shaped and the basal lobes point forwards; upper leaves clasp the stem. **Status** Widespread and common in suitable habitats, usually favouring well-drained acid soils; heaths and coastal cliffs and slopes are ideal.

Water Dock

Rumex hydrolapathum Polygonaceae Height to 2m

Large unbranched perennial. **Flowers** Borne in tall, dense spikes (Jul–Sep). **Fruits** Triangular, with few small teeth and 3 tubercles. **Leaves** are oval, up to 1m long and tapering at base. **Status** Widespread but absent from the N; commonest in S and E England. Associated with damp habitats such as ditches, river banks, canals and marshes.

Common Sorrel

Rumex acetosa Polygonaceae
Height to 60cm

Variable, usually upright perennial. The whole plant often turns red as it goes over. **Flowers** Reddish and borne in slender spikes (May–Jul). **Fruits** Nut-like with a small tubercle. **Leaves** Deep green, arrow-shaped and narrow; they taste mildly of vinegar. **Status** Widespread and common throughout Britain and Ireland, associated with a wide range of grassy habitats.

flowers

Curled Dock

Rumex crispus Polygonaceae
Height to 1m

Upright perennial. **Flowers** Flattened and oval; borne in dense leafless spikes that do not spread away from the stem (Jun–Oct). **Fruits** Oval and untoothed, usually with a single tubercle. **Leaves** Narrow, up to 25cm long and have wavy edges. **Status** Widespread and common, found in rough meadows and on disturbed ground.

fruit

Broad-leaved Dock

Rumex obtusifolius Polygonaceae
Height to 1m

Familiar wayside upright perennial. **Flowers** Borne in loose spikes that are leafy at the base (Jun–Aug). **Fruits** Have prominent teeth and 1 tubercle. **Leaves** Broadly oval, heart-shaped at the base and up to 25cm long. **Status** Common in disturbed meadows.

fruits

Wood Dock

Rumex sanguineus Polygonaceae
Height to 1m

Upright, straggly and branched perennial. **Flowers** Borne in spikes; leafy only at the base (Jun–Aug). **Fruits** Have a single elongated wart. **Leaves** Oval; the basal ones are heart-shaped at the base, sometimes red-veined and never waisted. **Status** Widespread and common; mainly absent from Scotland. Favours grassy woodland rides and shady meadows.

INSET: fruits

Blinks

Montia fontana Portulacaceae
Usually prostrate

Low-growing, sometimes mat-forming plant. The stems are sometimes reddish. **Flowers** Tiny and white; borne in terminal clusters (May–Oct). **Fruits** Rounded capsules. **Leaves** Narrow, oval and opposite. **Status** Widespread and common but least so in the S. Found on bare, damp ground; sometimes grows partly submerged in water.

Fat-hen

Chenopodium album
Chenopodiaceae Height to 1m

Upright branched annual. Often has a mealy appearance. **Flowers** Whitish green and borne in leafy spikes (Jun–Oct). **Fruits** Rounded and surrounded by 5 sepals, in a ring. **Leaves** Green, and matt-looking owing to a mealy coating; vary from oval to diamond-shaped. **Status** Widespread and common on disturbed arable land and waste places.

Clustered Dock

Rumex conglomeratus
Polygonaceae Height to 1m

Upright perennial with a zigzag stem and spreading branches. **Flowers** Borne in leafy spikes. **Fruits** Small, untoothed, with 3 elongated tubercles (Jun–Aug). **Leaves** Oval; basal ones heart-shaped at base and often waisted. **Status** Mostly common, but rare in Scotland. Found in meadows and woodland margins, often on damp soil.

fruits

Springbeauty

Claytonia perfoliata Portulacaceae
Height to 30cm

Annual, introduced from N America but now widely naturalised. **Flowers** White, 5-petalled and 5mm across; borne in loose spikes (Apr–Jul). **Fruits** Capsules. **Leaves** Oval and stalked at the base; flowering stems bear fused pairs of perfoliate leaves. **Status** Widespread and locally abundant on dry, sandy soil.

Identifying docks

Docks are extremely successful colonisers of bare and disturbed ground. Although British species are all superficially similar to the untrained eye, they each have distinctive leaves and fruits to aid identification. The common species are found in a range of disturbed situations, but a few docks are more specialised. **Fiddle Dock** *Rumex pulcher* (**1**), for example, thrives where grazing animals reduce competition from other plants and scuff the topsoil; its leaves are violin-shaped. **Golden Dock** *Rumex maritimus* (**2**) and **Marsh Dock** *R. palustris* (**3**) are both wetland species that occur sporadically; occasionally they appear in great profusion in drying mud during hot summers, but otherwise they can be completely absent from the same locations for years on end.

1

2

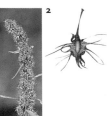

3

Flowers of east coast estuaries

Estuaries form where rivers meet the sea and deposits of silt settle, creating treacherous banks and miles of mudflats that are visible only at low tide. For the naturalist these vast expanses of mud are an exciting habitat that is full of wildlife interest throughout the year.

The higher reaches of an estuary are where some of the more interesting plants are found. Here, in the zone where sea meets land, we find communities of saltmarsh plants that colonise and stabilise the mud. Although they are represented by species from a range of plant families, they all have in common a tolerance of alternating exposure to air and inundation by water of varying salinities: saltmarshes are regularly submerged at high tide, and the salinity of the water is influenced by fresh water from the feed river.

Characteristic saltmarsh plants include Glasswort, Common Sea-lavender, Sea Purslane, Sea Aster and Golden Samphire; where conditions suit them, they grow in great abundance (*see* pp.42, 106 and 133 for descriptions). Although estuaries and saltmarsh communities can be found here and there around the entire coast of Britain and Ireland, those that are found on the east coast of England, from the Wash to Hampshire, arguably provide the greatest botanical variety and interest. Many of the speciality flowering plants found here grow nowhere else in our region. A keen eye is needed for many of the tinier species but some are showy and spectacular.

A north Norfolk saltmarsh in July, with abundant Common Sea-lavender.

Saltmarsh Goosefoot

Chenopodium chenopodioides
Chenopodiaceae Height to 30cm

Spreading annual that recalls Red Goosefoot (*see* p.43). The whole plant is often tinged red. **Flowers** Reddish green and borne in clusters (Jul–Sep). **Fruits** Reddish green. **Leaves** Diamond-shaped, toothed and always red below in maturity. **Status** Grows on drying mud in saltmarshes, with N Kent being the species' stronghold.

Grass-leaved Orache

Atriplex littoralis Chenopodiaceae
Height to 1m

Upright annual. **Flowers** Small, greenish and borne in spikes with small leaves (Jul–Sep). **Fruits** Greenish, toothed and warty. **Leaves** Long and narrow, sometimes with shallow teeth. **Status** Locally common on upper reaches of saltmarshes, particularly in E Anglia; also found on bare coastal ground.

Shrubby Sea-blite

Suaeda vera Chenopodiaceae
Height to 1m

Much-branched evergreen perennial with woody stems. **Flowers** Tiny and yellowish green; 1–3 appear in leaf axils (Jun–Oct). **Fruits** Contain rounded black seeds. **Leaves** Succulent, bluish green, blunt and almost semi-circular in cross-section. **Status** Restricted to coastal shingle and upper saltmarshes; locally common on the E Anglian coast.

Slender Hare's-ear

Bupleurum tenuissimum
Apiaceae Height to 50cm

Slender and easily overlooked annual, especially when not in flower. **Flowers** Yellow and borne in tiny umbels, 3–4mm across, surrounded by bracts and arising from leaf axils (Jul–Sep). **Fruits** Globular. **Leaves** Narrow and pointed. **Status** Local, entirely restricted to coastal grassland and upper reaches of saltmarshes, in S and E England only.

Sea-heath

Frankenia laevis Frankeniaceae
Prostrate

Branched, mat-forming and woody perennial. **Flowers** 5mm across with 5 pink and crinkly petals (Jun–Aug). **Fruits** Capsules. **Leaves** Small and narrow with inrolled margins; densely packed and opposite on side-shoots. **Status** Restricted to the drier, upper reaches of saltmarshes. Consequently, the species is local; it is found only from Hampshire to Norfolk.

Least Lettuce

Lactuca saligna Asteraceae
Height to 1m

Slender annual. Easy to overlook when its flower heads are closed; they open only on sunny days, at around 9–11am. **Flowers** In heads, around 1cm across, with yellow florets; at intervals up stem (Jul–Aug). **Fruits** Have unbranched hairs. **Leaves** Upper stem leaves are spear-shaped with clasping bases; lower leaves are increasingly divided. **Status** Grows on sea walls and upper saltmarshes; local along Thames Estuary and Rye Harbour, Sussex.

Matted Sea-lavender

Limonium bellidifolium
Plumbaginaceae Height to 25cm

Compact, wiry perennial. **Flowers** Pinkish lilac and borne in arching sprays, with many non-flowering shoots below (Jul–Aug). **Fruits** Capsules. **Leaves** Spoon-shaped and mainly basal, withering before flowers appear. **Status** Local and rather scarce. Grows in upper reaches of saltmarshes and is restricted mainly to the N Norfolk coast.

Estuarine bird life

Birdwatchers love estuaries because of the vast numbers of waders and wildfowl they support. The birds live there mainly from autumn to spring and feed on the wealth of invertebrate life that thrives in the oozing mud. Find out more about wading birds on pages 260–1.

Redshank

Flowers of sand & shingle

Sandy beaches, dunes and shingle beaches support hardy, resilient plants, specially adapted to cope with wind, inundation by salt water and sandy or pebbly ground. Just above the high-tide line, terrestrial plants such as Sea Sandwort, Sea-holly, Sea-milkwort and Marram Grass (*see* pp.44, 98, 101 and 169 for descriptions) are the first colonisers, starting the stabilising process that leads to the formation of sand dunes. In established dunes, a progression of vegetation can be followed inland, leading eventually to the creation of scrub and woodland. Wet dune hollows (slacks) are particularly rewarding areas for these plants. Many colonisers of sandy shores also grow on shingle, tolerating free drainage and little soil. Sea-kale, Yellow Horned-poppy and Sea Pea also favour shingle (*see* pp.61, 63 and 78).

Sea-holly growing on a Hampshire shingle beach.

Ray's Knotgrass
Polygonum oxyspermum
Polygonaceae Prostrate

Mat-forming coastal annual. **Flowers** Pinkish white and arise in leaf axils (Aug–Sep). **Fruits** Nut-like and protrude beyond the withering flower. **Leaves** Oval, leathery and alternate, sometimes with slightly inrolled margins. **Status** Grows on undisturbed coastal sand and shingle beaches. Local and commonest in the W.

Sea Knotgrass
Polygonum maritimum
Polygonaceae Prostrate

Similar to Ray's Knotgrass but perennial and woody at the base. **Flowers** Pinkish and arise in the leaf axils (Jul–Sep). **Fruits** Nut-like and protrude well beyond the perianth. **Leaves** Grey-green and rolled under at the margins. **Status** Grows on sand and shingle beaches in SW England and S Ireland.

Babington's Orache
Atriplex glabriuscula
Chenopodiaceae Prostrate

A spreading, mealy annual. Stems are usually reddish and the whole plant often turns red in autumn. **Flowers** Borne in leafy spikes (Jul–Sep). **Fruits** Diamond-shaped, maturing silvery white. **Leaves** Triangular or diamond-shaped. **Status** Locally common but restricted to stabilised shingle beaches and bare coastal ground.

Did you know?

In addition to the flowering plants that make stabilised sand dunes their home, a number of specialist fungi are unique, or nearly so, to this habitat. The dunes of the north Norfolk coast are a hotspot for unusual species – look out in particular for the **Winter Stalkball** *Tulostoma brumale* (*above*) and the **Tiny Earthstar** *Geastrum minimum* (*below*). Find out more about fungi on pp.184–97.

Sea Mouse-ear

Cerastium diffusum
Caryophyllaceae Height to 30cm

 Low annual that is sometimes prostrate. The plant is covered in sticky hairs. **Flowers** White and 3–6mm across, with notched petals (Apr–Jul). **Fruits** Capsules. **Leaves** Ovate and dark green; bracts do not have transparent margins (those of Common Mouse-ear do; *see* p.44). **Status** Locally common on sand near the coast; scarce inland.

Sand Catchfly

Silene conica Caryophyllaceae
Height to 35cm

 Upright, stickily hairy and greyish-green annual. **Flowers** 4–5mm across with 5 notched and pinkish petals; borne in clusters (May–Jul). **Fruits** Form within inflated flagon-shaped capsules. **Leaves** Narrow and downy. **Status** Local and scarce, restricted mainly to coastal SE England. Grows on sandy soils.

Childing Pink

Petrorhagia nanteuilii
Caryophyllaceae Height to 40cm

 Slender, wiry annual with greyish leaves and stem. **Flowers** 6–8mm across and borne terminally (Jul–Sep). **Fruits** Form within dry capsules. **Leaves** Narrow and wiry. **Status** Grows on stabilised coastal shingle. More widespread in the past, though nowadays found only at Pagham Harbour, Sussex, where it is very locally common.

Suffocated Clover

Trifolium suffocatum Fabaceae
Prostrate

 Unobtrusive annual, easily overlooked because it is low-growing and soon withers and dries. **Flowers** Whitish and in a stalkless central cluster with pointed bracts (Apr–May). **Fruits** Subterranean pods. **Leaves** Stalked and radiate from the centre of the plant. **Status** Scarce; grows on bare shingle and sand near the S coast.

Oysterplant

Mertensia maritima Boraginaceae
Prostrate

 A spreading blue-green plant of N coasts. **Flowers** Bell-shaped, pink in bud but soon turning blue (Jun–Aug). **Fruits** Nutlets. **Leaves** Fleshy, oval and blue-green. **Status** Local, confined to stony N beaches, growing around the high-tide mark. Shetland and Orkney are its best locations, where it is very locally common.

BELOW: Oysterplant

Lizard Orchid

Himantoglossum hircinum
Orchidaceae Height to 1m

 Extraordinary orchid. **Flowers** Smell of goats and have a greenish-grey hood with reddish streaks inside, and a long, twisted lip (up to 5cm); in tall spikes (May–Jul). **Fruits** Form and swell at the base of the flowers. **Leaves** Comprise oval basal leaves that wither, and smaller stem leaves that persist. **Status** Seen regularly in E Kent, on stabilised dunes.

Nesting seabirds

Human disturbance has all but excluded birds such as Ringed Plovers, Oystercatchers and terns from nesting on many beaches in southern England. However, in a few protected areas (particularly on the E Anglian coast), good colonies still exist in the south, while on remote beaches in northern Britain these birds fare better.

nesting Oystercatcher

Campions & catchflies

Many of the members of the pink family shown on these pages are colourful and showy, with flowers designed to attract pollinating insects. Some do have pink flowers, but white is a more typical petal colour among species within the group.

Moss Campion

Silene acaulis Caryophyllaceae
Prostrate

Charming cushion-forming perennial. **Flowers** Pink, 9–12mm across and 5-petalled (Jun–Aug). **Fruits** Capsules. **Leaves** Narrow and densely packed, creating a moss-like appearance to the cushion. **Status** Local on s uitable mountains, from Wales northwards. Typically found on mountain tops and rocky ledges, but also near the sea in the far N.

Greater Sea-spurrey

Spergularia media
Caryophyllaceae Height to 10cm

Robust fleshy perennial. **Flowers** Pinkish white and 7–12mm across, the 5 petals longer than the sepals (Jun–Sep). **Fruits** Capsules. **Leaves** Fleshy, bristle-tipped and semicircular in cross-section. **Status** Widespread and common around coasts, and typically associated with the drier upper reaches of saltmarshes.

Sea Campion

Silene uniflora Caryophyllaceae
Height to 20cm

Cushion-forming perennial that is confined to coastal habitats. **Flowers** White and 20–25mm across, with overlapping petals; borne on upright stems (Jun–Aug). **Fruits** Capsules. **Leaves** Grey-green, waxy and fleshy. **Status** Widespread and locally common around the coast, notably on cliffs and shingle beaches.

Sand Spurrey

Spergularia rubra Caryophyllaceae
Prostrate

Straggling, stickily hairy annual or biennial. **Flowers** Pink and 3–5mm across; the 5 petals are shorter than the sepals (May–Sep). **Fruits** Capsules. **Leaves** Grey-green, narrow and bristle-tipped; borne in whorls with silvery lanceolate stipules. **Status** Widespread and locally common. Found on dry sandy ground, often on heaths or on coastal dunes.

Bladder Campion

Silene vulgaris Caryophyllaceae
Height to 80cm

Upright showy perennial. **Flowers** White, drooping and 16–18mm across (Jun–Aug); petals are deeply divided and the calyx is swollen to form a purple-veined bladder. **Fruits** Capsules. **Leaves** Grey-green and oval; in opposite pairs. **Status** Widespread but common only in the S. Found on dry grassland on well-drained soil; often on chalk.

White Campion

Silene latifolia Caryophyllaceae
Height to 1m

Hairy, branched perennial. Sometimes hybridises with Red Campion. **Flowers** White, 5-petalled and 25–30mm across; separate-sex plants, the male flowers smaller than the females (May–Oct). **Fruits** Have erect teeth. **Leaves** Oval and borne in opposite pairs. **Status** Widespread and common throughout, found on disturbed ground, verges and grassy habitats.

Sea Mouse-ear

Cerastium diffusum
Caryophyllaceae Height to 30cm

 Low annual that is sometimes prostrate. The plant is covered in sticky hairs. **Flowers** White and 3–6mm across, with notched petals (Apr–Jul). **Fruits** Capsules. **Leaves** Ovate and dark green; bracts do not have transparent margins (those of Common Mouse-ear do; *see* p.44). **Status** Locally common on sand near the coast; scarce inland.

Childing Pink

Petrorhagia nanteuilii
Caryophyllaceae Height to 40cm

 Slender, wiry annual with greyish leaves and stem. **Flowers** 6–8mm across and borne terminally (Jul–Sep). **Fruits** Form within dry capsules. **Leaves** Narrow and wiry. **Status** Grows on stabilised coastal shingle. More widespread in the past, though nowadays found only at Pagham Harbour, Sussex, where it is very locally common.

Suffocated Clover

Trifolium suffocatum Fabaceae
Prostrate

 Unobtrusive annual, easily overlooked because it is low-growing and soon withers and dries. **Flowers** Whitish and in a stalkless central cluster with pointed bracts (Apr–May). **Fruits** Subterranean pods. **Leaves** Stalked and radiate from the centre of the plant. **Status** Scarce; grows on bare shingle and sand near the S coast.

Oysterplant

Mertensia maritima Boraginaceae
Prostrate

 A spreading blue-green plant of N coasts. **Flowers** Bell-shaped, pink in bud but soon turning blue (Jun–Aug). **Fruits** Nutlets. **Leaves** Fleshy, oval and blue-green. **Status** Local, confined to stony N beaches, growing around the high-tide mark. Shetland and Orkney are its best locations, where it is very locally common.

Lizard Orchid

Himantoglossum hircinum
Orchidaceae Height to 1m

 Extraordinary orchid. **Flowers** Smell of goats and have a greenish-grey hood with reddish streaks inside, and a long, twisted lip (up to 5cm); in tall spikes (May–Jul). **Fruits** Form and swell at the base of the flowers. **Leaves** Comprise oval basal leaves that wither, and smaller stem leaves that persist. **Status** Seen regularly in E Kent, on stabilised dunes.

Sand Catchfly

Silene conica Caryophyllaceae
Height to 35cm

 Upright, stickily hairy and greyish-green annual. **Flowers** 4–5mm across with 5 notched and pinkish petals; borne in clusters (May–Jul). **Fruits** Form within inflated flagon-shaped capsules. **Leaves** Narrow and downy. **Status** Local and scarce, restricted mainly to coastal SE England. Grows on sandy soils.

BELOW: Oysterplant

Nesting seabirds

Human disturbance has all but excluded birds such as Ringed Plovers, Oystercatchers and terns from nesting on many beaches in southern England. However, in a few protected areas (particularly on the E Anglian coast), good colonies still exist in the south, while on remote beaches in northern Britain these birds fare better.

nesting Oystercatcher

Goosefoots & oraches

Most plants thrive on stability and do not fare well when disturbed. However, a few hardy species can colonise agricultural land or, in the case of coastal species, grow on unstable and constantly changing shingle and sand. Through the very act of growing here, the plants help to stabilise their immediate environment.

Coastal stabilising plants

Goosefoot family members (Chenopodiaceae) are among the most characteristic plants of saltmarshes and sandy shores. The following seven species are among the most important stabilising influences on the coast. Their spreading roots help to bind the loose ground in which they grow.

Frosted Orache

Atriplex laciniata Chenopodiaceae
Usually prostrate

Distinctive silvery-grey plant with stems that are usually flushed with pink. **Flowers** Whitish and borne in clusters (Jul–Sep). **Fruits** Diamond-shaped and toothed. **Leaves** Fleshy, mealy, toothed and diamond-shaped. **Status** Widespread and locally common, but exclusively coastal; a characteristic plant of sandy beaches.

Sea Purslane

Atriplex portulacoides
Chenopodiaceae Height to 1m

Spreading, mealy perennial that sometimes forms rounded clumps. Entirely coastal, and restricted to the drier reaches of saltmarshes. **Flowers** Yellowish and borne in spikes (Jul–Oct). **Fruits** Lobed. **Leaves** Grey-green and oval at the base but narrow further up the stem. **Status** Widespread and locally common.

Perennial Glasswort

Sarcocornia perennis
Chenopodiaceae Height to 30cm

Branched and patch-forming succulent perennial with woody lower stems that turn orange with age. **Flowers** Small and yellow (Aug–Oct). **Fruits** Minute; they appear at stem junctions, in 3s, the central one being the largest. **Leaves** Small, paired and fleshy. **Status** Local around the coasts of S and E England and S Wales; restricted to the drier reaches of saltmarshes.

Common Glasswort

Salicornia europaea
Chenopodiaceae Height to 30cm

Fleshy, yellowish-green annual that recalls a miniature cactus. Often appears segmented. **Flowers** Tiny; they appear at stem junctions, are of equal size and occur in 3s (Aug–Sep). **Fruits** Minute seeds. **Leaves** Small, paired and fleshy. **Status** Locally abundant but entirely coastal: a classic saltmarsh plant.

flowers

Sea Beet

Beta vulgaris ssp. *maritima*
Chenopodiaceae Height to 1m

Sprawling, clump-forming coastal perennial. **Flowers** Green and borne in dense leafy spikes (Jul–Sep). **Fruits** Spiky, often sticking together in a clump. **Leaves** Dark green, glossy and leathery with reddish stems; their shape varies from oval to triangular. **Status** Locally common on cliffs, shingle beaches and other coastal habitats.

flowers

Prickly Saltwort

Salsola kali Chenopodiaceae
Height to 50cm

Spiky-looking, prickly annual. **Flowers** Tiny and yellowish, appearing at leaf bases (Jul–Oct). **Fruits** Similar to flowers. **Leaves** Swollen, flattened-cylindrical and spiny-tipped. **Status** Locally common on sandy beaches, usually growing near the strandline. Occasionally found beside regularly salted roads inland.

Annual Sea-blite

Suaeda maritima Chenopodiaceae
Height to 50cm

Much-branched annual. Forms small clumps that vary from yellowish green to reddish. **Flowers** Tiny and green; 1–3 appear in axils of upper leaves (Aug–Oct). **Fruits** Produce dark, flattish seeds. **Leaves** Succulent, cylindrical and pointed. **Status** Widespread and locally common in the upper reaches of saltmarshes around the coast.

Disturbed-ground specialists

Goosefoots and oraches are often found on cultivated ground, in gardens and on farms.

Red Goosefoot

Chenopodium rubrum
Chenopodiaceae
Height to 60cm

Variable upright annual. The stems often turn red in old or parched specimens. **Flowers** Small and numerous, borne in upright leafy spikes. **Fruits** Rounded and enclosed by 2–4 sepals (Jul–Oct). **Leaves** Shiny, diamond-shaped and toothed. **Status** Widespread and common in S, especially on manure-enriched land.

Common Orache

Atriplex patula Chenopodiaceae
Height to 60cm

Variable, branched annual; sometimes upright but often prostrate. **Flowers** Small, greenish and borne in leafy spikes (Jul–Sep). **Fruits** Diamond-shaped, toothless and lack warts. **Leaves** Toothed; the upper ones are lanceolate while the lower ones are triangular. **Status** Widespread and common.

Spear-leaved Orache

Atriplex prostrata
Chenopodiaceae Height to 70cm

Upright annual; stems are often tinged red. **Flowers** Borne in rather short spikes (Jul–Sep). **Fruits** Triangular and surrounded by green bracts. **Leaves** Triangular in outline and toothed; basal, largest teeth are at right angles to stalk. **Status** Widespread and locally common. Favours waste and bare ground near the sea.

Pinks

Pinks (Caryophyllaceae) have five-petalled flowers. The three illustrated here are rather delicate and have dainty white flowers (*see also* pp.44–7).

Thyme-leaved Sandwort

Arenaria serpyllifolia
Caryophyllaceae
Usually prostrate

Downy grey-green annual. **Flowers** White, 5-petalled and 5–7mm across (May–Sep); green sepals are shorter than petals. **Fruits** Pear-shaped. **Leaves** Oval; borne in opposite pairs on slender stems. **Status** Common on dry, bare soils.

Three-nerved Sandwort

Moehringia trinervia
Caryophyllaceae
Height to 40cm

Straggly, downy annual. **Flowers** White, 5-petalled and 14–16mm across; borne on long stalks (Apr–Jul). The white-margined green sepals are twice the length of the petals. **Fruits** Capsules. **Leaves** Ovate, with 3–5 obvious veins beneath. **Status** Fairly common in undisturbed woods.

Spring Sandwort

Minuartia verna Caryophyllaceae
Height to 10cm

Attractive, slightly downy perennial. **Flowers** White, 5-petalled and 7–9mm across (May–Sep). The green sepals are shorter than the petals. **Fruits** Capsules. **Leaves** Narrow, 3-veined and carried in whorls on slender stems. **Status** Local and extremely habitat-specific: characteristic of bare limestone soils or spoil from lead mines.

Stitchworts & chickweeds

Stitchworts, chickweeds, pearlworts and spurreys are pink family members (Caryophyllaceae) whose understated flowers have muted colours. What they lack in intensity, they make up for with abundance and many species flower in profusion (*see also* p.43).

Greater Stitchwort

Stellaria holostea Caryophyllaceae
Height to 50cm

 Familiar wayside perennial with rough-edged stems. **Flowers** White, with 5 notched petals; borne on the slender stems (Apr–Jun). **Fruits** Capsules. **Leaves** Narrow, fresh green, rough-edged and grass-like; easily overlooked in the absence of flowers. **Status** Widespread and common, found in open woodland and hedgerows.

Plants that tolerate salt

All flowering plants need water to survive but most will tolerate only fresh water, both surrounding their roots and on their foliage; even a brief inundation with sea water is enough to kill them. But there are a few species that thrive in saline soils and even tolerate salt spray and occasional inundation by the sea. These salt-tolerant plants are called halophytes, and **Sea Sandwort** *Honckenya peploides* (Caryophyllaceae) is one of the most regularly encountered of these around our shores. It is a prostrate, mat-forming perennial. The flowers are greenish white and 6–8mm across (May–Aug). The petals are slightly shorter than the sepals. The fruits are yellowish green and pea-like; and the leaves are oval and fleshy, borne as opposite pairs on creeping stems. It is a locally common species around coasts, particularly on stabilised coastal shingle and sandy beaches. Find out more about salt-tolerant plants on pp.38–43.

Lesser Stitchwort

Stellaria graminea
Caryophyllaceae Height to 50cm

 Perennial of open woodland, meadows and hedgerows, mainly on acid soils. Note the smooth-edged stems.
Flowers White and 5–15mm across, with 5 deeply divided petals (May–Aug). **Fruits** Capsules. **Leaves** Long, narrow, smooth-edged and grass-like. **Status** Widespread and common throughout.

Common Chickweed

Stellaria media Caryophyllaceae
Height to 30cm

 Sprawling annual; sometimes prostrate. The stems are hairy in lines on alternate sides between the leaf nodes. **Flowers** White, 5-petalled and 5–10mm across (Jan–Dec); 3–8 stamens. **Fruits** Capsules on long drooping stalks. **Leaves** Oval, fresh green and in opposite pairs; the upper ones are unstalked. **Status** Widespread and common on disturbed ground.

Common Mouse-ear

Cerastium fontanum
Caryophyllaceae Height to 30cm

 Hairy perennial with both flowering and non-flowering shoots. **Flowers** White, 5–7mm across, with 5 deeply notched petals (Apr–Oct). **Fruits** Capsules. **Leaves** Grey-green and borne in opposite pairs. **Status** Widespread and common in gardens and grasslands, and on disturbed ground.

Sticky Mouse-ear

Cerastium glomeratum
Caryophyllaceae Height to 40cm

 Annual with sticky glandular hairs. **Flowers** White, 10–15mm across, with 5 deeply notched petals; carried in compact clustered heads (Apr–Oct). **Fruits** Capsules. **Leaves** Pointed-ovate and borne in opposite pairs. **Status** Widespread and common throughout. Found mainly on dry, bare ground.

Water Chickweed

Myosoton aquaticum
Caryophyllaceae Height to 1m

Straggling perennial. **Flowers** White and 12–20mm across, with 5 deeply divided petals (much longer than the sepals) (Jun–Oct). **Fruits** Capsules. **Leaves** Heart-shaped with wavy edges. Borne in opposite pairs, the upper leaves unstalked. **Status** Common in England and Wales but scarce elsewhere; favours damp grassy ground and river margins.

Heath Pearlwort

Sagina subulata Caryophyllaceae
Height to 10cm

Mat-forming downy perennial with a basal rosette. **Flowers** Have 5 white petals that are equal in length to, or longer than, the sepals; borne on slender stickily hairy stalks (May–Aug). **Fruits** Capsules. **Leaves** Narrow, bristle-tipped and downy. **Status** Locally common in the N and W; scarce or absent elsewhere. Found on dry, sandy or gravelly soils.

Procumbent Pearlwort

Sagina procumbens
Caryophyllaceae Prostrate

Creeping perennial. Forms mats comprising a central rosette with radiating shoots that root at intervals, giving rise to erect flowering stems. **Flowers** Green, petal-less and borne on side-shoots (May–Sep). **Fruits** Capsules. **Leaves** Narrow and bristle-tipped but not hairy. **Status** Widespread and common on damp bare ground.

Procumbent Pearlwort

Corn Spurrey

Spergula arvensis Caryophyllaceae
Height to 30cm

Straggling stickily hairy annual. **Flowers** 4–7mm across and have 5 whitish petals (May–Aug). **Fruits** Capsules; longer than sepals and drooping at first. **Leaves** Narrow and borne in whorls along the stems. **Status** Widespread but not as common as it used to be, owing to herbicide use. A weed of arable land with sandy soils.

Rock Sea-spurrey

Spergularia rupicola
Caryophyllaceae Height to 20cm

Stickily hairy perennial, often with purplish stems. Sometimes forms clumps with woody bases. **Flowers** Pink, 5-petalled (petals and sepals equal) and 8–10mm across (Jun–Sep). **Fruits** Capsules. **Leaves** Narrow, flattened and fleshy; borne in whorls. **Status** Locally common in the W, on cliffs and rocky places near the sea.

Rock Sea-spurrey

Upright Chickweed

Moenchia erecta Caryophyllaceae
Height to 8cm

Tiny upright annual. **Flowers** White with 4 petals; they open only in bright sunshine (Apr–Jun). The sepals are white-edged and longer than petals. **Fruits** Capsules. **Leaves** Waxy grey-green, stiff and narrow. **Status** Local in England and Wales only, growing on short, dry grassland, typically on gravelly or sandy soils.

Knotted Pearlwort

Sagina nodosa Caryophyllaceae
Height to 12cm

Wiry perennial; the stems look 'knotted' owing to the clustered arrangement of leaves. **Flowers** White, 5-petalled (the petals are twice as long as the sepals) and 10mm across, with 5 styles (*cf.* Spring Sandwort, p.43) (Jul–Sep). **Fruits** Capsules. **Leaves** Short and clustered. **Status** Widespread but local. Found on damp sandy soils, often in coastal areas.

Campions & catchflies

Many of the members of the pink family shown on these pages are colourful and showy, with flowers designed to attract pollinating insects. Some do have pink flowers, but white is a more typical petal colour among species within the group.

Moss Campion
Silene acaulis Caryophyllaceae
Prostrate

 Charming cushion-forming perennial. **Flowers** Pink, 9–12mm across and 5-petalled (Jun–Aug). **Fruits** Capsules. **Leaves** Narrow and densely packed, creating a moss-like appearance to the cushion. **Status** Local on s uitable mountains, from Wales northwards. Typically found on mountain tops and rocky ledges, but also near the sea in the far N.

Greater Sea-spurrey
Spergularia media
Caryophyllaceae Height to 10cm

 Robust fleshy perennial. **Flowers** Pinkish white and 7–12mm across, the 5 petals longer than the sepals (Jun–Sep). **Fruits** Capsules. **Leaves** Fleshy, bristle-tipped and semicircular in cross-section. **Status** Widespread and common around coasts, and typically associated with the drier upper reaches of saltmarshes.

Sea Campion
Silene uniflora Caryophyllaceae
Height to 20cm

 Cushion-forming perennial that is confined to coastal habitats. **Flowers** White and 20–25mm across, with overlapping petals; borne on upright stems (Jun–Aug). **Fruits** Capsules. **Leaves** Grey-green, waxy and fleshy. **Status** Widespread and locally common around the coast, notably on cliffs and shingle beaches.

Sand Spurrey
Spergularia rubra Caryophyllaceae
Prostrate

 Straggling, stickily hairy annual or biennial. **Flowers** Pink and 3–5mm across; the 5 petals are shorter than the sepals (May–Sep). **Fruits** Capsules. **Leaves** Grey-green, narrow and bristle-tipped; borne in whorls with silvery lanceolate stipules. **Status** Widespread and locally common. Found on dry sandy ground, often on heaths or on coastal dunes.

Bladder Campion
Silene vulgaris Caryophyllaceae
Height to 80cm

 Upright showy perennial. **Flowers** White, drooping and 16–18mm across (Jun–Aug); petals are deeply divided and the calyx is swollen to form a purple-veined bladder. **Fruits** Capsules. **Leaves** Grey-green and oval; in opposite pairs. **Status** Widespread but common only in the S. Found on dry grassland on well-drained soil; often on chalk.

White Campion
Silene latifolia Caryophyllaceae
Height to 1m

 Hairy, branched perennial. Sometimes hybridises with Red Campion. **Flowers** White, 5-petalled and 25–30mm across; separate-sex plants, the male flowers smaller than the females (May–Oct). **Fruits** Have erect teeth. **Leaves** Oval and borne in opposite pairs. **Status** Widespread and common throughout, found on disturbed ground, verges and grassy habitats.

Red Campion

Silene dioica Caryophyllaceae
Height to 1m

Hairy biennial or perennial. **Flowers** Reddish pink and 20-30mm across; the male flowers are smaller than the females and occur on separate plants (Mar–Oct). **Fruits** Reveal 10 reflexed teeth when ripe. **Leaves** Hairy and borne in opposite pairs. **Status** Widespread and common in hedgerows, on grassy banks and in wayside places generally.

Ragged Robin

Lychnis flos-cuculi
Caryophyllaceae Height to 65cm

Delicate-looking perennial. **Flowers** Comprise 5 pink petals; each is divided into 4 'ragged' lobes (May–Aug). **Fruits** Capsules. **Leaves** Narrow, grass-like and rough, the upper ones in opposite pairs. **Status** Widespread and common in damp meadows, fens and marshes, but decreasing owing to agricultural changes, such as land drainage.

Maiden Pink

Dianthus deltoides
Caryophyllaceae Height to 20cm

Hairy perennial that sometimes forms clumps. **Flowers** 18-20mm across, with 5 pink petals that show white basal spots and have toothed margins (Jun–Sep). **Fruits** Capsules. **Leaves** Narrow, rough-edged and grey-green. **Status** Widespread but extremely local and declining. Associated with dry sandy soils, often on golf-course margins.

Corncockle

Agrostemma githago
Caryophyllaceae Height to 70cm

Distinctive downy annual. **Flowers** 30-45mm across, with 5 pinkish-purple petals and long, narrow, radiating sepals (May–Aug). **Fruits** Capsules. **Leaves** Narrow and grass-like. **Status** A plant of arable fields. Formerly widespread and common, but now extremely scarce and erratic owing to the use of agricultural herbicides.

Annual Knawel

Scleranthus annuus
Caryophyllaceae Height to 10cm

Yellowish-green sprawling annual. **Flowers** Comprise green pointed sepals and no petals; borne in clustered heads (May–Aug). **Fruits** Capsules. **Leaves** Narrow, pointed and borne in opposite pairs along the wiry stems. **Status** Widespread and locally common throughout. Associated with dry, bare soil and the margins of arable fields where they have not been sprayed.

flowers

Night and day

Many insect-pollinated flowers are bright and attention-grabbing, with petals that are designed to attract day-flying insects such as butterflies, bees and hoverflies, and arranged in a such a fashion as to ensure successful pollen transfer. Less is understood about the role of nocturnal insects in the process of pollination, but moths in particular are known to be important. Many flowers, especially those that are heavily scented, are pollinated by both day- and night-flying insects. **Night-flowering Catchfly** *Silene noctiflora* (**1**) and **Nottingham Catchfly** *S. nutans* (**2**) are more specific and rely on nocturnal insects exclusively; indeed, their flowers open only after dark, ensuring that moths are their sole pollinators. This strategy prevents the female part of the flower from being inundated by pollen from day-pollinated flowers and increases the chances of successful cross-pollination.

Night-flowering Catchfly is a scarce plant of arable fields, mainly on chalk or sandy soils. Its flowers, when open, recall those of White Campion. Nottingham Catchfly is a scarce plant of shingle beaches and calcareous grassland. In both species, the petals are inrolled during the daytime.

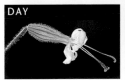

Common buttercups

At first glance, many of our common buttercup species (Ranunculaceae) look confusingly similar to one another. To identify them, pay attention to details of flower structure and leaf shape, and to where they are growing. In particular check whether or not the sepals are reflexed (folded back); whether the leaves and plant stalks are hairy or hairless; and the degree to which the leaves are dissected.

Meadow Buttercup

Ranunculus acris Ranunculaceae Height to 1m

Downy perennial. **Flowers** 18-25mm across, comprising 5 shiny yellow petals with upright sepals; borne on long, unfurrowed stalks (Apr–Oct). **Fruits** Hook-tipped, in a rounded head. **Leaves** Rounded and divided into 3-7 lobes; upper ones unstalked. **Status** Widespread and abundant throughout. Found in damp grassland.

Creeping Buttercup

Ranunculus repens Ranunculaceae Height to 50cm

Often unwelcome perennial whose long, rooting runners aid its spread. **Flowers** 20-30mm across, with 5 yellow petals and upright sepals; borne on furrowed stalks (May–Aug). **Fruits** In rounded heads. **Leaves** Hairy; divided into 3 lobes; the middle lobe is stalked. **Status** Widespread and common. Grows on lawns and other grassy places.

Bulbous Buttercup

Ranunculus bulbosus Ranunculaceae Height to 40cm

Hairy perennial. Note the swollen stem base. **Flowers** 20-30mm across with 5 bright yellow petals and reflexed sepals; borne on furrowed stalks (Mar–Jul). **Fruits** Smooth. **Leaves** Divided into 3 lobes, each of which is stalked. **Status** Widespread and often abundant. Found on dry grassland, including chalk downs.

Hairy Buttercup

Ranunculus sardous Ranunculaceae Height to 40cm

Hairy annual. **Flowers** 15-25mm across with 5 pale yellow petals and reflexed sepals (May–Oct). **Fruits** With a green border, inside which they are adorned with warts. **Leaves** Divided into 3 lobes; mainly basal. **Status** Local, mainly in the S and in coastal habitats. Favours grassy places, especially near the coast.

Corn Buttercup

Ranunculus arvensis Ranunculaceae Height to 40cm

Downy annual. **Flowers** 10-12mm across with 5 pale lemon-yellow petals (May–Jul). **Fruits** Distinctly spiny and bur-like. **Leaves** Divided into narrow lobes. **Status** Formerly widespread but now scarce, local and declining because of agricultural herbicide use; regular only in S England. Grows in arable fields.

Goldilocks Buttercup

Ranunculus auricomus Ranunculaceae Height to 30cm

Slightly hairy perennial. **Flowers** 15-25mm across with yellow petals, 1 or more of which is sometimes imperfect or absent (Apr–May). **Fruits** Roughly hairy. **Leaves** Rounded, 3-lobed basal leaves; narrowly lobed stem leaves. **Status** Widespread but local and declining. Found in damp, undisturbed woodland.

imperfect flower

Greater Spearwort

Ranunculus lingua Ranunculaceae Height to 1m

Robust, upright perennial. Plant has long runners. **Flowers** 20-40mm across with 5 yellow petals; on furrowed stalks (Jun–Sep). **Fruits** Rough, winged, with a curved beak. **Leaves** Narrow, 25cm long, sometimes toothed. **Status** Widespread but local. Grows in fens and the shallow margins of ponds and lakes.

Lesser Spearwort

Ranunculus flammula
Ranunculaceae Height to 50cm

 Upright or creeping perennial. Often roots where its leaf nodes touch the ground. **Flowers** 5–15mm across, usually solitary; borne on furrowed stalks (Jun–Oct). **Fruits** Beaked but not winged. **Leaves** Oval (basal leaves); the stem leaves are narrow. **Status** Widespread, commonest in the N. Favours damp ground, often beside rivers.

Lesser Celandine

Ranunculus ficaria Ranunculaceae
Height to 25cm

 Perennial that sometimes forms clumps or patches. **Flowers** 20–30mm across with 8–12 shiny yellow petals and 3 sepals (Mar–May); opening only in sunshine. **Fruits** In a rounded head. **Leaves** Heart-shaped, glossy, dark green. **Status** Widespread and common. Found in hedgerows, open woodland and bare ground.

Stream Water-crowfoot

Ranunculus penicillatus
Ranunculaceae Floating

 Annual or perennial. **Flowers** 15–25mm across with 5 white petals (May–Jul). **Fruits** In rounded heads. **Leaves** Lobed, rounded floating leaves and long, thread-like submerged ones that collapse out of water. **Status** Widespread and locally common, but found mainly in the S. Favours fast-flowing chalk streams and rivers.

Round-leaved Crowfoot

Ranunculus omiophyllus
Ranunculaceae Floating

 Creeping annual or biennial. **Flowers** 8–12mm across, with 5 white petals that are twice as long as the sepals (May–Aug). **Fruits** In rounded heads. **Leaves** Lobed and rounded. **Status** Rather local and restricted mainly to S and W England and Wales, and S Ireland. Favours damp, muddy places, especially water seepages.

Celery-leaved Buttercup

Ranunculus sceleratus
Ranunculaceae Height to 50cm

 Yellowish-green annual with hollow stems. **Flowers** 5–10mm across with pale yellow petals; in clusters (May–Sep). **Fruits** With elongated heads. **Leaves** Celery-like, divided into 3 lobes (basal leaves); stem leaves less divided. **Status** Widespread but only locally common in S. Favours marshes and wet grazing meadows, often on trampled ground.

Common Water-crowfoot

Ranunculus aquatilis
Ranunculaceae Floating

 Annual or perennial. **Flowers** 12–20mm across with 5 white petals (Apr–Aug). **Fruits** In rounded heads. **Leaves** Thread-like submerged leaves and floating ones that are entire but with toothed lobes. **Status** Widespread and common throughout. Found in both slow-flowing and still waters.

Pond Water-crowfoot

Ranunculus peltatus
Ranunculaceae Floating

 Annual or perennial. **Flowers** 15–30mm across with 5 white petals (May–Aug). **Fruits** In rounded, long-stalked heads. **Leaves** Lobed, rounded floating leaves and short, rigid, thread-like submerged ones. **Status** Widespread and common throughout. Grows in ponds, lakes and other areas of still water.

River Water-crowfoot

Ranunculus fluitans
Ranunculaceae Floating

 Robust perennial that forms extensive carpets in suitable habitats. **Flowers** 20–30mm across with 5 white overlapping petals (May–Aug). **Fruits** In rounded heads. **Leaves** Divided into narrow, thread-like segments; submerged leaves only. **Status** Widespread in England but scarce elsewhere. Favours fast-flowing streams and rivers.

Early-flowering buttercups

Although true buttercups are easy to recognise, not all members of the plant family to which they belong (Ranunculaceae) share these obvious characteristics, and the appearance of some of them is truly bizarre. Monk's-hood, for example, has inflated, helmeted flowers while those of Traveller's-joy are fine and frothy.

Early risers

Most of our insect-pollinated flowers appear in the months of April to July when most insects are active and on the wing. However, a select band of hardy plants flower much earlier – often in February and March – and on the face of it this may not seem like a sensible strategy. But on mild days in early spring there are indeed insects on the wing: these include the occasional butterfly, and significant numbers of bumblebees, and their activities are enough to serve the needs of the early flowerers.

Stinking Hellebore

Helleborus foetidus
Ranunculaceae Height to 75cm

Robust, strong-smelling perennial. When bruised, smells of rotting meat. **Flowers** Green, bell-shaped and 15–30mm across with purple margins; borne in clusters (Jan–May). **Fruits** Dry, many-seeded and splitting. **Leaves** Divided into toothed lobes; the lower ones persist through winter. **Status** Local and restricted to central and S England and Wales. Found in woodland on calcareous soils.

Green Hellebore

Helleborus viridis Ranunculaceae
Height to 60cm

Scentless perennial. **Flowers** Green (including the margins), with pointed sepals but no petals; borne in clusters (Feb–Apr). **Fruits** Dry, many-seeded and splitting. **Leaves** Divided into bright green elongate lobes; they are not evergreen. **Status** Local and scarce, least so in central and S England and Wales. Found in undisturbed woodland on calcareous soils and often associated with Beech.

Winter Aconite

Eranthis hyemalis Ranunculaceae
Height to 10cm

Attractive perennial that sometimes forms carpets on woodland floors. **Flowers** 12–15mm across, with 6 yellow sepals; borne on upright stems above the leaves (Jan–Apr). **Fruits** Dry, many-seeded and splitting. **Leaves** Spreading (3 per stem) and each divided into 3 lobes. **Status** Introduced to Britain but now widely planted and naturalised for its ornamental value and winter interest, and mostly found in England.

Marsh-marigold

Caltha palustris Ranunculaceae
Height to 25cm

Widespread perennial with stout hollow stems. **Flowers** 25–30mm across and comprise 5 petal-like sepals but no petals (Mar–Jul). **Fruits** Capsules. **Leaves** Kidney-shaped and shiny, up to 10cm across. **Status** Widespread and locally common, but its range is contracting. Found in damp woodland, marshes and wet meadows.

Globeflower

Trollius europaeus Ranunculaceae
Height to 60cm

Attractive perennial. **Flowers** Spherical, 30–40mm across, with 10–15 yellow sepals; borne on long, upright stems (May–Aug). **Fruits** Many-seeded and dry. **Leaves** Palmately divided into toothed lobes. **Status** Absent from the S but very locally common from N Wales to Scotland, also NW Ireland. Found in damp, upland and N meadows.

Monk's-hood

Aconitum napellus Ranunculaceae
Height to 1m

Dark green, almost hairless perennial. **Flowers** Bluish violet, 20mm across and helmeted; borne in upright spikes (May–Aug). **Fruits** Dry and many-seeded. **Leaves** Deeply divided into palmate lobes. **Status** Local, mainly in the S and SW. Found in damp woodland, often beside streams. Beware confusion with naturalised garden *Aconitum* species.

Columbine

Aquilegia vulgaris Ranunculaceae
Height to 1m

Familiar garden perennial but also a native plant. **Flowers** Nodding, purple and 30–40mm long, the petals with hooked-tipped spurs (May–Jul). **Fruits** Dry and many-seeded. **Leaves** Grey-green and comprise 3-lobed leaflets. **Status** Widespread but extremely local as a native species, favouring open woods on calcareous soils.

Traveller's-joy

Clematis vitalba Ranunculaceae
Length to 20m

Scrambling hedgerow perennial. **Flowers** Creamy, with prominent stamens; borne in clusters (Jul–Aug). **Fruits** Comprise clusters of seeds with woolly whitish plumes, hence the plant's alternative name of Old Man's Beard. **Leaves** Divided into 3–5 leaflets. **Status** Locally common in central and S England and in Wales, on chalky soils.

fruits

Mousetail

Myosurus minimus
Ranunculaceae Height to 10cm

Tufted and inconspicuous annual. **Flowers** long-stalked and 5mm across, with yellowish-green petals and sepals (Mar–Jul). **Fruits** Elongate, plantain-like and up to 7cm long. **Leaves** Narrow and grass-like. **Status** Mainly in S England, but scarce and declining. Found on arable field margins, often on sandy soil. (For more about arable field plants, *see* pp.54–5.)

Wood Anemone

Anemone nemorosa
Ranunculaceae Height to 30cm

Perennial that sometimes forms large carpets on suitable woodland floors. **Flowers** Solitary and comprise 5–10 white or pinkish petal-like sepals (Mar–May). **Fruits** Beaked and borne in rounded clusters. **Leaves** The stem leaves are long-stalked and divided into 3 lobes, each of these being further divided. **Status** Widespread and locally common.

Common Meadow-rue

Thalictrum flavum
Ranunculaceae Height to 1m

Upright perennial. **Flowers** Have small petals that drop, but showy yellow anthers; borne in dense clusters (Jun–Aug). **Fruits** Dry and papery. **Leaves** Fern-like and divided 2 or 3 times into toothed lobes. **Status** Widespread but local and common only in the S and E of England. Found in damp meadows, ditches and fens, and favours alkaline soils.

Lesser Meadow-rue

Thalictrum minus Ranunculaceae
Height to 1m

Variable, often short perennial. **Flowers** Yellowish, tinged purple, with prominent dangling stamens; borne in open clusters, the flowers drooping at first then erect (Jun–Aug). **Fruits** Dry and papery. **Leaves** Deeply divided 3 or 4 times. **Status** Widespread but local plant of dunes, dry grassland and rocky slopes; grows mainly in soils on chalk or limestone.

Farmland

The British and Irish countryside has been farmed for millennia – in fact, from the time when man first colonised these islands. It is hardly surprising then that, across much of the region, a patchwork of farmland dominates the landscape. Although intensive agriculture has now destroyed much intrinsic wildlife interest, a growing awareness of the need for environmentally sensitive farming is helping to reverse this trend.

Farmland birds have declined overall but there are some success stories. The **Common Buzzard** (*see* p.249) has increased in recent years thanks in part to a decline in persecution and increased Rabbit numbers. Woodpigeons have also fared well, as Rape – planted for oilseed – is a source of winter food.

Arable 'weeds' such as **Corncockle** (*see* p.47), once widespread and abundant, are now almost extinct in the wild, eliminated from the scene by agricultural herbicides. Find out about other arable field specialist plants such as Ground-pine, Pheasant's-eye and Corn Parsley, and their requirements, on pp.54–5.

Modern agricultural practices essentially exclude the **Harvest Mouse** (*see* p.203) from arable fields of wheat and barley. But it still lives in rough hay meadows and hedgerow margins. Field Voles, Common Shrews, Moles, Rabbits, Brown Hares and Foxes are among other frequently encountered farmland mammals.

Arable fields fall loosely into the category of grasslands (wheat, barley and oats are grasses, after all) but their botanical interest is often minimal. Even where organic methods are employed, invariably the intention is to grow the crop plant to the exclusion of all others. And where modern herbicides are employed, this outcome is usually achieved with frightening success: 'weed' plants are kept to an absolute minimum. Decades of chemical use have resulted in the soil's seed bank being depleted dramatically, so that many of the more delicate arable 'weed' species are essentially things of the past in many parts of Britain.

However, despite the best attempts of intensive agriculture, some species that were previously considered to be 'weeds'

have clung on, typically confined to scraps of marginal land that escape spraying either by luck or, in a few instances, through the foresight of enlightened farmers, of whom there is a growing number. And in areas where spraying ceases, for whatever reason, many of these arable plants do return in good numbers after a few years, which pleases conservationists although not necessarily farmers.

Of course, it is not just arable plants that have suffered from the intensification of farming in recent decades. Insecticides and toxic chemicals targeted at other invertebrates have depleted so-called pest animal species, and in many areas hedgerows – refuges for wildlife – are often grubbed out or badly managed. It is

little wonder then that most farmland birds – often used as barometers of the state of countryside – have suffered catastrophic decline. In many areas there is little for them to feed on, either in winter or summer, and nesting sites for many have either been removed or made inaccessible through changes in land use.

For many, the answer lies in a return to low-intensity mixed farming – hay meadows, grazing livestock and arable fields – with subsidies and grants targeted intelligently to encourage the production of good-quality food, to maximise biodiversity and, above all, to ensure a reasonable income for farmers. Changes in the way that we, collectively, fund farming subsidies could, and should, address all these issues.

If nutritional value rather than bulk is the aim, then botanical diversity is key to a good hay crop. Consequently, hay connoisseurs look for meadows that contain a wide range of native grass species, including **Cock's-foot** (*see* p.166), rather than the uniformity of grassland seeded with rank alien species.

A good hay meadow should be a floral delight, with native wildflower species in abundance. The presence of **Yellow-rattle** (*see* p.124), which is semi-parasitic on grass roots, is a sign of a mature, undisturbed meadow. Other hay meadow flowers include buttercups, Lesser Stitchwort, Bird's-foot Trefoil and knapweeds.

You can often detect a good hay meadow by the sound emanating from it – the songs of grasshoppers and **Roesel's Bush-crickets** (*see* p.379) fill the air in summer. Sweep a net through the vegetation and you will collect bush-crickets, beetles and moth caterpillars, as well as spiders and other invertebrates.

Flowers of arable fields

Arable fields are disturbed on a regular basis by ploughing. Most plants shun such conditions, but a group of specialists, which includes poppies for example, thrives here and indeed is seldom found elsewhere. These plants are adapted to regular disturbance of the soil and cannot compete with more vigorous plants that become established if the disturbance ceases. Most are annuals and have tough, resistant seeds that can remain dormant for decades but will germinate when the soil is tilled or rotavated (*see also* Corncockle (p.47) and Corn Marigold (p.135)).

Small-flowered Catchfly

Silene gallica Caryophyllaceae
Height to 40cm

Stickily hairy annual. **Flowers** 10–12mm across, pinkish or white and sometimes flushed red at the base; in 1-sided spikes (Jun–Oct). **Fruits** Inflated capsules. **Leaves** Hairy; the upper ones narrower than the basal ones. **Status** Widespread but local and scarce on arable land, mainly on sandy soil.

Pheasant's-eye

Adonis annua Ranunculaceae
Height to 40cm

Branched and hairless annual. **Flowers** Comprise 5–8 bright red petals that are blackish at the base (Jun–Aug). **Fruits** Long-stalked, elongate and wrinkled. **Leaves** Deeply divided and feathery, the upper ones partly shrouding the flowers. **Status** Mainly in the S, but local and rare in arable fields on chalky soils; badly affected increased use of agricultural herbicides.

Rough Poppy

Papaver hybridum Papaveraceae
Height to 40cm

Hairy annual. **Flowers** 2–5cm across with 4 crimson overlapping petals that have dark-blotched bases; borne on hairy stalks (Jun–Aug). **Fruits** Ovoid to spherical, with spreading yellowish hairs. **Leaves** Much divided and bristle-tipped. **Status** Mainly in S England; scarce and declining. Grows in arable fields, mainly on calcareous soils.

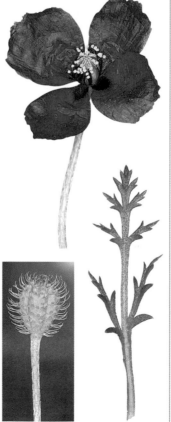

fruit

Prickly Poppy

Papaver argemone Papaveraceae
Height to 30cm

Delicate annual. **Flowers** 2–6cm across with 4 pale red petals that typically do not overlap but that do have a dark basal blotch (May–Aug). **Fruits** Narrow, elongated and ribbed with prickle-like bristles. **Leaves** Much divided and bristle-tipped. **Status** Mainly in S England; local and scarce. Grows on arable land, mainly on sandy soils.

fruit

Shepherd's-needle

Scandix pecten-veneris Apiaceae
Height to 50cm

Hairless annual that has spikes of distinctive fruit. **Flowers** Small, white and borne in umbels (May–Jul). **Fruits** Needle-like and 6–8cm long, much of the length forming a flattened 'beak'. **Leaves** Deeply divided 2–3 times. **Status** Found in S and E England, but scarce and declining. Grows in arable fields.

A hardy knotgrass

Unlikely though it may seem, some plants actually thrive on being trampled, or at least they tolerate a degree of trampling that excludes other species and hence minimises competition. One such plant is **Equal-leaved Knotgrass** *Polygonum arenastrum*, which favours farmland tracks and paths. It is similar in appearance to Knotgrass (*see* p.35), but all its leaves are of a similar size and it grows as prostrate mats on the ground.

Corn Parsley

Petroselinum segetum Apiaceae
Height to 60cm

Slender, wiry perennial that smells of Parsley. **Flowers** White and borne in umbels, 3–5cm across; these are open and irregular owing to the unequal length of the rays (Aug–Sep). **Fruits** Egg-shaped. **Leaves** Divided with ovate toothed leaflets. **Status** Extremely local, mainly the S, usually near the sea.

Spreading Hedge-parsley

Torilis arvensis Apiaceae
Height to 40cm

Wiry, branched and spreading annual. **Flowers** White and borne in long-stalked umbels, 2–4cm across, with 3–5 rays; bracts are absent (Jul–Sep). **Fruits** Egg-shaped with curved spines that lack hooks. **Leaves** Divided once or twice. **Status** Mainly in SE England but declining. Grows in arable fields on chalky soils.

Ground-pine

Ajuga chamaepitys Lamiaceae
Height to 20cm

Distinctive and unusual hairy annual. **Flowers** 8–15mm long and mainly yellow, with small purple markings; borne at leaf nodes (May–Aug). **Fruits** Nutlets. **Leaves** The stem leaves are deeply divided into 3 narrow lobes that smell of pine when rubbed. **Status** S England only; rare. Grows on dry, bare cultivated ground, on calcareous soils.

Field Gromwell

Lithospermum arvense Boraginaceae Height to 40cm

Upright downy perennial. **Flowers** 3–4mm across, 5-lobed and white; borne in clusters (May–Aug). **Fruits** Warty brown nutlets. **Leaves** Strap-shaped, blunter than those of Common Gromwell (*see* p.111), and without prominent side-veins. **Status** Local, mainly in S and E England; declining. Grows in arable fields and on dry, disturbed ground.

ABOVE: fruits
LEFT: flowers

Weasel's-snout

Misopates orontium
Scrophulariaceae Height to 25cm

Attractive and distinctive downy annual. **Flowers** 10–15mm long and pinkish purple, toadflax-like in shape but without a spur; they arise from leaf axils towards the top of the stem (Jul–Oct). **Fruits** Capsules. **Leaves** Narrow and linear. **Status** Mainly in S and E England; scarce and declining. Grows in arable fields, usually on sandy soils.

Broad-leaved Cudweed

Filago pyramidata Asteraceae
Height to 25cm

Similar to Common Cudweed (*see* p.132) but branches from base. **Flowers** In woolly clusters, 7–12mm across, consisting of 10–20 heads and overtopped by leaves; each head has yellow-tipped bract tips (Jul–Aug). **Fruits** Dry and do not split. **Leaves** Oval and sharp-pointed. **Status** S England only; rare. Grows in arable fields on chalky or sandy soils.

Cornflower

Centaurea cyanus Asteraceae
Height to 90cm

Creeping perennial with upright flowering stems. **Flowers** In heads, 15–30mm across, with bluish outer florets and reddish-purple inner florets (Jun–Aug). **Fruits** Hairless. **Leaves** Narrow; basal ones may be lobed. **Status** Formerly a common arable 'weed' prior to the use of modern agricultural herbicides; now local and occasional, though least so in the S.

Did you know?

Arable 'weeds' rely on soil disturbance for their survival and soon become crowded out by the vigorous growth of rank vegetation. There is a sad irony to the fact that grant-funded 'conservation' schemes designed to create wildlife headlands are often the final nail in the coffin at the local level for many scarce arable weeds: they cannot compete with the grasses and clovers with which such headlands are typically planted.

Flowers of the mountains

Britain's mountains are extremely inhospitable places for most plants to grow, being cold, wet and exposed, and often covered in snow during winter. But many moss and lichen species thrive here, and even a few hardy flowering species from a wide range of botanical families have become adapted to upland life. Those plants that can tolerate these conditions often flourish in the absence of competition from vigorous, less hardy species.

Mountain Sorrel

Oxyria digyna Polygonaceae
Height to 30cm

Distinctive, upright and hairless perennial. **Flowers** Greenish with red margins, and borne in loose upright spikes (Jul-Aug). **Fruits** Flat and broad-winged. **Leaves** Entirely basal, and rounded to kidney-shaped. **Status** Widespread and locally common in upland and northern areas. Grows on damp ground, often beside streams or in wet mountainside flushes. Commonest in Scotland.

Alpine Bistort

Persicaria vivipara Polygonaceae
Height to 30cm

Upright, unbranched perennial. **Flowers** Borne in terminal spikes; the upper ones are pale pink, while the lower ones are reddish-brown bulbils (a means of vegetative reproduction) (Jun-Aug). **Fruits** Nut-like. **Leaves** Narrow and grass-like. **Status** Locally common from N Wales northwards in upland grassland.

Cyphel

Minuartia sedoides
Caryophyllaceae
Prostrate

Distinctive and attractive cushion-forming perennial. **Flowers** Yellow, 4mm across and lack petals (Jun-Aug). **Fruits** Capsules. **Leaves** Narrow, fleshy and densely packed. **Status** Generally scarce, and restricted to damp stony ground on mountain tops in the Scottish Highlands and on a few Scottish islands.

Alpine Meadow-rue

Thalictrum alpinum
Ranunculaceae Height to 15cm

Short, easily overlooked perennial. **Flowers** Have purplish sepals and stamens and yellow anthers; borne in terminal clusters on slender stems (May-Jul). **Fruits** Dry and papery. **Leaves** Twice trifoliate, with dark green rounded leaflets. **Status** Local from N Wales to Scotland in upland grassland and on mountain ledges; scarce in W Ireland.

Purple Saxifrage

Saxifraga oppositifolia
Saxifragaceae Creeping

Mat-forming perennial with trailing stems. **Flowers** 10-15mm across and purple (Mar-Apr). **Fruits** Dry capsules. **Leaves** Small, dark green and have bristly margins; in opposite pairs. **Status** Locally common in N England and Scotland; rare in N Wales and NW Ireland. Found on mountains, but also on coastal rocks in the N.

Yellow Saxifrage

Saxifraga aizoides Saxifragaceae
Height to 20cm

Colourful clump-forming perennial. **Flowers** 10-15mm across with bright yellow petals; borne in clusters of 1-10 flowers (Jun-Sep). **Fruits** Dry capsules. **Leaves** Fleshy, narrow, toothed and unstalked. **Status** Locally common in N England, Scotland and N Ireland. Grows beside streams and on damp ground in mountains.

Alpine Saxifrage

Saxifraga nivalis Saxifragaceae
Height to 20cm

 A tough little perennial with glandular hairs on the purplish stems and margins of the leaves, which are purplish below. **Flowers** White or pink and borne in terminal clusters (Jul–Aug). **Fruits** Dry capsules. **Leaves** Fleshy and oval with deep rounded teeth. **Status** Rare in the Scottish Highlands; extremely rare in England, N Wales and W Ireland.

Sibbaldia

Sibbaldia procumbens Rosaceae
Prostrate

 Creeping tufted perennial. **Flowers** 5mm across with yellow petals that are sometimes absent (Jul–Aug). **Fruits** Dry and papery. **Leaves** Bluish green, divided into three leaflets, each of these ovate with 3 terminal teeth. **Status** Local in the Scottish Highlands, rare in N England. Grows on rocky ground and in short grassland in mountains.

Rock Speedwell

Veronica fruticans
Scrophulariaceae Height to 20cm

 Attractive and distinctive perennial with woody stems. **Flowers** 10–15mm across, 4-lobed and deep blue with a reddish centre; borne in open few-flowered terminal clusters (Jul–Sep). **Fruits** Flattened and hairy. **Leaves** Oval, unstalked and slightly toothed. **Status** Rare and restricted to a few mountain ledges high up in the Scottish Highlands.

Alpine Speedwell

Veronica alpina Scrophulariaceae
Prostrate

 Perennial with wiry creeping stems that root at the nodes, and short, upright flowering stems. **Flowers** 7–8mm across, 4-lobed and blue; borne on short stalks in crowded spikes (Jul–Aug). **Fruits** Flattened oval capsules. **Leaves** Oval and blunt-toothed. **Status** Scarce, and restricted to areas of short grass and rocks in the Scottish Highlands.

Alpine Lady's-mantle

Alchemilla alpina Rosaceae
Height to 25cm

 Tufted perennial. **Flowers** Tiny, yellowish green and borne in flat-topped clusters (Jun–Aug). **Fruits** Dry and papery. **Leaves** Lobed, the lobes typically divided to the base or nearly so; the undersurface of the leaf is silky-hairy. **Status** Widespread, but locally common in NW England and Scotland only. Grows in damp upland grassland.

Alpine Willowherb

Epilobium anagallidifolium
Onagraceae Height to 5cm

 Creeping hairless perennial with slender stems. **Flowers** 4–5mm across, pink and seldom open fully; borne on drooping stems (Jul–Aug). **Fruits** Long, erect red pods with cottony seeds. **Leaves** Oval, barely toothed and short-stalked. **Status** Local, restricted to damp ground in mountains, from N England northwards.

BELOW: Alpine Willowherb

Alpine Bartsia

Bartsia alpina Scrophulariaceae
Height to 25cm

 Upright downy perennial; semi-parasitic on the roots of other plants. **Flowers** 15–20mm long, purple and 2-lipped (the upper lip is longer than the lower); borne in spikes (Jul–Aug). **Fruits** Capsules. **Leaves** Oval and untoothed, the upper ones tinged purple. **Status** Rare in limestone grassland in N England and Scotland.

Early-flowering cabbages

Members of the cabbage family (Brassicaceae) are usually relatively easy to recognise: typically, their flowers have four even-sized petals, usually arranged in the shape of a cross. Their seeds are borne in pods, the shapes of which are useful for identification. The species here flower early in the year.

Hairy Bitter-cress

Cardamine hirsuta Brassicaceae
Height to 30cm

Upright annual with hairless stems. **Flowers** 2-3mm across (petals are sometimes absent) and borne terminally (Feb-Nov). **Fruits** Curved, up to 2.5cm long and overtop flowers. **Leaves** Deeply divided with rounded lobes; seen mainly as a basal rosette plus 1-4 stem leaves. **Status** Widespread and common. Found on damp, disturbed ground.

seeds and flowers

Wavy Bitter-cress

Cardamine flexuosa Brassicaceae
Height to 50cm

Similar to Hairy Bitter-cress but taller and with wavy hairy stems. **Flowers** 3-4mm across with 4 white petals (Mar-Sep). **Fruits** Curved and barely overtop flowers. **Leaves** Deeply divided with rounded lobes; seen as a basal rosette plus 4-10 stem leaves. **Status** Widespread and common. Favours damp and disturbed ground.

Large Bitter-cress

Cardamine amara Brassicaceae
Height to 60cm

Upright perennial. **Flowers** 12mm across with 4 white petals and violet anthers (Apr-Jun). **Fruits** Slender beaked pods, up to 4cm long. **Leaves** Deeply divided, with slightly toothed oval lobes. **Status** Widespread but local; scarce in, or absent from, W England and S Ireland. Found in damp, shady places in woods and marshes.

Cuckooflower

Cardamine pratensis Brassicaceae
Height to 50cm

Variable perennial, also known as Lady's-smock. **Flowers** 12-20mm across with 4 pale lilac or white flowers (Apr-Jun). **Fruits** Elongated and beaked. **Leaves** Seen mainly in a basal rosette of deeply divided leaves with rounded lobes; narrow stem leaves are also present. **Status** Widespread and locally common in damp grassy places.

Northern Rock-cress

Arabis petraea Brassicaceae
Height to 30cm

Variable perennial. **Flowers** 5-7mm across with 4 whitish or lilac petals (Jun-Aug). **Fruits** Curved and 4cm long. **Leaves** Appear as a basal rosette of deeply lobed stalked leaves, plus narrow toothed stem leaves. **Status** Local, restricted to rocky places in mountains from N Wales northwards; commonest in Scotland.

Thale Cress

Arabidopsis thaliana Brassicaceae
Height to 50cm

Distinctive annual. **Flowers** 3mm across with 4 white petals; borne in terminal clusters (Mar–Oct). **Fruits** Cylindrical and 20mm long. **Leaves** Broadly toothed, oval and form a basal rosette; upright flowering stems also bear a few small leaves. **Status** Widespread and fairly common in lowland areas. Grows on dry, sandy soils; often found on paths.

flowers and seeds

Hoary Whitlowgrass

Draba incana Brassicaceae
Height to 30cm

Upright hairy biennial. **Flowers** 3–5mm across with 4 slightly notched white petals (Jun–Jul). **Fruits** Cylindrical but twisted. **Leaves** Lanceolate; they appear as untoothed leaves in a basal rosette, plus toothed stem leaves. **Status** Local, restricted to N Wales, Scotland and NW Ireland, on upland limestone rocks and (rarely) on sand dunes in the N.

Common Whitlowgrass

Erophila verna Brassicaceae
Height to 20cm

Variable hairy annual. **Flowers** 3–6mm across and comprise 4 deeply notched whitish petals (Mar–May). **Fruits** Elliptical pods; borne on long stalks. **Leaves** Narrow and toothed; they form a basal rosette from the centre of which the flowering stalk arises. **Status** Common and widespread throughout, growing in dry, bare places.

Shepherd's-cress

Teesdalia nudicaulis Brassicaceae
Height to 25cm

Tufted, usually hairless annual. **Flowers** 2mm across with 4 white petals, 2 of which are much shorter than the others (Apr–Jun). **Fruits** Heart-shaped and notched. **Leaves** Deeply lobed and appear mainly as a basal rosette. **Status** Locally common in the S but rare elsewhere. Found on bare sandy ground or shingle.

Scurvygrass

As the name suggests, scurvygrass was once used to prevent scurvy, a disease caused by vitamin C deficiency – the plant's leaves are rich in it. Several species grow here, and most are salt-tolerant plants found on the coast. The most widely encountered species include: **Common Scurvygrass** *Cochlearia officinalis* (**1**), found on saltmarshes and sea cliffs, with flowers 8–10mm across, kidney-shaped basal leaves and clasping, arrow-shaped stem leaves; **Danish Scurvygrass** *C. danica* (**2**), a low-growing annual of sandy coastal soils, with flowers 4–6mm across, heart-shaped basal leaves and stalked ivy-shaped stem leaves; and **English Scurvygrass** *C. anglica* (**3**), a plant of estuaries, has flowers 10–14mm across, and long-stalked narrow basal leaves and clasping stem leaves.

Shepherd's-purse

Capsella bursa-pastoris
Brassicaceae Height to 35cm

Distinctive annual. **Flowers** 2–3mm across with 4 white petals; borne in terminal clusters (Jan–Dec). **Fruits** Green, triangular and notched. **Leaves** Vary from lobed to entire; the upper leaves are usually toothed and clasp the stem. **Status** Widespread and rather common throughout. Grows in arable fields, tracks, gardens and on wayside ground.

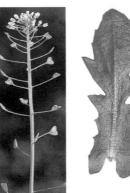

Field Penny-cress

Thlaspi arvense Brassicaceae
Height to 50cm

Annual that emits an unpleasant smell when crushed. **Flowers** 4–6mm across with 4 white petals (May–Sep). **Fruits** Rounded pods with a terminal notch. **Leaves** Comprise narrow, arrow-shaped leaves that clasp the upright stem; it lacks a basal rosette. **Status** Widespread and common throughout. Found in arable fields and on waste ground.

Cabbages

Although members of the cabbage family have shared flower characters, the plants themselves are tremendously varied in appearance and in their growing habits. They range from tiny species that grow, partly submerged, in acidic upland lakes to large bushy plants that adorn waysides and coast and happily survive being washed regularly with salt spray.

Awlwort

Subularia aquatica Brassicaceae
Height to 12cm

Aquatic annual. **Flower** Tiny, white and borne on short stems (Jun–Sep). **Fruits** Ovoid and 2mm long. **Leaves** Up to 6cm long, slender, cylindrical and pointed; they appear as a basal rosette. **Status** Local in the W and N, least so in Scotland. Found on the margins of gravelly upland lakes, usually submerged.

Hoary Cress

Lepidium draba Brassicaceae
Height to 60cm

Variable, often hairless perennial. **Flowers** Tiny and white; borne in large frothy terminal clusters (May–Oct). **Fruits** Heart-shaped and inflated. **Leaves** Grey-green, lance-shaped and toothed; stem leaves clasp the stem. **Status** Introduced; now locally common on disturbed ground.

Smith's Pepperwort

Lepidium heterophyllum
Brassicaceae Height to 40cm

Grey-green hairy branched perennial. **Flowers** 2–3mm across with 4 white petals and violet anthers; borne in dense heads (May–Aug). **Fruits** Oval and smooth. **Leaves** Comprise oval untoothed basal leaves and arrow-shaped clasping steam leaves. **Status** Locally common, especially in the S and W. Grows on dry, bare soil.

Swine-cress

Coronopus squamatus
Brassicaceae Prostrate

Creeping annual or biennial. **Flowers** 2–3mm across and white; borne in compact clusters (Jun–Sep). **Fruits** Knobbly and flattened. **Leaves** Deeply divided and toothed, sometimes forming a dense mat on the ground. **Status** Common only in S and E England. Grows on disturbed and waste ground, often near the sea.

Sea Radish

Raphanus raphanistrum ssp. *maritimus* Brassicaceae
Height to 60cm

Robust, roughly hairy annual. **Flowers** Yellow (May–Jul). **Fruits** Pods with up to 5 beaded segments. **Leaves** Comprise divided lower leaves and narrow entire upper leaves. **Status** Locally common on coasts in the S and SW only. **Similar species** Wild Radish ssp. *raphanistrum* has violet-veined white petals and is a relict of cultivation.

Garlic Mustard

Alliaria petiolata Brassicaceae
Height to 1m

Familiar wayside biennial. **Flowers** 6mm across with 4 white petals (Apr–Jun). **Fruits** Cylindrical, ribbed and 4–5cm long. **Leaves** Heart-shaped, toothed and borne up stem; they smell of garlic when crushed. **Status** Widespread and common, but least so in the N and W of region. Often found in hedgerows and on roadside verges.

BELOW: Sea Radish
INSET: Wild Radish

Sea-kale

Crambe maritima Brassicaceae
Height to 50cm

Robust perennial that forms domed and expansive clumps. **Flowers** 6-12mm across with 4 whitish petals; borne in flat-topped clusters (Jun-Aug). **Fruits** Oval pods. **Leaves** Fleshy with wavy margins; the lower ones are 25cm long and long-stalked. **Status** Very locally common on the coasts of England, Wales and Ireland. Grows on shingle and sandy beaches.

Rape

Brassica napus Brassicaceae
Height to 1.25m

Greyish-green annual or biennial. **Flowers** 25-35mm across with 4 yellow petals, often overtopped by buds (Apr-Sep). **Fruits** Cylindrical and 10cm long. **Leaves** Comprise deeply divided basal leaves that soon wither, and clasping stem leaves. **Status** Widely cultivated but also a casual occurrence in many places. Grows in arable fields and on waste ground.

Charlock

Sinapis arvensis Brassicaceae
Height to 1.5m

Wayside annual. Its stems are often dark reddish. **Flowers** 15-20mm across with 4 yellow petals (Apr-Oct). **Fruits** Cylindrical short-stalked pods with a rounded beak. **Leaves** Dark green, large and coarsely toothed; the lower leaves are stalked and hairy, while the upper ones are entire and unstalked. **Status** Widespread and common on both arable and waste ground.

Vegetable origins

Several wild-growing members of the cabbage family are ancestors of popular cultivated vegetables. For example, the turnip is descended from the **Wild Turnip** *Brassica rapa* ssp. *campestris* (**1**), a grey-green biennial with yellow flowers measuring 6-13mm across and cylindrical fruit pods. However, the most important ancestral *Brassica* is **Wild Cabbage** *B. oleracea* (**2**), a tough perennial of coastal chalk cliffs that has yellow flowers and grey-green fleshy leaves. It has given rise to cultivated forms of cabbage, plus kale, Brussels sprout, cauliflower and broccoli. Ancestral brassicas, although edible, are mostly unpalatable.

1

2

Black Mustard

Brassica nigra Brassicaceae
Height to 2m

Robust greyish annual. **Flowers** 12-15mm across with 4 yellow petals (May-Aug). **Fruits** Flattened and pressed close to stem. **Leaves** Stalked, the lower ones deeply lobed and bristly. **Status** Locally common in England and Wales; rather scarce elsewhere. Often found on sea cliffs, river banks and waste ground.

Sea Rocket

Cakile maritima Brassicaceae
Height to 25cm

Straggling fleshy hairless annual. **Flowers** 6-12mm across, pink or pale lilac and borne in terminal clusters (Jun-Sep). **Fruits** Waisted pods, the upper half being largest. **Leaves** Shiny and deeply lobed. **Status** Widespread and locally common around the coasts of Britain and Ireland. Grows on sandy and shingle beaches.

Cabbages & poppies

Members of the cabbage, fumitory (Fumariaceae) and poppy (Papaveraceae) families have attractive, distinctive flowers. They grow in a range of habitats, from arable fields and hedgerows to shingle beaches and freshwater marshes.

Sea Stock

Matthiola sinuata Brassicaceae
Height to 80cm

 Downy grey-green perennial, the base of which is not woody. **Flowers** Fragrant and 25–50mm across, with 4 pinkish petals (Jun-Aug). **Fruits** Narrow elongated pods. **Leaves** Narrow with toothed or lobed margins. **Status** Rare, growing on coastal dunes and sea cliffs in the SW of the region and in the Channel Islands.

Watercress cultivation

 These days, no green salad is complete without **Watercress** *Rorippa nasturtium-aquaticum*, the leaves of which have a strong peppery taste. This member of the cabbage family is a creeping perennial that grows to a height of 15cm or so. Its flowers have 4 white petals and are carried in terminal heads (May-Oct). Its fruits are pods, while the leaves are dark green and deeply divided. Watercress is widespread and common in shallow streams and ditches, and is widely cultivated in S England.

Hedge Mustard

Sisymbrium officinale
Brassicaceae Height to 90cm

 Tough upright annual or biennial. **Flowers** 3mm across with 4 yellow petals; borne in terminal clusters (May-Oct). **Fruits** Cylindrical, 1–2cm long and pressed close to the stem. **Leaves** Variable: the lower leaves are deeply divided while the stem leaves are narrow. **Status** Widespread and common throughout. Grows on waste ground and disturbed soil.

Winter-cress

Barbarea vulgaris Brassicaceae
Height to 80cm

 Upright hairless perennial. **Flowers** 7–9mm across with 4 yellow petals; borne in terminal heads (May-Aug). **Fruits** Long, narrow 4-sided pods. **Leaves** Dark green and shiny. The lower leaves are divided, the end lobe being large and oval, while the upper stem leaves are entire. **Status** Widespread but commonest in the S. Grows on damp ground.

Marsh Yellowcress

Rorippa palustris Brassicaceae
Height to 50cm

 Annual whose stems are upright, angular and hollow. **Flowers** 3mm across with 4 yellow petals, equal in length to the sepals; borne in terminal heads (Jun-Oct). **Fruits** Elliptical pods, 4-6mm long. **Leaves** Deeply lobed. **Status** Locally common throughout, except in the N. Grows in damp marshy hollows, sometimes in shallow water.

Common Fumitory

Fumaria officinalis Fumariaceae
Height to 10cm

Spreading or scrambling annual. **Flowers** 6-7mm long, pink with crimson tips, spurred and 2-lipped, the lower petal being paddle-shaped; borne in elongated spikes (Apr–Oct). **Fruits** Globular and 1-seeded. **Leaves** Grey-green and much divided; the lobes all lie in one plane. **Status** Widespread and common on well-drained arable soils.

Common Ramping-fumitory

Fumaria muralis Fumariaceae
Height to 10cm

Spreading or upright annual. **Flowers** 9-11mm long and pinkish purple with dark tips, the lower petal almost parallel-sided (not paddle-shaped) with erect margins; borne in spikes of 12-15 flowers (Apr–Oct). **Fruits** Globular and 1-seeded. **Leaves** Much divided. **Status** Widespread and fairly common on arable land, banks and walls.

Climbing Corydalis

Ceratocapnos claviculata
Fumariaceae Height to 70cm

Delicate climbing annual. **Flowers** Creamy white and 5-6mm long; borne in clusters (Jun–Sep). **Fruits** Capsules with 2-3 seeds. **Leaves** Much divided and end in tendrils that assist climbing. **Status** Widespread and common in W Britain; scarcer elsewhere and rare in Ireland. Favours woodland and scrub, mainly on acid soils.

Common Poppy

Papaver rhoeas Papaveraceae
Height to 60cm

Familiar wayside annual. **Flowers** 7-10cm across with 4 papery overlapping scarlet petals (often dark at the base); borne on slender stalks that have spreading hairs (Jun–Aug). **Fruits** Egg-shaped flat-topped capsules. **Leaves** Much divided into narrow segments. **Status** Widespread, commonest in S and E England. Grows on arable land and disturbed ground.

fruit

Yellow Horned-poppy

Glaucium flavum Papaveraceae
Height to 50cm

Blue-grey clump-forming perennial. **Flowers** 6-9cm across with over-lapping yellow petals (Jun–Sep). **Fruits** Elongated curved capsules up to 30cm long. **Leaves** Finely divided, the clasping upper ones having shallow toothed lobes. **Status** Locally common on most suitable shingle beaches although absent from the far N.

fruit

Long-headed Poppy

Papaver dubium Papaveraceae
Height to 60cm

Showy annual. **Flowers** 3-7cm across with 4 papery orange-red over-lapping petals that lack a dark basal blotch; borne on stalks that have close-pressed hairs (Jun–Aug). **Fruits** Narrow, elongated hairless capsules. **Leaves** Much divided into narrow segments. **Status** Widespread and fairly common, including in the N. Grows on arable land and disturbed ground.

fruit

Welsh Poppy

Meconopsis cambrica
Papaveraceae Height to 50cm

Showy perennial. **Flowers** 5-8cm across with 4 overlapping bright yellow petals; borne on slender stems (Jun–Aug). **Fruits** 4- to 6-ribbed capsules that split when ripe. **Leaves** Finely divided, toothed and stalked. **Status** Native to Wales, SW England and Ireland, growing in shady woods; naturalised as a garden escape elsewhere.

Greater Celandine

Chelidonium majus Papaveraceae
Height to 80cm

Tall, brittle-stemmed perennial. **Flowers** 2-3cm across and comprise 4 non-overlapping bright yellow petals (Apr–Oct). **Fruits** Narrow capsules that split from below when ripe. **Leaves** Grey-green and finely divided. **Status** Native in most areas, growing in hedgerows and along woodland rides; also naturalised as a garden escape.

Sundews & stonecrops

These plants are adapted to challenging habitats. Mignonette family members (Resedaceae) often colonise recently disturbed ground; sundews (Droseraceae) live in waterlogged, acid soils and supplement their nutrient intake by catching insects; and stonecrops (Crassulaceae) thrive on free-draining ground with little soil.

Carnivorous plants

Heath and moorland bogs are challenging places for plants to grow, as the ground is waterlogged for much of the time and conditions are often extremely acidic. Such habitats also lack nutrients essential to plants, but a select band of species has devised a novel way of overcoming this shortfall. Butterworts (*see* p.126) and sundews (family Droseraceae) are carnivorous: their sticky leaves trap insects, which are then digested and the resultant nutrients absorbed by the plants. Three sundew species are found in our region: **Round-leaved Sundew** *Drosera rotundifolia* (**1**), with rounded stalked leaves, which is widespread and locally common; **Oblong-leaved Sundew** *D. intermedia* (**2**), with 1cm-long, narrow, tapering leaves, which is commonest in the SW; and **Great Sundew** *D. anglica* (**3**), with 3cm-long, narrow, tapering leaves, which is locally common only in the N and NW.

1

2

2

3

Weld

Reseda luteola Resedaceae
Height to 1.2m

Upright hollow-stemmed biennial. **Flowers** Yellow-green with 4 petals; borne in tall spikes in the plant's second year (Jun–Aug). **Fruits** Globular pods. **Leaves** Narrow; they form a basal rosette in the plant's first year but appear as stem leaves in the second. **Status** Widespread and fairly common, except in the N and W. Grows on disturbed calcareous ground.

Navelwort

Umbilicus rupestris Crassulaceae
Height to 15cm

Distinctive perennial. **Flowers** Whitish, tubular and drooping; borne in spikes (Jun–Aug). **Fruits** Dry and splitting. **Leaves** Rounded and fleshy with a depressed centre above the leaf stalk. **Status** Widespread in W Britain and Ireland; scarce elsewhere. Grows on walls and banks, often in partial shade.

Wild Mignonette

Reseda lutea Resedaceae
Height to 70cm

Upright solid-stemmed biennial that is superficially similar to Weld. **Flowers** Yellowish green with 6 petals; borne in compact spikes (Jun–Aug). **Fruits** Erect oblong pods. **Leaves** Deeply divided with wavy edges. **Status** Widespread and locally common, but almost absent from Scotland. Grows on disturbed calcareous ground.

New Zealand Pygmyweed

Crassula helmsii Crassulaceae
Prostrate on land

Unwelcome creeping perennial. **Flowers** Tiny with 5 white or pink petals (Jun–Sep). **Fruits** Dry and splitting. **Leaves** Narrow, fleshy and 5–10mm long. **Status** An introduced species now widespread in England and expanding its range. It is a major threat to native aquatic plants: it colonises ponds and their margins, eventually excluding all native plant species.

Mossy Stonecrop

Crassula tillaea Crassulaceae
Prostrate

Tiny, often reddish annual that is easily overlooked. **Flowers** Tiny, whitish and arise from leaf axils (Jun–Sep). **Fruits** Dry and splitting. **Leaves** 1–2mm long, oval and densely crowded on the slender stems. **Status** Extremely local, mainly in S and E England. Found on bare and usually damp sandy soil and gravel.

Roseroot

Sedum rosea Crassulaceae
Height to 30cm

Distinctive greyish perennial. **Flowers** Yellow and 4-petalled; borne in rounded terminal clusters (May–Jul). **Fruits** Orange, superficially resembling the flowers. **Leaves** Succulent, oval and overlapping. **Status** Locally common in W Wales, N England, Scotland and Ireland. Grows on mountain ledges and sea cliffs.

English Stonecrop

Sedum anglicum Crassulaceae
Height to 5cm

Mat-forming perennial with wiry stems. **Flowers** Star-shaped and 12mm across with 5 white petals that are pink below (Jun–Sep). **Fruits** Dry and red. **Leaves** 3–5mm long, fleshy and often tinged red. **Status** Widespread and locally common, especially in W Britain and Ireland. Grows on rocky ground, shingle beaches and old walls.

White Stonecrop

Sedum album Crassulaceae
Height to 15cm

Mat-forming evergreen perennial. **Flowers** Star-shaped, 6–9mm across, and white above but often pinkish below; borne in terminal clusters (Jun–Sep). **Fruits** Dry and splitting. **Leaves** 6–12mm long, fleshy, shiny and green or reddish. **Status** Local, mainly in SW England; naturalised elsewhere. Grows on rocky ground and walls.

Hairy Stonecrop

Sedum villosum Crassulaceae
Height to 15cm

Upright unbranched biennial or perennial. **Flowers** 6–8mm across, pink and stalked; upright (not drooping) in bud (Jun–Aug). **Fruits** Dry and splitting. **Leaves** Fleshy, flat above and covered in sticky hairs; arranged spirally up the stems. **Status** Local from N England to central Scotland. Found on rocky ground and old walls.

Orpine

Sedum telephium Crassulaceae
Height to 50cm

Elegant perennial. **Flowers** Reddish purple and 5-petalled; borne in rounded terminal heads (Jul–Aug). **Fruits** Dry and splitting. **Leaves** Green, fleshy, oval and irregularly toothed; borne on reddish stems. **Status** Locally common in England and Wales but scarce elsewhere. Grows in shady woodland and scrub.

Biting Stonecrop

Sedum acre Crassulaceae
Height to 10cm

Distinctive attractive mat-forming perennial. **Flowers** Star-shaped and 10–12mm across, with 5 bright yellow petals (May–Jul). **Fruits** Dry and splitting. **Leaves** Fleshy, crowded and pressed close to stem; they taste peppery. **Status** Widespread and locally common. Found on well-drained ground such as sand dunes and old walls.

Saxifrages

Members of the saxifrage family (Saxifragaceae) are hardy plants, many of them thriving among stones and on gravel where other wildflowers would fail to gain a foothold. They can often tolerate free-draining growing conditions where water is at a premium and can conserve water in their rather fleshy leaves. They have flowers that comprise five petals, a characteristic they share with members of the rose family.

Opposite-leaved Golden Saxifrage

Chrysosplenium oppositifolium
Saxifragaceae Height to 15cm

Patch-forming perennial. **Flowers** 3–5mm across, yellow and lack petals; borne in flat-topped clusters with yellowish bracts (Mar–Jul). **Fruits** Dry capsules. **Leaves** Rounded, short-stalked and in opposite pairs. **Status** Locally common, mainly in the N and W. Grows in shady, damp places. **Similar species Alternate-leaved Golden Saxifrage** *C. alternifolium* has larger flowers and kidney-shaped long-stalked leaves that are sometimes alternate.

Opposite-leaved Golden Saxifrage

Alternate-leaved Golden Saxifrage

Meadow Saxifrage

Saxifraga granulata Saxifragaceae
Height to 45cm

Attractive hairy perennial. **Flowers** 20–30mm across with 5 white petals; borne in open clusters (Apr–Jun). **Fruits** Dry capsules. **Leaves** Kidney-shaped with blunt teeth; bulbils are produced at the leaf axils in autumn. **Status** Local and commonest in E England. Grows in grassy meadows, doing especially well on neutral or basic soils.

Mossy Saxifrage

Saxifraga hypnoides Saxifragaceae
Height to 20cm

Mat-forming upland perennial. **Flowers** 10–15mm across with 5 white petals; borne in clusters (May–Jul). **Fruits** Dry capsules. **Leaves** Pointed and lobed; non-flowering shoots are procumbent and leafy bulbils form in the leaf axils; the overall effect is of a moss-like plant. **Status** Locally common only in N England and Scotland. Grows on damp rocks.

Rue-leaved Saxifrage

Saxifraga tridactylites
Saxifragaceae Height to 15cm

Stickily hairy annual with reddish zigzagging stems. **Flowers** 4–6mm across with 5 white petals; borne in clusters (Jun–Sep). **Fruits** Dry capsules. **Leaves** Deeply divided into 1–5 finger-like lobes. **Status** Widespread and very locally common. Found on dry, bare ground and old walls, mainly on sandy or calcareous soils.

Starry Saxifrage

Saxifraga stellaris Saxifragaceae
Height to 25cm

Attractive perennial. **Flowers** Star-shaped with 5 white petals and red anthers; the petals have yellow basal spots but no red spots above; borne on slender stalks (Jun–Aug). **Fruits** Dry capsules. **Leaves** Oblong, toothed and form a basal rosette. **Status** Locally common in upland N Wales, N Britain and Ireland. Found on damp ground and along stream margins.

Grass-of-Parnassus

Parnassia palustris Saxifragaceae
Height to 25cm

Distinctive tufted, hairless perennial. **Flowers** 15–20mm across with 5 white petals and greenish veins; borne on upright stalks with clasping leaves (Jun–Sep). **Fruits** Dry capsules. **Leaves** Deep green; basal leaves are heart-shaped and stalked. **Status** Locally common only in N Britain and Ireland. Grows on damp, peaty grassland, marshes and moors.

Rose family

The rose family (Rosaceae) is a diverse group that includes herbaceous plants, shrubs and even trees (*see* pp.25–7). In most species, flower parts are arranged in 5s, although in burnets, lady's-mantles and some cinquefoils they are in 4s.

Lady's-mantle

Alchemilla vulgaris agg.
Rosaceae Height to 30cm

Variable grassland perennial, treated here as an aggregate of several native species. **Flowers** Yellowish green and borne in flat-topped clusters (May–Sep). **Fruits** Dry and papery. **Leaves** Rounded and variably lobed; leaf shape variation is used to separate the aggregated species. **Status** Widespread but local in the S in damp grassland (*see also* Alpine Lady's-mantle, p.57).

Great Burnet

Sanguisorba officinalis Rosaceae
Height to 1m

Elegant hairless perennial. **Flowers** Tiny and reddish purple; borne in dense ovoid heads on long stalks (Jun–Sep). **Fruits** Dry and papery. **Leaves** Deeply divided and comprise 3–7 pairs of oval toothed leaflets. **Status** Local and declining, and common only in central and N England. Grows in damp grassland and on river banks.

flowers

Salad Burnet

Sanguisorba minor Rosaceae
Height to 35cm

Perennial that smells of cucumber when crushed. **Flowers** Tiny and green with red styles; borne in dense rounded heads (May–Sep). **Fruits** Ridged and 4-sided. **Leaves** Deeply divided and comprise 4–12 pairs of rounded toothed leaflets; basal leaves form a rosette. **Status** Locally common. Found mainly in chalk grassland.

flowers

Parsley-piert

Aphanes arvensis Rosaceae
Creeping

Easily overlooked greyish-green downy annual. **Flowers** Minute, petal-less and green; borne in dense unstalked clusters along stems (Apr–Oct). **Fruits** Dry and papery. **Leaves** Fan-shaped, deeply divided into 3 lobes and Parsley-like. **Status** Widespread and generally common. Found on dry, bare ground and arable field margins.

Rose family

Mountain Avens

Dryas octopetala Rosaceae
Height to 6cm

Creeping perennial undershrub. **Flowers** 3–4cm across with 8 or more white petals and a mass of yellow stamens (Jun–Jul). **Flowers** Dry and 1-seeded, with feathery plumes. **Leaves** Dark green, oblong and toothed. **Status** Locally common in Scotland and W Ireland. Found on basic rocks, in mountains and, locally, at sea-level.

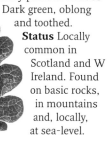

fruit

Wood Avens

Geum urbanum Rosaceae
Height to 50cm

Hairy perennial. **Flowers** 8–15mm across and comprise 5 yellow petals; upright in bud but dropping when fully open (May–Aug). **Fruits** Bur-like, with hooked red spines. **Leaves** Basal leaves have 3–6 pairs of side leaflets and a large terminal leaflet; stem leaves are 3-lobed. **Status** Widespread and common in hedgerows and woodland.

Water Avens

Geum rivale Rosaceae
Height to 50cm

Downy perennial. **Flowers** Nodding and bell-shaped, and comprise dark red sepals and pink petals (May–Sep). **Fruits** Bur-like but feathery. **Leaves** Comprise finely divided basal leaves and trifoliate stem leaves. **Status** Widespread and locally common, except in the S. Grows in damp meadows and marshes, and on mountain ledges, mostly on base-rich soils.

fruit

Hoary Cinquefoil

Potentilla argentea Rosaceae
Height to 50cm

Spreading perennial whose stems are coated in silky hairs. **Flowers** 8–12mm across with 5 yellow petals; borne in branched clusters (May–Jun). **Fruits** Dry and papery. **Leaves** Hand-like with narrow leaflets, the undersides of which are coated with silky hairs. **Status** Local, mainly in S and E England. Grows on dry, gravelly soils.

Marsh Cinquefoil

Potentilla palustris Rosaceae
Height to 40cm

Hairless perennial. **Flowers** Star-shaped and upright, and comprise 5 reddish sepals and smaller purple petals (May–Jul). **Fruits** Dry and papery. **Leaves** Greyish and deeply divided into 3 or 5 toothed oval leaflets. **Status** Widespread but local; common only in N England and Ireland. Grows in marshes and damp meadows.

Spring Cinquefoil

Potentilla neumanniana Rosaceae
Height to 15cm

Creeping, mat-forming perennial with woody stem bases. **Flowers** 1–2cm across with 5 yellow petals; borne in loose clusters (Apr–Jun). **Fruits** Dry and papery. **Leaves** Comprise palmate basal leaves with 5–7 leaflets, and trifoliate stem leaves. **Status** Widespread but extremely local. Found in dry, calcareous grassland.

Tormentil

Potentilla erecta Rosaceae
Height to 30cm

Creeping downy perennial. **Flowers** 7–11mm across with 4 yellow petals; borne on slender stalks (May–Sep). **Fruits** Dry and papery. **Leaves** Unstalked and trifoliate, but appear 5-lobed because of two large leaflet-like stipules at the base. **Status** Widespread and often abundant. Found in grassy places and on heaths and moors.

Creeping Cinquefoil

Potentilla reptans Rosaceae
Height to 20cm

Creeping perennial whose trailing stems root at the nodes (unlike Tormentil). **Flowers** 7–11mm across with 4 yellow petals (Jun–Sep). **Fruits** Dry and papery. **Leaves** Long-stalked, hairless and divided into 5–7 leaflets. **Status** Widespread and common throughout. Found in grassy places, including roadside verges.

Silverweed

Potentilla anserina Rosaceae
Creeping

Low-growing perennial with long, creeping stems. **Flowers** 15–20mm across with 5 yellow petals (May–Aug).

Fruits Dry and papery. **Leaves** Divided into a maximum of 12 pairs of leaflets (with tiny ones between them) that are covered in silvery silky hairs. **Status** Widespread and common. Found in damp grassy places and on bare ground.

Barren or fertile?

Tiny but delicious strawberries are sometimes found growing wild in hedgerows. Wild Strawberry is a common and widespread plant, but so is the closely related Barren Strawberry. This plant has similar flowers to Wild Strawberry but it does not produce edible fruits. Both species have similar leaves, but subtle differences allow separation. An ability to recognise Barren Strawberry plants can prevent literally fruitless searches for strawberries in areas where there is none to be found. **Wild Strawberry** *Fragaria vesca* (**1**) has leaves that comprise 3 oval toothed leaflets; the terminal tooth of the middle leaflet is usually longer than the adjacent ones. **Barren Strawberry** *Potentilla sterilis* (**2**) has similar leaves, but the terminal tooth of the middle leaflet is shorter than the adjacent ones.

Small-leaved Cotoneaster

Cotoneaster microphyllus Rosaceae Prostrate

Low-growing stiff evergreen undershrub with twigs that are downy when young. **Flowers** Small with white spreading petals; usually solitary (May–Jun). **Fruits** Spherical crimson berries. **Leaves** 5–10mm long, and oval but blunt or notched at the tip. **Status** Introduced and locally naturalised. Found on coastal limestone.

flowers and fruit

Roses & currants

Bramble

Rubus fruticosus agg.
Rosaceae Height to 3m

Scrambling shrub that is rather variable in appearance. Its arching stems are armed with variably shaped prickles and root when they touch the ground. **Flowers** 2–3cm across and white or pink (May–Aug). **Fruits** Familiar blackberries. **Leaves** 3–5 toothed leaflets. **Status** Widespread and common. Found in hedgerows and scrub.

flowers

fruits

Food for free

Many of the succulent, juicy soft fruits we relish eating belong to the gooseberry family (Grossulariaceae – Gooseberry, Black Currant and Red Currant) and rose family (Raspberry). Those cultivars commonly grown in British and Irish gardens all have their origins in the wild and are still locally common as wayside plants. **Gooseberry** *Ribes uva-crispa* (**1**) is a deciduous shrub with spiny stems, yellowish flowers and green hairy fruits; it is widespread and fairly common in woodlands and hedgerows. **Black Currant** *R. nigrum* (**2**) is a deciduous shrub of damp woodlands, with greenish bell-shaped flowers and familiar black fruits; the 5-lobed leaves are stickily hairy and aromatic when bruised. **Red Currant** *R. rubrum* (**3**) grows in damp woodland and beside streams, and has greenish bell-shaped flowers and shiny red berries; the rounded 5-lobed leaves are not aromatic when bruised. **Raspberry** *Rubus idaeus* (**4**) has arching biennial stems, flowers with 5 white petals and familiar red fruits; it grows in hedgerows and woodland margins.

Upland brambles

Bramble is a lowland plant, but two hardy counterparts favour upland areas in the west and north. **Stone Bramble** *Rubus saxatilis* (**1**) is locally common on rocky ground. Its flowers have 5 narrow white petals, its fruits are red and its leaves have toothed leaflets. **Cloudberry** *R. chamaemorus* (**2**) lacks prickles and has flowers that are 15–25mm across. Its fruits are red at first, ripening orange, and it has rounded lobed leaves. It grows on moorland.

Dewberry

Rubus caesius Rosaceae
Height to 10cm

Creeping perennial whose biennial stems bear weak prickles. **Flowers** 2–2.5cm across with 5 white petals (Jun–Aug). **Fruits** Bluish black, the large segments covered in a plum-like bloom. **Leaves** Trifoliate and toothed. **Status** Widespread and locally common. Found in dry grassy places, as well as in fens and dune slacks.

fruit

Dog-rose

Rosa canina Rosaceae
Height to 3m

Scrambling variable shrub whose long, arching stems bear curved thorns. **Flowers** 3–5cm across and fragrant, with 5 pale pink petals; borne in clusters of up to 4 flowers (Jun–Jul). **Fruits** Red egg-shaped hips that typically shed their sepals before they ripen. **Leaves** Comprise 5–7 hairless leaflets. **Status** Common in hedgerows and scrub.

fruits

Field-rose

Rosa arvensis Rosaceae
Height to 1m

Clump-forming shrub. The weak, trailing stems carry small numbers of curved thorns. **Flowers** 3–5cm across with 5 white petals; the styles form a column at least as long as the stamens (Jul–Aug). **Fruits** Rounded to ovoid red hips; the sepals do not persist. **Leaves** Have 5–7 oval leaflets. **Status** Locally common, except in the N. Grows in hedgerows and scrub.

fruit

Japanese Rose

Rosa rugosa Rosaceae Height to 1.5m

Showy shrub with upright stems that bear rather straight thorns. **Flowers** 6–9cm across with 5 pinkish-purple or white petals (Jun–Aug). **Fruits** Spherical red hips, 2–5cm across. **Leaves** Comprise 5–9 oval leaflets that are shiny above. **Status** Widely planted beside roads and often naturalised elsewhere.

fruit

Burnet Rose

Rosa pimpinellifolia Rosaceae
Height to 50cm

Clump-forming shrub. Its suckering stems have straight thorns and stiff bristles. **Flowers** 3–5cm across with 5 creamy white petals; usually solitary (May–Jul). **Fruits** Spherical, 5–6mm across and purplish black when ripe. **Leaves** Comprise 7–11 oval leaflets. **Status** Widespread, but only locally common. Found on sand dunes, calcareous grasslands and heaths.

fruit

Meadowsweet

Filipendula ulmaria Rosaceae
Height to 1.25m

Striking perennial. **Flowers** 4–6mm across, fragrant and creamy white; borne in sprays (Jun–Sep). **Fruits** Spirally twisted and 1-seeded. **Leaves** Dark green and comprise 3–5 pairs of oval leaflets with smaller leaflets between. **Status** Widespread and common throughout. Grows in damp meadows, marshes and stream margins.

Sweet-briar

Rosa rubiginosa Rosaceae
Height to 3m

Compact shrub with upright stems that bear short curved thorns, bristles and glands. **Flowers** 2–3cm across and pink; borne in clusters of up to 3 flowers (Jun–Jul). **Fruits** Ovoid red hips with persisting sepals. **Leaves** Have 5–7 oval toothed sweet-smelling leaflets. **Status** Locally common in the S. Found in hedgerows and scrub.

fruit

Harsh Downy-rose

Rosa tomentosa Rosaceae
Height to 2m

Dense shrub with arching stems that bear rather straight thorns. **Flowers** 3–4cm across with 5 pink or white petals; borne in clusters (Jun–Jul). **Fruits** Rounded red hips that are covered with bristles. **Leaves** Comprise 5–7 oval leaflets that are downy on both sides. **Status** Widespread and fairly common in the S. Grows in hedgerows and scrub.

Agrimony

Agrimonia eupatoria Rosaceae
Height to 50cm

Upright perennial. **Flowers** 5–8mm across with 5 yellow petals; borne in upright spikes (Jun–Aug). **Fruits** Bur-like and covered in spines. **Leaves** Comprise 3–6 pairs of oval toothed leaflets with smaller leaflets between. **Status** Widespread and generally common throughout. Grows in grassy places, hedgerows and roadside verges.

Dropwort

Filipendula vulgaris Rosaceae
Height to 50cm

Attractive perennial that is similar to Meadowsweet. **Flowers** 10–20mm across, unscented, and creamy white above and reddish below; borne in flat-topped sprays (May–Aug). **Fruits** Downy. **Leaves** Comprise 8–20 pairs of large leaflets with smaller leaflets between. **Status** Widespread, but local and restricted to calcareous grassland.

Flowers of chalk downlands

Soil chemistry has a profound effect upon the plants that will grow in any given area. Nowhere is this more noticeable than on the chalk downlands of southern England. Chalk landscapes are characterised by rolling open countryside where grassland dominates the scene. A closer look reveals a botanist's delight – a huge diversity of specialised plant species. Chalk soils are free-draining, alkaline and nutrient-poor, with little topsoil, and many of the plants growing here are found nowhere else.

Chalk downland is essentially a man-made habitat, resulting from forest clearance and centuries of grazing by animals. Continued grazing is essential to prevent scrub invasion and to maintain the habitat.

The flowering season is long, with a succession of colourful and often aromatic species blooming from early spring to late summer. Flowers such as Cowslip and Hairy Violet are first on the scene, but Kidney Vetch, knapweeds, Thyme, Marjoram, Yellowwort and Autumn Gentian follow on. But, chalk downland is probably best known for its orchids (*see also* pp.150–5).

Hampshire downland in July.

Chalk Milkwort

Polygala calcarea Polygalaceae
Height to 20cm

Small but attractive hairless perennial. **Flowers** 5-6mm long and usually bright blue; borne in short spikes of 6-20 flowers (May-Jun). **Fruits** Flattened. **Leaves** Blunt and narrowly ovate; they are crowded at the base and appear to form a rosette. **Status** Local in S and SE England only. Restricted to short calcareous grassland.

Pasqueflower

Pulsatilla vulgaris Ranunculaceae
Height to 25cm

Silky-hairy perennial. **Flowers** Purple and bell-shaped with 6 petal-like sepals; upright at first, then nodding (Apr-May). **Fruits** Comprise seeds with long silky hairs. **Leaves** Divided 2 or 3 times and comprise narrow leaflets. **Status** Rare and restricted to a few sites in S and E England. Found on dry calcareous grassland.

Wild Candytuft

Iberis amara Brassicaceae
Height to 30cm

Attractive downy annual. **Flowers** Comprise 4 white or mauve petals, 2 of which are much longer than the others (Jul-Aug). **Fruits** Rounded, winged and notched. **Leaves** Toothed and spoon-shaped, becoming smaller up the stem. **Status** Local, mainly in Chilterns. Favours disturbed chalky soil, often beside Rabbit burrows.

fruits

Wild Liquorice

Astragalus glycophyllos Fabaceae
Height to 30cm

Sprawling, hairless perennial with branched, zigzagging stems. **Flowers** 10-15mm long and yellowish green; borne in clusters (Jun-Aug). **Fruits** Curved and up to 4cm long. **Leaves** 15-20cm long with oval leaflets and large basal stipules. **Status** Local, mainly E England and S Scotland. Found in dry grassy places on calcareous soils.

Early Gentian

Gentianella anglica Gentianaceae
Height to 15cm

 Small and dainty biennial, superficially similar to the Autumn Gentian (*see* p.107) but much shorter and it flowers in spring. **Flowers** Reddish purple and borne in terminal clusters and from leaf axils (Apr–Jun). **Fruits** Capsules. **Leaves** Narrowly oval. **Status** Extremely local, growing in short chalk grassland in S England.

Field Fleawort

Tephroseris integrifolia Asteraceae
Height to 65cm

 Slender, unbranched, downy perennial. **Flowers** Borne in heads, 15–25mm across, with orange-yellow disc florets and yellow ray florets; borne in few-flowered clusters (May–Jul). **Fruits** Hairy. **Leaves** Oval, toothed at the base and forming a rosette; the stem leaves are few, narrow and clasping. **Status** Local; found mainly on downs in S and E England.

Late Spider-orchid

Ophrys fuciflora Orchidaceae
Height to 50cm

 Similar to both the Bee Orchid (*see* p.150) and Early Spider-orchid. **Flowers** Distinguished from the Bee Orchid by the broader, less rounded lip with its green upturned tip, and from the Early Spider-orchid by the pink (not green) outer flower segments. **Fruits** Egg-shaped. **Leaves** Narrowly oval. **Status** Rare on protected downland sites in Kent.

Military Orchid

Orchis militaris Orchidaceae
Height to 45cm

 Stately orchid, keenly sought by botanical enthusiasts. **Flowers** Pink and hooded, with the lower lip lobed into fanciful 'arms' and 'legs'; the flower spike opens from the bottom upwards (May–Jun). **Fruits** Egg-shaped. **Leaves** Oval and mainly basal. **Status** Formerly quite widespread in the Chilterns but now rare and restricted to just a few sites.

Chiltern Gentian

Gentianella germanica
Gentianaceae Height to 30cm

 Attractive and showy biennial, similar to Autumn Gentian (*see* p.107). **Flowers** Purple, 2cm across (larger than those of Autumn Gentian) and fringed with pale hairs (Aug–Oct). **Fruits** Capsules. **Leaves** Narrowly oval. **Status** Grows on chalk downland, mainly in the Chilterns, but with outposts southwards to N Hampshire.

Early Spider-orchid

Ophrys sphegodes Orchidaceae
Height to 35cm

 Perennial, often growing as a dwarf plant in grazed areas. **Flowers** Comprise green sepals and yellowish-green upper petals; lower lip is 12mm across, expanded, furry and maroon-brown, variably marked with a metallic blue H-shaped mark (Apr–May). **Fruits** Egg-shaped. **Leaves** Green and mainly basal. **Status** Local, in S England only. Found on dry chalk grassland.

Chalk Fragrant-orchid

Gymnadenia conopsea
Orchidaceae Height to 40cm

 Robust orchid that often grows in colonies. **Flowers** Fragrant, typically pink although variable, and with a 3-lobed lip and long spur; borne in dense cylindrical spikes (Jun–Aug). **Fruits** Egg-shaped. **Leaves** Oval and mostly basal. **Status** Widespread and locally common, particularly in S and SE England.

Monkey Orchid

Orchis simia Orchidaceae
Height to 45cm

 Attractive orchid. **Flowers** Pink and hooded, the lower lip with long lobes that fancifully resemble monkey limbs; borne in cylindrical heads, the flowers opening in succession from the top downwards (May–Jun). **Fruits** Egg-shaped. **Leaves** Mainly basal. **Status** Rare. Grows in a few chalk downland sites in Oxfordshire, Kent and Yorkshire.

Plants of the New Forest 'lawns'

Although the New Forest is known for its ancient trees and expanses of heathland, many of the botanical highlights are restricted to patches of close-cropped grassland. Referred to as 'lawns', these areas have been kept trimmed by generations of ponies and cattle, whose drinking pools are also populated with floral delights.

Flowers of the lawns

Closer examination of the lawn on your hands and knees reveals a huge variety of tiny plants. The species that succeed here cope with the constant nibbling and enrichment from the droppings of the grazing animals, as well as arid summer conditions. Flowers of Tormentil (*see* p.69) and Lesser Skullcap (*see* p.114) often adorn the turf, but in addition rarities such as Yellow Centaury and Chaffweed are waiting to be discovered.

Coral-necklace

Illecebrum verticillatum
Caryophyllaceae Prostrate

Charming and distinctive annual with square reddish stems. **Flowers** White and borne in discrete rounded clusters along the stems (Jun–Sep). **Fruits** Capsules. **Leaves** Small, oval and opposite. **Status** Grows in damp ground, often in trampled hollows at the margins of drying ponds and ruts. The New Forest is its stronghold.

Chaffweed

Anagallis minima Primulaceae
Height to 2cm

Tiny, hairless annual that is easily overlooked. **Flowers** Minute, pale pink and borne at the bases of stem leaves (Jun–Aug). **Fruits** Spherical and pinkish, like miniature apples. **Leaves** Oval and, uniquely among British species, they have a black line along the undersurface margin. **Status** Local on damp sandy ground. Easiest to find in the New Forest.

Grassland fungi

In addition to their plants, New Forest lawns are extremely important for grassland fungi, which put in an appearance here in the autumn. Members of the waxcap family are particularly well represented and the Parasol mushroom is locally common (*see* pp.184–97 for further information). Strangest of all is **Devil's Fingers** *Clathrus archeri* (**1**), an introduced species that is spreading and now locally common; it emits an unspeakable smell. Even the dung left behind by the grazing animals hosts an array of fungi, a New Forest speciality being the bizarre little **Nail Fungus** *Poronia punctata* (**2**), named after its resemblance to handmade nails, which typically have pitted heads.

Yellow Centaury

Cicendia filiformis Gentianaceae
Height to 5cm

Slender and extremely delicate plant that is easily overlooked when not in flower. **Flowers** 3–5mm across with 4 petal-like corolla lobes that open only in full sunshine (Jul–Aug). **Fruits** Capsules. **Leaves** Tiny, narrow and borne in opposite pairs. **Status** Extremely local on damp sandy ground. Easiest to find in the New Forest.

Flowers of damp hollows

During the summer months, the free-draining soils of the New Forest become parched and the lawn plants often turn decidedly yellow. By August there may be comparatively little to find in open areas of grassland, but this is just when damp hollows and the margins of drying drinking ponds come into their own. They, too, are subject to intense grazing pressure and trampling from hooves, but this suits specialist plants such as Pennyroyal, Hampshire-purslane and Small Fleabane.

Did you know?

The New Forest boasts one other botanical speciality that is not associated with its grazed lawns, namely Wild Gladiolus, which is found on bracken-covered slopes. You can learn out more about this species on p.148.

Hampshire-purslane

Ludwigia palustris Onagraceae
Height to 20cm

Typically a creeping perennial. Rather undistinguished but nevertheless quite distinctive, with a reddish appearance to the plant, particularly its stems. **Flowers** Tiny and inconspicuous (Jun–Aug). **Fruits** Capsules. **Leaves** Oval, opposite and strongly tinged reddish brown. **Status** Grows at the margins of ponds. Almost entirely confined to the New Forest.

Slender Marsh-bedstraw

Galium constrictum Rubiaceae
Height to 60cm

Similar to Common Marsh-bedstraw (*see* p.109) but more slender and with smooth stems. **Flowers** 2–3mm across, white and 4-petalled; borne in few-flowered clusters (May–Aug). **Fruits** Warty nutlets. **Leaves** Narrow; not bristle-tipped but with forward-pointing marginal bristles. **Status** Local, on marshy ground beside ponds; easiest to find in the New Forest.

Pennyroyal

Mentha pulegium Lamiaceae
Height to 30cm

Mint-scented, creeping, downy perennial that recalls Water Mint (*see* p.117). **Flowers** Mauve and borne in discrete whorls on upright flowering stems (Aug–Oct). **Fruits** Nutlets. **Leaves** Oval. **Status** Grows on damp grazed ground beside ponds, but also often thrives in ditches. Locally common only in the New Forest, which is therefore the best place in the region to see it.

Small Fleabane

Pulicaria vulgaris Asteraceae
Height to 40cm

Recalls Common Fleabane (*see* p.134) but is more branched and has much smaller flower heads. **Flowers** 1cm across, with short, yellow ray florets (Aug–Oct). **Fruits** Dry and hairy. **Leaves** Elongated oval. **Status** Grows around the trampled and grazed margins of pools. Extremely local; most easily seen in the New Forest; very locally common in some years.

Scented lawns

Wild Chamomile *Chamaemelum nobile* is another common component of many New Forest lawns, although it seldom gets a chance to flower. It can often be detected by the scent released by the leaves as you walk across them. Find out more about the species on p.131.

Chamomile lawn

Pea family

The pea family (Fabaceae) is a diverse group of plants with representatives that range in size from quite large shrubs, such as the gorses, to some of our smallest species, such as Spring Vetch. However, all share the same highly evolved flower shape and produce their seeds in pods.

Broom

Cytisus scoparius Fabaceae
Height to 2m

Deciduous, branched, spineless shrub with ridged, 5-angled green twigs. **Flowers** 2cm long and bright yellow; either solitary or borne in pairs (Apr–Jun). **Fruits** Oblong blackening pods that explode on dry, sunny days. **Leaves** Usually trifoliate. **Status** Widespread and common. Found on heaths and in hedgerows, favouring acid soils.

Common Gorse

Ulex europaeus Fabaceae
Height to 2m

Evergreen shrub with straight, grooved spines that are 15–25mm long. **Flowers** 2cm long, bright yellow and coconut-scented, with 4–5mm-long basal bracts (Jan–Dec, but mainly Feb–May). **Fruits** Hairy pods. **Leaves** Trifoliate when young. **Status** Widespread and common throughout. Grows on heaths and acid soils.

Dyer's Greenweed

Genista tinctoria Fabaceae
Height to 1m

Straggly deciduous grassland shrub that is superficially similar to Petty Whin but spineless. **Flowers** 15mm long, bright yellow and Broom-like; borne in leafy, showy, stalked spikes (Apr–Jul). **Fruits** Oblong, flat and hairless pods. **Leaves** Narrow and sometimes downy. **Status** Locally common in England, Wales and S Scotland. Usually grows on clay or calcareous soils. Intolerant of overgrazing and is easily wiped out at the local level by inappropriate management.

Western Gorse

Ulex gallii Fabaceae
Height to 1.5m

Dense evergreen shrub with spines that are almost smooth and 25mm long. **Flowers** 10–15mm long, bright yellow and with 0.5mm-long basal bracts (Jul–Sep). **Fruits** Hairy pods. **Leaves** Trifoliate when young. **Status** Restricted mainly to W Britain and Ireland; common on coastal cliffs in the W. Found on acid soils, often near coasts.

Dwarf Gorse

Ulex minor Fabaceae
Height to 1m

Spreading and typically rather low-growing evergreen shrub with spines that are soft, smooth and 10mm long. **Flowers** 10–15mm long, pale yellow and with 0.5mm-long basal bracts (Jul–Sep). **Fruits** Hairy pods. **Leaves** Trifoliate when young. **Status** Local and restricted mainly to acid soils on heaths; commonest in SE and E England.

Petty Whin

Genista anglica Fabaceae
Height to 1m

A rather spindly, hairless shrub that is armed with strong spines. **Flowers** 15mm long and deep yellow; borne in terminal clusters (Apr–Jun). **Fruits** Hairless and inflated. **Leaves** Narrow, hairless and waxy. **Status** Widespread but local in England, Wales and S Scotland. Found on heaths and moors, usually growing among Heather and Bell Heather.

Bush Vetch

Vicia sepium Fabaceae
Height to 1m

Scrambling, slightly downy perennial. **Flowers** 12–15mm long and pale lilac; borne in groups of 2–6 flowers (Apr–Oct). **Fruits** Hairless black pods. **Leaves** Comprise 6–12 pairs of broadly oval leaflets and end in branched tendrils. **Status** Common and widespread throughout Britain and Ireland. Grows in rough grassy places and scrub.

Common Vetch

Vicia sativa Fabaceae
Height to 75cm

Scrambling downy annual. **Flowers** Pinkish purple and 2–3cm long; they appear singly or in pairs (Apr–Sep). **Fruits** Pods that ripen black. **Leaves** Comprise 3–8 pairs of oval leaflets, ending in tendrils. **Status** Widespread and fairly common throughout. Grows in grassy places and hedgerows; sometimes seen as a relict of arable cultivation.

Tufted Vetch

Vicia cracca Fabaceae
Height to 2m

Slightly downy, scrambling perennial. **Flowers** 8–12mm long and bluish purple; borne in 1-sided spikes that are up to 8cm tall (Jun–Aug). **Fruits** Hairless pods. **Leaves** Comprise up to 12 pairs of narrow leaflets and end in a branched tendril. **Status** Widespread and common throughout. Found in grassy places, hedgerows and scrub.

Purple Milk-vetch

Astragalus danicus Fabaceae
Height to 30cm

Attractive downy, spreading perennial. **Flowers** 15–18mm long and purple; borne in stalked clusters (May–Jul). **Fruits** Pods, covered in white hairs. **Leaves** Hairy and deeply divided, comprising 6–12 pairs of oval leaflets. **Status** Local in E England and S Scotland; scarce in Ireland. Grows in dry calcareous grassland.

Wood Vetch

Vicia sylvatica Fabaceae
Height to 1.5m

Elegant straggling perennial. **Flowers** 12–20mm long, and white with purple veins; borne in spikes of up to 20 flowers (Jun–Aug). **Fruits** Hairless black pods. **Leaves** Comprise 6–12 pairs of oblong leaflets, ending in a branched tendril. **Status** Widespread but local. Grows in shady woods and on steep coastal slopes.

Spring Vetch

Vicia lathyroides Fabaceae
Height to 20cm

Rather delicate, spreading, downy annual. **Flowers** 5–8mm long, reddish purple and solitary (Apr–Jun). **Fruits** Hairless black pods. **Leaves** Comprise 2–4 pairs of bristle-tipped leaflets and unbranched tendrils. **Status** Local throughout the region. Found in short grassland, mainly on sandy soils and near the sea.

Peas & vetches

Many members of the pea family are associated with grassland habitats and tend to be rather straggly plants. However, what they lack in terms of robustness they make up for in tenacity, growing through, and using the support of, other plants. Twining tendrils often assist their progress.

Smooth Tare

Vicia tetrasperma Fabaceae
Height to 50cm

 Slender, scrambling annual that is easily overlooked. **Flowers** 4–8mm long and pinkish lilac; appear singly or in pairs (May–Aug). **Fruits** Smooth pods, usually containing 4 seeds. **Leaves** Comprise 2–5 pairs of narrow leaflets and end in tendrils. **Status** Locally common in England and Wales. Found in grassy places and hedgerows.

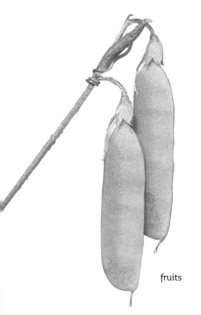

fruits

Hairy Tare

Vicia hirsuta Fabaceae
Height to 60cm

 Scrambling, downy annual. **Flowers** 2–4mm long and pale lilac; borne in groups of 1–9 flowers (May–Aug). **Fruits** Hairy 2-seeded pods. **Leaves** Comprise 4–10 pairs of leaflets and end in branched tendrils. **Status** Widespread and fairly common, except in the N and in Ireland. Found in grassy places, especially on neutral or calcareous soils.

BELOW: fruits

Yellow Vetch

Vicia lutea Fabaceae
Height to 50cm

 Rather prostrate, hairless, greyish-green annual. **Flowers** 25–35mm long and pale yellow; borne in groups of 1–3 (Jun–Sep). **Fruits** Hairy brown pods up to 4cm long. **Leaves** Have 3–10 pairs of bristle-tipped leaflets and tendrils. **Status** Very local, found mainly in the S. Grows in coastal grassy places and on stabilised shingle.

Meadow Vetchling

Lathyrus pratensis Fabaceae
Height to 50cm

 Scrambling perennial with long, angled stems. **Flowers** 15–20mm long and yellow; borne in open long-stalked terminal clusters of 4–12 flowers (May–Aug). **Fruits** 25–35mm-long pods that ripen black. **Leaves** Comprise a pair of narrow leaflets with a tendril and large stipules. **Status** Widespread and common. Favours grassy places.

Yellow Vetchling

Lathyrus aphaca Fabaceae
Height to 80cm

 Hairless, scrambling annual with angled stems and a waxy grey-green appearance. **Flowers** 12mm long and yellow; solitary and borne on long stalks (Jun–Aug). **Fruits** Curved brown pods. **Leaves** Reduced to tendrils but note the leaf-like stipules. **Status** Locally common in S England only. Restricted to chalk grassland.

Sea Pea

Lathyrus japonicus Fabaceae
Height to 12cm

 Spreading grey-green perennial with stems up to 1m long. **Flowers** 2cm long and purple, fading to blue; borne in heads of 2–15 flowers (Jun–Aug). **Fruits** Swollen pods, 5cm long. **Leaves** Comprise 2–5 pairs of oval leaflets and angular stipules. **Status** Mainly S and E England. Entirely restricted to coastal shingle and sand.

Bitter-vetch

Lathyrus linifolius Fabaceae
Height to 50cm

Upright, almost hairless perennial with winged stems. **Flowers** 15mm long, red fading to blue or green; borne in groups of 5-12 flowers (Apr-Jul). **Fruits** Brown pods, 4cm long. **Leaves** Have 2-4 pairs of narrow leaflets and no tendril. **Status** Widespread and locally common. Found in grassy places, heaths and woods, mainly on acid soils.

Marsh Pea

Lathyrus palustris Fabaceae
Height to 1m

Slender climbing perennial with winged stems. **Flowers** 2cm long and pinkish purple; borne in long-stalked groups of 2-8 flowers (May-Jul). **Fruits** Flat pods, 5cm long. **Leaves** Have 2-5 pairs of narrow leaflets and branched tendrils. **Status** Extremely local. Found in fens and damp grassy places on calcareous soils.

Grass Vetchling

Lathyrus nissolia Fabaceae
Height to 90cm

Upright, hairless or slightly downy perennial. Easily overlooked in its favoured grassy habitat when not in flower. **Flowers** 18mm long and crimson; solitary or paired on long, slender stalks (May-Jul). **Fruits** Narrow pods. **Leaves** Reduced to a pair of narrow, extremely grass-like leaflets. **Status** Locally common in meadows in SE.

Sainfoin

Onobrychis viciifolia Fabaceae
Height to 75cm

Distinctive straggly, slightly downy perennial. **Flowers** Pink with red veins; borne in conical spikes up to 8cm long (Jun-Aug). **Fruits** Warty oval pods. **Leaves** Comprise 6-14 pairs of oval leaflets. **Status** Possibly native in SE England but a relict of cultivation elsewhere. Found in dry calcareous grassland.

Mass flowering

In sites where growing conditions suit their needs, a few members of the pea family form sizeable carpets that thrive, at the local level, to the virtual exclusion of other plants, or at least in terms of their floral display. **Kidney Vetch** *Anthyllis vulneraria* (**1**) and **Horseshoe Vetch** *Hippocrepis comosa* (**2**), both characteristic of calcareous grassland, probably produce the most impressive displays. Kidney Vetch has yellow, orange or red flowers, borne in paired kidney-shaped heads that are 3cm across (May-Sep), and its leaves comprise pairs of narrow leaflets and a solitary terminal leaflet. It is widespread and locally common. Horseshoe Vetch has yellow flowers that are 7-10cm long and arranged in circular heads (May-Jul), and leaves that comprise pairs of narrow leaflets. It is locally common only in E and SE England.

Grassland peas

Common Restharrow

Ononis repens Fabaceae
Height to 70cm

Robust, creeping, woody perennial with hairy, spineless stems. **Flowers** 10–15mm long and pink, the wings and keel having a similar length; borne in clusters (Jul–Sep). **Fruits** Pods that are shorter than the calyx. **Leaves** Stickily hairy and trefoil with oval leaflets. **Status** Locally common on calcareous soils.

Spiny Restharrow

Ononis spinosa Fabaceae
Height to 70cm

Similar to Common Restharrow but upright and bushy, with spiny stems. **Flowers** 10–15mm long and deep pink, the wings shorter than the keel (Jul–Sep). **Fruits** Pods that are longer than the calyx. **Leaves** Trefoil with narrow oval leaflets. **Status** Local, mainly in England. Favours grassland on clay and heavy soils.

Ribbed Melilot

Melilotus officinalis Fabaceae
Height to 1.5m

Attractive upright, hairless biennial. **Flowers** Bright yellow and borne in spikes up to 7cm long (Jun–Sep). **Fruits** Wrinkled brown pods. **Leaves** Comprise 3 oblong leaflets. **Status** Locally common and possibly native in S England and Wales; introduced elsewhere. Found in grassy places and on waste ground.

White Melilot

Melilotus albus Fabaceae
Height to 1m

Distinctive upright, hairless biennial. **Flowers** White and borne in spikes up to 7cm long (Jun–Aug). **Fruits** Veined brown pods. **Leaves** Comprise 3 oblong leaflets. **Status** Introduced but now established locally in parts of S and E England. Found in grassy places and on disturbed soils on waste ground.

Common Bird's-foot Trefoil

Lotus corniculatus Fabaceae
Height to 10cm

Sprawling, solid-stemmed, usually hairless perennial. **Flowers** Red in bud but yellow and 15mm long when open; borne in heads on stalks to 8cm long (May–Sep). **Fruits** Slender pods; splayed like a bird's foot when ripe. **Leaves** Have 5 leaflets but appear trefoil (the lower pair is located at the stalk base). **Status** Common throughout in grassy places.

Greater Bird's-foot Trefoil

Lotus pedunculatus Fabaceae
Height to 50cm

Hairy, hollow-stemmed perennial. **Flowers** 15mm long and yellow; borne in heads on stalks up to 15cm long (Jun–Aug). Stalks are much longer than those of Common Bird's-foot Trefoil. **Fruits** Slender pods; splayed like a bird's foot when ripe. **Leaves** Have 5 dark green leaflets but appear trefoil (the lower pair is sited at the stalk base). **Status** Locally common throughout. Found in damp grassy places and fens, typically in wetter settings than Common Bird's-foot Trefoil.

Lucerne

Medicago sativa ssp. *sativa*
Fabaceae Height to 75cm

Downy or hairless perennial. **Flowers** 7-8mm long; borne in stalked heads of 5-40 flowers (Jun-Sep). **Fruits** Spirally twisted pods. **Leaves** Trefoil with narrow, toothed leaflets. **Status** A widely naturalised relict of cultivation. **Similar species Sickle Medick** (ssp. *falcata*) has yellow flowers and sickle-shaped pods, and is native to E Anglia.

Sickle Medick

Spotted Medick

Medicago arabica Fabaceae
Prostrate

Creeping annual. **Flowers** Small and yellow; borne in heads 5-7mm across comprising 1-6 flowers (Apr-Sep). **Fruits** Spirally coiled spiny pods (*see* box below). **Leaves** Trefoil, the heart-shaped leaflets bearing a dark central spot. **Status** Local in S and E England. Grows in dry grassy places and often found near the sea.

Bird's-foot Clover

Trifolium ornithopodioides
Fabaceae Prostrate

Low-growing, hairless annual that is easily overlooked. **Flowers** 5-8mm long and white or pale pink; borne in heads of 1-5 (May-Oct). **Fruits** Small pods. **Leaves** Clover-like and trefoil, with toothed oval leaflets. **Status** Very local in S England, Wales and Ireland. Found in dry grassy places, usually on sand or gravel.

Goat's-rue

Galega officinalis Fabaceae
Height to 1.5m

Bushy, much-branched perennial that is hairless or slightly downy. **Flowers** 10mm long and pale bluish lilac; borne in elongated spikes up to 5cm long (May-Aug). **Fruits** Cylindrical pods. **Leaves** Comprise 9-17 oval leaflets. **Status** Naturalised in central and S England on disturbed and waste ground; often found on roadside verges.

Black Medick

Medicago lupulina Fabaceae
Height to 20cm

Downy annual. **Flowers** Small and yellow; borne in dense spherical heads 8-9mm across comprising 10-50 flowers (Apr-Oct). **Fruits** Spirally coiled, spineless and black when ripe. **Leaves** Trefoil, each leaflet bearing a point at the centre of its apex. **Status** Widespread and rather common in short grassland and waste places throughout.

Seedpods of pea plants

Fruits of pea family members are known as pods, structures that protect and nourish the developing seeds. Some are miniature versions of their familiar cultivated counterparts, peas and beans, but others are more unusual in appearance. **Spotted Medick** (*see* above) (**1**) has spiny, spirally coiled pods that uncurl slightly and split when ripe. Although tiny, the pods of **Bird's-foot** *Ornithopus perpusillus* (**2**) are distinctive, being constructed and arranged like a bird's foot. The plant itself grows in dry sandy places and is locally common. Its flowers are 3-5mm long, creamy and red-veined, and are borne in heads of three to eight flowers (May-Aug).

Clovers, trefoils & Wood Sorrel

Clovers (Fabaceae) are a familiar sight in most meadows in Britain and Ireland and are among the classic components of lowland grassland. Red and White clovers are found in a wide range of grassy habitats, but several smaller, more delicate relatives, such as trefoils, also occur and are associated with open, bare habitats. The greatest variety of clover species is found near the coast on bare ground and short grassland. Clover flowers are often grouped together in conspicuous heads, favoured by pollinating insects, and their leaves are typically trefoil (*see* box), a characteristic shared by the unrelated Wood Sorrel.

White Clover
Trifolium repens Fabaceae
Height to 40cm

Creeping, hairless perennial that roots at its nodes. **Flowers** Creamy white, browning with age; borne in long-stalked, rounded heads, 2cm across (May–Oct). **Fruits** Concealed by the calyx. **Leaves** Trefoil; the rounded leaflets often bear white marks and translucent lateral veins. **Status** Widespread and common throughout in grassy places on a range of soil types.

Hop Trefoil
Trifolium campestre Fabaceae
Height to 25cm

Low-growing, hairy annual. **Flowers** 4–5mm long and yellow; borne in compact rounded heads, 15mm across (May–Oct). **Fruits** Pods, cloaked by brown dead flowers in hop-like heads. **Leaves** Trefoil; the terminal leaflet has the longest stalk. **Status** Widespread and generally common in the S. Found in dry grassland.

fruit

Lesser Hop Trefoil
Trifolium dubium Fabaceae
Height to 20cm

Low-growing annual that is similar to Hop Trefoil but hairless. **Flowers** 3–4mm long and yellow; borne in compact rounded heads, 8–9mm across (May–Oct). **Fruits** Pods, cloaked by brown dead flowers in hop-like heads. **Leaves** Trefoil with oval leaflets. **Status** Widespread and common throughout. Found in dry grassy places.

Hare's-foot Clover
Trifolium arvense Fabaceae
Height to 25cm

Charming, softly hairy annual. **Flowers** Pale pink and shorter than calyx teeth; borne in dense egg-shaped to cylindrical heads, 2–3cm long (Jun–Sep). **Fruits** Concealed by the calyx. **Leaves** Trefoil and comprise narrow leaflets. **Status** Locally common only in England and Wales. Grows in dry grassy areas, on sand or gravel.

Alsike Clover
Trifolium hybridum Fabaceae
Height to 60cm

Hairless perennial. Similar to White Clover, but it is upright, never roots at its nodes and has unmarked leaves. **Flowers** Whitish at first but turn pink or brown with age; borne in stalked heads, 2cm across (Jun–Oct). **Fruits** Concealed by the calyx. **Leaves** Trefoil with unmarked leaflets. **Status** Widespread and fairly common in grassy places; often a relict of cultivation.

Trefoil leaves

A characteristic of trefoils and clovers is that their leaves are divided into three leaflets; botanists describe this type of leaf arrangement as 'trefoil' or 'trifoliate'. In many of the clovers the leaflets are broadly oval, sometimes rounded, and are often marked in some way. However, possession of trefoil leaves is not an absolute guide to membership of the pea family – Wood Sorrel, which belongs to an unrelated family of plants, also shares this characteristic.

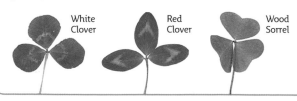

White Clover   Red Clover   Wood Sorrel

Red Clover

Trifolium pratense Fabaceae
Height to 40cm

 Familiar downy perennial. **Flowers** Pinkish red; borne in dense unstalked heads, 2–3cm across (May–Oct). **Fruits** Concealed by the calyx. **Leaves** Trefoil, the oval leaflets each bearing a white crescent-shaped mark; the stipules are triangular and bristle-tipped. **Status** Widespread and common throughout. Found in grassy places on a range of soil types.

Zigzag Clover

Trifolium medium Fabaceae
Height to 50cm

 Downy perennial. Similar to Red Clover but it has zigzagging stems. **Flowers** Reddish purple and borne in short-stalked heads that are 2–3cm across (May–Jul). **Fruits** Concealed by the calyx. **Leaves** Trefoil with narrow leaflets; the stipules are not bristle-tipped. **Status** Locally common in undisturbed grassy places.

Knotted Clover

Trifolium striatum Fabaceae
Height to 20cm

 Hairy annual. **Flowers** Pink and borne in unstalked egg-shaped heads that are 15mm long (May–Jul). **Fruits** Concealed by the calyx. **Leaves** Trefoil with spoon-shaped leaflets that are hairy on both sides but lack obvious lateral veins. **Status** Locally common, mainly in the S. Grows in dry grassy places, often on sand or gravel.

Knotted Clover

Rough Clover

Trifolium scabrum Fabaceae
Height to 15cm

 Downy annual. **Flowers** White and borne in unstalked heads that are 10mm long (May–Jul). **Fruits** Concealed by the calyx. **Leaves** Trefoil with oval leaflets that are hairy on both sides and have obvious lateral veins. **Status** Locally common in S England and S Wales but mainly coastal. Grows in bare grassland, often on gravel.

Sea Clover

Trifolium squamosum Fabaceae
Height to 30cm

 Downy annual. **Flowers** Pink and borne in rounded or egg-shaped heads that are 1cm long (Jun–Jul). **Fruits** Egg-shaped and resemble miniature Teasel heads. **Leaves** Trefoil with narrow leaflets. **Status** Locally common, mainly in the S. Confined to coastal grassland, including grassy sea walls that protect low-lying areas.

Strawberry Clover

Trifolium fragiferum Fabaceae
Height to 15cm

 Perennial with creeping stems that root at the nodes. **Flowers** Pink and borne in globular heads, 10–15mm across (Jul–Sep). **Fruits** Inflated pinkish heads that resemble pale berries. **Leaves** Trefoil, with oval unmarked leaflets. **Status** Local in the S and predominantly coastal. Found in grassy places, mainly on clay.

Wood Sorrel

Oxalis acetosella Oxalidaceae
Height to 10cm

Charming creeping perennial. **Flowers** 1cm across, bell-shaped and white or pale pink with lilac veins; borne on stalks (Apr–Jun). **Fruits** Hairless capsules. **Leaves** Trefoil, folding down at night, and borne on long stalks. **Status** Widespread and locally common. An indicator of undisturbed ancient woodlands.

Crane's-bills

Gardeners with a passion for herbaceous borders will be familiar with the range of colours and forms exhibited by cultivated varieties of crane's-bills (Geraniaceae). To the delight of anyone with an eye for colour, several of these garden favourites also occur in the wild as native species, along with a range of smaller and less showy relatives. Both crane's-bills and stork's-bills have five-petalled flowers that in most species are a shade of pink, and they have distinctive elongated fruits.

Meadow Crane's-bill

Geranium pratense Geraniaceae Height to 75cm

Hairy, clump-forming perennial. **Flowers** 3–3.5cm across with 5 rounded, bluish-lilac petals; borne in pairs on stalks (Jun–Aug). **Fruits** End in a long 'beak'. **Leaves** Deeply divided into 5–7 jagged lobes. **Status** Locally common, except in SE England, N Scotland and Ireland. Grows in meadows, mostly on base-rich soils.

Wood Crane's-bill

Geranium sylvaticum Geraniaceae Height to 60cm

Showy, tufted perennial. **Flowers** 2–3cm across with 5 reddish-purple petals (Jun–Aug). **Fruits** End in a long 'beak'. **Leaves** Deeply divided into 5–7 lobes but appear rather rounded overall. **Status** Absent from much of S England but locally common elsewhere. Grows in damp upland meadows and open woodlands, usually on base-rich soils.

Hedgerow Crane's-bill

Geranium pyrenaicum
Geraniaceae Height to 60cm

Upright, hairy perennial. **Flowers** 12–18mm across with deeply notched pink petals; borne in pairs (Jun–Aug). **Fruits** End in a long 'beak'. **Leaves** Rounded but divided halfway into 5–7 lobes. **Status** Probably introduced but locally common in S and E England. Grows in rough grassy places and on roadside verges.

Herb-Robert

Geranium robertianum
Geraniaceae Height to 30cm

Straggling, hairy annual. **Flowers** 12–15mm across with pink petals and orange pollen; borne in loose clusters (Apr–Oct). **Fruits** Have a long 'beak'. **Leaves** Hairy and deeply divided; often tinged red. **Status** Common in shady places.

Milkworts

Milkworts (family Polygalaceae) are delicate little grassland plants that are often associated with areas of good grassland, whose quality is reflected in the high milk yield of grazing cattle; this may explain the name. Three species are widespread in Britain. **Chalk Milkwort** *Polygala calcarea* grows on calcareous soils and is described on p.72. **Common Milkwort** *P. vulgaris* (**1**) is found on all but the most acid of soils. Its flowers are 6–8mm long, can be blue, pink or white and are borne in spikes of 10–40 flowers (May–Sep), and it has alternate basal leaves. **Heath Milkwort** *P. serpyllifolia* (**2**) grows on acid grassland and heaths, and its basal leaves grow in opposite pairs; the flowers are usually blue, and are borne in short spikes of 5–10 flowers (May–Sep).

1

2

Garden favourite

Most crane's-bills' flowers are subtle shades of pink or purple. But **Bloody Crane's-bill** *Geranium sanguineum* has garish reddish-purple flowers, 2–3cm across and solitary on long stalks (Jun–Aug). It is locally common on calcareous grassland and limestone pavements; it is also a popular rockery plant.

Long-stalked Crane's-bill

Geranium columbinum Geraniaceae Height to 60cm

 Elegant hairy annual that is sometimes tinged red. **Flowers** 12–18mm across with pink petals that are not notched; borne on long, slender stalks (Jun–Aug). **Fruits** Hairless and smooth. **Leaves** Divided to the base, the lower ones long-stalked. **Status** Local. Found on short, dry grassland, mainly on calcareous soils.

Dove's-foot Crane's-bill

Geranium molle Geraniaceae Height to 20cm

 Spreading, branched and extremely hairy annual. **Flowers** 5–10mm across with notched pink petals; borne in pairs (Apr–Aug). **Fruits** Hairless. **Leaves** Hairy and rounded, with the margins cut into 5–7 lobes. **Status** Common and widespread, especially in the S. Grows in dry grassy places, including roadside verges.

Round-leaved Crane's-bill

Geranium rotundifolium Geraniaceae Height to 40cm

 Downy annual that recalls Dove's-foot Crane's-bill. **Flowers** 10–12mm across with barely notched pink petals; usually borne in pairs (Jun–Jul). **Fruits** Hairy and not wrinkled. **Leaves** Rounded and only shallowly lobed. **Status** Distinctly local, mainly in S England. Found in dry grassy places, on calcareous and sandy soils.

Shining Crane's-bill

Geranium lucidum Geraniaceae Height to 30cm

 Branched, hairless annual that is sometimes tinged red. **Flowers** 10–15mm across; the petals are pink and are not notched, and the sepals are inflated (Apr–Aug). **Fruits** Hairless. **Leaves** Shiny and rounded, the margins cut into 5–7 lobes. **Status** Widespread but local. Found on shady banks and rocky slopes, mainly on limestone.

Cut-leaved Crane's-bill

Geranium dissectum Geraniaceae Height to 45cm

 Straggling, hairy annual. **Flowers** 8–10mm across with notched pink petals; borne on short stalks (May–Sep). **Fruits** Downy. **Leaves** Deeply dissected to the base, the lobes narrow and jagged. **Status** Generally common throughout, although scarce in N Scotland. Grows on disturbed ground and in cultivated soils.

Small-flowered Crane's-bill

Geranium pusillum Geraniaceae Height to 30cm

 Spreading, branched annual. Similar to Dove's-foot Crane's-bill but with shorter hairs. **Flowers** 4–6mm across with notched pink petals (Jun–Sep). **Fruits** Hairy. **Leaves** Deeply cut into 5–7 narrow lobes. **Status** Widespread but commonest in S and E England. Found in dry grassy places and on bare ground.

Stork's-bills

Many native plants have delightful common names, some relating to their medicinal properties. Another source for names is the (sometimes fanciful) resemblance of part of the plant to human or animal anatomical features or organs. Look at the fruits of stork's-bills (relatives of the crane's-bills), for example, and the origin of the common name is obvious. Three species are regularly found in Britain. **Common Stork's-bill** *Erodium cicutarium* (**1**) is a stickily hairy annual of bare grassy places, often near the coast. Its flowers are 8–14mm across with pink petals that are easily lost (May–Aug), and its leaves are finely divided and feathery. The larger, musk-smelling **Musk Stork's-bill** *E. moschatum* (**2**) grows in coastal sandy ground in the SW. Its flower measure 25–30mm (May–Sep), and its leaves are not feathery and have oval lobes. Unobtrusive by comparison, **Sea Stork's-bill** *E. maritimum* (**3**) grows on sandy ground, near the sea. Its tiny flowers often lack petals, and its leaves are oval and lobed.

1

2

3

Spurges

Members of the spurge family (Euphorbiaceae) are both widespread and well represented in Britain and Ireland. Their flowers lack petals and are either borne in tassel-like spikes, in the case of the mercuries *Mercurialis*, or in showy terminal heads with petal-like bracts, in the case of spurges *Euphorbia*.

Wood Spurge

Euphorbia amygdaloides Euphorbiaceae Height to 80cm

 Upright, downy and unbranched perennial that forms clumps. **Flowers** Yellow with petal-like bracts (true petals and sepals are absent); borne in umbel-like heads (Apr–Jun). **Fruits** Smooth. **Leaves** Dark green and 6cm long. **Status** Locally common in S England and Wales, but rare or absent elsewhere. Found in woodland and scrub.

Sea Spurge

Euphorbia paralias Euphorbiaceae Height to 60cm

 Upright perennial. **Flowers** Yellowish with petal-like bracts and horned lobes (true petals and sepals are absent); borne in umbel-like heads (Jun–Oct). **Fruits** Smooth. **Leaves** Grey-green, fleshy and closely packed up the stems. **Status** Widespread and locally common on sandy beaches and dunes in S and W England, Wales and Ireland.

BELOW: Sea Spurge

Sun Spurge

Euphorbia helioscopia Euphorbiaceae Height to 50cm

 Upright, hairless yellowish-green annual. **Flowers** Lack sepals and petals, and are yellow with green lobes; borne in flat-topped umbel-like clusters with 5 leaf-like basal bracts (May–Nov). **Fruits** Smooth. **Leaves** Spoon-shaped and toothed. **Status** Widespread and common. Found on disturbed ground and in cultivated soils.

Petty Spurge

Euphorbia peplus Euphorbiaceae Height to 30cm

 Upright, hairless annual that often branches from the base. **Flowers** Greenish with oval bracts (sepals and petals are absent); borne in flattish, umbel-like clusters (Apr–Oct). **Fruits** Smooth. **Leaves** Oval, blunt-tipped and stalked. **Status** Widespread and common almost throughout. Found on arable land and cultivated ground.

Dwarf Spurge

Euphorbia exigua Euphorbiaceae Height to 30cm

 Slender, low-growing grey-green annual that often branches from the base. **Flowers** Yellowish (sepals and petals are absent), the lobes bearing horns; borne in open, umbel-like clusters (Jun–Oct). **Fruits** Smooth. **Leaves** Untoothed and very narrow. **Status** Locally common in the S only. Found on cultivated ground, often on chalk.

Portland Spurge
Euphorbia portlandica
Euphorbiaceae Height to 40cm

Hairless greyish perennial that branches at the base. **Flowers** Have lobes with long, crescent-shaped horns (petals and sepals are absent); borne in umbel-like clusters (Apr–Sep). **Fruits** Rough. **Leaves** Spoon-shaped with a prominent midrib. **Status** Locally common in SW and W Britain, in grassland and on cliffs near the sea.

Irish Spurge
Euphorbia hyberna Euphorbiaceae Height to 55cm

Attractive tufted, hairless perennial. **Flowers** Yellow with rounded lobes (petals and sepals are absent); borne in flat-topped umbel-like clusters (May–Jul). **Fruits** Have long, slender warts. **Leaves** Oval, tapering and stalkless. **Status** Local, confined to SW England and SW Ireland. Found in shady woodland and hedgerows.

Flaxes

Members of the flax family (Linaceae) are widely cultivated and, depending on the species, are grown commercially either for linseed oil or as a source of fibre for the production of linen. Three native flax family members are also found in Britain. **Pale Flax** *Linum bienne* (**1**) is a wiry-stemmed plant, mainly of coastal calcareous grassland; its lilac-blue flowers are 12–18mm across and are borne in open clusters (Jun–Jul). Once valued for its purgative properties, **Fairy Flax** *L. catharticum* (**2**) is a delicate annual of calcareous grassland; its white flowers are 4–6mm across and are borne in loose terminal clusters (May–Sep). Least showy of all is **Allseed** *Radiola linoides* (**3**), a low-growing annual of peaty ground and acid soils that is local and most common in SW England; its minute white flowers are just 1–2mm across and are borne in dense terminal clusters (Jul–Aug).

Cypress Spurge
Euphorbia cyparissias
Euphorbiaceae Height to 45cm

Patch-forming, bushy perennial; often turns red late in the season. **Flowers** Yellow, the lobes have short, crescent-shaped horns; borne in umbel-like clusters (May–Aug). **Fruits** Warty. **Leaves** Narrow and linear. **Status** Local in S England. Grows in short grassland on calcareous soils, but also occurs occasionally as a garden escape.

Caper Spurge
Euphorbia lathyris Euphorbiaceae Height to 1.5m

Upright grey-green biennial. **Flowers** Fresh green with elongated heart-shaped bracts (petals and sepals are absent); borne in open clusters (Jun–Jul). **Fruits** 15–17mm across, green, 3-sided and caper-like. **Leaves** Narrow and borne in opposite pairs. **Status** Doubtfully native plant of woodland and hedgerows; often naturalised in the S.

Annual Mercury
Mercurialis annua Euphorbiaceae Height to 50cm

Hairless, branched and bushy annual. **Flowers** Yellowish green and borne in spikes on separate-sex plants (Jul–Oct). **Fruits** Bristly. **Leaves** Narrowly ovate, shiny and toothed. **Status** Locally common in the S. Often found growing on waste ground and in cultivated soils; especially common near the sea.

Dog's Mercury
Mercurialis perennis
Euphorbiaceae Height to 35cm

Hairy, creeping perennial with a foetid smell. **Flowers** Yellowish and rather tiny; borne in open spikes on separate-sex plants (Feb–Apr). **Fruits** Hairy. **Leaves** Oval, shiny and toothed. **Status** Widespread and generally common, but scarce in N Scotland and Ireland. Grows in woodlands (sometimes forming carpets) and also on limestone pavements.

St John's-worts & mallows

Members of the St John's-wort (Clusiaceae) and mallow (Malvaceae) families both have flowers that comprise five petals and are extremely colourful: St John's-worts are yellow while mallows are pink or purple, depending on the species. Both families include representatives that grow in habitats ranging from dry, free-draining areas to those that are waterlogged and boggy.

Identifying St John's-worts

The St John's-wort family is a challenging group to identify, and initially they can all appear rather similar. Features to look for include the presence or absence of pale and dark dots on the petals and leaves, the shape of the stem in cross-section, and the habitat in which they grow.

Tutsan

Hypericum androsaemum Clusiaceae Height to 80cm

Upright, hairless semi-evergreen shrub with 2-winged stems. **Flowers** 2cm across with 5 yellow petals (Jun–Aug). **Fruits** Berries that ripen from red to black. **Leaves** Oval, up to 15cm long and borne in opposite pairs. **Status** Locally common in S and W Britain and Ireland only. Found in shady woods and hedgerows.

Marsh St John's-wort

Hypericum elodes Clusiaceae Height to 15cm

Creeping, greyish-green and hairy perennial. **Flowers** 10–15mm across, yellow and terminal (Jun–Aug). **Fruits** Dry capsules. **Leaves** Rounded to oval, grey-green and clasp the stem. **Status** Rather local and confined mainly to SW England and Ireland. Grows on peaty ground and in marshes on acid soils.

Hairy St John's-wort

Hypericum hirsutum Clusiaceae Height to 1m

Downy perennial with round stems. **Flowers** 15mm across with pale yellow petals and pointed sepals that have marginal stalked black glands; borne in spikes (Jul–Aug). **Fruits** Dry capsules. **Leaves** Oval with translucent spots. **Status** Local in scrub on calcareous soils.

Perforate St John's-wort

Hypericum perforatum Clusiaceae Height to 80cm

Upright, hairless perennial with 2-lined stems. **Flowers** 2cm across, the deep yellow petals often with black marginal spots (Jun–Sep). **Fruits** Dry capsules. **Leaves** Oval with translucent spots; borne in opposite pairs. **Status** Widespread but commonest in the S. Found in meadows, scrub and woodland, usually on calcareous soils.

stem section

Trailing St John's-wort

Hypericum humifusum Clusiaceae Prostrate

Creeping, hairless perennial with trailing, 2-ridged stems. **Flowers** 8–10mm across with pale yellow petals (Jun–Sep). **Fruits** Dry capsules. **Leaves** Have translucent dots and are borne in pairs along the stems. **Status** Widespread but commonest in W Britain and W Ireland. Found on bare ground on heaths and moors with acid soils.

Slender St John's-wort

Hypericum pulchrum Clusiaceae Height to 60cm

Hairless perennial with rounded stems. **Flowers** 15mm across, the deep yellow petals marked with red spots and dark marginal dots (the latter also occur on the sepals) (Jul–Aug). **Fruits** Dry capsules. **Leaves** Paired and oval, with translucent spots. **Status** Widespread and common. Found in dry grassy places and heaths, mostly on acid soils.

Pale St John's-wort

Hypericum montanum Clusiaceae Height to 80cm

Perennial with round stems. Similar to Hairy St John's-wort but almost hairless. **Flowers** 10–15mm across with pale yellow petals and red sepals (Jul–Sep). **Fruits** Dry capsules. **Leaves** Lack translucent dots but have marginal black dots below. **Status** Local on calcareous soils.

Imperforate St John's-wort

Hypericum maculatum Clusiaceae
Height to 1m

 Upright, hairless perennial. Similar to Perforate St John's-wort but with square, unwinged stems. **Flowers** 2cm across with yellow petals (Jun–Aug). **Fruits** Dry capsules. **Leaves** Oval and lack translucent dots. **Status** Widespread and locally common throughout, except in the N. Found in woodland and scrub.

stem section

Square-stalked St John's-wort

Hypericum tetrapterum Clusiaceae
Height to 1m

 Upright, hairless perennial. Similar to Imperforate St John's-wort but the square stems are distinctly winged. **Flowers** 2cm across with yellow petals and undotted, pointed sepals (Jun–Sep). **Fruits** Dry capsules. **Leaves** Oval with translucent dots. **Status** Widespread and locally common, except in the N. Grows in damp ground.

stem section

Identifying mallows

Mallows are plants of grasslands and waysides whose flowers comprise five showy petals that are typically pink or purple. Identification of the species is made possible by looking at the size and shape of both the flowers and the leaves.

Musk-mallow

Malva moschata Malvaceae
Height to 75cm

 Branched perennial with hairy stems and downy, felty leaves. **Flowers** 3–6cm across and pale pink; borne in showy, terminal clusters of tightly packed flowers (Jul–Aug). **Fruits** Flat, round capsules. **Leaves** Rounded and 3-lobed at the base of the plant but becoming increasingly dissected up the stem. **Status** Widespread and locally common in England and Wales but scarce elsewhere. Found in dry grassy places, often in established meadows but also increasingly on mature roadside verges.

Common Mallow

Malva sylvestris Malvaceae
Height to 1.5m

 Upright or spreading perennial. **Flowers** 25–40mm across with 5 purple-veined pink petals that are much longer than the sepals; borne in clusters from the leaf axils (Jun–Oct). **Fruits** Flat, round capsules. **Leaves** Rounded at the base of the plant, 5-lobed on the stem. **Status** Widespread and common in the S; scarce elsewhere. Grows on grassy verges and disturbed ground.

fruit

Dwarf Mallow

Malva neglecta Malvaceae
Height to 1.5m

 Upright or spreading perennial. **Flowers** 1–2cm across with notched, purple-veined, pale lilac petals; borne in clusters along the stems (Jun–Sep). **Fruits** Flat, round capsules. **Leaves** Rounded with shallow lobes. **Status** Widespread and locally common in the S but scarce elsewhere. Grows on disturbed and waste ground.

Tree Mallow

Lavatera arborea Malvaceae
Height to 3m

 Imposing woody biennial, covered in starry hairs. **Flowers** 3–5cm across with dark-veined pinkish-purple petals; borne in terminal clusters (Jun–Sep). **Fruits** Flat, round capsules. **Leaves** 5–7 lobed. **Status** Locally common on W coasts of Britain and in S and W Ireland. Favours rocky ground and sea cliffs, often near seabird colonies.

Marsh-mallow

Althaea officinalis Malvaceae
Height to 2m

 Attractive downy perennial with starry hairs; very soft to touch. **Flowers** 35–40mm across and pale pink (Aug–Sep). **Fruits** Rounded flat capsules. **Leaves** Triangular with shallow lobes. **Status** Locally common on S coasts of Britain and Ireland. Found in coastal wetlands and often in upper reaches of saltmarshes.

Violets, winter shrubs & balsams

Some native plants have flowers whose petals are evenly sized, uniformly marked and arranged in a straightforward manner. This is not true of members of the violet (Violaceae) and balsam (Balsaminaceae) families. The former have five uneven petals adorned with a range of markings, while the latter have bizarre trumpet-shaped flowers. The flowers of both plants have long spurs, which are formed from one of the sepals in balsams, and from a petal in violets. At first glance, all our native violet species can look confusingly similar to one another. However, by paying attention to features such as flower and leaf shape, and growing habitat, most are relatively easy to identify.

Sweet Violet
Viola odorata Violaceae
Height to 15cm

Fragrant perennial herb. **Flowers** 15mm across and violet or white, with blunt sepals (Feb–May). **Fruits** Egg-shaped. **Leaves** Long-stalked and rounded in spring, with larger, heart-shaped leaves appearing in autumn. **Status** Widespread and locally common in England and Wales. Grows in woods and hedgerows, mostly on calcareous soils.

Hairy Violet
Viola hirta Violaceae
Height to 15cm

Similar to Sweet Violet but unscented and much more hairy. **Flowers** 15mm across, with pale violet petals and blunt sepals (Mar–May). **Fruits** Egg-shaped. **Leaves** Narrow and hairy. **Status** Widespread and locally common in England and Wales; absent from the N and scarce in Ireland. Found in dry grassland, mainly on calcareous soils.

Common Dog-violet
Viola riviniana Violaceae
Height to 12cm

Familiar perennial herb. **Flowers** 15–25mm across and bluish violet, with a blunt, pale spur that is notched at the tip, and with pointed sepals (Mar–May). **Fruits** Egg-shaped. **Leaves** Long-stalked, heart-shaped and mainly hairless. **Status** Widespread and locally common throughout. Grows in woodland rides and grassland.

Early Dog-violet
Viola reichenbachiana Violaceae
Height to 12cm

Similar to Common Dog-violet but with subtle differences in the flower and leaf form. **Flowers** 15–20mm across with narrow, pale violet petals and a spur that is darker than the petals and is not notched (Mar–May). **Fruits** Egg-shaped. **Leaves** Narrow heart-shaped. **Status** Locally common, mainly in the S and Ireland. Found in woods and hedgerows, mostly on chalk.

Heath Dog-violet
Viola canina Violaceae
Height to 30cm

Perennial herb that lacks a basal rosette of leaves. **Flowers** 12–18mm across with pale blue petals and a short greenish-yellow spur (Apr–Jun). **Fruits** Not inflated. **Leaves** Narrowly oval with a heart-shaped base. **Status** Widespread but only very locally common. Found on dry grassland, mainly on sandy soils.

Pale Dog-violet
Viola lactea Violaceae
Height to 12cm

Similar to Heath Dog-violet but note the differences in flower and leaf form. **Flowers** 12-18mm across with very pale petals and a short greenish spur (May-Jun). **Fruits** Not inflated.

Leaves Narrow and wedge-shaped at the base (not heart-shaped). **Status** Local on heathland.

Marsh Violet
Viola palustris Violaceae
Height to 15cm

Hairless, creeping perennial. **Flowers** 10-15mm across with rounded, dark-veined, pale lilac petals and a blunt, pale spur (Apr-Jul). **Fruits** Egg-shaped. **Leaves** Kidney-shaped and long-stalked.

Status Widespread but local; commonest in the N and W. Found in bogs and marshy places on acid soils.

Mountain Pansy
Viola lutea Violaceae
Height to 30cm

Almost hairless perennial. **Flowers** 15-30mm across and may be yellow, bluish violet or both (May-Aug). **Fruits** Egg-shaped. **Leaves** Narrow, with palmate basal stipules. **Status** Grows in upland calcareous grassland.

Field Pansy
Viola arvensis Violaceae
Height to 15cm

Variable annual. **Flowers** 10-15mm across and creamy white with an orange flush on the lower petal; the sepals are at least as long as the petals (Apr-Oct). **Fruits** Capsules. **Leaves**

Have deeply toothed stipules. **Status** Widespread and common throughout the region. Grows on arable land and cultivated ground.

Wild Pansy
Viola tricolor Violaceae
Height to 12cm

Also known as Heartsease. **Flowers** 15-25mm across; yellow and violet in ssp. *tricolor* but yellow in ssp. *curtisii* (Apr-Aug). **Fruits** Egg-shaped. **Leaves** Lanceolate with leaf-like stipules. **Status** Widespread and locally common. Ssp. *tricolor* is an annual of cultivated ground, whereas ssp. *curtisii* is a perennial of dry grassland.

Winter-flowering shrubs

Winter woodlands can often be bleak and uninspiring places. However, two native shrubs, that are found in a few mature and undisturbed sites on calcareous soils, provide splashes of colour from February onwards, well before the first primrose or daffodil appears. The most widespread of the two is **Spurge Laurel** *Daphne laureola* (**1**), an evergreen with clusters of flowers that each measure 8-12mm across, are yellowish and have four petal-like sepal lobes; later in the season, black berries form. More restricted in its range is the deciduous **Mezereon** *D. mezereum* (**2**), whose flowers are 8-12mm across with four pink, petal-like sepals, and which appear just before the leaves; red berries are formed later in the season.

Balsams

In addition to one scarce native species, several introduced balsams (family Balsaminaceae) are now widely established, if not altogether welcomed, residents of damp ground, often beside rivers. The most familiar is **Indian Balsam** *Impatiens glandulifera* (**1**), an impressive but invasive red-stemmed annual that reaches 2m in height; its flowers are 3-4cm long and pinkish purple, with a short, curved spur (Jun-Sep), and the fruits are club-shaped and explosive. Three other species are sometimes encountered: **Touch-me-not-Balsam** *I. noli-tangere*, a native species with 2-3cm-long yellow flowers that have a curved spur; **Orange Balsam** *I. capensis* (**2**), with 2-3cm-long orange flowers that have brown blotches and a curved spur; and **Small Balsam** *I. parviflora* (**3**), with 13-16mm-long yellow flowers that have a rather straight spur.

Rock-roses & willowherbs

The wildflowers shown here are found in a wide range of habitats. Some, including most willowherbs, thrive on disturbed and newly cleared ground; consequently, they are often found growing beside paths and tracks. Rock-roses tolerate dry, free-draining soils, while loosestrifes prefer damp ground and typically grow with their 'feet' in water.

Looking at willowherbs

Apart from a couple of showy, large-flowered species, willowherbs are generally difficult to identify, even for the experienced botanist. Key features to look for are the shape of the stigma - club-shaped (**1**) or four-lobed (**2**) in the centre - of the flower, whether the leaves are stalked or unstalked, and whether the leaves are alternate on the stem or arranged in opposite pairs.

Purple-loosestrife

Lythrum salicaria Lythraceae
Height to 1.5m

Downy, clump-forming perennial. **Flowers** 10–15mm across, reddish purple and 6-petalled; borne in tall spikes (Jun–Aug). **Fruits** Capsules. **Leaves** 4–7cm long, narrow, unstalked and borne in opposite pairs. **Status** Locally common except in the N. Grows in damp habitats such as river banks and fens.

Water-purslane

Lythrum portula Lythraceae
Prostrate

Low-growing, creeping and hairless annual. **Flowers** 1–2mm across, with 6 pinkish petals that are often absent or fallen; borne in leaf axils (Jun–Oct). **Fruits** Capsules. **Leaves** Oval and borne in opposite pairs. **Status** Locally common on damp bare ground and in shallow water, mainly on acid soils.

Rosebay Willowherb

Chamerion angustifolium
Onagraceae Height to 1.5m

Showy perennial. **Flowers** 2–3cm across with pinkish-purple petals; borne in tall spikes (Jul–Sep). **Fruits** Pods that contain cottony seeds. **Leaves** Lanceolate and arranged spirally up the stems. **Status** Widespread and common throughout. Grows on waste ground, cleared woodland and river banks, on a wide range of soil types.

Great Willowherb

Epilobium hirsutum Onagraceae
Height to 2m

Downy perennial with a round stem. **Flowers** 25mm across, pinkish purple with pale centres, and with a 4-lobed stigma; borne in terminal clusters (Jul–Aug). **Fruits** Pods that contain cottony seeds. **Leaves** Broadly oval, hairy and clasping. **Status** Widespread and common, except in the N. Favours damp habitats such as fens and river banks.

Broad-leaved Willowherb

Epilobium montanum Onagraceae
Height to 80cm

Upright perennial. Similar to Hoary Willowherb but almost hairless. **Flowers** 6–10mm across (drooping in bud), with notched pale pink petals and a 4-lobed stigma (Jun–Aug). **Fruits** Pods that contain cottony seeds. **Leaves** Oval, rounded at the base, toothed and opposite. **Status** Widespread and common. Found in woods and hedges.

A hardy cucumber

Edible marrows and cucumbers are tender plants that have exotic origins. But, there is a hardy native member of the family Cucurbitaceae, to which these vegetables belong, namely **White Bryony** *Bryonia dioica*. This climbing hedgerow perennial can reach a height of 4m and its progress is aided by long, unbranched tendrils. The flowers are greenish, are made up of five parts and are borne on separate-sex plants; they arise from leaf axils (May–Aug). The fruits are shiny red berries, and the leaves are 4–7cm across and divided into five lobes. It is common in England, although scarce elsewhere.

Rock-roses

Rock-roses belong to the family Cistaceae. They grow on dry soils, and their five showy petals are crinkly and easily dislodged. Only **Common Rock-rose** *Helianthemum nummularium* (**1**) is widespread. It is found on calcareous grassland and its flowers are 2.5cm across with five crinkly yellow petals (Jun–Sep); the leaves are narrow, oval, downy white below and paired. **White Rock-rose** *H. apenninum* (**2**) is a rarity of coastal calcareous grassland in Devon and Somerset; its flowers are 2.5cm across and have five crinkly white petals (May–Jul). **Hoary Rock-rose** *H. oelandicum* (**3**) is a rarity of limestone grassland, locally common in W Ireland but rare in N England and Wales. Its yellow flowers are 10–15mm across (May–Jul).

American Willowherb

Epilobium ciliatum Onagraceae
Height to 50cm

Upright perennial whose stems have 4 raised lines and spreading glandular hairs. **Flowers** 8–10mm across with notched pink petals and a club-shaped stigma (Jul–Sep). **Fruits** Pods that contain cottony seeds. **Leaves** Narrow, oval, toothed and short-stalked. **Status** Introduced but widely naturalised. Found on waste ground and in damp, shady places.

Marsh Willowherb

Epilobium palustre Onagraceae
Height to 50cm

Slender, upright perennial with a round, smooth stem. **Flowers** 4–7mm across and pale pink, with a club-shaped stigma (Jul–Aug). **Fruits** Pods that contain cottony seeds. **Leaves** Narrow, untoothed, unstalked and borne in opposite pairs. **Status** Widespread and locally common. Found in damp habitats, mainly on acid soils.

Hoary Willowherb

Epilobium parviflorum
Onagraceae Height to 75cm

Downy perennial. Similar to Great Willowherb but smaller, with non-clasping leaves. **Flowers** 12mm across with notched pale pink petals and a 4-lobed stigma (Jul–Sep). **Fruits** Pods that contain cottony seeds. **Leaves** Broadly oval; the upper ones are alternate. **Status** Widespread and common, except in the N. Found in damp habitats.

Spear-leaved Willowherb

Epilobium lanceolatum
Onagraceae Height to 80cm

Resembles a slender, grey-green form of Broad-leaved Willowherb with alternate leaves. **Flowers** 6–8mm across with a 4-lobed stigma; white at first, turning pink later (Jul–Sep). **Fruits** Pods that contain cottony seeds. **Leaves** Narrow and tapering to the base. **Status** Local, in S England and S Wales only. Found in shady places.

Square-stalked Willowherb

Epilobium tetragonum
Onagraceae Height to 1m

Upright, downy perennial with 4-ridged stems (sometimes winged). **Flowers** 6–8mm across (upright in bud) with pink petals and a club-shaped stigma (Jul–Aug). **Fruits** Pods (6–10cm long) that contain cottony seeds. **Leaves** Narrow and finely toothed. **Status** Common only in England and Wales. Found in damp woods and on river banks.

Chickweed Willowherb

Epilobium alsinifolium
Onagraceae Height to 20cm

Branched, usually upright perennial that is almost hairless. **Flowers** 8–11mm across, pinkish purple and seldom open fully; borne on drooping stalks (Jul–Aug). **Fruits** Long, green and erect. **Leaves** Ovate, short-stalked and slightly toothed. **Status** Local in mountains from N Wales northwards. Found on damp ground in upland areas.

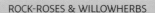

Carrot family

The term 'umbellifer' refers to members of the carrot family (Apiaceae). The group is diverse, with representatives in almost all habitats apart from the tops of mountains. Some are among the tallest of our native herbaceous plants, while others are tiny. All have flower heads shaped like an umbrella.

Pignut

Conopodium majus Apiaceae
Height to 25cm

Delicate, usually unbranched upright perennial with smooth, hollow stems. **Flowers** White and borne in umbels, 3–6cm across (Apr–Jun). **Fruits** Narrow and egg-shaped. **Leaves** Comprise finely divided basal leaves that soon wither, and narrow-lobed stem leaves. **Status** Locally common throughout in open woodland and grassland, mainly on dry acid soils.

Burnet-saxifrage

Pimpinella saxifraga Apiaceae
Height to 70cm

Downy branched perennial. **Flowers** White and borne in loose, open umbels (Jun–Sep). **Fruits** Egg-shaped and ridged. **Leaves** Deeply divided with oval leaflets at the base of the plant, and with stem leaves that are finely divided into narrow leaflets. **Status** Widespread and locally common in dry calcareous grassland; absent from NW Scotland.

Greater Burnet-saxifrage

Pimpinella major Apiaceae
Height to 1m

Branched perennial with hollow, ridged, hairless stems. **Flowers** White and borne in umbels, 3–6cm across (Jun–Sep). **Fruits** Egg-shaped and ridged. **Leaves** Deeply divided with toothed oval lobes. **Status** Widespread but distinctly local, commonest in central England and S Devon. Grows in shady and grassy places.

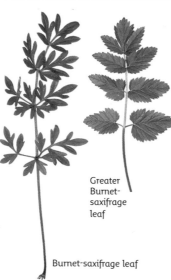

Greater Burnet-saxifrage leaf

Burnet-saxifrage leaf

Lesser Water-parsnip

Berula erecta Apiaceae
Height to 70cm

Spreading perennial. **Flowers** White; borne in short-stalked umbels, 3–6cm across (Jul–Sep). **Fruits** Ridged and spherical. **Leaves** Deeply divided with 7–14 pairs of oval, jagged-toothed leaflets. **Status** Locally common in damp ground.

Pepper-saxifrage

Silaum silaus Apiaceae
Height to 1m

Slender, hairless perennial with solid, ridged stems. **Flowers** Yellowish and borne in long-stalked umbels, 2–6cm across (Jul–Sep). **Fruits** Egg-shaped and ridged. **Leaves** Deeply divided with narrow, pointed leaflets. **Status** Locally common in England but scarce or absent elsewhere. Found in meadows on damp, heavy soils.

Lesser Marshwort

Apium inundatum Apiaceae
Creeping

Prostrate, hairless perennial with smooth stems. **Flowers** White and borne in small stalked umbels that have 2–4 rays (Jun–Jul). **Fruits** Narrow and egg-shaped. **Leaves** Divided with narrow, hair-like leaflets (*cf.* some water-crowfoot species, p.49). **Status** Widespread but local on the margins of ponds and marshes; sometimes partly submerged.

Hogweed

Heracleum sphondylium Apiaceae
Height to 2m

Robust, roughly hairy perennial with hollow, ridged stems. **Flowers** Off-white, with unequal petals; borne in umbels to 20cm across, with 40 or so rays (May–Aug). **Fruits** Elliptical, hairless and flattened. **Leaves** Up to 60cm long, broad, hairy and divided with ovate lobes. **Status** Widespread and common in meadows, open woodlands and on verges.

Giant Hogweed

Heracleum mantegazzianum
Apiaceae Height to 4m

Unmistakable size. The stems are hollow, ridged and purple-spotted. **Flowers** White; borne in umbels up to 50cm across (Jun-Jul). **Fruits** Narrow and oval. **Leaves** Deeply divided, to 1m long. **Status** Introduced and naturalised. Invasive and poisonous; now a notifiable weed.

Ground-elder

Aegopodium podagraria Apiaceae
Height to 1m

Patch-forming perennial, originally introduced as a culinary and medicinal herb but now rampantly invasive in many gardens; thrives on disturbance. **Flowers** White; borne in umbels, 2-6cm across with 10-20 rays (May-Jul). **Fruits** Egg-shaped and ridged. **Leaves** Fresh green, roughly triangular and twice trifoliate. **Status** Widespread in damp, disturbed soil.

Identifying Water-dropworts

Some water-dropworts prefer to have just their 'feet' in water, while others are truly aquatic, with leaves and stems that are almost entirely submerged. Several members of the group are rare and declining.

Hemlock Water-dropwort

Oenanthe crocata Apiaceae
Height to 1.25m

Highly poisonous plant, sometimes forming sizeable clumps. Its stems are hollow and grooved, and the plant smells of Parsley. **Flowers** White and borne in domed umbels, 5-10cm across, with 10-40 rays and numerous bracts (Jun-Aug). **Fruits** Cylindrical, with long styles. **Leaves** Divided 2-4 times and have toothed, tapering lobes. **Status** Fairly widespread, growing in marshes and damp ditches.

Tubular Water-dropwort

Oenanthe fistulosa Apiaceae
Height to 50cm

More delicate than Hemlock Water-droplet, with inflated, hollow stems. **Flowers** Pinkish white, in umbels, 2-4cm across, with 2-4 rays (Jul-Sep). **Fruits** Roughly cylindrical but angular; ripe umbels, which are globular. **Leaves** Have inflated stalks; leaflets of lower leaves are oval, those of upper ones are tubular. **Status** Local in damp ground and shallow water in England.

Parsley Water-dropwort

Oenanthe lachenalii Apiaceae
Height to 1m

Similar to Tubular Water-dropwort but has solid stems. **Flowers** White, in umbels, 2-6cm across, with 6-15 rays (Jun-Sep). **Fruits** Egg-shaped, ribbed, lacking swollen corky bases. **Leaves** Divided 2-3 times; leaflets of basal leaves are oval and flat, like fresh Parsley. **Status** Local in damp meadows and brackish marshes.

Corky-fruited Water-dropwort

Oenanthe pimpinelloides Apiaceae
Height to 1m

Has solid, ridged stems. **Flowers** White and borne in terminal, flat-topped umbels, 2-6cm across, with 6-15 rays (May-Aug). **Fruits** Cylindrical, with swollen corky bases. **Leaves** Divided 1-2 times, with narrow oval to wedge-shaped leaflets. **Status** Locally common in damp, mainly coastal grassland, particularly on clay soils.

Fine-leaved Water-dropwort

Oenanthe aquatica Apiaceae
Height to 1.3m

Bushy perennial. **Flowers** White and borne in flat-topped umbels, 2-5cm across, both terminal and arising opposite leaf stalks (Jun-Sep). **Fruits** Ovoid. **Leaves** Delicate-looking and divided 3-4 times; the submerged leaves have fine lobes and the aerial leaves have oval lobes. **Status** Extremely local and habitat-specific, growing in the margins of slow-flowing waters.

Carrot family

Wild Angelica

Angelica sylvestris Apiaceae
Height to 2m

 Robust and almost hairless plant with hollow, purplish stems. **Flowers** White (sometimes tinged pink); borne in robust, domed umbels up to 15cm across (Jun-Jul). **Fruits** Oval. **Leaves** Divided 2-3 times; the lower leaves are up to 60cm long; smaller upper leaves have inflated basal sheaths. **Status** Widespread throughout in damp meadows and woodlands.

Scots Lovage

Ligusticum scoticum Apiaceae
Height to 80cm

 Clump-forming perennial, once used as a culinary herb. **Flowers** White; borne in flat-topped umbels, 4-6cm across, on long, reddish stalks (Jun-Aug). **Fruits** Oval and flattened, with 4 wings. **Leaves** Shiny and divided 2 times, with oval leaflets and inflated, sheathing stalks. **Status** Locally common on sea cliffs and coastal grassland in the N.

Fool's-parsley

Aethusa cynapium Apiaceae
Height to 50cm

 Delicate, hairless and unpleasant-tasting annual. **Flowers** White and borne in umbels, 2-3cm across; the secondary umbels have a 'beard' of long upper bracts (Jun-Aug). **Fruits** Egg-shaped and ridged. **Leaves** Twice divided, flat and triangular in outline. **Status** Commonest in the S. Grows in disturbed soil in gardens and arable fields.

Fool's Watercress

Apium nodiflorum Apiaceae
Height to 20cm

 Creeping perennial whose leaves bear a passing resemblance to Watercress (see p.62). Roots at the nodes of its lower stems; the upright stems are hollow. **Flowers** White; borne in open umbels (Jul-Aug). **Fruits** Egg-shaped and ridged. **Leaves**

 Shiny and divided, with oval, toothed leaflets. **Status** Widespread and locally common in ditches and wet hollows.

Hemlock

Conium maculatum Apiaceae
Height to 2m

 Highly poisonous, hairless biennial, with hollow, purple-blotched stems and an unpleasant smell. **Flowers** White; borne in umbels, 2-5cm across (Jun-Jul). **Fruits** Globular with wavy ridges. **Leaves** Divided up to 4 times into fine leaflets. **Status** Widespread and locally common, except in the N. Grows on damp wayside ground, verges and riversides.

stem

Sweet Cicely

Myrrhis odorata Apiaceae
Height to 1.5m

 Downy perennial with hollow stems. The whole plant smells of aniseed when bruised and can be used in cooking. **Flowers** White, with unequal petals; borne in umbels up to 5cm across (May-Jun). **Fruits** Elongated and ridged. **Leaves** Fern-like, up to 30cm long, with basal sheaths. **Status** Introduced and naturalised on damp ground, mainly in the N.

Rock-samphire

Crithmum maritimum Apiaceae
Height to 40cm

 Branched perennial. The leaves are edible and formerly were much eaten. **Flowers** Greenish yellow; borne in umbels, 3-6cm across, with 8-30 rays and numerous bracts (Jun-Sep). **Fruits** Egg-shaped. **Leaves** Divided into narrow, fleshy lobes, triangular in cross-section. **Status** Widespread and common around the coasts of S and W Britain and Ireland.

Stone Parsley

Sison amomum Apiaceae
Height to 1m

 Upright, bushy perennial with an unpleasant smell (of nutmeg and petrol) when bruised. **Flowers** White and borne in open umbels, 1-4cm across, with unequal rays (Jul-Aug). **Fruits** Globular. **Leaves** Fresh green; the lower leaves have oval leaflets, while the upper ones have narrow leaflets. **Status** Local in grassy places on clay soils.

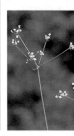 

Culinary ancestors

Many of the herbs and vegetables we commonly grow in our gardens are umbellifers, although harvesting often occurs before the plants come into flower and reveal their floral family ties. Some of these culinary delights are exotic imports, but many have their origins in the British countryside and their wild ancestors can still be found in meadows and waysides.

Wild Parsnip

Pastinaca sativa ssp. *sativa*
Apiaceae Height to 1m

 Upright, downy perennial that smells strongly when bruised. **Flower** Yellowish; borne in open, bractless umbels, 3–9cm across (Jun–Sep). **Fruits** Flattened and oval. **Leaves** Divided, with oval, toothed leaflets. **Status** Widespread and locally common in the S. Found mainly in dry calcareous grassland. **Similar species Garden Parsnip** *P. s. hortensis* is sometimes naturalised.

Wild Celery

Apium graveolens Apiaceae
Height to 1m

 Hairless biennial with a strong smell of celery. Its stems are solid and grooved. **Flowers** White; borne in short-stalked or unstalked umbels, 3–6cm across (Jun–Aug). **Fruits** Globular. **Leaves** Shiny and divided 1–2 times, with toothed, diamond-shaped lobes; the stem leaves look trifoliate. **Status** Rather local and mainly coastal, on rough, often saline, grassland and meadows.

Wild Carrot

Daucus carota ssp. *carota*
Apiaceae Height to 75cm

 Hairy perennial with solid, ridged stems. **Flowers** White (pinkish in bud) and borne in long-stalked umbels, to 7cm across; the central flower is red, and note the divided bracts beneath the umbel (Jun–Sep). **Fruits** Oval and spiny; the fruiting umbels are concave. **Leaves** Divided 2–3 times, with narrow leaflets. **Status** Locally common. Found inland in rough grassland, mostly on chalky soils or near the sea. **Similar species Sea Carrot** *D. c. gummifer* is similar but has more fleshy leaves and umbels that are flat or convex in fruit. It is found on cliffs, rocky slopes and dunes by the sea, mainly in SW England.

Fennel

Foeniculum vulgare Apiaceae
Height to 2m

 Grey-green, strong-smelling, hairless perennial with solid young stems and hollow older ones. **Flowers** Yellow and borne in open umbels, 4–8cm across (Jul–Oct). **Fruits** Narrow, egg-shaped and ridged. **Leaves** Feathery, comprising thread-like leaflets. **Status** Locally common in the S. Favours grassy places, mainly near the sea.

Garden Parsley

Petroselinum crispum Apiaceae
Height to 40cm

 Hairless biennial. Bright green, yellowing with age. **Flowers** Greenish yellow; borne in open umbels, 1–2cm across (Jun–Aug). **Fruits** Globular. **Leaves** Shiny, roughly triangular and divided 3 times; cultivars have variably crinkled leaflets. **Status** Rare native in dry coastal grassland. Widely cultivated and occasionally naturalised.

Carrot family

Marsh Pennywort

Hydrocotyle vulgaris Apiaceae
Creeping

Low-growing perennial and an atypical umbellifer. **Flowers** Tiny, pinkish and hidden by the leaves; borne in small umbels (Jun–Aug). **Fruits** Rounded and ridged. **Leaves** Round and dimpled with broad, blunt teeth. **Status** Widespread but commonest in W of region. Found in short, grassy vegetation on damp, mostly acid ground.

Sanicle

Sanicula europaea Apiaceae
Height to 50cm

Slender, hairless perennial. **Flowers** Pinkish and borne in small umbels on reddish stems (May–Aug). **Fruits** Egg-shaped with hooked bristles. **Leaves** Have 5–7 toothed lobes; the lower leaves are long-stalked. **Status** Very locally common throughout. Grows in deciduous woodland, mostly on neutral or basic soils and often under Beech.

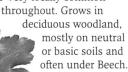

Sea-holly

Eryngium maritimum Apiaceae
Height to 60cm

Distinctive hairless perennial. **Flowers** Blue and borne in globular umbels that are up to 4cm long (Jul–Sep). **Fruits** Bristly. **Leaves** Waxy, blue-green and Holly-like, with spiny white margins and white veins. **Status** Widespread on shingle and sand around the coasts of England, Wales and Ireland; absent from N and E Scotland.

Did you know?

If you live in S Britain and travel the same country lane on a regular basis, you will probably have noticed that white umbellifer flowers are a hedgerow feature from early April right through to the end of August. Although the flowers seen in early spring may look superficially similar to those in bloom in summer, it is in fact three species, flowering in succession, that are responsible for the display. In chronological order of appearance, these are Cow Parsley, Rough Chervil and Upright Hedge-parsley.

Cow Parsley

Anthriscus sylvestris Apiaceae
Height to 1m

Downy herbaceous perennial with hollow, unspotted stems. **Flowers** White and borne in umbels up to 6cm across; bracts absent (Apr–Jun). **Fruits** Elongate and ridged. **Leaves** Deeply divided 2–3 times, only slightly hairy and fresh green. **Status** Widespread and common. Found in meadows and woodland margins, and on verges.

stem

Bur Chervil

Anthriscus caucalis Apiaceae
Height to 50cm

Delicate annual with hairless, hollow stems, flushed purple at the base. **Flowers** White and borne in umbels, 2–4cm across; bracts absent (May–Jun). **Fruits** Egg-shaped with hooked bristles. **Leaves** Finely divided, feathery and hairy below. **Status** Widespread and common, except in the N. Found on sandy ground, often coastal.

Alexanders

Smyrnium olusatrum Apiaceae
Height to 1.25m

Hairless biennial. **Flowers** Yellowish; borne in umbels, 4–6cm across, with 7–15 rays (Mar–Jun). **Fruits** Globular, ridged and black when ripe. **Leaves** Dark green, shiny, much divided. **Status** Introduced but widely naturalised, on S and SE coasts of England and Ireland. Grows in hedges and on verges mainly on calcareous soils.

Rough Chervil

Chaerophyllum temulum Apiaceae
Height to 1m

Biennial with solid, ridged, bristly, purple-spotted stems. **Flowers** are white; borne in umbels up to 6cm across (Jun-Jul). **Fruits** Elongate, tapering and ridged. **Leaves** Deeply divided 2-3 times, hairy and dark green. **Status** Common and widespread in England and Wales, in hedges and on verges.

Upright Hedge-parsley

Torilis japonica Apiaceae
Height to 1m

Slender annual with solid, unspotted, roughly hairy stems. **Flowers** White (or tinged pink); borne in long-stalked terminal umbels, 2-4cm across, with 5-12 rays (Jul-Aug). **Fruits** Egg-shaped with hooked purple bristles. **Leaves** Deeply divided 1-3 times and hairy. **Status** Widespread and common. Found in hedges and woodland margins.

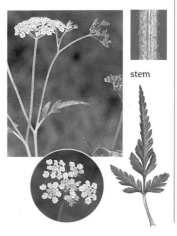

stem

Other species

Enchanter's Nightshade grows in shady woodland, while Dwarf Cornel is a hardy upland plant.

Enchanter's-nightshade

Circaea lutetiana Onagraceae
Height to 65cm

Delicate and slightly downy perennial. **Flowers** Small with white petals; borne in loose spikes above the leaves (Jun-Aug). **Fruits** Club-shaped and bristly. **Leaves** 10cm long, oval, heart-shaped at the base and gently toothed, with round stalks. **Status** Common and widespread, except in the N. Grows in woodland and hedgerows.

Dwarf Cornel

Cornus suecica Cornaceae
Height to 15cm

Creeping perennial. **Flowers** Small and purplish black; borne in dense umbels surrounded by 4 white bracts (Jun-Aug). **Fruits** Red berries. **Leaves** Ovate and pointed, with 3 main veins on both sides of the midrib. **Status** Scattered distribution in N England and locally common in parts of Scotland. Grows on upland moors and on some lower mountain slopes.

The importance of Ivy

Ivy *Hedera helix* is important for wildlife. Its flowers provide nectar for insects, its fruits are food for birds and the foliage provides safe roosting spots for birds and hibernating insects. Ivy is an epiphyte, not a parasite – a plant that grows on another plant, gaining support but not nutrition. Ivy's flowers are yellowish, with four parts, and are borne in globular heads (Sep-Nov). The fruits are berries that ripen purplish black, and the leaves have three or five lobes. It is common in wayside habitats.

fruits

Colourful alien colonisers

The majority of garden plants thrive in cultivation because they are given preferential treatment over native species, but in the wild most would not be able to compete with vigorous native plants. However, in milder parts of W Britain and S Ireland this is certainly not true of **Fuchsia** *Fuchsia magellanica* (**1**). This branched deciduous shrub, originally introduced for hedging, is now locally common as a naturalised plant of rocky ground near the sea. Its bell-shaped flowers, with red sepals and violet petals (Aug-Oct), are recognisable immediately. **Common Evening-primrose** *Oenothera biennis* (**2**) is altogether hardier and is now widely naturalised on verges, railway embankments and sand dunes. Its flowers, which are 4-5cm across, yellow and open only on dull days or evenings (Jun-Sep), are a colourful sight.

Primroses

Members of the primrose family (Primulaceae) include some of our most colourful and familiar spring flowers, several of which are associated with woodlands. Less familiar relatives are restricted to wetland habitats, while a few are specialists of seashore or upland habitats. All species characteristically have flowers that are fused at the base but with five obvious lobes.

Primrose

Primula vulgaris Primulaceae
Height to 20cm

Familiar herbaceous perennial. **Flowers** 2–3cm across, 5-lobed and pale yellow, usually with deep yellow centres; solitary, borne on hairy stalks that arise from the centre of the leaf rosette (Feb–May). **Fruits** Capsules. **Leaves** Oval, tapering and crinkly, to 12cm long; they form a basal rosette. **Status** Common throughout in hedgerows, woodlands and shady meadows.

Cowslip

Primula veris Primulaceae
Height to 25cm

Elegant downy perennial. **Flowers** 8–15mm across, fragrant, bell-shaped, stalked and orange-yellow; in rather 1-sided umbels of 10–30 flowers (Apr–May). **Fruits** Capsules. **Leaves** Tapering, wrinkled and hairy, forming a basal rosette. **Status** Widespread and locally common in dry, unimproved grassland, often on calcareous soils.

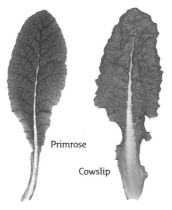

Primrose

Cowslip

Primrose

Oxlip

Primula elatior Primulaceae
Height to 20cm

Attractive perennial. **Flowers** 15–25mm across, pale yellow and 5-lobed, like small Primrose flowers; in 1-sided umbels of 10–20 flowers (Mar–May). **Fruits** Capsules. **Leaves** Oval, long-stalked and crinkly, and they end abruptly and do not taper; they form a basal rosette. **Status** Locally common in parts of E Anglia only. Grows in open woodland on clay soils.

False Oxlip

Primula veris × vulgaris
Primulaceae Height to 20cm

Natural hybrid between Primrose and Cowslip that superficially resembles Oxlip. **Flowers** 15–20mm across, yellow and 5-lobed; in many-sided umbels (Mar–May). **Fruits** Capsules. **Leaves** Oval, crinkly and tapering; they form a basal rosette. **Status** Local and scattered; absent from far N. Found in hedgerows and woodlands where both parents occur; always less common than the parent species.

Yellow Loosestrife

Lysimachia vulgaris Primulaceae
Height to 1m

Softly hairy perennial. **Flowers** 15–20mm across and yellow, with 5 pointed lobes; borne in terminal heads (Jun–Aug). **Fruits** Capsules. **Leaves** Narrow, ovate and borne in whorls of 3 or 4; often adorned with black dots. **Status** Locally common in damp grassland.

Creeping-Jenny

Lysimachia nummularia
Primulaceae Creeping

Low-growing, hairless perennial. **Flowers** 15–25mm across, yellow and bell-shaped, with 5 pointed lobes; borne on stalks arising from leaf axils (Jun–Aug). **Fruits** Capsules. **Leaves** Rounded or heart-shaped; borne in opposite pairs. **Status** Locally common in England; scarce or absent elsewhere. Grows in damp, grassy ground.

Yellow Pimpernel

Lysimachia nemorum Primulaceae
Creeping

 Hairless evergreen perennial. Similar to Creeping-Jenny but more delicate. **Flowers** 10–15mm across, yellow and star-shaped, with 5 lobes; borne on slender stalks arising from leaf axils (May–Aug). **Fruits** Capsules. **Leaves** Oval or heart-shaped; in opposite pairs. **Status** Common in damp, shady places.

Scarlet Pimpernel

Anagallis arvensis Primulaceae
Creeping

 Low-growing, hairless annual. **Flowers** 10–15mm across with 5 scarlet or pinkish-orange petals fringed with hairs; ssp. *foemina* has blue flowers fringed with hairs. The flowers open wide only in sunshine (Jun–Aug). **Fruits** Capsules. **Leaves** Oval and usually in pairs. **Status** Widespread and common on cultivated and disturbed ground.

Water-violet

Hottonia palustris Primulaceae
Aquatic

 Attractive and delicate plant. **Flowers** 20–25mm across, pale lilac with a yellow central 'eye'; borne in spikes on tall, hairless stems that rise clear of the water (May–Jun). **Fruits** Capsules. **Leaves** Feathery and divided into narrow lobes; both floating and submerged. **Status** Very locally common only in S and E England, in still or slow-flowing waters.

Chickweed Wintergreen

Trientalis europaea Primulaceae
Height to 20cm

 Delicate, spreading perennial. **Flowers** 12–18mm across and star-shaped, with 7 white petals; 1 or 2 are borne on long, slender stalks (Jun–Jul). **Fruits** Capsules. **Leaves** Oval; borne mainly in a whorl near the top of the stem. **Status** Very locally common in Scotland and scarce in N England. Grows in mature conifer forests.

Sea-milkwort

Glaux maritima Primulaceae
Height to 10cm

 Creeping, hairless plant. **Flowers** 5–6mm with 5 pink, petal-like sepals (May–Sep). **Fruits** Dark brown capsules. **Leaves** Oval and succulent in opposite pairs. **Status** Locally common in saltmarshes.

Bog Pimpernel

Anagallis tenella
Primulaceae Creeping

 Delicate, hairless perennial with trailing stems. **Flowers** Pink and funnel-shaped with 5 lobes, to 1cm long; borne on slender, upright stalks (Jun–Aug). **Fruits** Capsules. **Leaves** Rounded, short-stalked and paired. **Status** Widespread and locally common only in the W. Grows on damp ground, such as bogs and dune slacks, mainly on acid soils.

Brookweed

Samolus valerandi Primulaceae
Height to 12cm

 Hairless, pale green perennial. **Flowers** 2–3mm across with 5 white petals that are joined to the halfway point; borne in terminal clusters (Jun–Aug). **Fruits** Spherical capsules. **Leaves** Spoon-shaped; appearing mainly as a basal rosette. **Status** Widespread but local and mainly coastal in damp ground.

Jewels of the north

The Primrose *Primula vulgaris* is a common and widespread plant throughout much of Britain and Ireland. However, in a few favoured locations in N Britain it is joined by two tiny but beautiful relatives. **Bird's-eye Primrose** *Primula farinosa* (**1**) is seldom more than 12cm tall and is associated with damp limestone grassland in N England. Its flowers are 8–10mm across and pink, with a yellow central 'eye' (Jun–Jul), and its leaves are spoon-shaped and mealy white below. The tiny **Scottish Primrose** *Primula scotica* (**2**) is often less than 5cm tall and is very locally common in short coastal turf on W Mainland Orkney and the NE Scottish mainland; it grows nowhere else in the world. The leaves form a basal rosette from which arises a stalk bearing a terminal head of purplish, flowers with yellow centres (May–Jun and Aug).

 **1**

 **2**

Heathland

The heathlands of southern England are a naturalist's paradise, home to a fabulous array of specialist plants and animals. The habitat's name is clearly derived from the presence, and often dominance, of members of the heath family (*see* pp.104–5), all of which flourish on acid soils. For the ultimate visual display, visit an area of heathland in July, August or September, when the plants are in full bloom.

An interesting selection of specialist birds is found on good-quality heathland. Species such as the **Dartford Warbler** (*see also* p.300), Stonechat and Woodlark remain in our region throughout the year, while Nightjar and Hobby are migrants, present only during the breeding season.

Heathlands are excellent habitats for reptiles, and in particularly good locations all six of our native species can be found together. This includes not only relatively widespread species such as the Adder and Common Lizard, but also the rare and local **Sand Lizard** (*see also* p.328) and Smooth Snake.

All plants need water to survive but most cannot tolerate having their roots permanently waterlogged. Bog specialists such as sundews, Bog Pimpernel and cottongrasses are the exception to this rule, and **Sphagnum mosses** (*see also* p.180) will even carpet the margins of open water.

Heathland owes its existence to man and came about following forest clearance on acid sandy soils. Regimes of grazing, cutting and periodic burning in the past helped maintain the habitat, and today continued management is needed to ensure an appropriate balance between scrub encroachment and the continuation of open habitats. Ironically, man is also the biggest threat to heathland today: uncontrolled burns cause damage that takes decades for nature to repair, while the destruction of heathland for housing and road developments means the loss of this unique habitat for good.

Free-draining acid soils dominate most heathlands, and as a consequence heathlands support plants and animals that are adapted to seasonally arid conditions. However, if the landscape is undulating and the soil conditions are suitable, valley bottoms will collect water with the result that bogs develop. These wetland habitats are home to another unique set of species, able to tolerate saturated ground and extremely acid waters.

Heathlands are essentially restricted to southern England, with the majority of sites concentrated in Surrey, Hampshire and Dorset. However, further isolated examples of heathland can be found in south Devon and Suffolk, and in coastal districts of Cornwall and Pembrokeshire. This fragmented distribution adds to the problems that beset the habitat: 'island' populations of plants and animals have little chance of receiving genetic input from other sites, and they are prone to local extinctions following fires and other catastrophes, natural or otherwise.

Several species of gorse are well represented on heathlands, and shrubby forms such as **Common Gorse** (*see also* p.76) often dominate the landscape if left unchecked, creating a flowering spectacle in spring. All gorse species have yellow flowers and many are extremely spiny.

Walk across a southern heath in July and you will notice good numbers of **Silver-studded Blue butterflies** (*see also* p.349) on the wing and resting on sprays of heather. Other heathland butterflies include the Grayling and Small Copper. At dusk and after dark a good variety of moths can be seen.

Heathland hotspots
1. **New Forest National Park**, Hampshire; 2. **Kynance Cove**, Lizard Peninsula, Cornwall; 3. **Roydon Common National Nature Reserve (NNR)**, Norfolk; 4. **Arne Heath RSPB reserve**, Dorset; 5. **Thursley Common NNR**, Surrey.

Heathers

Heathers, heaths and their relatives belong to the family Ericaceae and are low-growing, shrubby plants associated with acid soils. Some have simple, almost needle-like leaves and typically their petals are fused, creating bell-shaped or tubular flowers. Although Crowberry is superficially similar to some members of the heather family, notably bilberries, botanists place it in a separate family, Empetraceae; because of similarities in its appearance and habitat preferences, it is included here.

The common four

Heathlands and heather moors are typically dominated by just three members of the heather family, each with a different tolerance to waterlogging: Bell Heather prefers the driest conditions, Cross-leaved Heath is found in the wettest, and Heather comes somewhere in between the two. Bilberry is also common and widespread, often occurring as an undershrub on heaths and in open woodland. Its absence from many seemingly suitable areas is hard to explain, although it does not tolerate grazing or burning to the extent of some of its relatives.

Bog specialists

Although heathers and their relatives are thoroughly adapted to the acid soils of heaths and moors, most find the waterlogged conditions of associated bogs too challenging for growth. However, two specialised members of the Ericaceae can tolerate permanently inundated peat bogs and thrive where conditions suit them. **Cranberry** *Vaccinium oxycoccos* (**1**) is a creeping evergreen undershrub with slender, trailing, wiry stems and dark green, narrow leaves that have inrolled margins. Its flowers are 8-10mm across and recall miniature Fuchsia flowers, being pink with reflexed petals (bent back beyond 90 degrees) and protruding stamens (May-Jul). The fruits are bright red berries. Cranberry is locally common only in N Wales, N England and E Ireland. **Bog-rosemary** *Andromeda polifolia* (**2**) is a hairless evergreen undershrub with narrow leaves that are bluish green above and whitish below. Its flowers are 8-10mm long, pink and urn-shaped; they are pendent on short stalks and are borne in small terminal clusters (May-Sep). It is local and declining, and is found from Wales and central England to S Scotland, as well as in Ireland.

ABOVE: fruit

Heather

Calluna vulgaris Ericaceae
Height to 50cm

Evergreen under-shrub. **Flowers** 4-5mm, bell-shaped, usually pink but sometimes white (Aug-Sep). **Fruits** Capsules. **Leaves** Short, narrow and borne in 4 rows along the stem. **Status** Locally abundant on most heaths and moors.

Bell Heather

Erica cinerea Ericaceae
Height to 50cm

Hairless evergreen. **Flowers** 5-6mm long, bell-shaped, purplish red, in spike-like groups (Jun-Sep). **Fruits** Capsules. **Leaves** Narrow, dark green, in whorls of 3 up stems. **Status** Locally common, some-times dominant on dry heaths.

Cross-leaved Heath

Erica tetralix Ericaceae
Height to 30cm

Downy undershrub. **Flowers** 6-7mm long, globular and pink; in terminal, 1-sided clusters (Jun-Oct). **Fruits** Downy capsules. **Leaves** Narrow, hair-fringed, in whorls of 4. **Status** Locally common in bogs.

Bilberry

Vaccinium myrtillus Ericaceae
Height to 75cm

Hairless deciduous undershrub with 3-angled green twigs. **Flowers** 5-6mm long, greenish pink and globular urn-shaped; pendent and borne on short stalks (Apr-Jun). **Fruits** Familiar and delicious black berries. **Leaves** Bright green, oval and finely toothed. **Status** Widespread and locally common on heathlands across much of the region; least numerous in the E.

fruit

Northern specialities

Moors and mountains in the north of Britain are home to a range of hardy specialist plants that are capable of withstanding harsh winter weather and the occasional cold snap in summer. Some of these plants have attractive flowers and most produce succulent berries that are feasted on by migrant birds in autumn.

Trailing Azalea

Loiseleuria procumbens Ericaceae
Creeping

Attractive low-growing perennial undershrub that forms prostrate mats. **Flowers** 5mm across, bell-shaped, deeply lobed and pink; either solitary or borne in clusters (May–Jun). **Fruits** Capsules. **Leaves** Thick, opposite and oblong, with downrolled margins. **Status** Locally common only in the Cairngorms on acid soils and stony ground on mountain plateaux.

Bog Bilberry

Vaccinium uliginosum Ericaceae
Height to 70cm

Rather straggly deciduous undershrub with round brown twigs. **Flowers** 6mm long, globular urn-shaped and pale pink; borne in clusters (May–Jun). **Fruits** Globular bluish-black berries. **Leaves** Ovate, bluish green and untoothed. **Status** Local, in N England and Scotland. Found on damp moorland and mountain ledges.

Bog Bilberry flowers and fruits

Bearberry

Arctostaphylos uva-ursi Ericaceae
Prostrate

Low-growing, mat-forming evergreen undershrub. **Flowers** 5–6mm long, urn-shaped and pink; borne on short stalks and in clusters (May–Aug). **Fruits** Shiny, bright red berries, 7–9mm across. **Leaves** Oval, untoothed and leathery; dark green and shiny above but paler below. **Status** Locally common only in Scotland. Grows on dry moorland and mountain slopes.

fruits

Arctic Bearberry

Arctostaphylos alpinus Ericaceae
Prostrate

Mat-forming deciduous undershrub, the stems often bearing the withered remains of the previous year's leaves. **Flowers** 4–5mm long, urn-shaped and white; borne in small clusters (May–Aug). **Fruits** Black berries up to 1cm across. **Leaves** Wrinkled and toothed, turning red in autumn. **Status** N Scotland only. Found on acid moorland.

Cowberry

Vaccinium vitis-idaea Ericaceae
Height to 20cm

Straggly evergreen undershrub. The twigs are round and downy when young. **Flowers** 5–6mm long, bell-shaped and pink; borne in drooping terminal clusters (May–Jun). **Fruits** Shiny, bright red berries. **Leaves** Leathery, oval and untoothed; dark green above but paler below. **Status** Locally common from N Wales northwards. Found on moors and in woodlands.

fruits

Crowberry

Empetrum nigrum Empetraceae
Height to 10cm

Mat-forming, Heather-like evergreen undershrub, whose young stems are reddish. **Flowers** Tiny and pinkish, with 6 petals; they arise at the base of the leaves (May–Jun). **Fruits** Shiny berries, 5–7mm across; green, ripening black. **Leaves** Narrow, shiny and dark green, with inrolled margins. **Status** Locally common only in N Britain on acid upland moors.

Regional specialities

Heather, Bell Heather and Cross-leaved Heath are common species that are found throughout the region where soil conditions suit them. However, three other members of the Ericaceae family are extremely local, two of them being named after the regions in which they are most common. **Cornish Heath** *Erica vagans* (**1**) forms low shrubs and, at suitable locations on the Lizard peninsula, it covers extensive areas; the display around Kynance Cove is particularly spectacular. The flowers are pale pink and small, but are borne in tall, dense spikes (Aug–Sep); the leaves are in fours or fives. Cornish Heath also occurs, very locally, in Ireland. **Dorset Heath** *E. ciliaris* (**2**) is a clump-forming evergreen undershrub whose flowers are 8–10cm long, elongated, egg-shaped and pinkish purple with projecting styles; they open in succession from the bottom and so the flower spikes taper towards the top (Jul–Sep). The leaves are narrow with bristly margins and are borne in whorls of three. It is locally common on Dorset heaths and also occurs in SW England and W Ireland. **St Dabeoc's Heath** *Daboecia cantabrica* (**3**) is an Irish speciality that is confined to heaths in the W. It is a hairy, straggly evergreen undershrub whose flowers are 10–14mm long, elongated, egg-shaped, pinkish purple and nodding, and are borne in open terminal spikes (Jun–Oct). The leaves are narrow with downrolled margins, and are dark green and hairy above but white below.

1

2

3

Wintergreens & gentians

Although many British wildflowers are widespread, those that are illustrated here have rather restricted distributions and well-defined habitat requirements. Some are found only within sight of the sea while others are confined to marshy ground or calcareous grassland, depending on the species concerned.

Wintergreens

Wintergreens (family Pyrolaceae) typically have five-petalled flowers and a rather waxy appearance. Despite its English name, **Common Wintergreen** *Pyrola minor* (**1**) is scarce, growing in open upland woodlands and moors. Its flowers are 5-6mm across, pinkish white and rounded, with a straight style that does not protrude beyond the petals (Jun–Aug). The oval, toothed leaves form a basal rosette.
Intermediate Wintergreen *P. media* is similar but has a protruding style; it is scarce, occurring mainly in E Scotland. Also local, **Round-leaved Wintergreen** *P. rotundifolia* (**2**) is regularly encountered growing in fens and coastal dune slacks. Its flowers are 8-12mm across, white and bell-shaped, and the style is S-shaped and protrudes beyond the petals (May–Aug). The leaves are rounded and form a basal rosette.
Serrated Wintergreen *Orthilia secunda* and **One-flowered Wintergreen** *Moneses uniflora* are rare plants of undisturbed Caledonian pine forests in the Scottish Highlands.

A strange lifestyle

The presence of chlorophyll and an ability to manufacture food by photosynthesis are defining features of almost all flowering plants. However, **Yellow Bird's-nest** *Monotropa hypopitys* breaks all the rules: it cannot photosynthesise and obtains its food from soil leaf mould. It is a member of the wintergreen family, has a waxy appearance and its flowers are 10-15mm long and bell-shaped (Jun–Sep). It is local in shady woodland, often under Beech.

Colourful coastal flowers

Members of the sea-lavender family (Plumbaginaceae) include some of our most colourful coastal species, and where conditions suit them they sometimes flower in great profusion, putting on spectacular displays in spring and summer. Several sea-lavenders occur in the region, and Thrift is also a member of this attractive group of plants.

Common Sea-lavender

Limonium vulgare
Plumbaginaceae Height to 30cm

Hairless, woody-based perennial. **Flowers** 6-7mm long and pinkish lilac; borne in branched, flat-topped heads on arching sprays (Jul–Sep). **Fruits** Capsules. **Leaves** Spoon-shaped with long stalks. **Status** Locally common only in S and SE England in saltmarshes.

Lax-flowered Sea-lavender

Limonium humile
Plumbaginaceae Height to 25cm

Similar to Common Sea-lavender but with subtle differences in the flower heads and leaves. **Flowers** 6-7mm long and pinkish lilac; borne in open, lax clusters with well-spaced flowers (Jul–Sep). **Fruits** Capsules. **Leaves** Narrow and long-stalked. **Status** Local in saltmarshes in England, Wales and S Scotland; also widespread and fairly common on Irish coasts.

Rock Sea-lavender

Limonium binervosum
Plumbaginaceae Height to 30cm

Hairless perennial. **Flowers** 6-7mm long and pinkish lilac; borne in small, well-spaced clusters on sprays that branch from below the middle (Jul–Sep). **Fruits** Capsules. **Leaves** Narrow, spoon-shaped and with winged stalks. **Status** Locally common on coastal cliffs and rocks, mainly in the W; also occasionally found on stabilised shingle beaches.

Thrift

Armeria maritima
Plumbaginaceae Height to 20cm

 Attractive cushion-forming perennial. **Flowers** Pink and borne in dense globular heads, 15–25mm across, that are carried on slender stalks (Apr–Jul). **Fruits** Capsules. **Leaves** Dark green, long and narrow. **Status** Widespread and locally abundant, often carpeting suitable coastal cliffs. Sometimes also found in saltmarshes and, rarely, on a few mountain tops.

Centauries & gentians

Members of the gentian family (Gentianaceae) have opposite, hairless leaves and colourful flowers that are trumpet-shaped, fused at the base and have four or five spreading lobes; in some species the flowers open fully only in bright sunshine.

Common Centaury

Centaurium erythraea
Gentianaceae Height to 25cm

 Variable hairless annual. **Flowers** 10–15mm across, unstalked and pink, with 5 petal-like lobes; borne in terminal clusters and on side shoots (Jun–Sep). **Fruits** Capsules. **Leaves** Grey-green and oval with 3–7 veins; the basal leaves form a rosette. **Status** Widespread and common, except in Scotland. Found in dry grassy places, including verges and sand dunes.

Lesser Centaury

Centaurium pulchellum
Gentianaceae Height to 15cm

 Slender annual; usually branches from near the base and lacks a basal rosette of leaves. **Flowers** 5–8mm across, short-stalked and dark pink; borne in open clusters (Jun–Sep). **Fruits** Capsules. **Leaves** Narrowly ovate with 3–7 veins and borne only on the stems. **Status** Widespread but local in England and Wales only; mainly coastal and most frequent in S.

Yellow-wort

Blackstonia perfoliata
Gentianaceae Height to 30cm

 Upright, grey-green annual. **Flowers** 10–15mm across with 6–8 bright yellow petal lobes, opening only in bright sunshine (Jun–Oct). **Fruits** Capsules. **Leaves** Ovate and waxy; the stem leaves are paired and fused at the base around the stem, while the basal leaves form a rosette. **Status** Locally common in S England and S Wales, mainly in calcareous grassland.

Yellow-wort flower

Common Centaury flower

Marsh Gentian

Gentiana pneumonanthe
Gentianaceae Height to 30cm

 Attractive hairless perennial. **Flowers** 25–45mm long, trumpet-shaped and bright blue, the outside marked with 5 green stripes; borne in terminal clusters (Jul–Oct). **Fruits** Capsules. **Leaves** Narrow, blunt, 1-veined and borne up the stem in opposite pairs. **Status** Local in England and Wales. Grows in bogs and damp grassy heaths on acid soils.

Spring Gentian

Gentiana verna Gentianaceae Height to 7cm

 Stunning perennial. **Flowers** 20–25mm across, 5-lobed and bright blue; borne on upright stems (May–Jun). **Fruits** Capsules. **Leaves** Oval and bright green; they form a basal rosette and are arranged in opposite pairs on the stem. **Status** Restricted to a few limestone grassland sites in Upper Teesdale (N England) and the Burren (W Ireland), but sometimes locally common there.

Field Gentian

Gentianella campestris
Gentianaceae Height to 10cm

 Biennial that is similar to Autumn Gentian, but note the differences in the flowers. **Flowers** 10–12mm across and bluish purple, sometimes creamy white, with 4 corolla lobes and extremely unequal calyx lobes (Jul–Oct). **Fruits** Capsules. **Leaves** Narrow and ovate. **Status** Locally common only in N England and Scotland. Found in neutral or acid grassland.

Autumn Gentian

Gentianella amarella
Gentianaceae Height to 25cm

 Variable, often purple-tinged biennial. **Flowers** 1cm across and purple, with 4 or 5 corolla lobes and equal calyx lobes; borne in upright spikes (Jul–Oct). **Fruits** Capsules. **Leaves** Form a basal rosette in the first year but wither before the flower stem appears in the second year. **Status** Locally common throughout in calcareous grassland and sand dunes.

Bedstraws

Members of the bedstraw family (Rubiaceae) are typically rather straggly plants that climb through wayside and grassland vegetation, assisted by their tough, often bristly stems and leaves. The flowers are four-lobed and generally small, and in some species they are produced in great profusion.

Field Madder
Sherardia arvensis Rubiaceae
Creeping

 Low-growing, hairy annual with square stems. **Flowers** 3–5mm across and pinkish with 4 corolla lobes; borne in small heads (May–Sep). **Fruits** Nutlets. **Leaves** Narrow, oval and arranged in whorls of 4–6 along the stems. **Status** Widespread; rather common in the S but becoming scarce further N. Found on arable and disturbed land.

Woodruff
Galium odoratum Rubiaceae
Height to 25cm

 Upright, hairless, square-stemmed perennial that smells of hay. **Flowers** 3–4mm across, white, 4-petalled and star-shaped; borne in clusters (May–Jun). **Fruits** Nutlets with hooked bristles. **Leaves** Lanceolate with bristly margins; borne in whorls of 6–8. **Status** Locally common except in the N. Found in shady woodlands, mostly on calcareous soils.

Lady's Bedstraw
Galium verum Rubiaceae
Height to 30cm

 Attractive branched perennial with square stems. The whole plant smells of hay. **Flowers** 2–3mm across, yellow and 4-petalled; borne in dense clusters (Jun–Sep). **Fruits** Smooth nutlets that ripen black. **Leaves** Narrow with down-rolled margins; borne in whorls of 8–12 and blackening when dry. **Status** Widespread and common throughout in dry grassland.

Hedge Bedstraw
Galium mollugo Rubiaceae
Height to 1.5m

 Scrambling perennial with smooth, square stems. **Flowers** 3mm across, white and 4-petalled; borne in large, frothy clusters (Jun–Sep). **Fruits** Wrinkled, hairless nutlets. **Leaves** Oval, 1-veined and bristle-tipped with forward-pointing marginal bristles. **Status** Widespread and fairly common, except in the N. Found in dry grassy places, mostly on base-rich soils.

Heath Bedstraw
Galium saxatile Rubiaceae
Height to 50cm

 Spreading, weak-stemmed perennial that blackens when dry. **Flowers** 3mm across, white and 4-petalled, with a sickly smell; borne in clusters (Jun–Aug). **Fruits** Hairless, warty nutlets. **Leaves** Narrow, ovate and bristle-tipped, with forward-pointing marginal bristles. **Status** Locally common throughout on heaths and grassland on acid soils.

BELOW: Heath Bedstraw

Squinancywort
Asperula cynanchica Rubiaceae
Height to 15cm

 Hairless, often prostrate perennial with 4-angled stems. **Flowers** 3–4mm across, pink, 4-petalled and star-shaped; borne in dense clusters (Jun–Sep). **Fruits** Warty nutlets. **Leaves** Narrow, variable in length and borne in whorls of 4. **Status** Locally common only in S England, in dry calcareous grassland; scarce or absent elsewhere.

Common Marsh-bedstraw

Galium palustre Rubiaceae
Height to 70cm

Delicate straggling perennial with rather rough stems. **Flowers** 3-4mm across, white and 4-petalled; borne in open clusters (Jun-Aug). **Fruits** Wrinkled nutlets. **Leaves** Narrow, widest towards the tip and not bristle-tipped; borne in whorls of 4-6. **Status** Widespread and common throughout the region. Grows in damp grassy places.

Fen Bedstraw

Galium uliginosum Rubiaceae
Height to 70cm

Straggly perennial. Its square stems have backward-pointing bristles along their edges. **Flowers** 2-3mm across, white and 4-petalled; borne in open, few-flowered clusters (Jun-Aug). **Fruits** Wrinkled brown nutlets. **Leaves** Narrow and spine-tipped, with backward-pointing marginal bristles; borne in whorls of 6-8. **Status** Local in damp grassy places on calcareous soils.

Northern Bedstraw

Galium boreale Rubiaceae
Height to 60cm

Rather robust form of bedstraw, and our only white-flowered bedstraw with 3-veined leaves. **Flowers** 4mm across, white and 4-petalled; borne in branched terminal clusters (Jun-Aug). **Fruits** Brown nutlets with hooked bristles. **Leaves** Blunt, leathery and dark green. **Status** Widespread only in N Britain and N Ireland in grassy uplands.

Wall Bedstraw

Galium parisiense Rubiaceae
Height to 40cm

Delicate annual. Its square stems have backward-pointing bristles along their edges. **Flowers** 2mm across, greenish white and 4-petalled; borne in small, open clusters (Jun-Jul). **Fruits** Warty, hairless nutlets. **Leaves** Narrow with forward-pointing marginal bristles. **Status** Local and rare although easily overlooked. Grows on old walls and on dry, sandy ground.

Cleavers

Galium aparine Rubiaceae
Height to 1.5m

Sprawling wayside annual. Its square stems have backward-pointing bristles along their edges that aid the plant's scrambling progress through vegetation. **Flowers** 2mm across and greenish white, with 4 petals; borne in clusters arising from leaf axils (May-Sep). **Fruits** Nutlets with hooked bristles. **Leaves** Have backward-pointing marginal bristles. **Status** Common throughout.

fruits and flower

Wild Madder

Rubia peregrina Rubiaceae
Height to 1m

Straggly, bedstraw-like perennial. Its 4-angled stems have backward-pointing bristles along their edges. **Flowers** 4-6mm across, yellowish green and 5-lobed; borne in clusters in leaf axils (Jun-Aug). **Fruits** Spherical black berries. **Leaves** Dark green, shiny and leathery, with marginal prickles. **Status** Local in the S, and its distribution is mainly coastal. Found in hedgerows and scrub.

fruits

Crosswort

Cruciata laevipes Rubiaceae
Height to 50cm

Attractive perennial with hairy, square stems. **Flowers** 2-3mm across, yellow and 4-petalled; borne in dense clusters arising from leaf axils (Apr-Jun). **Fruits** Smooth black nutlets. **Leaves** Oval, 3-veined and hairy; borne in whorls of 4. **Status** Widespread in England, Wales and S Scotland, though commonest in N and E England. Grows in grassy places, mainly on chalk or limestone.

Bindweeds & comfreys

Many of our native plants, such as bindweeds and comfreys, have extremely attractive flowers and over the centuries, gardeners have experimented with them in herbaceous borders. A few have proved to be so successful that they are probably now more familiar in cultivation than in the wild. Some, though attractive to look at, are invasive, and a small number of vigorous alien species have spread from the garden to the wild, damaging our native flora.

Parasitic plants

Common Dodder *Cuscuta epithymum* (family Cuscutaceae) is an intriguing plant: it is leafless and lacks chlorophyll, gaining its nutrition by parasitising host plants such as Heather and clovers. Consequently, it is found in grassy places and on heaths, where its slender, twining red stems form a seething mass among the vegetation. The flowers are 3–4mm across, pink and borne in dense clusters (Jul-Sep), and the plant is locally common only in the S.

Bogbean

Menyanthes trifoliata
Menyanthaceae Height to 15cm

Distinctive creeping aquatic perennial. **Flowers** 15mm across, star-shaped and pinkish white, with 5 fringed petal lobes; borne in spikes up to 25cm long (Mar-Jun). **Fruits** Capsules. **Leaves** Trifoliate; emergent ones have the texture and appearance of Broad Bean leaves. **Status** Locally common throughout in shallow waters of marshes, fens and bogs.

Sea Bindweed

Calystegia soldanella
Convolvulaceae Creeping

Prostrate coastal perennial. **Flowers** 3–5cm across, funnel-shaped and pink with 5 white stripes; borne on slender stalks (Jun-Aug). **Fruits** Capsules. **Leaves** Kidney-shaped, fleshy, up to 4cm long and long-stalked. **Status** Widespread on coasts but locally common only in the S. Grows on sand dunes, and occasionally on stabilised shingle.

Sea Bindweed

Field Bindweed

Convolvulus arvensis
Convolvulaceae
Creeping, or climbing to 3m

Familiar perennial that twines around other plants to assist its progress. **Flowers** 15–30mm across, funnel-shaped and either white or pink with broad white stripes (Jun-Sep). **Fruits** Capsules. **Leaves** Arrow-shaped, 2–5cm long and long-stalked. **Status** Common, except in N Scotland. Found in disturbed ground and arable land; a persistent garden weed.

Garden favourites

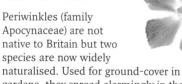  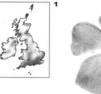

Periwinkles (family Apocynaceae) are not native to Britain but two species are now widely naturalised. Used for ground-cover in gardens, they spread alarmingly in the wild and smother and exclude less vigorous native species. **Lesser Periwinkle** *Vinca minor* (**1**) has five-lobed bluish violet flowers, 25–30mm across, the petals obliquely truncated on the outer margin (Feb-May); its leaves are shiny, dark green and arranged in opposite pairs. **Greater Periwinkle** *V. major* (**2**) is similar, but its flowers are 4–5cm across, with the petals acutely truncated on the outer margin (Mar-May). **Jacob's-ladder** *Polemonium caeruleum* (**3**) (family Polemoniaceae), another garden favourite, is native to Britain but is extremely local in the wild, confined to limestone grassland in N England; its flowers are 2–3cm across, are bright blue with five petal-like corolla lobes, and are borne in spikes (Jun-Jul).

Comfreys

Comfreys (family Boraginaceae) are hairy wayside plants that often grow beside rivers and ditches. Their flowers are tubular to bell-shaped. **Common Comfrey** *Symphytum officinale* (**1**) is a common native with strikingly winged stems, and its flowers, borne in curved clusters, are 12–18mm long and white, pink or purple (May–Jun); the hairy oval leaves have a stalk that runs down the main stem. **Russian Comfrey** *S.* × *uplandicum* (**2**) is a cultivated hybrid that is naturalised. It has only slightly winged stems and its flowers are usually bluish purple (May–Aug). The stalks of the upper leaves run only a short distance down the stem. **Tuberous Comfrey** *S. tuberosum* (**3**) is a scarce native species found mainly in the N, and has creamy-yellow flowers (Jun–Jul).

 1   **2** **3**

Bugloss

Anchusa arvensis Boraginaceae
Height to 50cm

 Roughly hairy annual. **Flowers** 5–6mm across and blue; borne in clusters (May–Sep). **Fruits** Egg-shaped nutlets. **Leaves** Narrow with wavy margins; the lower leaves are stalked, while the upper ones clasp the stem. **Status** Widespread and locally common in E England; becoming scarcer elsewhere. Grows on disturbed and often sandy soils.

Hound's-tongue

Cynoglossum officinale
Boraginaceae Height to 75cm

 Downy plant that smells strongly of mice. **Flowers** 5–7mm across, maroon and 5-lobed; borne in branched clusters (Jun–Aug). **Fruits** Comprise groups of 4 flattened oval nutlets with hooked bristles. **Leaves** Narrow and hairy; the lower ones are stalked. **Status** Commonest in S and E England. Grows in dry soil, often on chalk or near the coast.

fruits

Hedge Bindweed

Calystegia sepium
Convolvulaceae Climbing

Impressive, twining climber that smothers hedgerows by end of summer. **Flowers** 3–4cm across, white and funnel-shaped (Jun–Sep); 2 green bracts at base do not overlap one another. **Fruits** Capsules. **Leaves** Arrow-shaped, to 12cm long. **Status** Common wayside native.

Large Bindweed

Calystegia silvatica
Convolvulaceae Climbing

Introduced, twining perennial, similar to Hedge Bindweed. **Flowers** 6–7cm across, white and funnel-shaped (Jun–Sep); 2 basal green bracts overlap one another and conceal sepals. **Fruits** Capsules. **Leaves** Arrow-shaped, to 12cm long. **Status** Naturalised and now common in hedgerows.

Gromwells

Gromwells, members of the borage family (Boraginaceae), have delicate flowers and hard-cased fruits, known as nutlets. **Common Gromwell** *Lithospermum officinale* (**1**) is widespread but local. It grows in scrub on calcareous soils and has creamy white, 5-lobed flowers, 3–4mm across (Jun–Jul). The shiny white nutlets have a texture resembling glazed china. **Field Gromwell** *L. arvense*, a scarce arable 'weed', is similar and is discussed on p.55. **Purple Gromwell** *L. purpureocaeruleum* (**2**) is a scarce plant of scrub on calcareous soils. The flowers are 12–15mm across and funnel-shaped, and are pink at first but soon turn deep blue (Apr–Jun).

  1

ABOVE: fruits

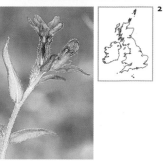 **2**

Lungworts & forget-me-nots

Most members of the borage family (Boraginaceae) have five-petalled blue flowers, including the charming little forget-me-nots. This complex group of similar-looking plants is a challenge to identify; features that are useful in distinguishing between them include flower size, the presence or absence of hairs on the stems, stalks and leaves, and whether these hairs are pressed close to the stem or are raised. Vervain is an unrelated species (family Verbenaceae) but is included here because of its passing resemblance to a tiny forget-me-not.

Lungworts

Lungworts are so called because their pale-spotted leaves bear a fanciful resemblance to lungs, and in the past they were thought to have medicinal properties that remedied complaints of these organs. **Narrow-leaved Lungwort** *Pulmonaria longifolia* (**1**) is a softly hairy native plant of woodland and scrub, usually on heavy soils; it is locally common only in the New Forest, Isle of Wight and E Dorset. Its flowers are 8mm across and bell-shaped, and are pink at first but soon turn blue; they are borne in terminal clusters (Apr–May). The basal leaves are narrowly oval and taper gradually to the base. **Lungwort** *P. officinalis* (**2**) is a closely related garden plant that is naturalised locally on verges and waysides. Its flowers are 1cm across and bell-shaped, and are pink at first but turn blue; they are borne in terminal clusters (Feb–May). The basal leaves taper abruptly to winged stalks.

Green Alkanet

Pentaglottis sempervirens
Boraginaceae Height to 60cm

Upright, bristly perennial. **Flowers** 8–10mm across and blue with a white centre; borne in clusters arising from the upper leaf axils (Apr–Jun). **Fruits** Rough nutlets. **Leaves** Oval, pointed and net-veined; the lower leaves are stalked and the upper ones are unstalked. **Status** Naturalised locally, mainly in England, and found in shady hedgerows and on roadside verges, favouring free-draining soils.

Viper's-bugloss

Echium vulgare Boraginaceae
Height to 80cm

Upright biennial covered in reddish bristles. **Flowers** 15–20mm long, funnel-shaped and bright blue with protruding purplish stamens; borne in tall spikes (May–Sep). **Fruits** Rough nutlets. **Leaves** Narrow and pointed; the basal leaves are stalked. **Status** Common only in England and Wales, in dry grassland. Found mainly on sandy and chalky soils, and often coastal.

Early Forget-me-not

Myosotis ramosissima
Boraginaceae Height to 10cm

Downy annual. **Flowers** 2–3mm across, 5-lobed and sky blue; borne in clusters (Apr–Oct). The corolla tube is shorter than the calyx tube. **Fruits** Nutlets. **Leaves** Oval, the basal ones forming a rosette. **Status** Widespread and common in most parts, except the far N. Grows in arable fields, bare grassy places and open woodland.

Changing Forget-me-not

Myosotis discolor Boraginaceae
Height to 20cm

 Branched, downy annual. **Flowers** 2–3mm across and 5-lobed, yellowish at first but soon changing to blue; borne in clusters (May–Sep). The mature corolla tube is longer than the calyx. **Fruits** Nutlets. The fruit stalks are shorter than the calyx. **Leaves** Oblong. **Status** Locally common in bare, dry and often disturbed ground, especially on sandy soil.

Water Forget-me-not

Myosotis scorpioides Boraginaceae
Height to 12cm

 Creeping perennial with angular stems that are covered in close-pressed hairs. **Flowers** 8–10mm across, 5-lobed and sky blue with a yellow 'eye'; borne in curved clusters (Jun–Sep). **Fruits** Nutlets. The fruit stalks are longer than the calyx. **Leaves** Narrow, oblong and unstalked. **Status** Widespread and locally common in watery habitats on neutral and basic soils.

Tufted Forget-me-not

Myosotis laxa Boraginaceae
Height to 12cm

 Branched perennial that lacks runners. Note the close-pressed hairs on the stems, leaves and calyx. **Flowers** 3–4mm across with rounded blue lobes, the calyx with pointed teeth; borne in clusters (May–Aug). **Fruits** Nutlets. The fruit stalks are 2–3 times the length of the calyx. **Leaves** Oblong. **Status** Common and widespread, growing in damp ground.

Alpine Forget-me-not

Myosotis alpestris Boraginaceae
Height to 25cm

Short, tufted perennial with spreading hairs on the stalks. **Flowers** Deep blue (not pale blue) (Jul–Sep). **Fruits** Black nutlets. The fruit stalks are hairy and the same length as the calyx. **Leaves** Oval and stalked. **Status** Scarce in the Scottish Highlands and rare in Upper Teesdale, growing mainly on calcareous soils.

Vervain

Verbena officinalis Verbenaceae
Height to 70cm

 Upright, hairy perennial with stiff, square stems. **Flowers** 4–5mm across and pinkish lilac with 2 lips; borne on slender spikes (Jun–Sep). **Fruits** Comprise a cluster of nutlets. **Leaves** Narrow and deeply lobed. **Status** Widespread and common in England and Wales; scarce elsewhere. Grows in dry, grassy places, especially on chalk and limestone.

Field Forget-me-not

Myosotis arvensis Boraginaceae
Height to 25cm

 Downy, branching annual. **Flowers** 5mm across, with 5 blue lobes; borne in forked clusters (Apr–Jun). The corolla tube is shorter than the calyx. **Fruits** Nutlets. The fruit stalks are longer than the calyx, which has spreading, hooked hairs. **Leaves** Oblong; basal ones form a rosette. **Status** Common, except in the N. Grows in dry grassland and on arable land.

Creeping Forget-me-not

Myosotis secunda Boraginaceae
Height to 12cm

 Creeping perennial. The stems have close-pressed hairs. **Flowers** 6–8mm across and blue with 5 slightly notched lobes; borne in clusters (Jun–Aug). The calyx is divided more than halfway into teeth. **Fruits** Nutlets. The fruit stalks are much longer than the calyx. **Leaves** Oblong. **Status** Common in W and N Britain and in Ireland. Grows in watery ground on acid soils.

Wood Forget-me-not

Myosotis sylvatica Boraginaceae
Height to 50cm

 Branched, hairy perennial with spreading hairs on the stems and leaves. **Flowers** 6–10mm across, 5-lobed and pale blue, the calyx with hooked hairs; borne in curved clusters (Apr–Jul). **Fruits** Brown nutlets. The fruit stalks are twice the length of the calyx. **Leaves** Oblong. **Status** Locally common in SE and E England only. Grows in damp, shady woodlands.

Mint family

Mints and their relatives (family Lamiaceae) comprise a group of herbaceous plants that have opposite leaves and square stems. Their unusual-looking flowers are, in most species, strikingly two-lipped; typically, they are borne in whorls up the flowering stems. Members of the Lamiaceae include several widespread and common wayside plants.

Bugle

Ajuga reptans Lamiaceae
Height to 20cm

Upright perennial with stems that are hairy on 2 opposite sides. The leafy, creeping runners root at intervals. **Flowers** 15mm long and bluish violet; the lower lip has pale veins (May–Jun). **Fruits** Nutlets. **Leaves** Ovate; the lower leaves are stalked and the upper ones unstalked in opposite pairs. **Status** Commonest in the S. Grows in woods and meadows on damp, heavy soils.

Skullcap

Scutellaria galericulata Lamiaceae
Height to 40cm

Creeping downy or hairless perennial with upright flowering stalks. **Flowers** 10–15mm long and bluish violet; borne in pairs (Jun–Sep). **Fruits** Nutlets. **Leaves** Oval, stalked and toothed. **Status** Locally common throughout, except in Ireland and N Scotland, in damp ground.

Lesser Skullcap

Scutellaria minor Lamiaceae
Height to 15cm

Creeping, hairless perennial with upright flowering stalks. **Flowers** 6–10mm long and pink, the lower lip marked with purplish spots; borne on leafy, upright stems (Jul–Oct). **Fruits** Nutlets. **Leaves** Lanceolate to oval and usually almost untoothed. **Status** Widespread and locally common, mainly in the S. Grows in damp grassy places on acid soils.

Selfheal

Prunella vulgaris Lamiaceae
Height to 20cm

Downy perennial with rooting runners and upright flowering stems. **Flowers** 10–15mm long and bluish violet; in cylindrical heads with purplish bracts and calyx teeth (Apr–Jun). **Fruits** Nutlets. **Leaves** Paired and oval. **Status** Common and widespread, but mostly on calcareous and neutral soils.

Wood Sage

Teucrium scorodonia Lamiaceae
Height to 40cm

Downy perennial. **Flowers** 5–6mm long and yellowish; unlike most other members of the mint family, they lack an upper lip; borne in leafless spikes (Jun–Sep). **Fruits** Nutlets. **Leaves** Oval, heart-shaped at the base, and wrinkled. **Status** Locally common. Grows in woodland rides and on heaths and coastal cliffs, mainly on acid soils.

Ground-ivy

Glechoma hederacea Lamiaceae
Height to 15cm

Softly hairy, aromatic perennial with creeping, rooting runners and upright flowering stems. **Flowers** 15–20mm long and bluish violet; borne in open whorls arising from leaf axils. **Fruits** Nutlets. **Leaves** Kidney-shaped to rounded, toothed and long-stalked. **Status** Widespread and common, except in the far N. Found in woods, hedgerows and meadows.

Bastard Balm

Melittis melissophyllum Lamiaceae
Height to 60cm

Hairy, strong-smelling perennial. **Flowers** 25–40mm long, fragrant, mainly white and variably adorned with pink or purple; borne in whorls (May–Jul). **Fruits** Nutlets. **Leaves** Oval, toothed and stalked. **Status** Local in SW along woodland rides and in shady hedgerows and areas of scrub.

Yellow Archangel

Lamiastrum galeobdolon
Lamiaceae Height to 45cm

Hairy perennial. **Flowers** 17–20mm long, rich yellow with reddish streaks, the lip divided into 3 equal lobes; borne in whorls (Apr–Jun). **Fruits** Nutlets. **Leaves** Oval to triangular and toothed; cultivated variegated forms are sometimes naturalised. **Status** Locally common in England and Wales. Found in woods and hedgerows, mainly on basic soils.

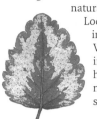

variegated form

Betony

Stachys officinalis Lamiaceae
Height to 50cm

Downy or hairless perennial. **Flowers** 12–18mm long and reddish purple; borne in showy terminal heads that are reminiscent of marsh-orchid spikes (Jun–Sep). **Fruits** Nutlets. **Leaves** Stalked and oblong; the basal ones are typically heart-shaped. **Status** Fairly common in England and Wales. Grows on grassy waysides and in hedgerows, usually on sandy or chalky soils.

White Dead-nettle

Lamium album Lamiaceae
Height to 40cm

Downy, slightly aromatic, patch-forming perennial. **Flowers** 25–30mm long and white, with a hairy upper lip and toothed lower lip; borne in whorls (Mar–Nov). **Fruits** Nutlets. **Leaves** Toothed, ovate to triangular with a heart-shaped base. Similar to those of Common Nettle but lacking stinging hairs. **Status** Common except in N Scotland. Found in hedges and meadows.

Hemp-nettles

Hemp-nettles are downy or hairy annual plants of disturbed ground and cultivated soils, whose flowers are borne in whorls. As its English name suggests, **Common Hemp-nettle** *Galeopsis tetrahit* (**1**) is a widespread and locally common species (height to 50cm), with flowers that are 15–20mm long and pinkish (Jul–Sep); the leaves are ovate, toothed and stalked. By comparison, the showy **Large-flowered Hemp-nettle** *G. speciosa* (**2**) (height to 50cm) is much more local and common only in the N on peaty soil; its flowers are 25–35mm long and yellow with purple on the lower lip (Jul–Sep), and the leaves are ovate, toothed and stalked. **Red Hemp-nettle** *G. angustifolia* (**3**) (height to 30cm) is a scarce plant of arable fields and coastal shingle; its flowers are 15–25mm long and reddish pink, with a hooded upper lip and a two-lobed lower lip (Jul–Sep).

1

2

3

Red Dead-nettle

Lamium purpureum Lamiaceae
Height to 30cm

Downy, pungently aromatic annual that is sometimes tinged purplish. **Flowers** 12–18mm long and purplish pink, with a hooded upper lip and a lower lip that is toothed at the base; borne in whorls on upright stems (Mar–Oct). **Fruits** Nutlets. **Leaves** Heart-shaped to oval, round-toothed and stalked. **Status** Common throughout on disturbed ground and in cultivated soils.

Henbit Dead-nettle

Lamium amplexicaule Lamiaceae
Height to 20cm

Trailing, straggly annual. **Flowers** 15–20mm long and pinkish purple with a hairy lip and long corolla tube; borne in widely spaced whorls (Mar–Nov). Only a few flowers in a given whorl open at any one time. **Fruits** Nutlets. **Leaves** Rounded and blunt-toothed. **Status** Locally common, least so in the N. Found in cultivated soil and on disturbed ground, typically in dry locations.

Aromatic mints

A characteristic of many members of the mint family (Lamiaceae) is that their leaves are aromatic when bruised. In the majority of species the fragrance produced is pleasing, especially in the case of the true mints (*Mentha* species). However, in one or two members of the family the aroma released is extremely disagreeable.

Black Horehound

Ballota nigra Lamiaceae
Height to 50cm

 Straggly, hairy perennial that releases a pungent, unpleasant smell when bruised. **Flowers** 12–18mm long and pinkish purple; borne in whorls with striking calyx teeth (Jun–Sep). **Fruits** Nutlets. **Leaves** Stalked and ovate or heart-shaped at the base. **Status** Locally common in England and Wales but scarce elsewhere. Grows on disturbed ground, beside tracks and on roadside verges, favouring free-draining soils.

Marsh Woundwort

Stachys palustris Lamiaceae
Height to 1m

 Robust, non-smelling perennial with creeping stems and unbranched flowering stalks. **Flowers** 12–15mm long and pinkish purple with white markings; borne in elegant open spikes (Jun–Sep). **Fruits** Nutlets. **Leaves** Toothed, narrow and oblong, often heart-shaped at the base. **Status** Locally common throughout. Found on damp ground in marshes, and beside ditches and rivers.

Field Woundwort

Stachys arvensis Lamiaceae
Height to 25cm

 Creeping or upright, slightly hairy annual. **Flowers** 15–20mm long and pinkish, with purple markings; borne in open leafy spikes (Apr–Oct). **Fruits** Nutlets. **Leaves** Oval, bluntly toothed and short-stalked. **Status** Locally common only in the S and W of the region. Grows in arable fields and on disturbed ground, mainly on sandy, acid soils.

Hedge Woundwort

Stachys sylvatica Lamiaceae
Height to 75cm

 Roughly hairy perennial that releases an unpleasant smell when bruised. **Flowers** 12–18mm long and reddish purple with white markings on the lower lip; borne in open spikes (Jun–Oct). **Fruits** Nutlets. **Leaves** Ovate to heart-shaped, toothed and long-stalked. **Status** Widespread and common in hedgerows, waysides and on verges, often on disturbed ground.

Meadow Clary

Salvia pratensis Lamiaceae
Height to 1m

 Downy and slightly aromatic, upright perennial. **Flowers** 2–3cm long and bluish violet; borne in whorls on upright, showy spikes (Jun–Jul). **Fruits** Nutlets. **Leaves** Narrow, bluntly toothed and wrinkled; heart-shaped at the base. **Status** Rare and restricted to a few sites in S England. Grows in dry grassland on chalk and limestone soils.

Wild Clary

Salvia verbenaca Lamiaceae
Height to 80cm

 Downy perennial. The upper part of the plant is often tinged purple. **Flowers** 8–15mm long and bluish violet, the calyx sticky and coated with long, white hairs; borne in whorls in compact spikes (May–Aug). **Fruits** Nutlets. **Leaves** Oval with jagged teeth, and mainly basal. **Status** Local in S and E England only. Found in dry grassland, often on chalky soils and near the coast.

Common Calamint

Clinopodium ascendens Lamiaceae
Height to 50cm

 Hairy perennial that branches from the base and smells of mint. **Flowers** 12–16mm long and pinkish lilac, with darker spots on the lower lip; borne in heads comprising dense whorls (Jun–Sep). **Leaves** Rounded and long-stalked. **Status** Very locally common only in the S. Grows in dry grassland, hedgerows and on verges, often on chalk or limestone soils.

Cat-mint

Nepeta cataria Lamiaceae
Height to 50cm

 Greyish, downy perennial with a minty smell that cats find alluring. **Flowers** 8–12mm long and white with purple spots; borne in whorls and terminal heads (Jul–Sep). **Fruits** Nutlets. **Leaves** Heart-shaped, toothed, downy below and woolly above. **Status** Local in S England and S Wales in dry, grassy places; naturalised as a garden escape elsewhere.

Water Mint

Mentha aquatica Lamiaceae
Height to 50cm

 Robust, hairy, reddish-stemmed perennial that smells strongly of mint. **Flowers** 3–4mm long and lilac-pink; borne in dense terminal heads up to 2cm long, and in a few sub-terminal whorls Jul–Oct). **Fruits** Nutlets. **Leaves** Oval and toothed. **Status** Widespread and common on damp ground, sometimes growing in shallow standing water.

Corn Mint

Mentha arvensis Lamiaceae
Height to 30cm

 Straggly, hairy perennial with a strong and rather pungent smell of mint. **Flowers** 3–4mm long and lilac; borne in dense whorls at intervals along the stem, not terminally (May–Oct). **Fruits** Nutlets. **Leaves** Toothed, oval and short-stalked. **Status** Widespread and generally common in damp arable land, and on paths and disturbed ground.

Lesser Calamint

Clinopodium calamintha
Lamiaceae Height to 50cm

 Tufted, greyish perennial. Similar to Common Calamint but downier and more branched. **Flowers** 10–15mm long, pale pinkish mauve and almost unspotted; borne in whorls (Jul–Sep). After flowering, the calyx tube has hairs projecting from its mouth. **Fruits** Nutlets. **Leaves** Oval and almost untoothed. **Status** Local, restricted to S and E England.

Cultivated mints

Mint is such a popular culinary herb that it is not surprising to learn that the different forms have been widely cultivated for their subtly distinct aromas. **Round-leaved Mint** *Mentha suaveolens* (**1**) (height to 70cm) is extremely aromatic and smells distinctly of apples. Its lilac flowers are borne in dense spikes (Jul–Sep), and the leaves are oval to rounded, wrinkled and coated in woolly hairs. It is native in the W, but often escapes from cultivation elsewhere, and it grows in damp ground. **Spear Mint** *M. spicata* (**2**) (height to 75cm) is the most popular cultivated culinary mint, with lilac flowers that are borne in whorled terminal spikes (Jul–Oct). Its leaves are narrowly ovate, toothed and almost unstalked, and the plant grows in damp ground locally throughout the region and often escapes from cultivation. **Peppermint** *M. × piperata* (**3**) (height to 1m), a robust hybrid between Spear and Water mints, has a strong peppermint smell and pinkish-lilac flowers that are borne mainly in terminal spikes with a few whorls below (Jul–Sep). Its leaves are narrow, ovate and stalked, and it is naturalised occasionally in damp ground across the region.

  **1**   **2**   **3**

Mints

Basil-thyme

Clinopodium acinos Lamiaceae
Height to 20cm

Downy annual. **Flowers** 7-10mm long and bluish violet with a white patch on the lower lip; borne in few-flowered whorls along much of the length of the stems (May-Aug). **Fruits** Nutlets. **Leaves** Oval, stalked and only slightly toothed. **Status** Locally common only in S and E England; scarce or absent elsewhere. Grows in dry, grassy habitats on calcareous soils.

Wild Basil

Clinopodium vulgare Lamiaceae
Height to 35cm

Hairy, pleasantly aromatic perennial. **Flowers** 15-22mm long and pinkish purple; borne in whorls with bristly purple bracts and arising from the axils of the upper leaves (Jul-Sep). **Fruits** Nutlets. **Leaves** Ovate, toothed and stalked. **Status** Locally common in S and E England; scarce or absent elsewhere. Grows in dry, grassy places, on calcareous soils.

Wild Marjoram

Origanum vulgare Lamiaceae
Height to 50cm

Downy, pleasantly aromatic perennial; a popular culinary herb. **Flowers** Maroon when in bud but pinkish purple and 6-8mm long when open; borne in dense terminal clusters with purplish bracts (Jul-Sep). **Fruits** Nutlets. **Leaves** Oval and pointed; arranged in opposite pairs. **Status** Locally common only in the S. Grows in dry grassland on calcareous soils.

Gipsywort

Lycopus europaeus Lamiaceae
Height to 75cm

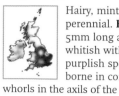

Hairy, mint-like perennial. **Flowers** 5mm long and whitish with small purplish spots; borne in compact whorls in the axils of the upper leaves (Jul-Sep). **Fruits** Nutlets. **Leaves** Yellowish green and deeply cut into lobes. **Status** Common in central and S England but scarce or absent elsewhere. Grows in damp ground beside ditches and ponds.

White Horehound

Marrubium vulgare Lamiaceae
Height to 50cm

Upright, downy and aromatic perennial. **Flowers** 12-15mm long and white; borne in whorls, only a few flowers appearing at any one time (Jun-Oct). **Fruits** Nutlets. **Leaves** Oval, toothed and wrinkled. **Status** Local, mainly near the S coast of England; occasional elsewhere. Grows on dry, often disturbed ground, mainly on chalky soil.

Thyme

Famed for the flavour their small oval leaves impart in cooking, thymes are mat-forming perennial herbs that grow wild in the British countryside. The most widely encountered species is **Wild Thyme** *Thymus polytrichus* (**1**) (height to 5cm), which has slender, woody runners and pinkish-purple flowers that are 3-4mm long, borne in dense terminal heads and have dark purplish calyx tubes (Jun-Sep); the four-angled stems are hairy on two opposite sides. Wild Thyme is widespread and common throughout the region, growing on dry grassland and heaths, as well as coastal cliffs and dunes. **Large Thyme** *T. pulegioides* (**2**) (height to 15cm) is common only in the S and lacks woody runners; its flowers are similar to, but larger than, those of Wild Thyme (Jun-Aug), and the stems have rows of hairs along the four edges and downy hairs on two opposite sides. It grows on chalk downland and dry heaths.

Nightshades

Although a few members of the mint family (Lamiaceae) have an unpleasant taste, none is actually harmful and some are invaluable additions to the culinary palette. However, the same is not true of members of the potato family (Solanaceae): all representatives described here are, to varying degrees, poisonous and some are even deadly.

Thorn-apple

Datura stramonium Solanaceae
Height to 1m

 Unmistakable branched annual. The whole plant is poisonous. **Flowers** 7–10cm long, white and trumpet-shaped with 5 lobes (Jun–Oct). **Fruits** Distinctive green capsules, up to 5cm long and armed with strong spines. **Leaves** Long-stalked, to 20cm long and ovate to triangular with toothed lobes. **Status** Widely naturalised on cultivated ground.

fruit

Deadly Nightshade

Atropa belladonna Solanaceae
Height to 1m

 Robust, downy perennial. **Flowers** 25–30mm long, purplish and bell-shaped; pendent from leaf axils (Jun–Aug). **Fruits** Globular, deadly poisonous black berries, 15–20mm across. **Leaves** Broadly oval, pointed and stalked. **Status** Locally common only in S and E England. Grows in scrub and on disturbed ground, mostly on calcareous soils.

fruit

Henbane

Hyoscyamus niger Solanaceae
Height to 75cm

 Stickily hairy, unpleasant-smelling plant. The whole plant is poisonous. **Flowers** 2–3cm across, funnel-shaped and creamy yellow with a purplish centre and purple veins (Jun–Aug). **Fruits** Capsules. **Leaves** Oval and pointed. **Status** Locally common only in S and E England; scarce elsewhere. Grows on disturbed ground, often on sandy soils.

Duke of Argyll's Teaplant

Lycium barbarum Solanaceae
Height to 1.5m

 Deciduous perennial with spiny, woody stems. **Flowers** 8–10mm long, purplish and 5-lobed, with projecting yellow anthers; borne in groups of 1–3 flowers arising from leaf axils (Jun–Sep). **Fruits** Egg-shaped red berries. **Leaves** Narrow and grey-green. **Status** Introduced from China and naturalised, often near the sea.

Bittersweet

Solanum dulcamara Solanaceae
Height to 1.5m

 Scrambling perennial with a woody base. **Flowers** 10–15mm across with 5 purple, petal-like corolla lobes and projecting yellow anthers; borne in hanging clusters on purple stems (May–Sep). **Fruits** Poisonous, egg-shaped red berries, to 1cm long. **Leaves** Oval and pointed. **Status** Common, except in the N and in Ireland. Found in hedgerows and scrub.

Black Nightshade

Solanum nigrum Solanaceae
Height to 60cm

 Straggly annual whose stems are sometimes blackish. **Flowers** 7–10mm across with white corolla lobes and projecting yellow anthers; borne in pendent clusters of 5–10 flowers (Jul–Sep). **Fruits** Spherical berries, green at first but ripening black. **Leaves** Oval and toothed. **Status** Locally common only in the S. Grows in cultivated and disturbed soils.

fruits

Mulleins & toadflaxes

Members of the figwort family (Scrophulariaceae) are a varied group, with representatives that range from tiny, insignificant plants to those that are tall and impressive. Their flower structure also varies, ranging from simple and basic, as in speedwells and mulleins, to extremely elaborate, as in figworts and toadflaxes.

Fluellens

Fluellens are intriguing creeping plants that are associated with cultivated soils and arable fields, mainly in S England. Two species are found in Britain; while superficially similar, these are distinguishable by studying their flowers and leaves. **Round-leaved Fluellen** *Kickxia spuria* (**1**) has 8–15mm-long flowers that have a curved spur and are mainly yellow but with a purple upper lip (Jul–Oct); the leaves are oval or slightly rounded.

Sharp-leaved Fluellen *K. elatine* (**2**) has similar flowers but a straight spur, and its leaves are triangular to arrow-shaped.

Unusual mulleins

Mulleins are popular garden plants whose wild ancestors grow in the countryside. **Moth Mullein** *Verbascum blattaria* (**1**) (height to 1.3m) has angled stickily hairy stems. Its solitary flowers are 2–3cm across, usually yellow but sometimes pinkish white, with pink-haired stamens (Jun–Sep). It is occasional in S and E England on waste ground. **Hoary Mullein** *V. pulverulentum* (**2**) (height to 2m) is branched with woolly white hairs. Its flowers are 15–35mm across and yellow, with white-haired stamens (Jul–Sep). It is local in E Anglia only, in dry calcareous grassland. **Twiggy Mullein** *V. virgatum* (**3**) (height to 1.5m) has yellow flowers 1–2cm across, and heart-shaped clasping leaves; it is restricted to SW England. **White Mullein** *V. lychnitis* (**4**) (height to 1.5m) has white flowers, 15–20mm across, in branched spikes (Jul–Aug); it is local in S England, mainly in calcareous grassland.

Dark Mullein

Verbascum nigrum
Scrophulariaceae Height to 1m

Upright, unbranched biennial with ridged, purplish stems. **Flowers** 1–2cm across and yellow, the stamens coated in purple hairs; borne in elongated spikes (Jun–Aug). **Fruits** Capsules. **Leaves** Dark green and oval; the lower ones are long-stalked. **Status** Locally common in S and E England only. Found on verges and disturbed ground, on calcareous and sandy soils.

Mudwort

Limosella aquatica
Scrophulariaceae Height to 10cm

Rosette-forming annual. **Flowers** 3–5mm across and bell-shaped, with 5 pinkish-white lobes; borne on slender stalks arising from leaf axils (Jun–Oct). **Fruits** Capsules. **Leaves** Narrow, ovate and long-stalked. **Status** Rare, on the margins of drying ponds.

Great Mullein

Verbascum thapsus
Scrophulariaceae Height to 2m

Upright biennial with coating of woolly white hairs. **Flowers** 15–35mm across, 5-lobed and yellow; borne in tall, dense, usually unbranched spikes (Jun–Aug). **Fruits** Egg-shaped capsules. **Leaves** Ovate and woolly; they form a basal rosette in the first year, from which tall, leafy stalks arise in the second. **Status** Widespread and locally common in dry, grassy places. Commonest on free-draining soils.

Common Toadflax

Linaria vulgaris Scrophulariaceae Height to 75cm

Greyish-green, hairless, branched perennial. **Flowers** 15–25mm long and yellow, with orange centres and long spurs; borne in tall cylindrical spikes (Jun–Oct). **Fruits** Capsules. **Leaves** Narrow and linear. **Status** Locally common, although scarce in Ireland. Grows in dry grassland, on roadside verges; sometimes found on the margins of arable fields.

Pale Toadflax

Linaria repens Scrophulariaceae
Height to 75cm

Greyish-green, creeping perennial with upright leafy stems. **Flowers** 7–14mm long and lilac with dark veins; borne in terminal spikes (Jun–Sep). **Fruits** Capsules. **Leaves** Narrow; borne in whorls on the lower part of the stem. **Status** A garden escape but also possibly native in England and Wales. Grows in dry, grassy places and on disturbed ground.

Small Toadflax

Chaenorhinum minus Scrophulariaceae Height to 25cm

Upright, downy and slightly sticky annual. **Flowers** 6–8mm long and pinkish lilac with a yellow patch and short spur; borne on long stalks that arise from leaf axils (May–Oct). **Fruits** Capsules. **Leaves** Narrow. **Status** Fairly common in arable fields and railway tracks, favouring calcareous soils.

Ivy-leaved Toadflax

Cymbalaria muralis Scrophulariaceae Trailing

Hairless perennial with purplish stems. **Flowers** 10–12mm across and lilac, with yellow and white at the centre, and a curved spur; borne on long stalks (Apr–Nov). **Fruits** Capsules; their stalks become recurved with maturity, forcing the fruit into crevices. **Leaves** Long-stalked and ivy-shaped. **Status** Widely naturalised on rocks and walls.

Cornish Moneywort

Sibthorpia europaea Scrophulariaceae Prostrate

Hairy, mat-forming perennial with rooting, creeping stems. **Flowers** Tiny, with 2 yellow lobes and 3 pink ones; solitary, borne on short, slender stalks (Jul–Oct). **Fruits** Capsules. **Leaves** 2cm across, long-stalked and kidney-shaped, with 5–7 lobes. **Status** Local in SW England, Sussex, S Wales and SW Ireland. Found on damp, shady banks and beside streams.

Figworts

Figworts are robust plants with square stems. Although their flowers are small, they are beautiful when viewed in close-up, being greenish with a maroon upper lip. Their fruits bear a fanciful resemblance to miniature figs, hence their name. Three species are regularly encountered, although one of these is distinctly local. **Common Figwort** *Scrophularia nodosa* (**1**) (height to 70cm) is widespread and common, and has unwinged stems and pointed oval leaves with sharp teeth. **Water Figwort** *S. auriculata* (**2**) (height to 70cm) is common in damp ground, in woodlands and beside fresh water. Its stems have prominent wings, and its leaves are oval but blunt-tipped, with rounded teeth. **Balm-leaved Figwort** *S. scorodonia* (**3**) (height to 70cm) is locally common only in SW England. It is downy and grey, with stems that are square and angled; the leaves are oval, toothed and wrinkled.

1

2
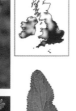

3

Exotics

Introduced from North America, members of the genus *Mimulus* have escaped and thrive in damp ground. **Monkeyflower** *M. guttatus* (height to 50cm) has flowers (*below*), 25–45mm across, that are yellow with small red spots (Jun–Sep). **Blooddrop Emlets** *M. luteus* (height to 50cm) has similar flowers, measuring 25–45mm across, but with much larger red blotches on the throat.

Speedwells & foxgloves

These members of the figwort family (Scrophulariaceae) vary enormously in terms of their size and structure. At one extreme is the tall, imposing Foxglove, and at the other the delicate speedwells. Speedwell species, with their four-lobed flowers, all look rather similar to one another at first glance. However, careful examination of the flowers, fruits and leaves, and of the growing habitat, makes identification relatively easy.

Thyme-leaved Speedwell

Veronica serpyllifolia
Scrophulariaceae Height to 20cm

Downy perennial with creeping, rooting stems and hairless, upright flowering stems. **Flowers** 5-7mm across and pale blue or white; borne on short stalks in loose spikes (Apr-Oct). **Fruits** Flattened, oval capsules. **Leaves** Thyme-like, small, oval and untoothed. **Status** Common throughout on bare and disturbed ground, and in short grassland.

Wall Speedwell

Veronica arvensis
Scrophulariaceae Height to 20cm

Softly hairy annual. **Flowers** 2-4mm across and blue, partly concealed by the bracts; borne in dense, leafy spikes (Mar-Oct). **Fruits** Flattened, hairy, heart-shaped capsules with a projecting style. **Leaves** Oval and toothed; the lower leaves are short-stalked and the upper ones unstalked. **Status** Common throughout in dry, bare locations, including banks and old walls.

Germander Speedwell

Veronica chamaedrys
Scrophulariaceae Height to 20cm

Delicate perennial with creeping, rooting stems and upright flowering stems that have 2 lines of hairs. **Flowers** 10-12mm across and blue with a white centre; borne on slender stalks in open terminal spikes (Apr-Jun). **Fruits** Flattened, hairy, heart-shaped capsules. **Leaves** Oval, toothed, hairy and short-stalked. **Status** Common throughout in grassy places.

Ivy-leaved Speedwell

Veronica hederifolia
Scrophulariaceae Prostrate

Creeping, hairy, branched annual. **Flowers** 4-5mm across and pale lilac-blue; borne on short stalks arising from leaf axils (Mar-Aug). **Fruits** Flattened, broadly rounded, hairless capsules. **Leaves** 10-12mm across, kidney-shaped to rounded, deeply lobed and Ivy-like. **Status** Common on bare ground and in fields.

Wood Speedwell

Veronica montana
Scrophulariaceae Height to 20cm

Delicate perennial with creeping, rooting stems and upright flowering stems that are hairy all round. **Flowers** 7-9mm across and lilac; borne on slender stalks in open spikes (Apr-Jul). **Fruits** Flattened, heart-shaped capsules. **Leaves** Oval, toothed, hairy and stalked. **Status** Locally common in damp woodland soil.

Slender Speedwell

Veronica filiformis
Scrophulariaceae Prostrate

Mat-forming, downy perennial with creeping stems. **Flowers** 8-10mm across and bluish with a white lip; borne on long, slender stalks arising from leaf axils (Apr-Jul). **Fruits** Seldom produced. **Leaves** 5-10mm across, rounded to kidney-shaped, blunt-toothed and short-stalked. **Status** Introduced; now locally common in the S in short grassland.

Heath Speedwell

Veronica officinalis
Scrophulariaceae Height to 10cm

Mat-forming, hairy perennial with creeping, rooting stems and upright flowering stems. **Flowers** 6-8mm across and lilac-blue with darker veins; borne in cylindrical, often conical spikes (May-Aug). **Fruits** Flattened, heart-shaped capsules. **Leaves** Oval, toothed, unstalked and hairy on both sides. **Status** Widespread and common in grassland and on heaths.

Wetland speedwells

Speedwells are found in a range of British habitats, from the tops of mountains (see p.57) to arable fields. Those shown here are associated with marshes and riversides. **Marsh Speedwell** *Veronica scutellata* (**1**) (height to 20cm) is a delicate plant of boggy ground, especially on acid soils. Its flowers are 6-7mm across and pale pink or white, with dark lines (Jun–Aug), and the leaves are narrow, lanceolate and 2-4cm long. **Brooklime** *V. beccabunga* (**2**) (height to 30cm) is a more robust plant that roots at the nodes and grows in shallow standing water and on damp ground. Its flowers are 7-8mm across, blue and arise from leaf axils (May–Sep), and its leaves are oval, fleshy and short-stalked. **Blue Water-speedwell** *V. anagallis-aquatica* (**3**) (height to 25cm) grows in damp ground, often beside streams and ponds. Its flowers are 5-6mm across, pale blue and borne in dense spikes with flower stalks that are as long as the bracts (Jun–Aug). Its leaves are narrow, oval, pointed, toothed and up to 12cm long. **Pink Water-speedwell** *V. catenata* (**4**) (height to 20cm) is similar to, and sometimes hybridises with, Blue Water-speedwell; it grows in similar habitats, but its flowers are pink and the flower stalks are shorter than the bracts.

Foxglove

Digitalis purpurea
Scrophulariaceae Height to 1.5m

Downy biennial. **Flowers** 4-5cm long, pinkish purple (sometimes white) with darker spots in the throat; borne in tall spikes (Jun–Sep). **Fruits** Green capsules. **Leaves** 20-30cm long, downy, oval and wrinkled; they form a rosette in the first year, from which the flowering spike appears in the second. **Status** Common throughout in woods and moors and on cliffs, especially on acid soils.

Eyebright

Euphrasia officinalis agg.
Scrophulariaceae Height to 25cm

Variable semi-parasite of the roots of other plants. Experts recognise many species but here these are treated as one. **Flowers** 5-10mm long and 2-lipped (the lower lip has 3 lobes), whitish (sometimes tinged pink) with purple veins and a yellow throat (May–Sep).

Fruits Capsules. **Leaves** Oval and toothed. **Status** Widespread and locally common in undisturbed grassland.

Red Bartsia

Odontites vernus
Scrophulariaceae Height to 40cm

Downy semi-parasite. **Flowers** 8-10mm long, pinkish purple and 2-lipped, the lower lip with 3 lobes; borne in 1-sided, elongated, slightly curved spikes (Jun–Sep). **Fruits** Capsules. **Leaves** Narrow, toothed, unstalked and paired. **Status** Common on disturbed ground and in arable field margins.

Field-speedwells

Plants of arable land and ploughed fields, three species of field speedwells are regularly encountered in the region. **Common Field-speedwell** *Veronica persica* (**1**) is a straggling, prostrate and hairy annual with reddish stems that is found throughout the region. Its flowers (8mm across) are pale blue, but with white on the lower lip (Jan–Dec), and its paired, oval leaves are pale green. **Grey Field-speedwell** *V. polita* (**2**) is similar but is restricted to chalky soil and is less common in the N; its flowers are entirely blue (Mar–Nov) and the leaves are grey-green. **Green Field-speedwell** (**3**) *V. agrestis*, widespread on sandy soils, has tiny flowers by comparison (3-4mm), and these are extremely pale and have a white lower lip; its leaves are fresh green.

fruit

Cow-wheats & broomrapes

Undisturbed grassland supports a great diversity of wildflowers, all growing alongside one another and competing for the same soil nutrients and water. Some gain an advantage by growing more vigorously than their neighbours, while others are more cunning, parasitising the plants around them to gain part or all of their nutrition. Some parasitic species obtain only part of their nutrition from their host (semi-parasitic) while others are wholly dependent (parasitic).

Marsh Lousewort
Pedicularis palustris
Scrophulariaceae Height to 60cm

Upright, hairless semi-parasite with a single branching stem. **Flowers** 20–25mm long, pinkish purple and 2-lipped, the upper lip with 4 teeth; borne in open leafy spikes (May–Sep). **Fruits** Inflated capsules. **Leaves** Feathery and deeply divided into toothed lobes. **Status** Widespread and locally common in marshes, bogs and fens, not exclusively on acid soils.

Yellow Bartsia
Parentucellia viscosa
Scrophulariaceae Height to 40cm

Stickily hairy, unbranched semi-parasitic annual. **Flowers** 15–35mm long, bright yellow and 2-lipped, the lower lip with 3 lobes; borne in leafy spikes (Jun–Sep). **Fruits** Capsules. **Leaves** Narrow and unstalked. **Status** Very locally common near the coasts of S and SW England and W Ireland. Grows in damp grassy places, often in dune slacks.

Yellow-rattle
Rhinanthus minor
Scrophulariaceae Height to 45cm

Variable, almost hairless semi-parasitic annual. **Flowers** 10–20mm long, yellow, 2-lipped and rather tubular and straight; borne in spikes that have green, triangular, toothed bracts (May–Sep). **Fruits** Inflated capsules, inside which the ripe seeds rattle. **Leaves** Oblong with rounded teeth. **Status** Common throughout in undisturbed meadows and on stabilised dunes.

Lousewort
Pedicularis sylvatica
Scrophulariaceae Height to 20cm

Spreading, branched semi-parasite. **Flowers** 20–25mm long, pale pink and 2-lipped, the upper lip with 2 teeth; borne in few-flowered leafy spikes (Apr–Jul). **Fruits** Inflated capsules. **Leaves** Feathery and divided into toothed leaflets. **Status** Widespread and locally common throughout. Found on damp heaths and moors, and in bogs, usually on acid soils.

Cow-wheats

Cow-wheats are annual members of the figwort family (Scrophulariaceae), and are semi-parasitic on the roots of other plants. Their leaves are narrow and are borne in opposite pairs. Of our four species, only **Common Cow-wheat** *Melampyrum pratense* (**1**) (height to 35cm) is widespread and common, growing on heaths and woodland rides, mainly on acid soils. Its flowers are 10–18mm long, pale yellow, rather tubular but flattened laterally, and two-lipped but with the mouth almost closed (May–Sep). **Small Cow-wheat** *M. sylvaticum* (height to 25cm) is a scarce plant of upland birch or pine woodlands. Its flowers are similar to those of Common Cow-wheat but are smaller (8–10mm long), with the mouth opening widely and the lower lip curved down (Jun–Aug). **Crested Cow-wheat** *M. cristatum* (**2**) (height to 50cm) is a rare, striking plant of verges and grassy woodland rides. Its flowers are 12–16mm long, yellow and purple, and two-lipped; they are borne in four-sided spikes with triangular, toothed bracts that are tinged purple at the base (Jun–Sep). **Field Cow-wheat** *M. arvense* (**3**) (height to 50cm) is a rare plant of arable fields and grassland. Its flowers are 20–25mm long, yellow and pink, and two-lobed; they are borne in cylindrical spikes with numerous toothed bracts that are tinged reddish purple (Jun–Sep).

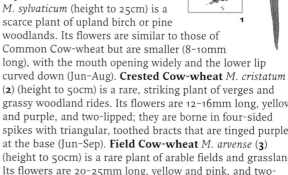
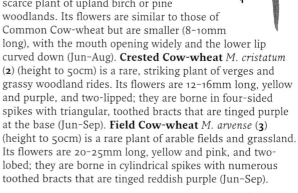

1

2

3

Broomrapes

Broomrapes (family Orobanchaceae) lack chlorophyll, cannot photosynthesise, and parasitise other plants to obtain their nutrition. Their flowers are borne in open, upright spikes. Several species occur in our region, some are local, and many are host-specific. **Common Broomrape** *Orobanche minor* (**1**) (height to 40cm) is the commonest and grows in grassy places in central and S England, Wales and S Ireland. It parasitises clovers and related plants and often has purplish-tinged stems. Its flowers are 10-18mm long, pinkish yellow with purple veins, tubular with a smoothly curved dorsal surface, and two-lipped (Jun-Sep). More impressive but distinctly local

is **Knapweed Broomrape** *O. elatior* (**2**) (height to 70cm), a parasite of knapweeds with a thick, yellowish-brown stem that is slightly swollen at the base. It grows in chalk grassland in S and E England only, and its flowers are 18-25mm long, yellow tinged purple, the filaments hairy at the base (Jun-Jul). **Greater Broomrape** *O. rapum-genistae* (**3**) (height to 80cm) is found on heaths mainly in England and Wales, and parasitises Broom and gorses. Its stems are yellowish and distinctly swollen at the base, and the flowers are 20-25mm long, yellow tinged purple, with filaments that are hairless at the base (May-Jul). Ivy Broomrape *O. hederae* (height to 60cm), found mainly in S and W Britain, resembles Common Broomrape but parasitises Ivy growing on chalky soil. Thyme Broomrape *O. alba* (height to 25cm) is a scarce, red-tinged plant of SW England, W Scotland and Ireland; it parasitises thymes and related plants.

Moschatel

Adoxa moschatellina
Adoxaceae Height to 10cm

Hairless, carpet-forming perennial. **Flowers** 6-8mm across and yellowish green; borne in long-stalked heads of 5 flowers, 4 of these facing outwards and the fifth facing upwards (Apr-May). **Fruits** Spongy. **Leaves** Pale green and fleshy; the basal leaves are twice 3-lobed and the stem leaves are 3-lobed. **Status** Locally common in damp, shady woodlands and hedgerows.

Toothwort

Lathraea squamaria
Orobanchaceae Height to 25cm

Native plant that parasitises the roots of Hazel and other shrubs. **Flowers** 15-18mm long, tubular and pinkish lilac to creamy white, borne in one-sided spikes (Apr-May).

Fruits Capsules contained within the dead flower. **Leaves** Creamy white and scale-like. **Status** Absent from N Scotland and W Ireland; only locally common, growing in woodlands on base-rich soils.

Purple Toothwort

Lathraea clandestina
Orobanchaceae

Locally naturalised, mainly subterranean plant. Parasitises the roots of trees such as poplars, willows and alders. **Flowers** Plant visible above ground only when flowers appear; these are 4-5cm long, purple and held erect (Mar-May). **Fruits** Capsules. **Leaves** Purplish and scale-like. **Status** Introduced and naturalised in damp woodland.

Valerians

Valerians belong to the family Valerianaceae, whose members are characterised by having leaves borne in opposite pairs and flowers arranged in umbel-like heads. **Red Valerian** *Centranthus ruber* (**1**) (height to 75cm) is a familiar grey-green garden perennial that is now widely naturalised on rocky ground. Its flowers are 8-10mm long, reddish or pink (sometimes white), and are borne in dense terminal heads (May-Sep). **Common Valerian** *Valeriana officinalis* (**2**) (height to 1.5m) is a widespread native of grassy wayside places. Its flowers are 3-5mm long, funnel-shaped, five-lobed and pale pink, and are borne in dense terminal umbels (Jun-Aug). **Marsh Valerian** *V. dioica* (**3**) (height to 30cm) is a locally common perennial of damp grassland and fens. Its flowers are pale pink (May-Jun).

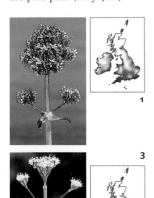

Butterworts & plantains

All British plants face challenges when it comes to their growth and survival. Tough and resilient leaves allow plantains (family Plantaginaceae) to grow in harsh environments, some species trampled by feet in lawns and on paths, others growing in well-drained soils by the sea. Cornsalads (family Valerianaceae) thrive in arable fields while butterworts (family Lentibulariaceae) overcome acid, nutrient-poor soils by supplementing their diet with carnivorous habits – their leaves catch and digest insects.

Cornsalads

Cornsalads are members of the valerian family (Valerianaceae) and are so named because typically they grow in arable fields, and because their small, oblong leaves can be used as a salad vegetable. Several species occur in Britain but only three are regularly encountered; these are best distinguished by studying the fruits. Although it is nothing like as numerous as it was once, **Common Cornsalad** *Valerianella locusta* (**1**) (height to 30cm) is still widespread and is found in dunes as well as on farmland. Its flowers are 1–2mm across, pinkish lilac, five-lobed and borne in flat-topped terminal clusters that are 1–2cm across (Apr–Aug), and its fruits are flattened and rather oval. **Broad-fruited Cornsalad** *V. rimosa* (**2**) (height to 30cm) is similar, but is restricted to S England and is less leafy. Its flowers are borne in few-flowered clusters (Jul–Aug) and its fruits are egg-shaped, not flattened, and grooved on one side. **Narrow-fruited Cornsalad** *V. dentata* (**3**) (height to 30cm) is confined to arable fields on chalky soils in SE England; its flowers are 1–2mm across, lilac and borne in few-flowered clusters (Jun–Jul), and its fruits are narrowly ovoid, flat on one side and round on the other.

Greater Plantain

Plantago major Plantaginaceae
Height to 20cm

Persistent, usually hairless perennial. **Flowers** 3mm across, pale yellow with initially purple anthers that turn yellow later; borne on slender spikes, 10–15mm long (Jun–Oct). **Fruits** Capsules. **Leaves** Broad, oval, to 25cm long, and with 3–9 veins and a distinct, narrow stalk. **Status** Very common on lawns and disturbed grassland.

Large-flowered Butterwort

Pinguicula grandiflora
Lentibulariaceae Height to 20cm

Irish speciality. **Flowers** 25–30mm across, violet with a purple-streaked white throat, a 10–12mm-long spur and overlapping lip lobes (May–Jul). **Fruits** Capsules. **Leaves** Sticky; trap and digest insects. **Status** Native to SW Ireland, introduced to SW England.

Common Butterwort

Pinguicula vulgaris
Lentibulariaceae Height to 15cm

Carnivorous plant. **Flowers** 12–14mm across, violet with a white throat, funnel-shaped with a spreading lower lip and 4–7mm-long spur (May–Aug). **Fruits** Capsules. **Leaves** Sticky; trap and digest insects. **Status** Widespread and locally common in N and W Britain and Ireland.

Pale Butterwort

Pinguicula lusitanica
Lentibulariaceae Height to 10cm

Dainty carnivorous plant. **Flowers** 7–9mm across, pale pinkish lilac with a short spur (Jul–Sep). **Fruits** Capsules. **Leaves** Narrow and sticky; trap and digest insects. **Status** Restricted to SW and NW Britain and Ireland.

Hoary Plantain

Plantago media Plantaginaceae
Height to 25cm

Persistent, downy perennial. **Flowers** 2mm across, whitish with lilac anthers; borne on slender spikes that are up to 20cm long (May–Aug). **Fruits** Capsules. **Leaves** Greyish and narrowly ovate, tapering gradually to broad stalks. **Status** Widespread and common only in England. Grows in lawns and trampled and compressed grassland, doing best mainly on calcareous soils.

flowers

Buck's-horn Plantain

Plantago coronopus
Plantaginaceae Height to 15cm

Downy, greyish-green perennial. **Flowers** 2mm across, with a brownish corolla and yellow stamens; borne in slender spikes, 2–4cm long (May–Jul). **Fruits** Capsules. **Leaves** 20cm long, 1-veined and deeply divided. **Status** Widespread and common on coastal grassland, disturbed ground and cliffs in England and Ireland; also occurs inland in SE England.

Sea Plantain

Plantago maritima
Plantaginaceae Height to 15cm

Coastal perennial, tolerant of salt spray and occasional immersion in seawater. **Flowers** 3mm across and brownish with yellow stamens; borne in slender spikes, 2–6cm long (Jun–Aug). **Fruits** Capsules. **Leaves** Narrow, strap-like and untoothed, with 3–5 faint veins. **Status** A mainly coastal plant that is widespread and common in saltmarshes, but also found on coastal cliffs.

Ribwort Plantain

Plantago lanceolata
Plantaginaceae Height to 15cm

Persistent perennial. **Flowers** 4mm across and brownish with white stamens; borne in compact 2cm-long heads on furrowed stalks that are up to 40cm long (Apr–Oct). **Fruits** Capsules. **Leaves** Narrow, to 20cm long and with 3–5 distinct veins. **Status** Widespread and common throughout. Found on disturbed grassland, cultivated ground and tracks.

Shoreweed

Littorella uniflora
Plantaginaceae Creeping

Aquatic perennial with creeping runners, sometimes forming patches. **Flowers** Greenish; the stalked males have long stamens, while the female flowers are stalkless and basal. **Fruits** Capsules. **Leaves** 7–10cm long, semicircular in cross-section, narrow and spongy. **Status** Local, mainly in the N and W. Grows on the margins of lakes and ponds that have acid waters.

Arrowgrasses

Arrowgrasses belong to the family Juncaginaceae and have wiry, superficially grass-like leaves that are borne in a basal rosette. Their flowers, which are rather insignificant, are borne in long spikes that bear a passing resemblance to those of the unrelated plantains. Two species are found in the region. **Sea Arrowgrass** *Triglochin maritimum* (**1**) (height to 50cm) is a tufted saltmarsh perennial, widespread around all coasts, whose flowers are 3–4mm across, three-petalled and green, edged with purple; they are borne in a long-stalked spike (May–Sep). The fruits are egg-shaped with six segments and the leaves are not furrowed. **Marsh Arrowgrass** *T. palustre* (**2**) (height to 50cm) is similar but it grows in freshwater habitats, notably marshy meadows, and is locally common throughout. Its flowers are smaller (Jun–Aug), the fruits are club-shaped with three segments and the leaves are furrowed.

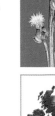

Honeysuckles, teasels & bellflowers

Members of the bellflower family and most members of the honeysuckle family attract insects by producing blooms that are relatively large. By contrast, the individual flowers of the teasel family are tiny but, because they appear together in large, showy heads, the same dramatic effect is produced.

The honeysuckle family

Honeysuckle and related species belong to the family Caprifoliaceae, and they range in size from small herbs to large shrubs and woody climbers. Their leaves are arranged in opposite pairs and typically their flowers are tubular or bell-shaped.

Honeysuckle

Lonicera periclymenum
Caprifoliaceae Height to 5m

Woody climber that twines clockwise through shrubs and trees. **Flowers** 3–5cm long, scented, trumpet-shaped, 2-lipped and creamy yellow to white; borne in whorled heads (Jun–Aug). **Fruits** Red berries. **Leaves** Grey-green, oval and in opposite pairs. **Status** Common in woodland and hedgerows.

Twinflower

Linnaea borealis Caprifoliaceae
Height to 7cm

Charming, creeping, mat-forming perennial. **Flowers** 5–9mm long, pink and bell-shaped; borne in pairs on upright, slender stalks (Jun–Aug). **Fruits** Dry and papery. **Leaves** Oval to rounded and borne in pairs on wiry stems. **Status** Rare and restricted to undisturbed pine forests in NE Scotland.

Dwarf Elder

Sambucus ebulus Caprifoliaceae
Height to 2m

Unpleasant-smelling deciduous shrub. **Flowers** 3–5mm across and pinkish white; borne in flat-topped clusters, 8–15cm across (Jun–Aug). **Fruits** Poisonous black berries, borne in clusters. **Leaves** Divided into 7–13 narrow leaflets. **Status** Widespread but local, mainly in the S. Grows in hedgerows and scrub, and on roadside verges.

fruits

The teasel family

Teasels and closely related species belong to the family Dipsacaceae. They are upright plants with opposite leaves and their small flowers are borne in dense, compact heads; in some species these heads are adorned with sharp bracts.

Small Scabious

Scabiosa columbaria Dipsacaceae
Height to 65cm

Upright, branching perennial. **Flowers** Bluish violet and borne in compact heads, 2–3cm across, the outer flowers larger than the inner ones (Jun–Sep). **Fruits** Dry and papery. **Leaves** Comprise deeply lobed basal leaves in a rosette, and shallow-lobed stem leaves. **Status** Locally common in England and Wales. Grows in calcareous grassland.

Field Scabious

Knautia arvensis Dipsacaceae
Height to 75cm

Robust, hairy biennial or perennial. **Flowers** Bluish violet and borne in heads, 3–4cm across, the outer flowers larger than the inner ones (Jun–Oct). **Fruits** Dry and papery. **Leaves** Comprise lobed, spoon-shaped basal leaves in a rosette and stem leaves that are less divided. **Status** Widespread and common, except N Scotland. Grows in dry grassland.

Devil's-bit Scabious

Succisa pratensis Dipsacaceae
Height to 75cm

Upright perennial. **Flowers** Pinkish lilac to violet-blue, with projecting anthers that resemble tiny mallets; borne in dense, domed, long-stalked terminal heads, 15–25mm across (Jun–Oct). **Fruits** Dry and papery. **Leaves** Comprise spoon-shaped basal leaves and narrow stem leaves. **Status** Widespread and common in damp grassland, woodland rides and marshes.

Wild Teasel

Dipsacus fullonum Dipsacaceae
Height to 2m

 Biennial with prickly stems. **Flowers** Pinkish purple; borne in egg-shaped heads, 6–8cm long, that are adorned with spiny bracts (Jul–Aug). **Fruits** Dry and papery. **Leaves** Spine-coated; they appear as rosettes in the first year and in the second year are seen as opposite stem leaves that are fused at the base and collect water. **Status** Common in damp grassland on heavy soils.

Small Teasel

Dipsacus pilosus Dipsacaceae
Height to 1.25m

 Upright biennial with hairy stems. **Flowers** White; borne in spherical to egg-shaped heads, 15–20mm across, with spiny bracts (Jul–Sep). **Fruits** Dry and papery. **Leaves** Oval; the long-stalked basal leaves form a rosette and the stem leaves sometimes have 2 basal lobes. **Status** Local, in England and Wales only. Grows along woodland margins and on banks.

The bellflower family

Bellflowers belong to the family Campanulaceae, and include plants that have opposite and undivided leaves and bell-shaped flowers, typically with five pointed lobes at the mouth. Their flowers are usually a shade of blue or pinkish lilac, depending on the species.

Round-headed Rampion

Phyteuma orbiculare
Campanulaceae Height to 50cm

 Hairless perennial. **Flowers** Bluish violet; borne in rounded heads, 10–15mm across, on long, slender stems (Jun–Aug). **Fruits** Dry capsules. **Leaves** Comprise oval basal leaves and narrow, unstalked stem leaves. **Status** Local, restricted to a few areas of undisturbed chalk grassland in S England.

Sheep's-bit

Jasione montana Campanulaceae
Height to 30cm

 Downy biennial. **Flowers** Sky blue; borne in rounded heads, 30–35mm across, on slender stalks (May–Sep). The anthers do not project (cf. Devil's-bit Scabious). **Fruits** Dry capsules. **Leaves** Wavy-edged; the basal leaves form a rosette and the stem leaves are narrow. **Status** Local, commonest in the W and near the sea. Found in dry grassland, heaths and dunes, on acid soils.

Harebell

Campanula rotundifolia
Campanulaceae Height to 40cm

 Hairless perennial with wiry stems. **Flowers** 15mm long, blue and bell-shaped, with sharp, triangular teeth (Jul–Oct). **Fruits** Dry capsules. **Leaves** Comprise rounded basal leaves that wither, and narrower stem leaves that persist. **Status** Widespread and common throughout except, perhaps surprisingly, in the SW. Grows in dry, grassy places, on both calcareous and acid soils.

Clustered Bellflower

Campanula glomerata
Campanulaceae Height to 25cm

 Upright, hairy perennial. **Flowers** 15–25mm long, violet-blue and bell-shaped, with blunt teeth; borne mainly in terminal clusters (Jun–Oct). **Fruits** Dry capsules. **Leaves** Long-stalked and heart-shaped at the base but narrower and clasping on the stem. **Status** Locally common in S and E England only. Grows in grassland and on verges, on calcareous soils.

Nettle-leaved Bellflower

Campanula trachelium
Campanulaceae Height to 75cm

 Roughly hairy perennial with sharply angled stems. **Flowers** 3–4cm long, bluish violet and bell-shaped, with flared, triangular lobes (Jul–Aug). **Fruits** Dry capsules. **Leaves** Stalked; heart-shaped at the base of the plant and nettle-like on the stem. **Status** Locally common in S and E England only. Grows in hedgerows and scrub on calcareous soils.

Bellflowers & daisies

Most members of the bellflower family have rather symmetrical flowers that look distinctly bell-shaped. Those of lobelias are the exception, with lobes that differ in size and shape from one another. The way in which daisy family members arrange their flowers is entirely different: individual flowers are tiny but they are grouped together in large, showy heads.

Giant Bellflower

Campanula latifolia
Campanulaceae Height to 1m

Upright perennial with bluntly angled stems. **Flowers** 4–5.5cm long, pale blue (sometimes white) and bell-shaped, with triangular lobes (Jul–Aug). **Fruits** Dry capsules. **Leaves** Oval to lanceolate and toothed; the lower ones have winged stalks. **Status** Locally common in central and N England. Grows in shady woods and hedgerows.

Ivy-leaved Bellflower

Wahlenbergia hederacea
Campanulaceae Creeping

Delicate, trailing perennial. **Flowers** 5–10mm long, pale blue and narrowly bell-shaped, with flared, triangular lobes at the mouth (Jul–Aug). **Fruits** Dry capsules. **Leaves** 5–10mm across, pale green and lobed, sometimes resembling tiny Ivy leaves. **Status** Locally common in SW on damp acid moors and heaths.

Venus's Looking-glass

Legousia hybrida Campanulaceae Height to 40cm

Hairy, straggly perennial. **Flowers** 5–10mm across; purple when fully open, which happens only in bright sunshine (May–Aug). **Fruits** Narrow, tapering capsules. **Leaves** Oblong and wavy; the upper ones are stalkless and the lower ones are short-stalked. **Status** Local, almost entirely restricted to S and E England (formerly much more common). Grows in arable fields on chalky soils.

Water Lobelia

Lobelia dortmanna
Campanulaceae Aquatic

Hairless perennial. **Flowers** 15–20mm long, lilac and 2-lipped, the upper lip with 2 narrow lobes and the lower one with 3 narrow lobes; borne in spikes on slender stalks (Jul–Sep). **Fruits** Capsules. **Leaves** Narrow and fleshy; borne in rosettes on lake beds. **Status** Locally common in acid, gravelly lakes.

Heath Lobelia

Lobelia urens Campanulaceae Height to 50cm

Upright, hairless perennial. **Flowers** 10–15mm long, bluish purple and 2-lipped, with 2 narrow upper lobes and 3 narrow lower lobes; borne in open spikes (Jul–Aug). **Fruits** Capsules. **Leaves** Dark green, oval at the base of the plant and narrow on the stem. **Status** Local and scarce, found mainly from Sussex to Devon. Grows on damp, grassy heaths on acid soils.

The daisy family

The flowers of members of the daisy family (Asteraceae) are varied in appearance, but several have superficially very similar, daisy-like compound flower heads comprising two distinct types of florets: yellow disc florets are found in the centre, surrounded by an array of white ray florets.

Daisy

Bellis perennis Asteraceae Height to 10cm

Familiar downy perennial. **Flowers** Solitary heads, 15–25mm across, borne on slender stems, and comprising yellow disc florets and white (often faintly crimson-tipped) ray florets (Mar–Oct). **Fruits** Dry and papery. **Leaves** Spoon-shaped and form prostrate rosettes from which the flower stalks arise. **Status** Common throughout in lawns and other areas of short grass.

Oxeye Daisy

Leucanthemum vulgare
Asteraceae Height to 60cm

 Downy or hairless perennial. **Flowers** Borne in solitary heads, 30-50mm across, with yellow disc florets and white ray florets (May-Sep). There are no scales between the disc florets. **Fruits** Dry and papery. **Leaves** Dark green and toothed; the lower spoon-shaped leaves form a rosette and the stem leaves are lobed. **Status** Common throughout in dry, grassy places, often on disturbed ground.

Stinking Chamomile

Anthemis cotula Asteraceae
Height to 50cm

 Similar to Scented Mayweed but hairless and unpleasantly scented. **Flowers** Borne in solitary heads, 20-35mm across, with yellow disc florets and white ray florets (Jul-Sep). Scales are present between the disc florets. **Fruits** Dry and papery. **Leaves** Feathery and much divided. **Status** Common only in the S. Grows in disturbed ground.

Scentless Mayweed

Tripleurospermum inodorum
Asteraceae Height to 75cm

 Scentless, hairless, straggly perennial. **Flowers** Borne in clusters of stalked heads, 20-40mm across, with yellow disc florets and white ray florets (Apr-Oct). There are no scales between the disc florets and the base of the flower head is solid. **Fruits** Dry and papery, tipped with black oil glands. **Leaves** Feathery. **Status** Common throughout in disturbed and cultivated ground.

Scented Mayweed

Matricaria recutita Asteraceae
Height to 60cm

 Similar to Scentless Mayweed but scented and aromatic. **Flowers** Borne in clusters of long-stalked heads, 20-30mm across, with yellow disc florets and white ray florets (Jun-Aug). There are no scales between the disc florets and the flower head is hollow. **Fruits** Dry and papery, without black oil glands. **Leaves** Feathery. **Status** Common only in the S. Found on disturbed ground.

Wild Chamomile

Chamaemelum nobile Asteraceae
Height to 25cm

 Creeping, greyish perennial. Pleasantly aromatic. **Flowers** Borne in solitary heads, 18-24mm across, with yellow disc florets and white ray florets (Jun-Aug). Scales are present between the disc florets. **Fruits** Dry and papery. **Leaves** Feathery, with fine, bristle-tipped lobes. **Status** Local, mainly in the S. Grows in short, often grazed, grassland on sandy soils.

Corn Chamomile

Anthemis arvensis Asteraceae
Height to 50cm

 Pleasantly aromatic annual with downy stems. **Flowers** Borne in solitary heads, with yellow disc florets and white ray florets (Jun-Jul). Scales are present between the disc florets. **Fruits** Dry and papery. **Leaves** Much divided; the lobes are broader than in other mayweeds, and are downy below. **Status** Locally common in the S. Grows in cultivated calcareous ground.

Sea Mayweed

Tripleurospermum maritimum
Asteraceae Height to 60cm

 Similar to Scentless Mayweed but more branched. **Flowers** Borne in clusters of solitary long-stalked heads, 20-40mm across, with yellow disc florets and white ray florets (Apr-Oct). Scales between disc florets are absent and base of flower head is solid. **Fruits** Dry and papery. **Leaves** Much divided, with fleshy segments. **Status** Common on coastal shingle and sand.

Pineapple Mayweed

Matricaria discoidea Asteraceae
Height to 12cm

 Bright green, hairless perennial. Smells of pineapple when crushed. **Flowers** Yellowish-green disc florets only (no ray florets); borne in hollow, rounded to conical heads, 8-12mm long (May-Nov). **Fruits** Dry and papery. **Leaves** Finely divided and feathery. **Status** Widespread and common on disturbed ground, paths and tracks.

Fleabanes & cudweeds

The individual flowers of members of the daisy family (Asteraceae) are always tiny, but when grouped together in compound heads they are much more effective at attracting insect pollinators. Some of the species shown on these pages have flowers that, to the human eye at least, are understated, while others are showy and colourful.

Canadian Fleabane

Conyza canadensis Asteraceae
Height to 1m

 Upright, hairy annual. **Flowers** Grouped in heads, 5-8mm long, with pinkish or white florets; the heads are borne in dense, much-branched inflorescences (Jul-Oct). **Fruits** Dry and papery. **Leaves** Narrow and strap-like. **Status** Introduced and increasing. Found on disturbed and bare ground, and often seen growing beside roads.

Blue Fleabane

Erigeron acer Asteraceae
Height to 30cm

 Hairy annual or biennial with stiff, reddish stems. **Flowers** Borne in clusters of heads, each 12-18mm across; bluish-purple ray florets largely conceal yellow disc florets (Jun-Aug). **Fruits** Dry and papery. **Leaves** Spoon-shaped at the base of the plant but narrow on the stem. **Status** Common in England and Wales only. Grows in dry, grassy places and on coastal shingle.

Common Cudweed

Filago vulgaris Asteraceae
Height to 25cm

 Annual with a white-woolly coating; it branches towards the top of the plant. **Flowers** Borne in rounded, woolly clusters, 10-12mm across, of 20-35 heads, not overtopped by leaves; the heads have yellow florets and yellowish-tipped, straight bract tips (Jul-Aug). **Fruits** Dry and papery. **Leaves** Narrow, woolly and wavy. **Status** Locally common in the S. Found in dry, grassy places.

Small Cudweed

Filago minima Asteraceae
Height to 20cm

 Upright, greyish, woolly annual. **Flowers** Borne in clusters of 3-6 conical or egg-shaped heads, 3-4mm long, with bracts that are woolly at the base only and tipped yellow (Jul-Sep). **Fruits** Dry and papery. **Leaves** Narrow and woolly. **Status** Very locally common in England and Wales only. Found on grassy heaths, on sandy, acid soils.

Marsh Cudweed

Gnaphalium uliginosum
Asteraceae Height to 20cm

 Greyish green, woolly, branched annual. **Flowers** Borne in clusters of unstalked heads, 3-4mm long, with yellow disc florets and brown bracts (Jul-Oct). **Fruits** Dry and papery. **Leaves** Narrow and woolly on both sides; the upper ones sometimes overtop the flower heads. **Status** Widespread and common. Grows in damp, disturbed ground and on tracks.

Heath Cudweed

Gnaphalium sylvaticum
Asteraceae Height to 50cm

Greyish, leafy perennial with prostrate non-flowering stalks and upright flowering stems. **Flowers** Borne in clusters of heads, each 5–7mm long, with yellow-brown florets and brown-tipped bracts (Jul–Sep). **Fruits** Dry and papery. **Leaves** Green and hairless above but white-woolly below. **Status** Locally common in dry, grassy places on heaths and in woods.

Ploughman's-spikenard

Inula conyzae Asteraceae
Height to 1m

Downy biennial or perennial, often with red-tinged stems. **Flowers** Borne in clusters of egg-shaped heads, 8–10mm long, with yellow florets and purplish and green bracts (Jul–Sep). **Fruits** Dry and papery. **Leaves** Comprise oval basal leaves and narrower stem leaves. **Status** Locally common in England and Wales only. Found in dry grassland on calcareous soils.

Upland specialists

Most cudweeds and similar related species are lowland plants, but two are restricted to upland heaths and moors, often on acid soils where few other flowering plants can thrive apart from certain grass, sedge and rush species. **Dwarf Cudweed** *Gnaphalium supinum* (**1**) is a tiny (height to 10cm), compact, greyish-green perennial of the Scottish Highlands, whose flowers are borne in clusters of small, compact, brown heads that are fringed by bracts (Jul–Aug); its leaves are narrow and woolly on both sides, the upper ones surrounding the flower heads. **Mountain Everlasting** *Antennaria dioica* (**2**) (height to 20cm) is a more widespread and showy plant, a downy perennial with leaf rosettes from which flower stems arise. The flowers themselves are borne in umbel-like clusters of compact, woolly, separate-sex heads, the male flowers with white-tipped bracts and the larger female heads with pink-tipped bracts (Jun–Aug).

Marshy ground specialists

Most plants grow best where soil drainage is good and shun permanently waterlogged ground. The bur-marigolds are exceptions and love conditions where they can grow with their 'feet' in the water; pond margins and marshes are ideal. Two species are found in our region, both of them annuals and locally common in the S; although superficially similar, they can be distinguished by looking at the flowers and leaves. **Trifid Bur-marigold** *Bidens tripartita* (**1**) (height to 60cm) has almost hairless reddish stems, and its flowers are grouped in heads, 10–25mm across, with yellow disc florets (no ray florets) and five to eight leaf-like bracts below (Jul–Oct); its leaves are stalked and three-lobed. **Nodding Bur-marigold** *B. cernua* (**2**) (height to 70cm) has hairy stems and flowers grouped in nodding heads, 15–30mm across with yellow disc florets and five to eight leaf-like bracts below (Jul–Oct); its leaves are narrow, unstalked and undivided.

Coastal highlights

If you visit the coast and especially saltmarshes and sea cliffs in late summer you should be rewarded with a colourful display put on by two charming species of the daisy family. **Golden Samphire** *Inula crithmoides* (**1**) (height to 75cm) has bright yellow and orange flowers arranged in clusters of heads, 15–30mm across, with spreading, yellow ray florets and orange-yellow central disc florets (Jul–Sep). Its leaves are bright green, narrow and fleshy, and the plant is found around the coasts of SW Britain and Ireland. The flowers of the more widespread **Sea Aster** *Aster tripolium* (**2**) (height to 75cm) are reminiscent of those of the cultivated Michelmas-daisy and comprise umbel-like clusters of heads, each 1–2cm across, with yellow disc florets and bluish-lilac ray florets (Jul–Sep); its leaves are fleshy and narrow, with a prominent midrib.

Wayside daisies

It is usually possible to find at least one representative species of the daisy family (Asteraceae) in flower in any month from midwinter right through to late autumn. Though some are distinctly seasonal, blooming for just a few weeks, a few have a prolonged flowering period, producing a succession of flowers, or appearing as a succession of plants, over a time span of six months or more.

Common Fleabane

Pulicaria dysenterica Asteraceae
Height to 50cm

Creeping perennial with upright, woolly flowering stems. **Flowers** Borne in clusters of heads, 15–30mm across, with spreading yellow ray florets and deeper yellow central disc florets (Jul–Sep). **Fruits** Dry and papery. **Leaves** Heart-shaped; the stem leaves are clasping and the basal ones soon wither. **Status** Common (except in Scotland) in damp meadows on heavy soils.

Gallant-soldier

Galinsoga parviflora Asteraceae
Height to 75cm

Upright, branched, hairless annual. **Flowers** Borne in much-branched clusters of untidy-looking heads, 3–5mm across, with yellow disc florets and 4–5 white rays (May–Oct). **Fruits** Have long hairs. **Leaves** Oval, toothed, stalked and paired. **Status** Introduced and naturalised, mainly in S and SE England. Found on waste ground and in cultivated fields.

Goldenrod

Solidago virgaurea Asteraceae
Height to 75cm

Upright, slightly downy perennial. **Flowers** Borne in spikes of heads, each 5–10mm across, with yellow ray and disc florets (Jun–Sep). **Fruits** Have long hairs. **Leaves** Stalked and spoon-shaped at the base; the stem leaves are narrow and unstalked. **Status** Widespread and locally common in woods and grassland, and on rocky banks; tolerates a wide range of soil types.

Early risers

Three frost-tolerant members of the daisy family are familiar harbingers of spring, putting in a floral appearance before much else is in flower. All are plants of damp, often disturbed ground. **Butterbur** *Petasites hybridus* (**1**) (height to 50cm) is an impressive riverside plant that forms patches; it is locally common throughout the region. Its flowers are borne in tall spikes that comprise pinkish-red heads on separate-sex plants; the male flower heads are 7–12mm across and the females 3–6mm across (Mar–May). The flowers are finishing as the heart-shaped leaves, measuring 1m across, appear. **Winter Heliotrope** *P. fragrans* (**2**) (height to 20cm) is a patch-forming perennial garden plant that is now widely naturalised. Its vanilla-scented flowers are borne in spikes of pinkish-lilac heads, each 10–12mm across (Dec–Mar); the leaves are rounded, 20cm across, long-stalked and present all year. The widespread and common **Colt's-foot** *Tussilago farfara* (**3**) (height to 20cm) is a charming, creeping perennial that appears in early spring as leafless stalks with purplish bracts. On these are borne heads of yellow flowers, 15–35mm across (Feb–Apr); the leaves, which are rounded, heart-shaped and 10–20cm across, appear after flowering has finished.

Canadian Goldenrod

Solidago canadensis Asteraceae
Height to 2m

Upright, downy perennial. **Flowers** Borne in branching clusters of crowded, arching, 1-sided sprays of numerous yellow individual flower heads (Jul–Oct). **Fruits** Hairy. **Leaves** Oval, toothed and 3-veined. **Status** A familiar introduced garden plant that is naturalised locally in damp wayside ground, hedgerows and rough grassland.

Yarrow

Achillea millefolium Asteraceae
Height to 50cm

Downy, aromatic perennial with creeping stems and upright flowering stalks. **Flowers** Borne in flat-topped clusters of heads, each 4–6mm across, with yellowish disc florets and pinkish-white ray florets (Jun–Nov). **Fruits** Dry and papery. **Leaves** Dark green and feathery. **Status** Common throughout in meadows, verges and hedgerows, and on waste ground.

Sneezewort

Achillea ptarmica Asteraceae
Height to 60cm

Upright perennial with stiff, angular stems. **Flowers** Borne in open clusters of heads, each 1–2cm across, with greenish-yellow disc florets and white ray florets (Jul–Sep). **Fruits** Dry and papery. **Leaves** Narrow, undivided, untoothed and stalkless. **Status** Locally common throughout in damp meadows, woodland rides and clearings, mainly on acid soils.

Tansy

Tanacetum vulgare Asteraceae
Height to 75cm

Upright, aromatic perennial. **Flowers** Borne in flat-topped, umbel-like clusters of up to 70 golden-yellow, button-like heads, 7–12mm across, with disc florets only (Jul–Oct). **Fruits** Dry and papery. **Leaves** Yellowish green and deeply divided, with lobes that themselves are further divided. **Status** Common and widespread on roadside verges, and in hedgerows and on disturbed ground.

A thing of the past?

In the days before intensive agriculture came to dominate the farming scene, arable 'weeds' were very much a feature of the British and Irish landscape. Nowadays, most areas are, to a degree, sterilised by the use of herbicides, but in places like the Isles of Scilly, where more traditional approaches to agriculture persist, you can get a glimpse of what things must have been like elsewhere in the past. **Corn Marigold** *Chrysanthemum segetum* (height to 50cm) is one of the more showy arable plants and, when conditions are suitable, it appears in huge numbers. It is widespread, although occasional, elsewhere, turning up in corners of fields that have been missed by sprayers or on organically farmed arable land. This annual's flowers are grouped in heads, 3–6cm across, with orange-yellow disc florets and yellow ray florets (Jun–Oct); its leaves are narrow, deeply lobed or toothed, and slightly fleshy.

Feverfew

Tanacetum parthenium Asteraceae Height to 50cm

Branched, downy, aromatic perennial. **Flowers** Borne in open clusters of daisy-like heads, 1–2cm across, with yellow disc and white ray florets (Jul–Aug). **Fruits** Dry and papery. **Leaves** Yellowish green and divided. **Status** This widely naturalised garden plant is often found near habitation in disturbed ground, and on verges and old walls.

Ragworts, groundsels & knapweeds

Many daisy family members (Asteraceae) can survive regular grassland cutting and colonise freshly disturbed ground. They appear out of the blue because they produce large quantities of seed, dispersed by wind or animal carriers.

Ragworts

Ragworts have bright yellow flower heads; three species are widespread. **Common Ragwort** *Senecio jacobaea* (**1**) (height to 1m), the commonest, is often reviled because it is poisonous to grazing animals, although the living plant is avoided. But it is a fantastic source of nectar for insects. Flowers in flat-topped clusters of heads, each 15–25mm across (Jun–Nov); leaves deeply divided with a blunt end lobe.

Hoary Ragwort *S. erucifolius* (**2**) (height to 1.5m) is similar but downy; flowers in clusters of pale yellow heads, 15–20mm across (Jul–Aug); leaf lobes narrower and more pointed than Common Ragwort. It is locally common in England and Wales only.

Oxford Ragwort *S. squalidus* (**3**) (height to 50cm) is naturalised on disturbed ground; flowers in clusters of heads (May–Dec), each 15–25mm across (with black-tipped bracts); divided leaves have a pointed end lobe.

Marsh Ragwort *S. aquaticus* (**4**) (height to 80cm), is a local marshland plant whose few-flowered clusters of heads are not flat-topped (Jul–Aug).

Hemp-agrimony

Eupatorium cannabinum
Asteraceae Height to 1.5m

Tall, downy perennial. **Flowers** Dull pinkish lilac; borne in clusters of heads, each 2–5mm across, comprising 5–6 florets (Jul–Sep). **Fruits** Hairy. **Leaves** Trifoliate and borne in opposite pairs up the stem. **Status** Widespread and common, except in the N. Grows mainly in damp grassland and marshes, and (perhaps surprisingly) also in scrub on chalky, dry, free-draining soils.

Mugwort

Artemisia vulgaris Asteraceae
Height to 1.25m

Aromatic, branched plant with ribbed, reddish stems. **Flowers** Borne in reddish heads, 2–3mm across; carried in tall, branched spikes (Jul–Sep). **Fruits** Dry and papery. **Leaves** Dark green and hairless above, but with silvery-down below; the lower leaves are stalked, while the upper ones are unstalked. **Status** Widespread and common on roadside verges, disturbed land and waste ground.

Groundsels

A successful group of plants associated with recently disturbed ground or unstable, changing soils; groundsels have deeply divided leaves. Three species are widespread: **Groundsel** *Senecio vulgaris* (**1**) (height to 40cm) is a familiar sight in cultivated garden soil. Its flowers comprise cylindrical heads, 10mm long, of yellow disc florets only, with black-tipped greenish bracts (Jan–Dec). **Heath Groundsel** *S. sylvaticus* (**2**) (height to 70cm) is superficially similar but more robust, and is locally common on sandy soils on heaths; it often appears in the wake of fires or gorse clearance. Its flowers comprise stickily hairy conical heads, 10mm long, of yellow disc florets, recurved ray florets and bracts that are not black-tipped (Jun–Sep). **Sticky Groundsel** *S. viscosus* (**3**) (height to 60cm) resembles Heath Groundsel but the whole plant (not just the flower heads) is stickily hairy and pungent; it is locally common in dry, bare places, often near the coast, and its flowers comprise conical heads, 12mm long, of yellow disc florets, recurved ray florets and bracts that are not black-tipped (Jul–Sep).

Wormwood

Artemisia absinthium Asteraceae
Height to 80cm

Aromatic perennial with silky-hairy stems. **Flowers** Borne in yellowish heads that are 3–5mm across, bell-shaped and nodding; carried in tall, branched spikes (Jul–Sep). **Fruits** Dry and papery. **Leaves** Divided into deeply cut lobes, with silvery hairs on both surfaces. **Status** Locally common in England and Wales only. Grows in disturbed grassland, often on coasts and on roadside verges.

Common Knapweed

Centaurea nigra Asteraceae
Height to 1m

Downy, branched perennial with stiff, grooved stems that are swollen beneath the base of the flowers. **Flowers** Borne in solitary heads, 2–4cm across, with reddish-purple florets; the swollen base is covered in brown bracts (Jun–Sep). **Fruits** Hairless. **Leaves** Narrow; the basal leaves are slightly lobed. **Status** Common throughout in a range of different soil types.

Greater Knapweed

Centaurea scabiosa Asteraceae
Height to 1m

Elegant plant with stiff, grooved stems that are swollen beneath the base of the flower. **Flowers** Borne in solitary heads, 3–5cm across, with reddish-purple disc florets (the outer ones elongated and spreading); the swollen base has brown bracts (Jun–Sep). **Fruits** Hairless. **Leaves** Oblong and divided. **Status** Locally common in S and E England only. Grows in calcareous grassland.

Greater Burdock

Arctium lappa Asteraceae
Height to 1m

Downy wayside plant. **Flowers** with purplish florets and hooked spiny bracts; in egg-shaped heads 2–4cm across.

Leaves Heart-shaped, longer than they are wide, with solid stems. **Fruits** have hooked hairs that catch in animal fur and clothing. **Status** Locally common in a variety of habitats. **Similar species Lesser Burdock** *A. minus* is similar but with smaller flowers and leaves wider than long, with hollow stems.

Sea Wormwood

Seriphidium maritimum
Asteraceae Height to 65cm

Aromatic, much-branched perennial. **Flowers** Borne in egg-shaped, slightly nodding, yellow heads, 1–2mm across; carried in dense, branched, leafy spikes (Aug–Oct). **Fruits** Dry and papery. **Leaves** Deeply divided and downy on both sides. **Status** Locally common in saltmarshes and on sea walls in England and Wales; tolerates occasional inundation.

Seed dispersal

The dispersal of seeds is desirable for two reasons: it minimises the possibility that the next generation will compete with its parents for space or food resources; and it allows the colonisation of new sites. Some plants, including members of the daisy family, have evolved an intriguing number of ways to disperse their seeds. Goat's-beard, for example, has seeds with hairy 'parachutes' that aid wind dispersal. In contrast, burdock seeds have hook-tipped spines that trap whole seed heads in the fur of animals; the seeds are transported until the seed head disintegrates.

ABOVE: Burdock seed head caught in dog's fur
RIGHT: Goat's-beard seed head

Goat's-beard

Tragopogon pratensis Asteraceae
Height to 60cm

Upright plant that is easily overlooked when not in flower. **Flowers** Borne in heads, 3–4cm across, with yellow florets and long, narrow bracts; they close by midday and remain closed on dull mornings (May–Aug). **Fruits** White 'clocks', 8–10cm across. **Leaves** Narrow, grass-like and clasping or sheathing. **Status** Locally common in the S only. Grows in grassy places.

Thistles

Although not the only plants to have spines, thistles have perfected this protective defence against grazing animals. With some species, the coating of spines is so complete that it is impossible to handle them without gloves. Most thistles have colourful flowers that are extremely attractive to pollinating insects.

Chicory

Cichorium intybus Asteraceae
Height to 1m

Branched perennial with stiff, grooved stems. **Flowers** Borne in heads, 3–4cm across, with sky-blue florets; open only in the morning and in sunny weather (Jun–Sep). **Fruits** Dry and papery. **Leaves** Stalked and lobed at the base of the plant; the upper leaves are narrow and clasping. **Status** Locally common in S England only. Found in bare, grassy places, often on calcareous soil.

Saw-wort

Serratula tinctoria Asteraceae
Height to 75cm

Spineless perennial with grooved stems. **Flowers** In open clusters of heads, each 15–20mm long, with pinkish-purple florets and purplish bracts (Jul–Oct). **Fruits** Have unbranched hairs. **Leaves** Variable but edges are always saw-toothed. **Status** Locally common in damp meadows.

Musk Thistle

Carduus nutans Asteraceae
Height to 1m

Biennial with cottony stems, spiny except below the flowers. **Flowers** Solitary, nodding heads, 3–5cm across, with reddish-purple florets and spiny bracts (Jun–Aug). **Fruits** Have unbranched hairs. **Leaves** Lobed and spiny. **Status** Locally common in England and Wales only. Found in dry, grassy places.

Welted Thistle

Carduus crispus Asteraceae
Height to 1.3m

Branched biennial. The stems are cottony, with spiny wings except just below the flower heads. **Flowers** Borne in clusters of egg-shaped heads, 2–3cm long, with reddish-purple florets and woolly green bracts (Jun–Aug). **Fruits** Have unbranched hairs. **Leaves** Oblong, 3-lobed and spiny at the base of the plant. **Status** Common, except in Ireland and N Scotland. Found in grassland.

Slender Thistle

Carduus tenuiflorus Asteraceae
Height to 1m

Upright, greyish biennial. Similar to Welted Thistle but the entire stem is spiny-winged. **Flowers** Borne in egg-shaped heads, 5–10mm across, with pinkish-red florets; carried in dense terminal clusters (Jun–Aug). **Fruits** Have unbranched hairs. **Leaves** Divided, spiny and cottony below. **Status** Locally common in coastal grassland, except in the N.

Carline Thistle

Carlina vulgaris Asteraceae
Height to 60cm

Upright biennial with stiff spines. Dead plants persist through winter. **Flowers** Borne in clusters of golden-brown, rayless heads, 15–40mm across, surrounded by spreading, straw-coloured bracts (Jul–Sep); dead flower heads persist. **Fruits** Have feathery hairs. **Leaves** Oblong with wavy margins and spiny lobes; the lower leaves are downy. **Status** Locally common in dry calcareous grassland.

Spear Thistle

Cirsium vulgare Asteraceae
Height to 1m

Upright biennial. The stems are downy and spiny-winged between the leaves. **Flowers** Borne in solitary heads, or small clusters of heads, each 2–4cm across, with purple florets topping a basal ball that is coated with spiny bracts (Jul–Sep). **Fruits** Have feathery hairs. **Leaves** Deeply lobed and spiny. **Status** Common throughout in grassland and on disturbed ground.

Melancholy Thistle

Cirsium heterophyllum Asteraceae
Height to 1m

Unbranched perennial. The stems are grooved, cottony, spineless and unwinged. **Flowers** Borne in solitary heads, or small clusters of heads, each 3–5cm across, with reddish-purple florets (Jun–Aug). **Fruits** Have feathery hairs. **Leaves** Oval, toothed and barely spiny, hairless above but white-felted below. **Status** Locally common only in the N. Grows in grassland.

Meadow Thistle

Cirsium dissectum Asteraceae
Height to 75cm

Creeping perennial. The stems are unwinged, downy and ridged. **Flowers** Borne in solitary heads, 20–25mm across, with reddish-purple florets and darker bracts (Jun–Jul). **Fruits** Have feathery hairs. **Leaves** Oval, toothed, green and hairy above, and white and cottony below. **Status** Locally common in S and central England, Wales and Ireland. Favours grassy places.

Dwarf Thistle

Cirsium acaule Asteraceae
Height to 5cm

Creeping, flattened perennial with a rosette of extremely spiny leaves. **Flowers** Borne in usually stalkless heads, 3–5cm across, with reddish-purple florets (Jun–Sep). **Fruits** Have feathery hairs. **Leaves** Deeply divided with wavy, spiny lobes. **Status** Locally common in S and E England and S Wales only. Grows in short grassland on calcareous soils.

Creeping Thistle

Cirsium arvense Asteraceae
Height to 1m

Creeping plant with upright, unwinged and mostly spineless stems. **Flowers** Borne in clusters of heads, 10–15mm across, with pinkish-lilac florets and darker bracts (Jun–Sep). **Fruits** Have feathery hairs. **Leaves** Deeply lobed and spiny; the upper leaves are clasping. **Status** Widespread and common throughout in disturbed ground and grassy areas.

Woolly Thistle

Cirsium eriophorum Asteraceae
Height to 1.5m

Upright biennial with furrowed, cottony, unwinged stems. **Flowers** Borne in solitary heads, 6–7cm across, comprising reddish-purple florets topping a ball that is coated with cottony bracts (Jul–Sep). **Fruits** Have feathery hairs. **Leaves** Pinnate, spiny and cottony below. **Status** Local, in calcareous grassland.

Cotton Thistle

Onopordum acanthium
Asteraceae Height to 2.5m

Has strongly winged, spiny stems coated in cottony down. **Flowers** Borne in clusters of heads, 30–35mm across, with reddish-purple florets and a globular base covered in spine-tipped bracts (Jul–Sep). **Fruits** Have unbranched hairs. **Leaves** Have wavy, spiny lobes; cottony on both surfaces. **Status** Possibly introduced; local in the S on disturbed ground and verges.

Marsh Thistle

Cirsium palustre Asteraceae
Height to 1.5m

Upright, branched biennial that is often tinged reddish. The stems have continuous spiny wings. **Flowers** Borne in clusters of heads, 10–15mm across, with dark reddish-purple florets (Jul–Sep). **Fruits** Have feathery hairs. **Leaves** Deeply divided, lobed and spiny. **Status** Widespread and common throughout. Grows in damp grassland.

Yellow daisies

Superficially similar daisy family members (Asteraceae) are a familiar sight in the countryside. Many are easy to identify from leaf and flower shape but some groups – notably dandelions and hawkweeds – baffle even experienced botanists who usually treat them as super species.

Dandelions

Dandelion classification is complex and here it is treated as a super species. Dandelions have a basal rosette of toothed leaves, and flower heads borne on hollow stems that yield a milky latex. **Common Dandelion** *Taraxacum officinale* agg. (height to 35cm) is common in grassy places. Its flower heads are 3–6cm across (Mar–Oct). The fruits have a hairy 'parachute', arranged as a white 'clock'.

Sow-thistles

The hollow stems of sow-thistles yield a milky sap if broken; their fruits have feathery hairs and form a 'clock'. Three species are common. **Smooth Sow-thistle** *Sonchus oleraceus* (**1**) (height to 1m) grows in disturbed ground and its flowers are borne in clusters of heads, each 20–25mm across, with pale yellow florets (May–Oct); the matt leaves are divided, with triangular lobes, spiny margins and clasping, pointed basal auricles. **Perennial Sow-thistle** *S. arvensis* (**2**) (height to 2m) grows in damp, disturbed ground. Its flowers are borne in umbel-like clusters of heads, each 4–5cm across, with yellow florets (Jul–Sep); the leaves are narrow, shiny, dark green above and greyish below, with lobes, soft marginal spines and clasping, rounded basal auricles. **Prickly Sow-thistle** *S. asper* (**3**) (height to 1m) grows on cultivated land. Its flowers are borne in umbel-like clusters of heads, each 20–25mm across, with rich yellow florets (Jun–Oct); the leaves are glossy green above, with wavy, crinkly and sharp-spined margins, and clasping, rounded auricles at the base.

Wild lettuces

Three species of wild lettuce, related to the familiar salad vegetable, grow in our region. Their stems yield a milky sap if broken and their small, compound flowers comprise ray florets. **Wall Lettuce** *Mycelis muralis* (**1**) (height to 1m) is locally common, hairless and with purple-tinged stems; it grows on shady banks, usually on chalky soils. Its flowers are borne in clusters of heads, each 7–10mm across, with five yellow ray florets (Jun–Sep), and the leaves are deeply lobed, the end lobe triangular. **Prickly Lettuce** *Lactuca serriola* (**2**) (height to 1.75m) is a stiff, branched biennial of disturbed ground. Its flowers are borne in heads, 11–13mm across, with yellow florets, and are carried in open, branched inflorescences (Jul–Sep). The fruits are black, and its leaves are grey-green and held stiffly erect, with pointed clasping bases and weak spines on the margins and lower midrib. **Great Lettuce** *L. virosa* (height to 2m) is similar to Prickly Lettuce but is taller. Its heads of flowers, 9–11mm across, have yellow florets and are carried in open, branched inflorescences (Jul–Sep). Its fruits are maroon, and the leaves are dark green and spreading, with rounded, clasping bases.

Nipplewort

Lapsana communis Height to 1m

Much-branched annual with stiff stems that do not produce latex when broken. **Flowers** Borne in heads, 1–2cm across, with yellow florets, and carried in open clusters (Jul–Oct); the buds are nipple-like. **Fruits** Hairless. **Leaves** Oval to lanceolate, toothed and short-stalked. **Status** Common throughout in cultivated and disturbed ground, and often found in gardens.

Mouse-ear Hawkweed

Pilosella officinarum Height to 25cm

Hairy, mat-forming perennial. The stems produce milky latex when broken. **Flowers** Borne in solitary heads, 2–3cm across, with pale yellow florets that have a red stripe below (May–Oct). **Fruits** Have unbranched hairs. **Leaves** Spoon-shaped, green and hairy above, and downy white below; they form a basal rosette. **Status** Common in a wide range of dry, grassy places.

Bristly Oxtongue

Picris echioides Height to 80cm

Bristly annual or biennial. **Flowers** Borne in clusters of heads, each 20–25mm across, with pale yellow florets (Jun–Oct). **Fruits** Have feathery hairs. **Leaves** Oblong (the upper ones are clasping), with swollen-based bristles and pale spots. **Status** Locally common in dry grassland.

Hawkweeds

Experts recognise hundreds of superficially similar species of Hawkweeds (genus *Hieracium*) and their identification is beyond the scope of this book; here they are treated as a single super species. They grow in grassy places and attain a height of 80cm; their stems produce a milky latex when broken. Their flowers are borne in heads, 2–3cm across, with yellow florets, and are carried on hairy stalks in clusters (Jul–Sep). The leaves are ovate and toothed; the basal, rosette leaves are stalked while stem leaves are unstalked.

Hawkweed Oxtongue

Picris hieracioides Asteraceae Height to 70cm

Branched, bristly perennial. **Flowers** Borne in heads 20–25mm across (Jul–Sep). **Fruits** Have feathery hairs. **Leaves** Narrowly oblong, toothed and covered in bristles that are not swollen at the base.

Status Locally common in SE England only. Grows in rough grassland, often in coastal areas.

Cat's-ear

Hypochaeris radicata Height to 50cm

Tufted perennial with hairless stems. **Flowers** Borne in solitary heads, 25–40mm across, with yellow florets that are much longer than the bristly, purple-tipped bracts; the flower stalks branch 1–2 times and are swollen beneath the flower heads (Jun–Sep). **Fruits** Beaked with feathery hairs. **Leaves** Oblong, bristly and wavy-edged; they form a basal rosette. **Status** Common throughout in dry grassland.

Hawkbits

Hawkbits are herbaceous grassland plants whose leaves form a basal rosette. Their flowers are borne mostly in solitary heads; unlike Cat's-ear (see below), these do not have scales between the florets. Three species are common. **Rough Hawkbit** *Leontodon hispidus* (**1**) (height to 35cm) is coated in rough white hairs, favours calcareous soils, and is locally common. Its flower heads are 25–40mm across, with golden-yellow florets (Jun–Oct), and the leaves are wavy-lobed and very hairy.
Autumn Hawkbit *L. autumnalis* (**2**) (height to 25cm) is slightly hairy and found mostly on acid soils. Its flower heads are 15–35mm across, with yellow florets; scale-like bracts are seen on the stem below the head (Jun–Oct); the leaves are oblong and deeply lobed. **Lesser Hawkbit** *L. saxatilis* (**3**) (height to 25cm) is similar to both both Autumn and Rough hawkbits; its stems are hairless above but bristly below. Its solitary flower heads droop in bud and are 20–25mm across, with yellow florets (Jun–Oct); the flower stalk lacks scale-like bracts. Its leaves are deeply lobed and sparsely hairy. It is common except in the N.

Hawk's-beards

Hawk's-beards grow in grassy places and have branched stems and alternate leaves. Three species are common within their ranges. **Smooth Hawk's-beard** *Crepis capillaris* (**1**) (height to 80cm) has flowers borne in clusters of heads, each 15–25mm across, with yellow florets and two rows of bracts, the outer ones spreading (Jun–Oct); its leaves are deeply cut with pointed lobes, the upper leaves with clasping, arrow-shaped bases. **Rough Hawk's-beard** *C. biennis* (**2**) (height to 1.2m) has stems that are roughly hairy and purplish towards the base, and flowers that are borne in heads, 25–30mm across, with yellow florets and two rows of bracts, the outer ones unequal and spreading (Jun–Sep); the leaves are deeply cut, the upper leaves lacking clasping, arrow-shaped bases. **Beaked Hawk's-beard** *C. vesicaria* (**3**) (height to 1.2m) has flowers borne in heads, 15–25mm across with orange-yellow florets, the outer ones striped red (Jun–Oct); the leaves are deeply cut and have a large end lobe.

Lily family

Members of the lily family (Liliaceae) are a varied group of herbaceous plants whose summer growth typically arises from bulbs or tubers. In most species the leaves are narrow with parallel veins, and their flowers comprise six segments that are equal, or nearly so.

Bog Asphodel

Narthecium ossifragum Liliaceae
Height to 20cm

Tufted, hairless perennial. The whole plant turns orange-brown when in fruit. **Flowers** 12–15mm across, yellow and star-like, with woolly orange anthers; carried in spikes (Jun–Aug). **Fruits** Splitting capsules. **Leaves** Narrow, basal and borne in a flat fan. **Status** Widespread in the N and W; much more local in the S and E. Grows in boggy heaths and moors.

Scottish Asphodel

Tofieldia pusilla Liliaceae
Height to 20cm

Delicate, hairless perennial. **Flowers** 2–3mm across and greenish white with 3 blunt lobes; carried in dense, rounded spikes on slender stems (Jun–Aug). **Fruits** Capsules. **Leaves** Iris-like and borne in a flat basal fan. **Status** Widespread in the Scottish Highlands and local in Upper Teesdale. Grows in damp ground and bogs, mainly in mountains.

Meadow Saffron

Colchicum autumnale Liliaceae
Height to 10cm

Bulbous perennial. It flowers in autumn, long after the leaves have withered. Can be confused with Autumn Crocus (*see* p.149). **Flowers** Pinkish purple and 6-lobed, with orange anthers and 6 stamens; borne on slender stalks (Aug–Oct). **Fruits** Capsules. **Leaves** Long and ovate, seen in spring. **Status** Local in meadows.

Star-of-Bethlehem

Ornithogalum angustifolium Liliaceae Height to 25cm

Bulbous perennial. **Flowers** 3–4cm across and star-like (they open only in sunshine), with 6 white petals that have a green stripe on the back; borne in umbel-like clusters (May–Jun). **Fruits** Capsules with 6 ridges. **Leaves** Narrow with a white stripe down the centre. **Status** Possibly native in E England; naturalised elsewhere. Grows in dry grassland.

Garden favourites

The colourful and ornate flowers of ornamental lilies have long been a favourite with gardeners who specialise in creating stunning herbaceous borders. The two species described below in particular have a long heritage in British horticulture and have become naturalised in a few locations.

Martagon Lily

Lilium martagon Liliaceae
Height to 1.8m

Upright, bulbous perennial. **Flowers** 4–5cm across, reddish purple with darker spots and extremely recurved flower lobes; borne in open spikes (Jun–Jul). **Fruits** Capsules. **Leaves** Ovate, glossy dark green and whorled. **Status** A familiar garden plant, naturalised and possibly native in a few places in the S. Grows in woodland and scrub.

Pyrenean Lily

Lilium pyrenaicum Liliaceae
Height to 1.5m

Upright, bulbous perennial. **Flowers** 4–5cm across, yellow with dark spots, and with recurved flower lobes; borne in spikes (May–Jul). **Fruits** Capsules. **Leaves** Narrow and alternate up the stem. **Status** A familiar garden plant that is naturalised in a few locations in SW England, Wales and W Ireland. It grows on banks and in hedgerows.

Spiked Star-of-Bethlehem

Ornithogalum pyrenaicum
Liliaceae Height to 80cm

 Upright perennial that is also known as Bath Asparagus. **Flowers** 2cm across and greenish white; borne in tall, drooping-tipped spikes (May-Jul). **Fruits** Capsules. **Leaves** Grey-green, narrow and basal; they soon wither. **Status** Very local, and found mainly from Bath eastwards along the M4 corridor. Grows in scrub and open woodland.

Spring Squill

Scilla verna Liliaceae
Height to 5cm

 Resilient, hairless perennial. **Flowers** 10-15mm across, bell-shaped and lilac-blue with a purplish bract; borne in upright terminal clusters (Apr-Jun). **Fruits** Capsules. **Leaves** Wiry, curly, basal and 4-6 in number; they appear in early spring, before the flowers. **Status** Locally common in short coastal grassland in W Britain and E Ireland.

Spring woodland star attractions

Spring woodlands are renowned for the mass displays put on by wildflowers such as Bluebells and Ramsons, and these are discussed more fully on pp.144-5. However, a careful search among the massed ranks of commoner species may reveal a couple of floral gems. Their discovery will indicate that you are probably in a special site with an ancient ancestry.

Herb-Paris

Paris quadrifolia Liliaceae
Height to 35cm

 Unusual and distinctive perennial. **Flowers** Comprise narrow greenish-yellow petals and sepals, topped by a dark purplish ovary and yellow stamens; solitary and terminal (May-Jun). **Fruits** Black berries. **Leaves** Broad, oval and borne in whorls of 4. **Status** Widespread but always extremely local. Grows in damp woodland, and usually found on calcareous soils.

Yellow Star-of-Bethlehem

Gagea lutea Liliaceae
Height to 15cm

 Delicate perennial. Easily overlooked when not in flower. **Flowers** 2cm across, yellow and star-like; borne in umbel-like clusters of 1-7 flowers (Mar-May). **Fruits** 3-sided capsules. **Leaves** Comprise a single narrow basal leaf with a hooded tip and 3 ridged veins. **Status** Local, least uncommon in central England. Usually found on calcareous or heavy soils.

Wild Tulip

Tulipa sylvestris Liliaceae
Height to 40cm

 Bulbous perennial. **Flowers** 3-4cm across when fully open, fragrant and with yellow petals that are sometimes tinged green below; solitary, borne on slender stems (May-Jun). **Fruits** Capsules. **Leaves** Grey-green, narrow and up to 25cm long. **Status** Introduced and now widely, but locally, naturalised, mainly in S England. Grows in meadows and grassy woodland rides.

Autumn Squill

Scilla autumnalis Liliaceae
Height to 7cm

 Hairless perennial. **Flowers** 10-15mm across, bell-shaped and bluish purple; borne in compact terminal clusters on a slender stalk, the flowers lacking an accompanying bract (Jul-Sep). **Fruits** Capsules. **Leaves** Wiry and basal, appearing in autumn. **Status** Very locally common in short coastal grassland; restricted to SW England.

Herb-Paris

Woodland lilies

Ever since people first colonised Britain, its woodlands have been exploited for timber and managed to varying degrees to maximise their productivity. Coppicing – the cutting back of a tree or shrub to the base, from which fresh shoots arise – has been, and still is, a widespread form of management. The practice suits a select band of woodland flowers and Bluebell, Ramsons, Wild Daffodil and Snowdrop (all lily family members) can often carpet the ground. But why do they thrive under this form of management? All four produce leaves and flowers early in the year, before the deciduous trees and shrubs under which they grow are in leaf themselves. As a result, they can take advantage of high light levels with little competition from other species. By late spring the woodland canopy has closed in so much that most herbaceous plants on the woodland floor cannot photosynthesize properly. But by then the lily family members have already finished growing for the season and survive until the following spring as dormant, underground bulbs.

Bluebells under Hazel coppice and Oak.

Woodland management – past and present

Initially, woodland was probably exploited on a rather ad hoc basis. However, because trees are valuable and uncontrolled exploitation is damaging, most significant areas of woodland have been managed at least since medieval times as a sustainable resource. Although these woodlands are heavily influenced by man, the resulting habitat continuity and stability has had an unintentional, but markedly beneficial, impact on many woodland plants and animals. Coppicing - a common form of woodland management - involves cutting back the tree or shrub to a stump, or stool. From this base arise fresh shoots that typically are straight and so are ideal for use as poles. Once the shoots have been cut, the coppicing process can be repeated almost indefinitely if performed sensitively, the stool increasing in girth over time. Hazel is often used for coppicing, its poles being harvested every seven to ten years. It is an ideal companion shrub to grow under oak. Larger trees such as Alder and Ash are also frequently coppiced; these produce much thicker poles that are harvested at more extended intervals than with Hazel, and these are often used for firewood. The sensitive coppicing of Ash can actually prolong the life of the tree considerably, so much so in fact, that Ash stools occasionally reach 7m in diameter and 800 years or more in age.

Ancient coppiced Ash stool with Bluebells and Early Purple Orchids.

Bluebell

Hyacinthoides non-scripta
Liliaceae Height to 50cm

Hairless, bulbous perennial. **Flowers** Bell-shaped, with 6 recurved lobes at the mouth, and bluish purple (very occasionally pink or white); borne in 1-sided drooping-tipped spikes (Apr-Jun). **Fruits** Capsules. **Leaves** Long, 15mm wide, glossy green and all basal. **Status** Widespread and locally abundant. Grows in woodland and also on coastal cliffs.

Ramsons

Allium ursinum Liliaceae
Height to 35cm

Bulbous perennial that smells strongly of garlic. **Flowers** 15-20cm across, white and bell-shaped; borne in spherical terminal clusters on slender, 3-sided, leafless stalks (Apr-May). **Fruits** Capsules. **Leaves** Ovate, up to 7cm wide and 25cm long, and all basal. **Status** Widespread and locally abundant. Grows in damp woodland, mainly on calcareous soils.

Wild Daffodil

Narcissus pseudonarcissus
Liliaceae Height to 50cm

Bulbous woodland perennial. Often flourishes immediately after coppicing. **Flowers** 5-6cm across, with 6 pale yellow outer segments and a deep yellow trumpet (Mar-Apr). **Fruits** Capsules. **Leaves** Grey-green, narrow and all basal. **Status** A locally common native in parts of England and Wales; naturalised elsewhere. Grows in open woods and meadows.

Snowdrop

Galanthus nivalis Liliaceae
Height to 25cm

Familiar spring perennial. **Flowers** 15-25cm long and nodding, the 3 outer segments pure white and the inner 3 white with a green patch; solitary and nodding (Jan-Mar). **Fruits** Capsules. **Leaves** Grey-green, narrow and all basal. **Status** Possibly native in S Britain but widely naturalised. Grows in damp woodland and is locally abundant.

Threats from cultivation

Spanish Bluebell *Hyacinthoides hispanica* (height to 40cm) is a popular garden plant that, unfortunately, is often dumped in the countryside and spreads into woodlands from neighbouring gardens. Although charming to look at, it is a problem in conservation terms because, being vigorous, it often crowds out native Bluebells and, furthermore, it hybridises with Bluebells to produce an even more robust plant. It is similar in appearance to Bluebell, but the lobes of the bell-shaped flowers are not recurved at the mouth and, unlike Bluebell, the flower spikes are not 1-sided (May-Jun). The long leaves are up to 35mm wide. The hybrid *H. non-scripta × hispanicus* (right) is also popular in gardens and, in many areas, is more widely naturalised than Spanish Bluebell.

Garlic, onions & lilies

One of the more pleasurable aspects of studying wildflowers is the chance to appreciate the subtly attractive scents with which many of their blooms are perfumed. However, many members of the lily family (Liliaceae) that grow wild in countryside have pungent smells – akin to the aroma of culinary onion and garlic – associated with their foliage. While these odours are not unpleasant, they are certainly not something that most people would describe as alluring!

Babington's Leek

Allium ampeloprasum var. *babingtonii* Liliaceae
Height to 2m

 Showy, bulbous perennial. **Flowers** 6–8mm long and purplish; borne in spherical heads, to 9cm across (Jun–Aug); bulbils may be present. **Fruits** Capsules. **Leaves** Flat, narrow, to 50cm long, waxy and with finely toothed margins. **Status** Rare, in coastal grassland in the SW. **Similar species Wild Leek** *A. ampeloprasum* has more compact flower heads and favours similar habitats.

Sand Leek

Allium scorodoprasum Liliaceae
Height to 1m

 Upright, bulbous perennial. **Flowers** Ovoid, stalked and purplish, the stamens not projecting; borne in rounded umbels, 3–4cm across, with purple bulbils and 2 papery bracts (Jul–Aug). **Fruits** Capsules. **Leaves** Narrow, keeled and rough-edged. **Status** Local, mainly in N England and S Scotland. Grows in dry grassland on sandy soils.

Three-cornered Garlic

Allium triquetrum Liliaceae
Height to 45cm

 Bulbous perennial that smells strongly of garlic when bruised. **Flowers** 2cm long, bell-shaped and white, with narrow green stripes; borne in drooping umbels on 3-sided stems (Mar–Jun). **Fruits** Capsules. **Leaves** Narrow and keeled; 3 per plant. **Status** An introduced garden plant, naturalised locally in the SW. Grows in hedges and on disturbed ground.

Rosy Garlic

Allium roseum Liliaceae
Height to 70cm

 Elegant and showy bulbous perennial. **Flowers** Pink, stalked and borne in heads 4–5cm across (May–Jun); the stamens do not protrude and bulbils are sometimes present. **Fruits** Capsules. **Leaves** Greyish green and flat. **Status** A popular introduced garden plant that is naturalised locally in the S. Grows in dry grassland, often on calcareous soils.

Chives

Allium schoenoprasum Liliaceae
Height to 40cm

 Tufted, strong-smelling bulbous perennial. **Flowers** Purplish and borne in heads, 2–4cm across, comprising 10–30 flowers and 2 papery bracts; the stamens do not project (Jun–Sep). **Fruits** Capsules. **Leaves** Grey-green, hollow and cylindrical. **Status** Widely cultivated, and local as a native plant, mainly in the W. Found in damp, grassy places on limestone rocks.

Wild Onion

Allium vineale Liliaceae
Height to 60cm

 Bulbous perennial. **Flowers** Pink or white, long-stalked and borne in umbels along with greenish-red bulbils and a papery bract; the proportion of flowers to bulbils varies considerably (Jun–Jul). **Fruits** Capsules. **Leaves** Grey-green, hollow and semicircular in cross-section. **Status** Common in the S. Grows in dry grassland and on roadside verges.

Keeled Garlic

Allium carinatum Liliaceae
Height to 80cm

Bulbous perennial. **Flowers** Long-stalked and pink, with protruding stamens; borne in open heads, with bulbils and 2 long bracts (Jul–Aug). **Fruits** Capsules. **Leaves** Slender, and semicircular or rounded in cross-section. **Status** Naturalised locally in dry grassland. **Similar species Field Garlic** *A. oleraceum* has white flowers and the stamens do not protrude. It is a rare native.

Angular Solomon's-seal

Polygonatum odoratum Liliaceae
Height to 50cm

Creeping perennial with angled, arching stems. **Flowers** Bell-shaped, white and not waisted; borne in clusters of 1–2, arising from leaf axils (May–Jun). **Fruits** Blackish berries. **Leaves** Ovate and alternate. **Status** Local, mainly in N and NW England. Grows on rocky ground, mainly on limestone and sometimes on limestone pavements.

Lily-of-the-valley

Convallaria majalis Liliaceae
Height to 20cm

Creeping, patch-forming perennial. **Flowers** Bell-shaped and white; borne in 1-sided spikes, the flowers stalked and nodding (May–Jun). **Fruits** Red berries. **Leaves** Oval and basal; 2 or 3 per plant. **Status** Occurs locally in England and Wales as a native plant; sometimes naturalised as a garden escape elsewhere. Grows in dry woodland, often on calcareous soils.

May Lily

Maianthemum bifolium Liliaceae
Height to 20cm

Attractive, creeping perennial. The upright stalks each bear a pair of leaves and a flower spike. **Flowers** 2–5mm across, white and consist of 4 parts; borne in spikes, 3–4cm long (May–Jun). **Fruits** Red berries (rarely produced). **Leaves** Heart-shaped and shiny, the lower one long-stalked. **Status** Local, in N England only. Found in mature woodlands, often on acid soils.

Common Solomon's-seal

Polygonatum multiflorum
Liliaceae Height to 60cm

Creeping perennial with rounded, arching stems. **Flowers** White, bell-shaped and waisted in the middle; borne in clusters of 1–3, arising from leaf axils (May–Jun). **Fruits** Bluish-black berries. **Leaves** Ovate and alternate. **Status** Locally common in S England only; scarce or absent elsewhere. Grows in dry woodland, often on calcareous soils.

Snowflakes

Snowflakes are relatives of the Snowdrop, and two members of the group (genus *Leucojum*) are found in Britain. They are attractive bulbous perennials that are grown in gardens and are naturalised on occasions; both species also occur as rare natives growing in damp woodland. **Spring Snowflake** *Leucojum vernum* (**1**) has solitary flowers that are 20–25mm long, bell-shaped and mainly white, but with green marks on the six pointed segments (Apr–Jun); its leaves are bright green and strap-like. **Summer Snowflake** *L. aestivum* (**2**) is similar but has slightly smaller flowers (15–20mm long) that are borne in umbels of two to five flowers with a greenish spathe (May–Jun).

Lilies, irises & arums

Gardeners prize plants of the lily (Liliaceae), iris (Iridaceae) and arum (Araceae) families for the bold statements their striking foliage and flowers make in the herbaceous border. Several popularly cultivated species, alien in origin, have become naturalised in the countryside at large, and even horticultural favourites that remain confined to the garden have native relatives growing wild.

Asparagus

Asparagus officinalis Liliaceae
Height to 1.5m

 Branched, hairless perennial. **Flowers** 4–6mm long, greenish and bell-shaped; borne in leaf axils, on separate-sex plants (Jun–Sep). **Fruits** Red berries. **Leaves** Reduced to tiny bracts; the feathery 'leaves' are, in fact, slender stems. **Status** Garden Asparagus (ssp. *officinalis*) is locally naturalised on free-draining soils, while the prostrate Wild Asparagus (ssp. *prostratus*) grows on sea cliffs in the SW.

fruits

Butcher's-broom

Ruscus aculeatus Liliaceae
Height to 1m

Atypical evergreen member of the lily family. **Flowers** Tiny and solitary; borne on the upper surface of leaf-like structures (Jan–Apr). **Fruits** Red berries. **Leaves** Minute; the plant's oval, spine-tipped leaf-like structures are flattened branches. **Status** Locally common native in the S; naturalised elsewhere. Grows in shady woods, often on chalky soils.

Wild Gladiolus

Gladiolus illyricus Iridaceae
Height to 90cm

 Upright, hairless perennial that is easily overlooked when not in bloom. **Flowers** Striking and pinkish purple, 3–4cm long; borne in tall spikes (Jun–Jul). **Fruits** Capsules. **Leaves** Slender and extremely grass-like. **Status** Confined to the New Forest in Hampshire, where it grows among Bracken on heaths and in the margins of open woodland.

Yellow Iris

Iris pseudacorus Iridaceae
Height to 1m

 Familiar and robust perennial. **Flowers** 8–10cm across and bright yellow with faint purplish veins; borne in clusters of 2–3 flowers (May–Aug). **Fruits** Oblong and 3-sided. **Leaves** Grey-green, sword-shaped and often wrinkled. **Status** Widespread and common throughout, found growing in pond margins and marshes, and on river banks.

The decline of the Fritillary

The **Fritillary** *Fritillaria meleagris* (height to 30cm) is a perennial member of the lily family (Liliaceae) that grows only in undisturbed water meadows. In the days before land drainage and intensive industrial agriculture changed the face of much of the British countryside, this habitat and its floral emblem were widespread in central and S England. However, the species' range and abundance have declined catastrophically and are symptomatic of the fate of undisturbed grassland generally. Today, although it is still locally abundant in a handful of sites, the Fritillary is generally considered to be a rare and vulnerable plant. Its charming flowers are 3–5cm long, nodding, bell-shaped and usually pinkish purple with dark chequerboard markings (sometimes pure white), and its leaves are grey-green, narrow and grass-like. Probably the best place to see Fritillaries is at North Meadow in Cricklade, Wiltshire.

white form

Stinking Iris

Iris foetidissima Iridaceae
Height to 60cm

 Tufted perennial. **Flowers** 7-8cm across, purplish and veined (May-Jul). **Fruits** Green, oblong and 3-sided, splitting to reveal orange seeds. **Leaves** Dark green and sword-shaped, with a rich, unpleasant smell. **Status** Locally common in S England and S Wales only. Grows in scrub and woodlands, mostly on calcareous soils.

fruits

Autumn Crocus

Crocus nudiflorus Iridaceae
Height to 20cm

 Similar to Spring Crocus but it flowers in autumn; also beware confusion with Meadow Saffron (*see* p.142). **Flowers** 3-4cm across (when fully open), violet and veined, with 3 stamens (Meadow Saffron has 6) (Sep-Oct). **Fruits** Capsules. **Leaves** Narrow with a white midrib stripe. **Status** Locally naturalised in grassland. **Similar species Bieberstein's Crocus** *C. speciosus* has darker-veined flowers.

Bieberstein's
Crocus

Black Bryony

Tamus communis Dioscoreaceae
Height to 3m

 Twining perennial. Similar to the unrelated White Bryony (*see* p.92), but note different leaf shape and lack of tendrils. **Flowers** Tiny, yellowish green and 6-petalled; borne on separate-sex plants (May-Aug). **Fruits** Red berries. **Leaves** Heart-shaped, glossy and net-veined. **Status** Widespread in England and Wales. Grows in hedgerows and scrub.

fruits

Italian Lords-and-ladies

Arum italicum Araceae
Height to 50cm

 Robust perennial. **Flowers** Comprise a yellowish spathe (not purple-margined) that is cowl-shaped with an overhanging tip, and a yellow spadix (May-Jun). **Fruits** Red berries, borne in a spike. **Leaves** Arrowhead-shaped but blunter than those of Lords-and-ladies; some appear in autumn. **Status** Common in S England only, in shady places; seen elsewhere as a garden escape.

Spring Crocus

Crocus vernus Iridaceae
Height to 20cm

 Perennial that resembles many cultivated forms of garden crocuses. **Flowers** 3-4cm across (when open fully), and pinkish purple with pale stripes (Mar). **Fruits** Capsules. **Leaves** Long, narrow and linear, with a white midrib stripe on the upper side. **Status** A garden escape and deliberately planted; now naturalised locally in undisturbed grassland.

Montbretia

Crocosmia × crocosmiiflora
Iridaceae Height to 70cm

 Showy, spreading and patch-forming perennial. **Flowers** Reddish orange and borne in 1-sided spikes (Jul-Aug). **Fruits** Capsules. **Leaves** Narrow, linear, flat and superficially iris-like. **Status** A cultivated hybrid that has become naturalised as a garden escape, mainly in the S. Grows in hedgerows, and on roadside verges and disturbed ground.

Lords-and-ladies

Arum maculatum Araceae
Height to 50cm

 Distinctive perennial. **Flowers** Comprise a pale green, purple-margined, cowl-shaped spathe that partly shrouds a club-shaped, purplish-brown spadix; borne on slender stalks (Apr-May). **Fruits** Red berries, borne in a spike. **Leaves** Arrowhead-shaped, stalked and sometimes dark-spotted; they appear in spring. **Status** Commonest in the S. Grows in woods and hedges.

fruits

Orchids

Orchids (family Orchidaceae) occupy a unique place in British botany: countless books have been devoted to these intriguing plants and undoubtedly they have a bigger fan club than any other group of wildflowers. Their flowers are undeniably fascinating to look at, but at least part of their allure comes from the rarity value associated with some species. Generally, they are associated with pristine, undisturbed habitats that are a pleasure to visit in their own right.

Insect mimics

Members of the orchid genus *Ophrys* have intriguing flowers that bear a fanciful resemblance to small furry bees or wasps and emit a convincing copy of the female insect's pheromone. Male insects are often so convinced that they try to mate with the flowers, inadvertently picking up a ball of pollen in the process; this is then transferred to the next flower they visit. Two *Ophrys* species are fairly common and widespread in the S. **Bee Orchid** *O. apifera* (**1**) (height to 30cm) grows in dry grassland, mainly on calcareous soils. Its flowers are vaguely bumblebee-like, with pink sepals and green upper petals; the lower petal is 12mm across, expanded, furry and maroon, with variable pale yellow markings. **Fly Orchid** *O. insectifera* (**2**) (height to 40cm) grows in dry grassland and scrub on calcareous soils. Its flowers resemble tiny wasps, with greenish sepals and thin, brown, antennae-like upper petals; the lower petal is 6–7mm across, velvety maroon with a metallic blue patch, and is elongated with two side lobes. Two other *Ophrys* species are found in Britain: Early Spider-orchid and Late Spider-orchid, both of which are rare chalk downland specialists (*see* p.73).

male Digger Wasp attempting to mate with Fly Orchid flower

Early Purple Orchid
Orchis mascula Height to 40cm

 Attractive perennial. **Flowers** Pinkish purple, with a 3-lobed lower lip (8–12mm long) and a long spur; borne in tall spikes (Apr–Jun). **Fruits** Egg-shaped. **Leaves** Glossy and dark green with dark spots; they appear initially as a rosette, from January onwards. **Status** Locally common and widespread in woodland, scrub and grassland areas, mainly on neutral or calcareous soil.

Lady Orchid
Orchis purpurea Height to 75cm

 Impressive orchid. **Flowers** Have a dark red hood and a pale pink, red-spotted lip; borne in a cylindrical spike, 10–15cm tall, with flowers opening from the bottom up (Apr–Jun). **Fruits** Egg-shaped. **Leaves** Broad and oval; they form a basal rosette and loosely sheathe the stem. **Status** Confined to S England; locally common in Kent only, in woodland on chalk soil.

Man Orchid
Orchis anthropophorum Height to 30cm

 Distinctive orchid. **Flowers** Fancifully man-like, with a green hood, an elongated 4-lobed lip (12–15mm long) and a spur; borne in tall, dense spikes (May–Jun). **Fruits** Egg-shaped. **Leaves** Oval and fresh green; they form a basal rosette and sheathe the lower stem. **Status** Local, with isolated colonies on chalk grassland and scrub in SE England.

Burnt Orchid
Neotinea ustulata Height to 15cm

 Charming orchid. **Flowers** Dark maroon in bud, but they have a red-spotted white lip and a red hood when open; borne in compact cylindrical spikes reminiscent of intact cigar ash (May–Jul). **Fruits** Egg-shaped. **Leaves** Dull green; they form a basal rosette and also sheathe the stem. **Status** Very locally common in S England only. Grows on chalk downland.

Pyramidal Orchid

Anacamptis pyramidalis
Height to 30cm

Distinctive orchid. **Flowers** Deep pink with a 3-lobed lip and a long spur; borne in dense conical or domed heads (Jun–Aug). **Fruits** Egg-shaped. **Leaves** Grey-green and narrow, partially sheathing the stem. **Status** Locally common on dry chalk grassland and stabilised dunes.

Bird's-nest Orchid

Neottia nidus-avis Height to 35cm

Bizarre-looking orchid. It lacks chlorophyll and obtains its food from soil fungi. **Flowers** Brownish, with a hood and a 2-lobed lip that is 10mm long; borne in spikes (May–Jul). **Fruits** Egg-shaped. **Leaves** Reduced to brownish scale-like structures. **Status** Widespread except in N Scotland, but always local. Grows in undisturbed woodland, usually under Beech trees.

Lesser Twayblade

Neottia cordata Height to 20cm

Charming little orchid that is easily overlooked because of its small size. **Flowers** 2–3mm long and reddish, with a hood and forked lip; borne in a loose spike on a reddish stem (Jun–Aug). **Fruits** Egg-shaped. **Leaves** Comprise a pair of oval basal leaves, to 4cm long. **Status** Widespread but local, mainly in the N and NW. Grows on moorland and in conifer forests.

Common Twayblade

Neottia ovata Height to 50cm

Distinctive orchid. **Flowers** Yellowish green, with a hood and deeply forked lower lip (10–15mm long); borne in loose spikes (May–Jul). **Fruits** Egg-shaped. **Leaves** Comprise a pair of broad, oval basal leaves, up to 16cm long, that appear before the flower stem emerges. **Status** Widespread and common in woodlands and grassland, on a range of soil types.

Green-winged Orchid

Anacamptis morio Height to 40cm

Attractive orchid. **Flowers** Vary from plant to plant, ranging from pinkish purple to almost white; the upper petals are marked with dark veins and often suffused green, and the lip has a red-dotted pale central patch; borne in spikes (Apr–Jun). **Fruits** Egg-shaped. **Leaves** Glossy green. **Status** Locally common in dry grassland, mainly in the S.

The rarest of the rare

Orchid-hunting can become a compulsive habit and, once bitten by the bug, enthusiasts often try to see all the British species. Two orchids are a real challenge to find because they are so rare and elusive, and their whereabouts are often shrouded in secrecy. The Holy Grail for orchid enthusiasts is the enigmatic **Ghost Orchid** *Epipogium aphyllum* (**1**) (height to 15cm), which at the time of writing has not been seen since the 1980s. It has just a few sites to its name (shady Beech woodlands mainly in the Chilterns), it appears only infrequently and it seldom flowers in precisely the same place in succession. To make matters worse, Ghost Orchids can appear at any time from June to September and seldom last more than a few days, often being consumed by slugs. The flowers are pinkish yellow and have a porcelain-like texture. Count yourself extremely lucky if you see one. The **Lady's-slipper Orchid** *Cypripedium calceolus* (**2**) (height to 50cm) is at the other end of the spectrum in terms of showiness, having magnificent flowers that comprise maroon outer segments and an inflated yellow lip, 4–5cm across (May–Jun). It grows on limestone soils in N England, where its few known sites are heavily protected. Occasionally, it is discovered at new sites in N England, but the native origins of these plants is doubtful.

1

2

Orchids

Frog Orchid
Dactylorhiza viride Orchidaceae
Height to 20cm

Compact orchid. **Flowers** Fancifully frog-like; the sepals and upper petals form a greenish hood and the lip is 6–8mm long and yellowish brown; borne in open spikes (Jun–Aug). **Fruits** Egg-shaped. **Leaves** Broad and oval; they form a basal rosette and partially sheathe the lower stem. **Status** Locally common on chalk downs in the S, and on N upland pastures.

Common Spotted-orchid
Dactylorhiza fuchsii Orchidaceae
Height to 60cm

Robust orchid. **Flowers** Vary in colour from pale pink to pinkish purple; darker streaks and spots adorn the lower lip, which has 3 even-sized lobes; borne in spikes (May–Aug). **Fruits** Egg-shaped. **Leaves** Glossy green and dark-spotted. **Status** Locally common in grassland, mostly on calcareous or neutral soils.

Heath Spotted-orchid
Dactylorhiza maculata
Orchidaceae Height to 50cm

Similar to Common Spotted-orchid but it grows in different habitats. **Flowers** Pale, sometimes almost white, but with darker streaks and spots; the central lobe of the lower lip is smaller than the outer 2; borne in open spikes (May–Aug). **Fruits** Egg-shaped. **Leaves** Narrow and dark-spotted. **Status** Locally common throughout in damp, acid soils on heaths and moors.

Early Marsh-orchid
Dactylorhiza incarnata
Orchidaceae Height to 60cm

Subtly attractive orchid. **Flowers** Usually flesh-pink, but also creamy white or reddish purple in certain subspecies; the 3-lobed lip is folded back along the mid-line; borne in spikes (May–Jun). **Fruits** Egg-shaped. **Leaves** Yellowish-green, unmarked and narrow. **Status** Local in damp meadows, usually on calcareous soils.

Southern Marsh-orchid
Dactylorhiza praetermissa
Orchidaceae Height to 70cm

Robust orchid. **Flowers** Pinkish purple with a broad 3-lobed lip, 11–14mm long, the lobes shallow and blunt; borne in tall spikes (May–Jun). **Fruits** Egg-shaped. **Leaves** Narrowly oval, glossy dark green and usually unmarked. **Status** Common only in the S. Grows in water meadows, fens and wet dune slacks.

Did you know?

Spotted-orchids and Marsh-orchids, which belong to the genus *Dactylorhiza*, are a complex group of superficially rather similar species that are the subject of much debate among botanists concerned with taxonomy. The species described here are reasonably easy to identify: they are often separated geographically from one another, or favour distinctly different growing habitats. Failing that, flower structure is a particularly useful feature to examine.

Northern Marsh-orchid

Dactylorhiza purpurella Orchidaceae Height to 60cm

Northern counterpart of Southern Marsh-orchid. **Flowers** Reddish purple, the lip broadly diamond-shaped and with indistinct lobes and dark streaks; borne in spikes (Jun–Jul). **Fruits** Egg-shaped. **Leaves** Green and narrow, including those at the base. **Status** Widespread in N England, N Wales and N Ireland. Grows in damp meadows.

Irish Marsh-orchid

Dactylorhiza majalis Orchidaceae Height to 60cm

Robust orchid. **Flowers** Magenta to pinkish purple, with a broad lip, the side lobes of which are broader than the central one; borne in spikes (May–Jul). **Fruits** Egg-shaped. **Leaves** Narrowly oval and usually dark-spotted; the basal leaves are broadest and the stem leaves are sheathing. **Status** Local and mainly coastal in Ireland. Grows in damp meadows.

Narrow-leaved Marsh-orchid

Dactylorhiza traunsteineri Orchidaceae Height to 50cm

Delicate orchid. **Flowers** Pinkish purple, with a well-marked 3-lobed lip whose central lobe is longer than the side ones; borne in short, open, few-flowered spikes (May–Jun). **Fruits** Egg-shaped. **Leaves** Narrowly oval, keeled and usually unspotted. **Status** Very local in England, Wales and Ireland. Grows in fens and marshes with calcereous soils.

Small-white Orchid

Pseudorchis albida Orchidaceae Height to 30cm

Slender, elegant orchid. **Flowers** 2–3mm across and greenish white; borne in dense cylindrical spikes (May–Jul). **Fruits** Egg-shaped. **Leaves** Narrowly oval and pointed. **Status** Very locally common in Scotland; increasingly scarce, or absent, further S. Grows in grassland and on rock ledges, in mountains and upland regions.

Greater Butterfly-orchid

Platanthera chlorantha Orchidaceae Height to 50cm

Elegant orchid. **Flowers** Greenish white with a long, narrow lip, a long spur (15–25mm) and pollen sacs that form an inverted 'V'; borne in open spikes (Jun–Jul). **Fruits** Egg-shaped. **Leaves** Comprise a single basal pair and a few smaller stem leaves. **Status** Local in woodland and grassland on calcareous soils.

Lesser Butterfly-orchid

Platanthera bifolia Orchidaceae Height to 40cm

Attractive orchid. **Flowers** Greenish white with a long, narrow lip, a long spur (25–30mm) and parallel pollen sacs; borne in open spikes (May–Jul). **Fruits** Egg-shaped. **Leaves** Comprise a pair of large, oval, basal leaves and much smaller, scale-like stem leaves. **Status** Widespread but local in undisturbed grassland, moors and heaths.

Coralroot Orchid

Corallorrhiza trifida Orchidaceae Height to 25cm

Intriguing yellowish-green plant that lacks leaves and feeds on decaying plant matter. **Flowers** Greenish white, the lip often tinged and streaked with red; borne in open spikes (Jun–Jul). **Fruits** Egg-shaped. **Leaves** Absent. **Status** Very locally common only in N England and Scotland. Grows in damp, shady woodland and in dune slacks.

Fen Orchid

Liparis loeselii Orchidaceae Height to 20cm

Unusual and distinctive orchid. **Flowers** Yellowish green with narrow, spreading lobes; borne in spikes (Jun–Jul). **Fruits** Egg-shaped. **Leaves** Form a cup-like arrangement comprising the basal pair of leaves. **Status** Extremely local and generally rare. Grows in a few fen locations in Norfolk, with additional coastal dune sites in S Wales and N Devon.

Helleborines & lady's-tresses

Helleborines are an unusual group of orchids with bell- or trumpet-shaped flowers. Lady's-tresses are dainty by comparison, with flowers that are arranged spirally.

White Helleborine

Cephalanthera damasonium Orchidaceae Height to 50cm

 Attractive orchid. **Flowers** 15-20mm long, creamy white, bell-shaped, only partially open and with a leafy bract; borne in tall spikes (May–Jul). **Fruits** Egg-shaped. **Leaves** Broad and oval at the base, but becoming smaller up the stem. **Status** Locally common only in S England, in woods on calcareous soils, often under beech.

Sword-leaved Helleborine

Cephalanthera longifolia Orchidaceae Height to 50cm

 Elegant orchid, similar to the White Helleborine. **Flowers** 20mm long, pure white, bell-shaped and open more fully than those of White Helleborine; borne in tall spikes (May–Jun). **Fruits** Egg-shaped. **Leaves** Long and narrow (narrower than those of the White Helleborine), largest at the base. **Status** Local and scarce, mainly in SE England in scattered colonies. Grows in woodlands and scrub only on calcareous soils.

Red Helleborine

Cephalanthera rubra Orchidaceae Height to 50cm

 Attractive orchid whose beauty is often hard to appreciate since it grows in deep shade. **Flowers** Reddish pink, bell-shaped and open wide at the mouth; borne in open spikes (Jun–Jul). **Fruits** Egg-shaped. **Leaves** Long and narrow. **Status** Rare and local in S England only. Grows in woods on calcareous soils, usually under Beech.

Marsh Helleborine

Epipactis palustris Orchidaceae Height to 50cm

 Attractive wetland orchid. **Flowers** Have reddish-green sepals, whitish upper petals marked with red, and a frilly, whitish lip with red streaks; borne in open spikes (Jul–Aug). **Fruits** Pear-shaped. **Leaves** Broad, oval, basal leaves, narrower up the stem. **Status** Locally common in marshes, fens and dune slacks.

Dune Helleborine

Epipactis dunensis Orchidaceae Height to 65cm

 Upright perennial. **Flowers** Comprise narrow, greenish-white sepals and upper petals, and a broadly heart-shaped, greenish-white lip that is sometimes tinged pink towards the centre and curves under at the tip; borne in open spikes, the flowers seldom opening fully (Jun–Jul). **Fruits** Pear-shaped. **Leaves** Oval. **Status** Rare, restricted to dunes in Anglesey and N England.

Dark-red Helleborine

Epipactis atrorubens Orchidaceae Height to 60cm

 Upright, downy perennial. **Flowers** Dark reddish purple with broad sepals and upper petals, and a broad, heart-shaped lip; borne in spikes (Jun–Jul). **Fruits** Pear-shaped and downy. **Leaves** Oval and tinged purple. **Status** Very local and restricted to N England, N Wales, NW Scotland and W Ireland. Grows on limestone soils.

Autumn Lady's-tresses

Spiranthes spiralis Orchidaceae Height to 15cm

 Charming little orchid. **Flowers** White and downy; borne in a distinct spiral up the grey-green stem (Aug–Sep). **Fruits** Egg-shaped and downy. **Leaves** Appear as a basal rosette of oval leaves that wither before the flower stem appears. **Status** Locally common in S England, Wales and SW Ireland. Found in short, dry grassland inland and on coastal turf and dunes.

Helleborines of deep shade

Most plants require strong sunlight in order to photosynthesise and will become sickly if shaded by overhanging branches. However, four superficially similar species of helleborines find deep shade just to their liking; they can be separated by studying their flower structure. The most common and widespread is **Broad-leaved Helleborine** *Epipactis helleborine* (**1**) (height to 75cm), a clump-forming plant with downy stems. Its flowers comprise broad, greenish sepals tinged purple around the margins, broad upper petals that are strongly tinged purple, and a heart-shaped purplish lip, the tip of which is usually curved under; they are borne in spikes of up to 100 flowers (Jul–Sep), and the leaves are broadly oval and strongly veined. **Violet Helleborine** *E. purpurata* (**2**) (height to 75cm) is similar but has violet-tinged stems and grows on chalky soils. Its flowers comprise narrow sepals and upper petals that are greenish white inside, and a heart-shaped whitish lip that is tinged purplish towards the centre; they are borne in spikes (Aug–Sep), and the leaves are narrow and parallel-sided. **Narrow-lipped Helleborine** *E. leptochila* (**3**) (height to 70cm) is a local plant of Beech woodlands on chalk soils. Its flowers comprise narrow, greenish-white sepals and upper petals, and a narrow, heart-shaped, greenish-white lip that is sometimes tinged pink towards the centre and has a tip that does not curve under; they are borne in open spikes (Jun–Jul), and the leaves are narrowly oval. **Green-flowered Helleborine** *E. phyllanthes* (**4**) (height to 50cm) grows on chalky soils and has rather insignificant-looking flowers that comprise yellowish-green sepals and petals. The flowers are pendent, invariably do not open fully and are borne in open spikes (Jul–Sep), and the leaves are narrowly oval and strongly veined.

1

2

3

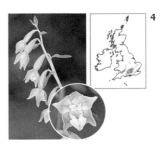

4

Irish Lady's-tresses

Spiranthes romanzoffiana
Orchidaceae Height to 25cm

Upright perennial. **Flowers** Greenish white; borne in a triple spiral up the stem, creating a rather conical spike (Jul–Aug). **Fruits** Egg-shaped. **Leaves** Appear as a basal rosette of narrowly oval leaves, and as narrow leaves up the stem. **Status** Rare, found only in SW and N Ireland, NW Scotland and Dartmoor. Grows in waterlogged grassland and marshes.

Creeping Lady's-tresses

Goodyera repens Orchidaceae Height to 25cm

Creeping perennial with upright flowering stalks. **Flowers** White, slightly sticky and do not open fully, arranged in a spiral in open spikes (Jul–Aug). **Fruits** Egg-shaped. **Leaves** Oval, stalked and evergreen. **Status** Locally common only in Scotland and N England, in mature conifer forests (mainly Scots Pine).

Musk Orchid

Herminium monorchis
Orchidaceae Height to 15cm

Delicate, yellowish perennial. **Flowers** Tiny, yellowish and smelling of honey, with narrow lobes to the lip; borne in dainty-looking spikes, 2–4cm long (Jun–Jul). **Fruits** Egg-shaped. **Leaves** Oval at the base of the plant but small and bract-like up the stem. **Status** Locally common only in S England in calcareous grassland.

Bog Orchid

Hammarbya paludosa
Orchidaceae Height to 8cm

Delicate, yellowish-green perennial. Its stems are bulbous at the base. **Flowers** Tiny, with a narrowly ovate lip; borne in spikes, to 5cm long (Jul–Sep). **Fruits** Egg-shaped. **Leaves** Rounded oval, and mainly basal. **Status** Widespread but local in Scotland; absent elsewhere, although locally common in the New Forest. Grows among *Sphagnum* moss in floating bogs.

Grassland & meadows

What could be more beautiful than a grassy meadow full of wildflowers and native grass species? Many rural examples are a naturalist's paradise, but as long as the vegetation is cut sympathetically and not sprayed, even the most unpromising urban grassland is usually full of wildlife and a variety of plants.

Kestrels (*see also* p.250) and Barn Owls feed on Field Voles, which can be abundant prey in good years. Depending on where you are in Britain and Ireland, and the management regime of the habitat in question, other grassland birds that can be found include Meadow Pipits, Skylarks and partridges.

The **Marsh Fritillary** (*see also* p.341) is one of many specialist butterflies which can be found in grasslands. Members of the brown family are also well represented and include Meadow Brown, Gatekeeper, Marbled White and Ringlet. Skippers are also considered to be archetypal meadow butterflies.

Slow-worms (*see also* p.328) make a good living feeding on invertebrates, such as slugs and earthworms. Depending on the geographical location and management, Grass Snakes and Common Lizards may also be present, the diet of the former including larger prey animals than just invertebrates.

As with many other habitats, the species you will find in any unimproved meadow depends partly on its geographical location and the drainage of the site. Perhaps more importantly though, the variety of different species – or biodiversity – depends on the soil chemistry. Those grassland habitats found on basic soils on limestone and chalk tend to have greater biodiversity than those on neutral or acid soils, and certainly harbour many species found nowhere else. This is especially evident in the range of wildflower species to be found. Biodiversity generally diminishes the further north

you travel in Britain, but visually, good northern meadows are often just as colourful in floral terms as their southern counterparts.

As recently as 50 years ago, few people would have given a second thought to the protection of lowland meadows: they were so widespread and common. Today, however, their fate is a real conservation issue. Modern farming has destroyed the intrinsic wildlife value of many sites through the application of selective herbicides, and seeding with rank grasses, in a process euphemistically known as 'improvement' aimed at increasing crop yields and productivity.

You would be hard pushed to find many meadows that have not suffered from damaging agricultural practices to some degree, but a few do survive more-or-less intact, and when you come across one you will recognise it instantly. Instead of uniformity you will discover diversity, most notably in the form of a kaleidoscope of floral colour and an abundance of invertebrate life. The ideal management regime for most lowland grasslands is to cut the vegetation in late summer for hay, once flowering has finished.

The **Common Green Grasshopper** (*see also* p.377) is well represented in meadow habitats, and a widespread species in southern Britain. You are also likely to come across Meadow, Common Green and Common Field grasshoppers. Other insects and invertebrate groups are also well represented in grasslands.

Meadow Buttercup (*see also* p.48) is a characteristic grassland flower. Other common species include Ox-eye Daisy, Common Knapweed, Red Bartsia, Creeping Thistle, Bird's-foot Trefoil and Hogweed. There are fewer species as you travel north, but a lack in diversity is usually made up for by abundance.

Anthills, such as this one covered in **Wild Thyme** (*see also* p.118), are very much a feature of undisturbed grassland, particularly where grazed rather than cut. The larger they are, the less disturbed the habitat has been. Excessive grazing, or cutting using harsh mechanical means, destroys these fragile structures.

Grassland habitats & hotspots

Grassland varies greatly across Britain, so it can be divided up into more specialised habitats: **water meadows**, **machair**, **upland grassland**, **limestone** and **chalk downlands**. Some of the best examples or 'hotspots' of each of these are given below.

Water meadows

Although typically rather difficult to negotiate and damp underfoot, wet grazing meadows often support interesting plants such as Marsh Marigold, various marsh-orchid species, Ragged Robin and Water Avens, not to mention a wealth of different sedges and rushes. Wetland invertebrates abound and in areas that are not disturbed ground-nesting waders such as Snipe and Redshank sometimes nest.

Hotspot *see map:*
1 North Meadow, Cricklade, Wiltshire – extensive water meadows with large numbers of Snake's-head Fritillaries

Water meadow, Old Basing, Hampshire.
INSET: Water Vole

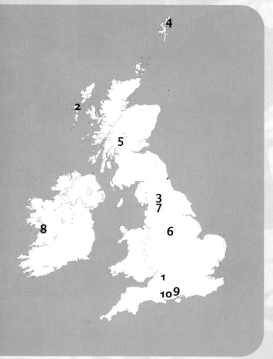

Machair

A unique type of grassland, called machair, can be found on the coasts of north-west Ireland, particularly Co. Mayo, and on the Outer Hebrides. Here, wind-blown sand has been colonised and stabilised by low-growing grassland plants. This fragile habitat is vulnerable to disturbance and only where it is sensitively managed by grazing or cutting does it survive intact. Machair is extremely important for ground-nesting birds including many wader species and the Corncrake.

Hotspot *see map:*
2 Balranald, North Uist – wonderful machair habitat www.rspb.org.uk/reserves/guide/b/balranald

Balranald North Uist.
INSET: Yellow-rattle

Heather and grass moor, lowland Scotland.
INSET: Yorkshire Fog grass

Upland grassland

Although heather moorland predominates in acid upland areas, grasses, rushes and sedges often flourish where soils are neutral or slightly basic. A handful of species, notably Purple Moor-grass and Deer Grass, tend to dominate, so that such areas lack both the floral and invertebrate interest found in southern, lowland counterparts. Nevertheless, in a few northern locations, two speciality butterflies – Mountain Ringlet and Large Heath – can be found.

Hotspots *see map:*
3 Swaledale Hay Meadows – superb upland hay meadows, at their most colourful in May and June www.yorkshiredales.org.uk
4 Fetlar, Shetland – unspoilt hay meadows
5 Ben Lawers, Killin, Scotland – vast tracts of upland grassland www.nts.org.uk

Limestone

Chalk is the dominant calcareous rock in southern Britain, being replaced by limestone further north. Subtle differences in the flora can be observed between the two underlying bedrocks, but generally speaking, many calcareous-loving flower species have a distinctly southern bias. Yet this does not mean that limestone grassland is any less colourful than chalk downland, or that it lacks speciality plants. You will find many plants in the Peak District and Yorkshire Dales that you can't see on the South Downs, such as the Bird's-eye Primrose.

Hotspots *see map:*
6 Lathkill Dale, Derbyshire – superb limestone grassland in the Derbyshire Dales
7 Malham, Yorkshire – limestone grassland and limestone pavement www.nationaltrust.org.uk
8 The Burren, Ireland – limestone grassland and pavement with a unique flora

Peak District. INSET: Globeflower

Chalk specialists

Centuries of sheep grazing on chalk soils has created a unique grassland habitat, the best examples of which are found on the North and South Downs of southern England. The free-draining, impoverished soil may seem like an unpromising botanical location, but the close-cropped grass harbours an incredible diversity of herbaceous plants. Many of these are entirely restricted to chalk soils as they are unable to compete with the lush growth of grasses in typical hay meadows.

Hotspots *see map:*
9 Old Winchester Hill, Hampshire – fantastic views with Round-headed Rampion being a speciality
10 Martin Down, Hampshire – one of the most extensive areas of unspoilt chalk grassland remaining, with abundant butterfly specialities. For further details of all these sites visit www.english-nature.org.uk

Chalkland.
INSET: Common Spotted-orchid

👓 Find out more about...

- Chalk downland flowers on **pp.72–3**
- Limestone pavements on **p.177**
- The specialist 'lawn' flora of the New Forest on **pp.74–5**
- 'Weeds' of arable fields on **pp.54–5** (arable crops are grasses after all)
- Saltmarshes, where specialised grass species predominate, on **pp.38–9**
- Sand dunes, and their colonisation by grasses, on **pp.40–1**

Rushes

The name rush is applied to a number of different groups of tough, grass-like plants that are usually found in damp and waterlogged areas. *Juncus* species are the true rushes and are widely represented in Britain, being found in suitable habitats in all areas.

Toad Rush

Juncus bufonius Juncaceae
Height to 40cm

 Tufted, red-tinged annual. **Flowers** Greenish white; borne in branched clusters, topped by a sharp spine (May–Sep). **Fruits** Brown, egg-shaped and shorter than the sepals. **Leaves** Narrow and grooved. **Status** Common. Grows on damp, bare ground, including ruts along tracks, and on the margins of ponds.

Saltmarsh Rush

Juncus gerardii Juncaceae
Height to 50cm

 Often covers extensive areas. **Flowers** Dark brown, borne in loose clusters and flanked by leaf-like bracts (Jun–Jul). **Fruits** Brown, egg-shaped, glossy and equal in length to the sepals. **Leaves** Dark green, arising at the base of the plant and on the stems. **Status** Locally common in coastal saltmarshes.

Soft Rush

Juncus effusus Juncaceae
Height to 1.5m

 Perennial with stems that are yellowish green, glossy and smooth. **Flowers** Pale brown; in loose or open clusters near the stem tops (Jun–Aug). **Fruits** Yellow-brown, egg-shaped, indented at the tip and shorter than the sepals. **Leaves** Absent. **Status** Widespread and common. Found on overgrazed grassland, mostly on acid soils.

BELOW: Soft Rush flower

Great Wood-rush

Luzula sylvatica Juncaceae
Height to 80cm

 Robust tufted perennial. **Flowers** Brown; borne in heads of 3 in branched, open clusters (Jun–Jul). **Fruits** Brown and egg-shaped. **Leaves** 5–20mm across, hairy and glossy. **Status** Widespread and locally common but predominantly in the N and W. Found in woodlands and rocky, upland terrain, mainly on acid soils.

Hard Rush

Juncus inflexus Juncaceae
Height to 1.2m

 Tufted perennial with stiff, ridged, bluish- or greyish-green stems. **Flowers** Brown; borne in loose clusters below a long bract (Jun–Aug). **Fruits** Brown, and egg-shaped with a tiny point. **Leaves** Absent. **Status** Widespread and common except in the N. Grows in damp grassy places but avoids acid soils.

Compact Rush

Juncus conglomeratus Juncaceae
Height to 1m

 Similar to a compact-flowered form of Soft Rush, but the stems are darker, not glossy, ridged and rough. **Flowers** Brown; borne in compact clusters (May–Jul). **Fruits** Dark brown, egg-shaped and as long as the sepals. **Leaves** Absent. **Status** Widespread and locally common throughout. Found on damp, grazed grassland, mainly on acid soils.

Field Wood-rush

Luzula campestris Juncaceae
Height to 25cm

Tufted perennial. Also known as Good Friday Grass. **Flowers** Brown with yellow anthers; borne in heads (1 unstalked, several stalked) and arranged in clusters (Apr–May). **Fruits** Brown and globular. **Leaves** Grass-like, fringed with long, white hairs. **Status** Locally common throughout in dry grassland, especially on calcareous soils.

Heath Wood-rush

Luzula multiflora Juncaceae
Height to 30cm

Tufted perennial. **Flowers** Brown and borne in stalked heads of 5–12 flowers, the heads in clusters of 3–10 (May–Jun). **Fruits** Brown, globular and shorter than the sepals. **Leaves** Grass-like, fringed with white hairs. **Status** Locally common on southern heaths, and in woodlands on acid soils. Much more widespread on moors in the N and W.

Common Club-rush

Schoenoplectus lacustris
Cyperaceae Height to 3m

Tall, impressive grey-green perennial with narrow stems. **Flowers** Stalked, egg-shaped, brown spikelets; borne in clusters (Jun–Aug). **Fruits** Greyish brown. **Leaves** Narrow and submerged. **Status** Widespread but only locally common. Grows along river margins and in fresh and brackish marshes.

Common Spike-rush

Eleocharis palustris Cyperaceae
Height to 50cm

Creeping, hairless perennial with green, leafless stems. **Flowers** Brown; borne in terminal egg-shaped spikelets of 20–70 flowers (May–Jul). **Fruits** Yellowish brown. **Leaves** Reduced to brownish basal sheaths on the stems. **Status** Widespread and locally common throughout. Grows in marshes and pond margins.

Jointed Rush

Juncus articulatus Juncaceae
Height to 60cm

Creeping or tufted and upright perennial. **Flowers** Brown and borne in open, branched clusters (Jun–Aug). **Fruits** Brown, egg-shaped and abruptly pointed at the tip. **Leaves** Curved, narrow and flattened with a transverse joint. **Status** Locally common in marshes, damp heaths, moors and dune-slacks.

Sea Club-rush

Bolboschoenus maritimus
Cyperaceae Height to 1.25m

Creeping, robust perennial. The stems are rough and are triangular in cross-section. **Flowers** A tight terminal cluster of egg-shaped spikelets, flanked by a long, leafy bract (Jul–Aug). **Fruits** Dark brown. **Leaves** Rough and keeled. **Status** Locally common. Grows at the margins of brackish water near the sea.

Common Club-rush

Sedges

Although they are not showy or colourful, sedges and their allies (family Cyperaceae) are true flowering plants. Their flowers may lack petals and sepals, but the basic reproductive structures are still there; these are encased in scaly features: such flowers are known as spikelets. In some species, spikelets are solitary but in others they are grouped together into heads or spikes. The shape of the swollen fruits is often important in identification.

White Beak-sedge

Rhynchospora alba Cyperaceae
Height to 40cm

 Tufted perennial. **Flowers** Spikelets that are whitish at first; borne in flat-topped terminal clusters with a narrow bract (Jun–Sep). **Fruits** Appear similar to the flowers. **Leaves** Pale green; they arise at the base of the plant and on the stem. **Status** Very local in S Britain but widespread in Scotland and W Ireland. Grows in bogs and wet heaths on acid soils.

Carnation Sedge

Carex panicea Cyperaceae
Height to 30cm

 Greyish, tufted perennial. **Flowers** Borne in inflorescences that comprise a single terminal male spike above 1–3 female spikes (May–Jun). **Fruits** Greenish, pear-shaped and not beaked. **Leaves** Greyish green and rough. **Status** Widespread and locally common throughout. Grows in damp ground, avoiding acid soils.

Stiff Sedge

Carex bigelowii Cyperaceae
Height to 30cm

 Creeping perennial with stiff, sharply 3-sided stems. **Flowers** Borne in inflorescences comprising a single male spike above 2–3 female spikes (Jun–Jul). **Fruits** Short-beaked and green grading to brown. **Leaves** Ridged and curved, with inrolled margins. **Status** Locally common only in Scotland. Grows on mountains and upland moors.

Sand Sedge

Carex arenaria Cyperaceae
Height to 35cm

 Creeping perennial. Progress of its underground stems can be detected by aerial shoots, which appear in straight lines. **Flowers** Comprise pale brown spikes borne in a terminal head, the male flowers above the females (May–Jul). **Fruits** Yellowish brown and beaked. **Leaves** Wiry. **Status** Locally common in sand dunes on most suitable coasts.

Yellow Sedge

Carex viridula Cyperaceae
Height to 40cm

 Tufted perennial. Several subspecies occur, each with subtly different growing requirements. **Flowers** Comprise a terminal, stalked male spike above rounded to egg-shaped clusters of yellowish female flowers (May–Jun). **Fruits** Yellow and beaked. **Leaves** Narrow, curved and longer than the stems. **Status** Common throughout in damp ground.

Glaucous Sedge

Carex flacca Cyperaceae
Height to 50cm

 Common grassland sedge with 3-sided stems. **Flowers** Comprise an inflorescence with 1–3 brown male spikes above 2–5 brown female spikes (Apr–May). **Fruits** Greenish, flattened and with only a tiny beak. **Leaves** Pale green and stiff. **Status** Widespread and locally common throughout. Usually found growing on calcareous soils.

Oval Sedge

Carex ovalis Cyperaceae
Height to 60cm

 Tufted sedge with 3-sided stems that are rough at the top. **Flowers** Yellowish brown and borne in compact, rather egg-shaped terminal clusters (Jun–Jul). **Fruits** Brownish and beaked. **Leaves** Green and rough-edged. **Status** Widespread and locally common. Grows in rough grassland, mainly on acid soils.

Remote Sedge

Carex remota Cyperaceae
Height to 60cm

 Slender, rather distinctive sedge. **Flowers** Arranged in an extended inflorescence, 10–20cm long, with widely spaced spikes and long, leafy bracts (Jun–Aug). **Fruits** Greenish, egg-shaped and flattened, with a short beak. **Leaves** Pale green, narrow and rough-edged. **Status** Widespread and locally common in damp, shady places.

False Fox-sedge

Carex otrubae Cyperaceae
Height to 80cm

 Tufted perennial with robust, rough, 3-sided stems. **Flowers** Greenish brown and borne in a dense head with long bracts (Jun–Jul). **Fruits** Smooth, beaked and ribbed. **Leaves** Stiff, upright and 5–10mm wide. **Status** Widespread but locally common only in S England. Grows in damp, grassy ground, doing particularly well on heavy soils.

Wood Sedge

Carex sylvatica Cyperaceae
Height to 50cm

 Elegant and tufted sedge with smooth, 3-sided, arched stems. **Flowers** Borne in nodding inflorescences comprising a single terminal male spike and 3–4 slender, long-stalked female spikes (May–Jun). **Fruits** Green, 3-sided and beaked. **Leaves** Pale green, 3–6mm across and drooping. **Status** Locally common throughout Britain and Ireland. Grows in damp woods.

Deergrass

Trichophorum cespitosum
Cyperaceae Height to 20cm

Tufted perennial with round, smooth stems. Sometimes forms small tussocks. **Flowers** Comprise a single egg-shaped, terminal brown spikelet (May–Jun). **Fruits** Matt brown. **Leaves** Consist of a single strap-like leaf near the base. **Status** Locally very common in Britain and Ireland, mainly in the N and W. Grows in boggy moorland terrain, on acid soils.

Hairy Sedge

Carex hirta Cyperaceae
Height to 70cm

Distinctive sedge, recognised by its very hairy leaves. **Flowers** Borne in inflorescences of 2–3 brown male spikes above 2–3 yellowish female spikes (Apr–Jun). **Fruits** Green, beaked and ridged. **Leaves** Long, grey-green and covered in long, white hairs. **Status** Widespread and locally common throughout. Grows in damp grassland.

Pendulous Sedge

Carex pendula Cyperaceae
Height to 1.5m

 Clump-forming sedge whose stems are tall, arching and 3-sided. **Flowers** Borne in inflorescences that comprise 1–2 male spikes above 4–5 long, drooping, unstalked female spikes (Jun–Jul). **Fruits** Flattened, greyish and short-beaked. **Leaves** Long, yellowish and up to 2cm wide. **Status** Locally common in damp woodlands on heavy soils.

Pill Sedge

Carex pilulifera Cyperaceae
Height to 25cm

Tufted sedge with 3-sided stems. **Flowers** Borne in inflorescences that comprise a single male spike above 2–4 egg-shaped female spikes (May–Jun). **Fruits** Green, rounded, downy and ribbed. **Leaves** Yellowish green, narrow, wiry and flaccid. **Status** Widespread but local, except in the N and W. Grows on heaths and dry grassland with acid soils.

Did you know?

Although some sedge species are rather catholic in their habitat requirements, many are extremely specific and will grow only if soil conditions suit them precisely and the habitat in which they are growing is undisturbed. It is for this reason that ecologists place great emphasis on the importance of the discovery of certain sedges and refer to them as 'indicator species', meaning the habitat in which they are growing is important in conservation terms.

Grasses

In ecological terms, grasses (family Poaceae) are extremely important, partly because many species have binding roots and a creeping habit that allows them to consolidate unstable or shifting soils. They have relatively simple, wind-pollinated flowers and parallel-veined leaves, and many of them spread vegetatively. Their role in meadows is obvious – they are the dominant plants – but certain grass species are also important in wetlands.

Common Bent

Agrostis capillaris Poaceae
Height to 70cm

Creeping perennial. **Flowers** Greenish-brown spikelets that are borne in heads with spreading, whorled branches (Jun–Aug). **Fruits** Small, dry nutlets. **Leaves** Narrow, with blunt ligules. **Status** Widespread and locally common throughout the region. Found in stable grassland, mainly on acid soils.

Purple Moor-grass

Molinia caerulea Poaceae
Height to 80cm

Tussock-forming perennial. **Flowers** Purplish-green spikelets, borne in long, branched, spike-like heads (Jul–Sep). **Fruits** Small, dry nutlets. **Leaves** Grey-green and 3–5mm wide, with purplish leaf sheaths. **Status** Widespread and locally common. Usually associated with damp ground on acid heaths and grassy moors.

Black Bent

Agrostis gigantea Poaceae
Height to 1.5m

Widespread, creeping perennial. **Flowers** Comprise numerous 1-flowered purplish-brown spikelets that are borne in branching clusters (Jul–Aug). **Fruits** Small, dry nutlets. **Leaves** Dark green, up to 6mm across and with long ligules. **Status** Commonest in central and S England. Grows on waste ground, verges and arable fields.

Creeping Bent

Agrostis stolonifera Poaceae
Height to 1m

Hairless perennial with creeping runners and upright stems. **Flowers** Comprise yellowish-green spikelets that are borne in dense, tall, branched and whorled heads (Jun–Aug). **Fruits** Small, dry nutlets. **Leaves** Narrow, with bluntly pointed ligules. **Status** Widespread and common, especially in the S. Grows in meadows and on waste ground.

Bristle Bent

Agrostis curtisii Poaceae
Height to 50cm

Tufted, clump-forming perennial. **Flowers** Yellowish-green spikelets that are borne in tall, dense, spike-like heads (Jun–Aug). **Fruits** Small, dry nutlets. **Leaves** Grey-green, narrow and bristle-like, with pointed ligules. **Status** Locally common in SW Britain only. Grows in damp heaths and moors, mainly on acid soils.

Reedbeds

Arguably the most significant plant associated with wetland habitats is **Common Reed** *Phragmites australis* (height to 2m). Its spreading roots allow it to grow in soft, yielding mud, so that before long it often covers vast areas, lending its name to a habitat type – reedbeds. In ecological terms, the species is important because the binding effect of the roots eventually leads to the formation of dry ground, on which colonising woodland can grow. These days, reedbeds are often managed to ensure that they remain as such, and the reeds are still cut for roof thatching. The flowers of Common Reed are spikelets, purplish brown at first and then fading, and are borne in branched, one-sided terminal clusters (Aug–Sep). The leaves are broad and long, and the whole plant turns golden in autumn and winter.

False Brome

Brachypodium sylvaticum Poaceae
Height to 1m

Tufted, softly hairy perennial. **Flowers** Borne in unbranched, slightly nodding heads, with long spikelets that are short-stalked (Jul–Aug). **Fruits** Small, dry nutlets. **Leaves** Long, 10–12mm wide and hairy. **Status** Widespread and common throughout. Usually found growing in shady woods and hedgerows.

Perennial Rye-grass

Lolium perenne Poaceae
Height to 90cm

Tufted, hairless perennial with wiry stems. **Flowers** Borne in unbranched heads, the spikelets green and lacking awns (May–Aug). **Fruits** Small, dry nutlets. **Leaves** Deep green and often folded when young. **Status** Common in meadows but also widely cultivated on farmland.

Wall Barley

Hordeum murinum Poaceae
Height to 30cm

Tufted annual. **Flowers** Borne in unbranched spikes, 9–10cm long, with spikelets in 3s, each with 3 stiff awns (May–Jul); the flowering stems are prostrate at the base. **Fruits** Small, dry nutlets. **Leaves** 7–8mm wide, with short, blunt ligules. **Status** Widespread and common. Found in bare ground and waste places, often near the sea.

Marsh Foxtail

Alopecurus geniculatus Poaceae
Height to 50cm

Tufted, grey-green perennial. The stems are bent at sharp angles at the base and at joints. **Flowers** Borne in cylindrical, purplish heads, 5–6cm long, the spikelets with long awns and blunt ligules (Jun–Sep). **Fruits** Small, dry nutlets. **Leaves** Smooth below. **Status** Widespread and locally common. Grows in damp grassland.

stem joint

Italian Rye-grass

Lolium multiflorum Poaceae
Height to 90cm

Tufted annual or biennial with rough stems and leaves. **Flowers** Borne in unbranched heads, the spikelets brown with long awns (May–Aug). **Fruits** Small, dry nutlets. **Leaves** Often rolled when young. **Status** Widespread and common, except in the N. Commonly cultivated on farmland to provide grazing for livestock.

Crested Dog's-tail

Cynosurus cristatus Poaceae
Height to 50cm

Distinctive, tufted perennial with wiry stems. **Flowers** Borne in compact, flat heads, the spikelets usually greenish (Jun–Aug). **Fruits** Small, dry nutlets. **Leaves** Narrow and short, with narrow, blunt ligules. **Status** Widespread and common throughout. Found in grassland and on roadside verges.

Meadow Foxtail

Alopecurus pratensis Poaceae
Height to 1m

Tufted, hairless perennial. **Flowers** Borne in smooth, cylindrical, purplish-grey heads, 7–9cm long, of 1-flowered spikelets, with pointed glumes and long awns (Apr–Jun). **Fruits** Small, dry nutlets. **Leaves** Rough, 5–8mm wide and with blunt ligules. **Status** Widespread and common throughout. Grows in meadows and on verges.

Did you know?

Because the leaves of grasses are simple and their flowers lack obvious colourful parts that correspond to petals or sepals, botanists use special names to help describe their structures. An awn is a stiff, bristle-like projection; a ligule is a membraneous flap seen at the base of a leaf; and a glume is a pair of scaly bracts at the base of a flower head.

Grasses

Sweet Vernal-grass
Anthoxanthum odoratum Poaceae
Height to 50cm

Tufted, downy perennial that has a sweet smell when dried. **Flowers** Borne in relatively dense spike-like clusters, 3–4cm long, of 3-flowered spikelets, each with 1 straight and 1 bent awn (Apr–Jul). **Fruits** Small, dry nutlets. **Leaves** Flat with blunt ligules. **Status** Widespread and common. Grows in a wide range of grassland.

Yellow Oat-grass
Trisetum flavescens Poaceae
Height to 50cm

Slender, softly hairy perennial. **Flowers** Borne in an open inflorescence of yellowish, 2–4-flowered spikelets, each with a bent awn (Jun–Jul). **Fruits** Small, dry nutlets. **Leaves** Narrow and flat, with a blunt ligule. **Status** Widespread and locally common. Grows in dry grassland, usually on calcareous soils.

Quaking-grass
Briza media Poaceae
Height to 40cm

Distinctive perennial. **Flowers** Borne in an open inflorescence, the dangling spikelets resembling miniature hops or cones and carried on wiry stalks (Jun–Sep). **Fruits** Small, dry nutlets. **Leaves** Pale green, forming loose tufts. **Status** Widespread and locally common. Grows in dry grassland, usually on calcareous soils.

Tufted Hair-grass
Deschampsia cespitosa Poaceae
Height to 1.5m

Tufted, clump-forming perennial. **Flowers** Borne in a long-stemmed inflorescence comprising spreading clusters of 2-flowered, silvery-purple spikelets (Jun–Jul). **Fruits** Small, dry nutlets. **Leaves** Dark green, wiry and narrow, with rough edges. **Status** Widespread and common in damp grassland, woodland rides and marshes.

False Oat-grass
Arrhenatherum elatius Poaceae
Height to 1.5m

Tall perennial. **Flowers** Borne in an open inflorescence comprising numerous 2-flowered spikelets, 1 floral element with a long awn (May–Sep). **Fruits** Small, dry nutlets. **Leaves** Broad and long, with a blunt ligule. **Status** Common in disturbed grassland, verges and waysides; sometimes sown in meadow-seeding mixes.

Wild-oat
Avena fatua Poaceae
Height to 1m

Distinctive annual of arable crops. **Flowers** Borne in an inflorescence comprising an open array of stalked, dangling spikelets, each of which is shrouded by the glumes and has a long awn (Jun–Aug). **Fruits** Small, dry nutlets. **Leaves** Dark green, broad and flat. **Status** Widespread and common in arable fields and on waste ground.

Cock's-foot
Dactylis glomerata Poaceae
Height to 1m

Tufted, tussock-forming perennial. **Flowers** Borne in an inflorescence of long-stalked, dense, egg-shaped heads that spread and then fancifully resemble a bird's foot (Jun–Jul). **Fruits** Small, dry nutlets. **Leaves** Rough, with slightly inrolled margins. **Status** Widespread and common in grassland and woodland rides.

Wavy Hair-grass
Deschampsia flexuosa Poaceae
Height to 1m

Tufted perennial. **Flowers** Borne in inflorescences comprising open clusters of purplish spikelets with a long, bent awn (Jun–Jul). **Fruits** Small, dry nutlets. **Leaves** Inrolled and hair-like. **Status** Locally common in Britain; scarce in Ireland. Grows in dry ground on heaths and moors, usually on acid soils.

Red Fescue

Festuca rubra Poaceae
Height to 50cm

Clump-forming perennial. **Flowers** Borne in an inflorescence, the spikelets 7-10mm long and usually reddish (May-Jul). **Fruits** Small, dry nutlets. **Leaves** Either narrow, wiry and stiff, or flat (on flowering stems). **Status** Common in grassy places; often particularly abundant in coastal grassland and the dry upper margins of saltmarshes.

Meadow Fescue

Festuca pratensis Poaceae
Height to 1m

Tufted, hairless perennial. **Flowers** Borne in a 1-sided inflorescence, with open spikelets carried on unequal, paired branches, the shorter one with a single spikelet (Jun-Aug). **Fruits** Small, dry nutlets. **Leaves** Flat and 4mm across. **Status** Widespread and common throughout, growing in damp meadows.

Sheep's Fescue

Festuca ovina Poaceae
Height to 30cm

Variable, tufted, hairless perennial. **Flowers** Borne in branched but compact heads of grey-green spikelets, each with a short awn (May-Jul). **Fruits** Small, dry nutlets. **Leaves** Short, narrow, inrolled, hair-like and waxy. **Status** Widespread and locally common throughout, growing in dry grassland on chalk and limestone.

Rough Meadow-grass

Poa trivialis Poaceae
Height to 90cm

Loosely tufted perennial with creeping runners. **Flowers** Borne in a pyramidal inflorescence, the purplish-brown spikelets carried on whorls of stalks (May-Jul). **Fruits** Small, dry nutlets. **Leaves** Pale green, and soft with a pointed ligule and a rough sheath. **Status** Common in damp places.

Mountain grasses

Mountain plants have to be hardy, and upland grass species are no exception. Given the often damp, rainy summer weather experienced in such habitats, wind pollination could be a problem, something that two species overcome in unusual ways. **Alpine Meadow-grass** *Poa alpina* (**1**) (height to 20cm) does not produce flowers but instead forms little plantlets that are ready to grow once they come into contact with the ground. In a similar manner, **Viviparous Fescue** *Festuca vivipara* (**2**) (height to 20cm) produces green bulbils, not flowers.

Annual Meadow-grass

Poa annua Poaceae
Height to 25cm

Annual or short-lived perennial. **Flowers** Borne in an inflorescence that is roughly triangular in outline, comprising branches with oval spikelets at their tips (Jan-Dec). **Fruits** Small, dry nutlets. **Leaves** Pale green, blunt-tipped and often wrinkled. **Status** Widespread and common in bare grassland and disturbed ground.

Flattened Meadow-grass

Poa compressa Poaceae
Height to 50cm

Upright, tufted perennial with flattened stems. **Flowers** Borne in a pyramidal inflorescence, the purplish spikelets carried in stalked whorls (May-Jun). **Fruits** Small, dry nutlets. **Leaves** Have a hooked tip and blunt ligule. **Status** Widespread but local, growing in dry, grassy places.

Grasses

Common Couch

Elytrigia repens Poaceae
Height to 1.2m

 Tough, creeping perennial. **Flowers** Borne in a stiff, unbranched inflorescence, with many-flowered, yellowish-green, alternately arranged spikelets (Jun–Aug). **Fruits** Small, dry nutlets. **Leaves** Flat, green and downy above. **Status** Widespread and common throughout, growing in cultivated and disturbed ground.

Bearded Couch

Elymus caninus Poaceae
Height to 1m

 Tufted perennial with downy stem joints. **Flowers** Borne in a rather lax, unbranched inflorescence, the spikelets alternate and with a long, straight awn (Jun–Aug). **Fruits** Small, dry nutlets. **Leaves** Flat. **Status** Widespread and locally common in England and Wales. Grows in damp, shady places in woods and hedgerows.

Soft-brome

Bromus hordaceus Poaceae
Height to 1m

 Softly downy annual or biennial. **Flowers** Borne in moderately compact heads with short-stalked, hairy, egg-shaped spikelets (Jun–Aug). **Fruits** Small, dry nutlets. **Leaves** Greyish green and rolled when young, with hairy sheaths. **Status** Widespread and common, especially in the S. Found in meadows and on verges.

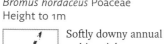

Hairy Brome

Bromopsis ramosus Poaceae
Height to 1.8m

 Stout, elegant, tufted perennial. **Flowers** Borne in open, arching and drooping inflorescences with long-stalked, slender, purplish spikelets, each 2cm long and with a long awn (Jul–Aug). **Fruits** Small, dry nutlets. **Leaves** Dark green and drooping. **Status** Widespread and common. Grows in woods and shady places.

Wood Melick

Melica uniflora Poaceae
Height to 50cm

 Delicate, creeping perennial. **Flowers** Borne in a loose, open inflorescence with brown, egg-shaped, stalked spikelets along the side branches (May–Jul). **Fruits** Small, brown nutlets. **Leaves** Pale green and lax. **Status** Locally common in England and Wales only. Grows in dry, shady woodland, often on chalk and under Beech.

Timothy

Phleum pratense Poaceae
Height to 1.5m

 Robust, hairless, tufted perennial. **Flowers** Borne in dense, rough, cylindrical heads, 13–16cm long, with purplish-green spikelets, each with a short awn (Jun–Aug). **Fruits** Small, dry nutlets. **Leaves** Broad, rough and flat, with blunt ligules. **Status** Widespread and common in meadows, agricultural land and waysides.

Smaller Cat's-tail

Phleum bertolonii Poaceae
Height to 60cm

Tufted perennial that is similar to Timothy but smaller in all respects. **Flowers** Borne in dense cylindrical heads, 6–8cm long (Jun–Jul). **Fruits** Small, dry nutlets. **Leaves** Rough and flat with pointed ligules. **Status** Widespread and locally common, except in the N. Grows in grassy places, often on calcareous soils.

Coastal colonisers

Coastal erosion and the need for sea defences are a constant concern. Surprisingly, certain grass species are just as important in protecting the coast as steel and concrete reinforcements: they are the unsung heroes of coastal consolidation and their colonisation can lead, eventually, to the creation of stable, dry land where once there was sea. **Marram** *Ammophila arenaria* (**1**) (height to 1m) is probably the most familiar of the coastal colonisers, a plant of dunes that stabilises shifting sands by means of its underground stems; its flowers are borne in dense spikes, with one-flowered, straw-coloured spikelets (Jul–Aug), and its leaves are tough, grey-green, rolled and sharply pointed. Sometimes growing alongside it is **Lyme-grass** *Leymus arenarius* (**2**) (height to 1.5m), a blue-grey perennial whose flowers are borne in tall heads of paired, grey-green spikelets (Jun–Aug), and whose leaves are up to 15mm wide, with inrolled margins.

Saltmarshes are also dependent upon specific grass species for their creation and maintenance. **Common Cord-grass** *Spartina anglica* (**3**) (height to 1.3m) is one of several closely related species found in such habitats; its flowers are borne in a stiff inflorescence of elongated clusters of between three and six yellowish flower heads, up to 35cm long (Jul–Sep), and its leaves are grey-green and tough. **Common Saltmarsh-grass** *Puccinellia maritima* (**4**) (height to 30cm) is a tufted, sward-forming perennial whose flowers are borne in spike-like heads, and with spikelets that occur along all the branches and are mainly upright, not spreading (Jul–Aug); its leaves are grey-green and inrolled. **Reflexed Saltmarsh-grass** *P. distans* (**5**) (height to 60cm) is similar but its inflorescence branches are reflexed in fruit. Although formerly this species was restricted to the coast, it has now spread inland along the verges of winter-salted roads.

Yorkshire Fog

Holcus lanatus Poaceae
Height to 1m

Variable, tufted perennial with grey-green, downy stems. **Flowers** Borne in heads that are tightly packed at first but then spread; comprise reddish-tipped, grey-green, 2-flowered spikelets (May–Aug). **Fruits** Small, dry nutlets. **Leaves** Grey-green and downy. **Status** Widespread and common. Grows in meadows, woods and waste ground.

not fully opened

Creeping Soft-grass

Holcus mollis Poaceae
Height to 60cm

Similar to Yorkshire Fog but more slender, and the stems are downy only at the joints. **Flowers** Borne in heads, compact at first then spreading, with purplish-green spikelets, each with a bent awn (Jun–Aug). **Fruits** Small, dry nutlets. **Leaves** Grey-green. **Status** Widespread and common. Grows in woods and on heaths, mainly on acid soils.

Blue Moor-grass

Sesleria caerulea Poaceae
Height to 45cm

Blue-green, tufted, patch-forming perennial. **Flowers** Borne in dense, egg-shaped heads with bluish-green spikelets (Apr–Jun). **Fruits** Small, dry nutlets. **Leaves** Rough-edged and blunt with a fine point at the tip. **Status** Locally common only in N England, S Scotland and W Ireland. Found in dry limestone grassland.

Mat-grass

Nardus stricta Poaceae
Height to 20cm

Densely tufted, wiry perennial that often forms large patches. **Flowers** Borne in narrow, 1-sided spikes, the spikelets in 2 rows (Jun–Aug). **Fruits** Small, dry nutlets. **Leaves** Grey-green and wiry. **Status** Locally common in upland regions; scarce or absent elsewhere. Grows on moors, mountains and (more rarely) heaths.

Plants of mires, bogs & fens

A mire (often popularly referred to as a marsh) is a habitat in which vegetation has colonised the margins of open water but the ground remains waterlogged; if the water is acid, then technically it is a bog, while if it is alkaline it is a fen. Although representatives from most flowering plant families can be found in these wetland habitats, sedges, grasses and other closely related species are the plants that dominate and often define them.

Common Sedge

Carex nigra Cyperaceae
Height to 50cm

Variable, creeping sedge with 3-angled stems that are rough at the top. **Flowers** Borne in inflorescences of 1–2 thin male spikes above 1–4 female spikes, the latter with black glumes (May-Jun). **Fruits** Short-beaked and green grading to blackish. **Leaves** Long, narrow and appear in tufts. **Status** Widespread and common in damp grassland and marshes.

Greater Tussock-sedge

Carex paniculata Cyperaceae
Height to 1m

Distinctive, coarsely hairy perennial, recognised throughout the year by the large tussocks it forms. **Flowers** Brown and borne in a dense terminal spike, 5–15cm long (May-Jun). **Fruits** Ribbed and with a winged beak. **Leaves** Narrow and up to 1.2m long. **Status** Widespread and locally common, mainly in the S. Found in marshes and fens.

Black Bog-rush

Schoenus nigricans Cyperaceae
Height to 50cm

Tufted perennial with rigid, smooth, round stems. **Flowers** Dark brown spikelets, arranged in a terminal head, with a leaf-like bract (May-Jul). **Fruits** Whitish. **Leaves** Long, green and arise at the base of the plant. **Status** Commonest in N and W Britain, and in Ireland. Grows in dune slacks and marshes, usually on alkaline soils.

Lesser Pond-sedge

Carex acutiformis Cyperaceae
Height to 1.2m

Creeping perennial. **Flowers** Comprise 2–3 brownish male spikes above 3–4 yellowish-green female spikes (Jun-Jul). **Fruits** Green, egg-shaped but flattened. **Leaves** Bluish grey and rough. **Status** Locally common in marshes and on the margins of ponds and streams.

Lesser Pond-sedge

Great Fen-sedge

Cladium mariscus Cyperaceae
Height to 2.5m

Imposing plant, still cut commercially in E Anglia. **Flowers** Glossy reddish-brown spikelets (Jul-Aug). **Fruits** Shiny and dark brown. **Leaves** Long, saw-edged and often bent at an angle. **Status** Locally common in E Anglia and W Ireland, in fens and lake margins.

Great Fen-sedge

Cottongrasses

Despite their name, cottongrasses are not really grasses at all, but members of the sedge family (Cyperaceae). Several species occur in our region and two are common and widespread. They are found in muddy areas and tend to be inconspicuous except in summer, when their cottony seed heads are visible.

Common Cottongrass *Eriophorum angustifolium* (**1**) (height to 75cm) grows in very boggy ground with peaty, acid soils; its flowers are borne in inflorescences of drooping, stalked spikelets (Apr–May), while the fruits are dark brown with cottony hairs and the fruiting heads resemble balls of cotton wool. **Hare's-tail Cottongrass** *E. vaginatum* (**2**) (height to 50cm) is a tussock-forming perennial of moors and heaths on acid, peaty soil; its flowers comprise a terminal spikelet emerging from an inflated sheath (Apr–May), while the fruits are yellowish brown with tufts of cottony hairs.

Water margins

A specialised group of grass-like plants can be found growing with their 'feet' permanently in still or slow-flowing water. Although their roots and stems (and, in some cases, their leaves) are inundated, their flowering stems are typically emergent and conspicuous. Included in this group are members of the grass family (Poaceae), the bur-reed family (Sparganiaceae) and the bulrush family (Typhaceae).

Floating Sweet-grass

Glyceria fluitans Poaceae
Floating

Aquatic grass that sometimes covers the water surface. **Flowers** Borne in an open inflorescence comprising an array of narrow spikelets (Jul–Aug). **Fruits** Small, dry nutlets. **Leaves** Broad, green and are usually seen floating at the water's surface. **Status** Locally common in still and slow-flowing fresh water in lowland regions.

Reed Sweet-grass

Glyceria maxima Poaceae
Height to 2m

Impressive aquatic grass that often forms large patches. **Flowers** Borne in a large, branched inflorescence with narrow spikelets (Jul–Aug). **Fruits** Small, dry nutlets. **Leaves** Bright green, long and 2cm wide, with a dark mark at the junction. **Status** Locally common only in SE England. Grows in shallow fresh water and marshy ground, often on margins of ponds and ditches.

Bulrush

Typha latifolia Typhaceae
Height to 2m

Sedge-like plant, formerly known as Great Reedmace. **Flowers** Borne in spikes, and comprise a brown, sausage-like array of female flowers and a narrow terminal spire of male flowers, the two contiguous (Jun–Aug). **Fruits** Have cottony down. **Leaves** Grey-green, long and 1–2cm wide. **Status** Widespread and common. Grows in freshwater margins.

Bulrush

Lesser Bulrush

Typha angustifolia Typhaceae
Height to 2m

Similar to Bulrush; formerly known as Lesser Reedmace. **Flowers** Borne in spikes and comprise a brown, sausage-like array of female flowers that are separated from a narrow terminal spire of male flowers by a gap (Jun–Aug). **Fruits** Have cottony down. **Leaves** Dark green, long and 3–6mm wide. **Status** Locally common in the S. Grows in freshwater margins.

Lesser Bulrush

Branched Bur-reed

Sparganium erectum
Sparganiaceae Height to 1m

Upright, hairless, sedge-like perennial. **Flowers** Borne in spherical heads; the yellowish male heads are separate from, and carried above, the green female heads (Jun–Aug). **Fruits** Beaked and borne in spherical heads. **Leaves** Bright green, linear, keeled and 3-sided. **Status** Locally common throughout. Grows in still or slow-flowing fresh water.

NON-FLOWERING PLANTS

Often seen as the poor relations of the plant world, non-flowering plants lack the stature and impact of their showy cousins. But if you take a closer look they reveal themselves to be as intricate and varied as their flowering relatives. Non-flowering plants range from the simplest, single-celled algae to ferns, which have roots and an internal vascular transportation system. As in more advanced plants, internal chemical reactions allow non-flowering plants to photosynthesise, trapping sunlight energy using pigments, including chlorophyll, to create food. But their reproductive structures are primitive and they do not produce flowers or seeds. In evolutionary terms, non-flowering plants are the primitive ancestors from which more complex forms evolved. At the bottom of the evolutionary ladder are algae, and from these humble origins evolved plants called bryophytes, a group that includes the liverworts and mosses. The pinnacle of complexity in non-flowering plants is seen in the pteridophytes, which include ferns and their allies, the clubmosses and horsetails.

Serrated Wrack *Fucus serratus*, a common rocky shore seaweed.

Algae

Most algae are aquatic, with both marine and freshwater representatives. All species are tied to water for their reproduction. Algae range in size from single-celled organisms (visible individually only under the microscope, although en masse they can sometimes be seen as a green 'bloom' in water) to large seaweeds, some exceeding 2m in length. Some free-living algae comprise chains of cells or form discrete colonies. Reproduction varies from simple cell division to sexual reproduction: in seaweeds this involves the release of male and female sex cells into the sea, where fertilisation takes place. Algae, particularly microscopic free-living species, form the basis of most aquatic food chains and are extremely important to the world's ecology.

LEFT: *Spirogyra* is a microscopic alga comprising chains of linked cells.
BELOW: *Trentepohlia* sp. is a bark-encrusting alga that stains tree trunks orange.

Mosses

Mosses are primitive land plants whose structure is marginally more complex than that of liverworts. Their stems bear scale-like leaves but they lack proper roots, fibres at the base of the plant serving merely to anchor it to the substrate on which it grows. Mosses lack a proper cuticle and are prone to desiccation. Most thrive only in damp situations with high humidity and a good supply of water. Some species grow directly on the ground, at times forming mounds or carpets, while others grow on dead logs, living tree branches or rock surfaces. Sexual reproduction is similar to that of liverworts, and stalked spore capsules are found in many species.

ABOVE: White Fork Moss *Leucobryum glaucum* growing on a woodland floor.

Common Haircap.

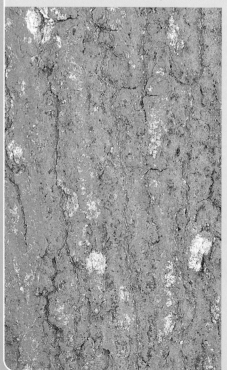

Liverworts

Pellia epiphylla.

Liverworts are a varied group of usually prostrate primitive plants, the body of which is called the thallus. Some are flat and almost seaweed-like, while others are leafy in appearance. Root-like outgrowths on their undersides anchor liverworts to the substrate on which they grow, and the plants lack a proper cuticle, making them prone to desiccation. The greatest diversity of liverworts is found in habitats that are damp and humid.

Sexual reproduction involves a two-stage process called alternation of generations. The first stage, called the gametophyte, produces male sex cells that swim in a film of water and fertilise female sex cells. The resulting fertilised embryo develops into a second stage, the sporophyte, which grows on the gametophyte and produces a capsule from which airborne spores are liberated and dispersed. These then develop into gametophytes.

Ferns

Most ferns are relatively large and robust plants. Some grow to form sizeable clumps, while others (notably Bracken) spread extensively by vegetative growth. Like mosses and liverworts, fern reproduction involves alternation of generations. The stage in the life cycle that most people regard as typically fern-like is the spore-producing sporophyte, the spores themselves being produced in structures on the fronds. Liberated spores give rise eventually to a tiny liverwort-like plant that produces sex cells, the males of which must swim in a film of water to reach the anchored female cells. So, although some ferns grow in seemingly dry habitats, they cannot survive if, for at least part of the year, there is not enough water to allow fertilisation to take place.

Common Polypody.

Damp woodlands provide an ideal growing environment for ferns such as Woodland Lady-fern *Athyrium filix femina.*

Horsetails

Field Horsetail.

These perennial plants have upright, jointed aerial stems that appear in spring with whorls of radiating branches. Spore-producing organs are arranged in cones, either on vegetative shoots or special reproductive ones, depending on species. Liberated spores grow into small single-sex plants, and fertilisation involves male cells swimming in a surface film of water and finding female cells on another plant. Horsetails spread and colonise new ground partly using a food-storing underground stem called a rhizome. They favour damp, sometimes disturbed ground, are locally abundant and can be invasive.

Clubmosses

Members of this intriguing group of non-flowering plants may look like mosses but they are more closely related to ferns. Unlike their more primitive relatives the mosses, these perennial evergreens have true roots, which can collect moisture and nutrients from the soil. The leaves lack the structure seen in those of flowering plants, but their tough stems possess proper vascular tissues for transporting water and food around the plant. As with liverworts, reproduction involves alternation of generations, the spores being produced in cylindrical terminal cones. Clubmosses grow in damp, undisturbed ground.

Marsh Clubmoss.

Lichens: a special case

Defining lichens has always been difficult. They may look like primitive, non-flowering plants, but in reality they are created by mutually beneficial relationships between fungi (which are not plants) and tiny photosynthetic partners that live inside the fungal body. The appearance of any given lichen species is determined by the species of fungus involved and the partner, which can be either an alga or a cyanobacterium (a bacterium capable of photosynthesising), and some fungi play host to either form of partner. Lichen identification is covered on pp.182–3 and you can find out more about fungi on pp.184–97.

RIGHT: Map Lichen *Rhizocarpon geographicum* growing on a mountain boulder in Snowdonia.
BELOW: A colourful lichen community (with English Stonecrop) growing on a rocky coast.

Ferns

Beloved of Victorian botanists and plant collectors, ferns have fallen out of fashion somewhat in recent years. These venerable and attractive non-flowering plants are worth looking for as they are beautiful in their own special way.

Bracken

Pteridium aquilinum
Dennstaedtiaceae Frond to 2m

The commonest fern in the region. **Fronds** Compact, curled-tipped fronds appear in spring. Mature fronds are green and are repeatedly divided 3 times. **Spore cases** Borne around frond margins. **Status** Widespread and often abundant in woodland and on hillsides, favouring dry, acid soils.

young fronds

Royal Fern

Osmunda regalis Osmundaceae
Frond to 3m

Large and impressive fern. **Fronds** Triangular overall, repeatedly divided 2 times into oblong lobes. Separate, central fertile fronds are covered in golden spores, Jun-Aug. **Status** Widespread but only locally common. Found in damp, shady places, mostly on acid soils.

Broad Buckler-fern

Dryopteris dilatata
Dryopteridaceae Frond to 1m

Robust fern. **Fronds** Dark green and repeatedly divided 3 times; the stalks have dark-centred scales; Apr-Nov. **Status** Widespread and common in Britain and Ireland. Favours damp woods, heaths and mountain slopes, usually on acid soils.

Broad Buckler-fern

Narrow Buckler-fern

Marsh Fern

Thelypteris palustris
Thelypteridaceae Frond to 30cm

Attractive fern that is restricted to waterlogged ground. **Fronds** Elongated and triangular, the lobes divided twice; the stalks are hairless. **Spore cases** Round and closely packed under the margins of the frond lobes. **Status** Local, typically found on base-rich soils in fens.

Narrow Buckler-fern

Dryopteris carthusiana
Dryopteridaceae Frond to 1m

Attractive fern of damp woodland and bogs, similar in appearance to Broad Buckler-fern. **Fronds** Parallel-sided, narrower than those of Broad Buckler-fern and yellowish green; the scales on the stalks do not have dark centres. **Status** Widespread and fairly common in suitable habitats.

Hay-scented Buckler-fern

Dryopteris aemula
Dryopteridaceae Frond to 50cm

Fern whose fresh green fronds smell of hay when crushed, hence its common name. **Fronds** Stay green throughout winter; repeatedly divided 3 times and have pale brown scales on the stalk. **Status** Locally common only in W Britain and Ireland.

Male-fern

Dryopteris filix-mas
Dryopteridaceae Frond to 1.25m

Large, clump-forming fern. **Fronds** Remain green throughout winter; broadly oval in outline, repeatedly divided 2 times and with pale brown scales on the stalks. **Spore cases** Round; spores produced Aug-Oct. **Status** Common and widespread throughout, growing in woods and on banks.

Scaly Male-fern

Dryopteris affinis Dryopteridaceae
Frond to 1m

Clump-forming, deciduous fern. **Fronds** Yellow-green in colour; they do not overwinter. The stalks have orange-brown scales; the margins of the smaller frond lobes look as though they have been cut neatly with scissors. **Status** Locally common in N and W Britain and Ireland. Favours shady woods, usually on acid soils.

Lady-fern

Athyrium filix-femina
Woodsiaceae Frond to 1.5m

Large but rather delicate fern, forming sizeable and attractive clumps. **Fronds** Pale green and repeatedly divided 2 times. **Spore cases** Curved and ripen in autumn. **Status** Widespread and fairly common throughout although scarce in central E England. Found in damp woods and on banks and hillsides.

Hart's-tongue

Phyllitis scolopendrium
Aspleniaceae Frond to 60cm

Evergreen fern. **Fronds** Fresh green, undivided and strap-like, forming clumps. **Spore cases** Dark brown and borne in rows on the underside of the fronds. **Status** Fairly widespread but commonest in the W of the region. Found in damp, shady woods and on banks, mainly on calcareous soil.

Polypody

Polypodium vulgare
Polypodiaceae Frond to 50cm

Distinctive clump-forming fern. **Fronds** Dark green and leathery, divided simply and borne on slender stalks. They appear in May and persist over winter. **Status** Commonest in W Britain and Ireland. Typical fern of damp, shady gorges and wooded valleys, mostly on acid soils; often epiphytic on tree branches.

Lemon-scented Fern

Oreopteris limbosperma
Thelypteridaceae Frond to 80cm

Yellowish-green fern that smells of lemon when crushed. **Fronds** Superficially similar to those of Male-fern but they taper top and bottom. **Spores** Black and produced around the margins of the frond lobes. **Status** Widespread on moorland slopes and heaths on acid soils.

Hard-fern

Blechnum spicant Blechnaceae
Frond to 60cm

Distinctive fern of woods and shady heaths. **Fronds** The bright green, sterile, overwintering fronds are divided simply and form spreading clumps, while the fertile fronds are borne upright and have very narrow lobes. **Status** Locally common throughout in suitable habitats; favours acid soils.

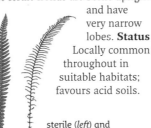

sterile (*left*) and fertile (*right*) frond

Shield-ferns

Members of this group of tufted ferns have frond lobes that end in a bristle. Their name derives from the fact that the spores are protected by a shield-like scale during their development. Two species are widespread in our region.
Soft Shield-fern *Polystichum setiferum* (**1**) (frond to 1.5m) is yellowish green, feels soft to the touch and has a stalk that is covered in golden scales; it favours damp woodland.
Hard Shield-fern *P. aculeatum* (**2**) (frond to 1m) is darker green, feels tough, leathery and prickly to the touch, and has a stalk that is covered in reddish-brown scales; it favours wooded stony ground.

Ferns of rocks & stone walls

Stony ground is a feature of western and upland Britain and Ireland, and here ferns thrive on, or among, the rocks. These habitats provide shelter and protection from the elements, increased humidity and a firm footing for fern roots. Stone walls are man-made substitutes for natural rock faces.

Brittle Bladder-fern

Cystopteris fragilis
Dryopteridaceae Frond to 40cm

A delicate fern that grows in tufts. **Fronds** Repeatedly divided 2–3 times; they appear in Apr–Oct. **Spore cases** Rounded. **Status** Locally common in NW England and NW Scotland. Found growing from crevices in rocks and stone walls, mostly on limestone.

Mountain specialists

In botanical terms, mountains are inhospitable environments where heavy rain or snow, cold temperatures and strong winds can hamper growth. However, two fern species do well if they can find the right foothold. **Parsley Fern** *Cryptogramma crispa* (**1**) (frond to 25cm) is a Parsley-like species with fronds that form clustered tufts; it grows mostly on acid rocks, and is generally scarce, although locally common in Snowdonia and the Lake District. **Holly-fern** *Polystichum lonchitis* (**2**) (frond to 50cm) is a tufted plant with stiff, leathery, evergreen fronds whose lobes are curved and spiny; it is confined mainly to calcareous rocks.

Filmy-ferns

Filmy-ferns superficially resemble flattened mosses or miniature seaweeds, with almost translucent fronds. Prone to desiccation, they grow only on shady rocks where spray or seepage ensures high humidity. They favour acid rocks and where conditions suit them they form matted carpets. Two species are found locally in our region, although they occur mainly in the wetter western parts. **Tunbridge Filmy-fern** *Hymenophyllum tunbrigense* (**1**) (frond to 3cm) has dull green, short-stalked and flattened fronds whose veins do not reach the tip. **Wilson's Filmy-fern** *H. wilsonii* (**2**) has darker, narrower fronds with longer stalks and veins that do reach the tip.

Spleenworts

These tufted, evergreen ferns (genus *Asplenium*) like rock crevices and stone walls. **Black Spleenwort** *A. adiantum-nigrum* (**1**) (frond to 40cm) is the most widespread; its triangular fronds taper basally and have a dark brown stalk base; spore cases coat the frond underside and the plant favours acid soils. **Lanceolate Spleenwort** *A. obovatum* (**2**) (frond to 25cm) is similar, but its narrow triangular fronds do not taper basally; its spore cases are marginal and it is a W coast species. **Maidenhair Spleenwort** *A. trichomanes* (**3**) (frond to 20cm) has fronds with blackish-brown stalks and oval lobes; basal scales have a dark central stripe. **Green Spleenwort** *A. viride* (**4**) (frond to 15cm) is similar to Maidenhair Spleenwort but fronds have toothed lobes and a green stalk; basal scales lack a dark central stripe. **Sea Spleenwort** *A. marinum* (**5**) (frond to 30cm) grows in sea caves in W Britain; the shiny, bright green fronds taper at both ends and have oblong lobes and a green midrib.

Rustyback

Ceterach officinarum Aspleniaceae
Frond to 20cm

 Distinctive fern of stone walls and rocks. **Fronds** Dark green, divided simply into rounded lobes and forming tufted clumps; the lobe undersides are covered in rusty-brown scales. **Status** Widespread in suitable habitats, but common only in SW England, W Wales and Ireland. Today, it is mostly commonly seen on undisturbed stone walls.

Wall-rue

Asplenium ruta-muraria
Aspleniaceae Frond to 12cm

 Delicate little fern that often looks densely tufted. **Fronds** Evergreen, dull green in colour and divided simply into oval lobes. **Spore cases** On underside of lobes. **Status** Widespread but commonest in W Britain and Ireland. Found on stone walls and rocks, often in areas of limestone.

Oak Fern

Gymnocarpium dryopteris
Woodsiaceae Frond to 40cm

 Distinctive, long-stalked fern. **Fronds** Triangular, appearing singly from creeping roots and borne flat and level; they are bright green and divided 3 times. **Status** Local, commonest in the N. Favours neutral or acid soils and grows on rocky ground with woodland cover.

Beech Fern

Phegopteris connectilis
Thelypteridaceae Frond to 30cm

 Small, rather delicate fern. **Fronds** Triangular and superficially similar to those of Oak Fern. They are yellow-green and are usually held upright, but with the lowest lobe usually bent downwards. **Status** Local, commonest in the N. Favours wooded stony hillsides, mainly on acid soils.

Limestone pavements

Britain and Ireland have some of the finest examples of limestone pavements in the world, and these fascinating natural rock plateaux are among our most precious habitats. Laid bare originally by the scouring action of glaciers, the flat limestone outcrops have been weathered over the millennia by the erosive action of mildly acid rainwater. The result is a network of deeply eroded gullies (called grykes) on the surface, the overall appearance resembling massive and closely aligned building blocks. Although little actually grows on the surface of these pavements, their crevices and gullies harbour a phenomenal diversity of plant life, some species of which are seldom found elsewhere. Ferns do particularly well in limestone pavement gullies and several rare species are unique, or nearly so, to this habitat. Sadly, limestone pavements have suffered greatly from the removal of rocks for the gardening industry; fortunately, however, the best remaining areas are now protected by law.

Limestone pavement near Orton, Cumbria.

Atypical ferns

With their divided and sometimes feathery fronds, most ferns are instantly recognisable as such. However, among their number are some strange and atypical species, two of which are described here. **Adder's-tongue** *Ophioglossum vulgatum* (**1**) (frond to 10cm) grows in damp, undisturbed grassland and dune slacks, and is widespread but only locally common; its frond is bright green, oval and borne upright on a short stalk, and its spores are carried terminally on a tall fertile spike. **Moonwort** *Botrychium lunaria* (**2**) (frond to 15cm) grows on grassy moors, mountain slopes and undisturbed meadows, and is widespread but seldom common; its single stalk bears a solitary frond, divided into 3–9 rounded lobes, while the spores are carried on a separate fertile spike.

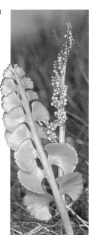

Fern allies & mosses

Most fern allies and mosses are very low-growing and easy to overlook. However, if you take the trouble to seek them out you will find that their representatives grow almost everywhere and that there are few terrestrial or freshwater habitats in which examples of their kind do not flourish.

Aquatic fern allies

Two unusual fern relatives grow around lake and pond margins; they look more like grasses than ferns. **Quillwort** *Isoetes lacustris* (**1**) (height to 20cm) grows as tufts of narrow, quill-like leaves on the beds of gravelly mountain lakes. **Pillwort** *Pilularia globulifera* (**2**) (frond to 10cm) grows locally in the shallow margins of acid ponds; the plant's spores are borne in small, basal, pea-like 'pills'.

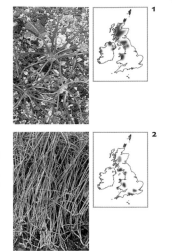

Clubmosses

Looking more like miniature conifer seedlings, clubmosses are often associated with moors and mountains. **Stag's-horn Clubmoss** *Lycopodium clavatum* (**1**) (height to 10cm) is evergreen with trailing and upright stems, both cloaked in pointed, scale-like leaves; its cones are borne on long stalks and it grows in N Wales, N England and Scotland. **Fir Clubmoss** *Huperzia selago* (**2**) (height to 10cm) is a tufted plant whose stems are cloaked in green, needle-like leaves; its spore cases are borne on the stem and it grows on dry, grassy moors in Scotland. **Alpine Clubmoss** *Diphasiastrum alpinum* (**3**) (height to 5cm) grows in short, upland grassland, while **Marsh Clubmoss** *Lycopodiella inundata* (**4**) (height to 3cm) is a mainly creeping plant of lowland bogs, commonest in the New Forest in Hampshire.

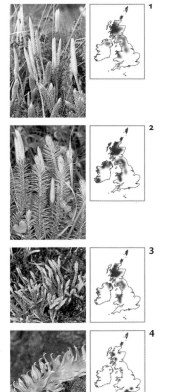

Silky Forklet-moss

Dicranella heteromalla
Dicranaceae Height to 3cm

 Widespread and often extremely common moss. **Leaves** Narrow, slightly curved and pointed. **Spore capsules** The ripe capsule is brown and is held at an angle. **Status** Common throughout. Found on bare ground on tracks and along woodland rides; favours neutral to acid soils.

Pointed Spear-moss

Calliergonella cuspidata
Amblystegiaceae Height to 4cm

 Typically yellow-green moss, with pointed tips to the shoots. **Stems** Reddish. **Leaves** Pressed together when young, which can give it a rather bedraggled appearance. **Status** Widespread and common. Found on dry chalk grassland and in damp ground such as lawns.

Capillary Thread-moss

Bryum capillare Bryaceae
Height to 3cm

 Forms compact and distinctive cushions. **Leaves** Oval and tipped with a fine point. **Spore capsules** Elongated, ovoid, long-stalked and drooping; green ripening to brown. **Status** Common and widespread. Found on roofs and walls.

Common Feather-moss

Kindbergia praelonga
Brachytheciaceae Spreading

 Forms tangled, spreading mats. **Stems** Much branched. **Leaves** Yellowish and feather-like. **Status** Common in shady woodlands.

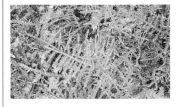

Rough-stalked Feather-moss

Brachythecium rutabulum
Brachytheciaceae Height to 4cm

 Shiny, carpet-forming moss. **Stems** Branching. **Leaves** Shiny, oval and pointed, covering the stems; usually dark green but sometimes tinged yellow. **Spore capsules** Curved and long-stalked. **Status** Very common in damp lawns, woodland and on banks.

Common Smoothcap

Atrichum undulatum
Polytrichaceae Height to 3cm

Familiar woodland-floor moss. **Leaves** Long, narrow and dark green, with wavy, toothed-edged margins. **Spore capsules** Curved, brown, borne on long stalks and held at an angle. **Status** Common and widespread. Found in woodlands, on most soil types except chalk.

Common Pocket-moss

Fissidens taxifolius Fissidentaceae
Height to 2cm

Forms rather tangled and straggly masses. **Leaves** Oval with pointed tips, carried on the stems usually more or less in one plane, creating a Yew-like appearance. **Spore capsules** Narrow and are borne on stalks that arise near the plant base. **Status** Common throughout. Found in damp, shady places.

Horsetails

These perennial plants (genus *Equisetum*), belonging to their own family (Equisetaceae), are seen mainly in the spring, often dying back and withering by late summer. Several species are found in Britain and Ireland, most of them associated with damp ground. **Field Horsetail** *E. arvense* (**1**) (height to 75cm) is the commonest species, forming spreading patches in grassy places and on waste ground; sterile shoots with ridged stems carry whorls of unbranched branches, while fertile stems appear in early spring and ripen in May. **Wood Horsetail** *E. sylvaticum* (**2**) (height to 50cm), an elegant species of shady woodland and moors, is widespread but commonest in N Britain; sterile stems resemble the growing tips of conifers and carry whorls of slender branches that are further branched, and fertile stems ripen in May. **Water Horsetail** *E. fluviatile* (**3**) (height to 1m) grows in marshes and along the margins of ponds and lakes, and is widespread and locally common; the tall, unbranched stems are jointed and thin, with whorls of narrow, jointed branches, and the spores are borne in blunt, cone-like structures at the ends of some stems. **Marsh Horsetail** *E. palustre* is similar but has pointed cones.

Grey-cushioned Grimmia

Grimmia pulvinata Grimmiaceae
Height to 3cm

Forms compact cushions. **Leaves** Narrow and grey-green; the greyish, pointed tips can sometimes give the whole cushion a silvery appearance, especially in dry weather. **Status** Common and widespread, especially in limestone areas. Grows on walls and roofs and hence a common garden species.

Silky Wall Feather-moss

Homalothecium sericeum
Brachytheciaceae Height to 2cm

Forms creeping, tangled mats. **Stems** Branched and mat-forming; in dry weather, they curl and turn brown. **Leaves** Narrow, finely tipped and glossy, covering the stems. **Status** Common and widespread. Found in a wide range of sites, including old brick and stone walls and at the base of tree trunks.

Swan-neck Mosses

Campylopus sp. Dicranaceae
Height to 4cm

A group of tufted mosses that grow on damp rocks, wet peat or rotting wood. **Leaves** Very narrow, typically tapering to a fine, hair-like point. **Status** Native species have a mainly W range, but the introduced *C. introflexus* is now widespread and sometimes grows on acid roof tiles in urban and suburban settings.

Mosses & liverworts

As a testament to the diversity and colonising powers of mosses and liverworts, few areas of lawn, open woodland floor or tree trunk do not have at least one species of these plants growing on them.

Cypress-leaved Plait-moss

Hypnum cupressiforme Hypnaceae
Spreading

Typically forms flattened clusters or mats. **Leaves** Curved, oval and pointed, overlapping each other to cover the stems. **Spore capsule** Borne on a short stalk. **Status** Widespread and common. Found at the base of tree trunks, and on walls and boulders.

Large White-moss

Leucobryum glaucum
Leucobryaceae Height to 4cm

Distinctive moss that forms cushions on the ground. Larger specimens of cushions may become eroded towards the centre, exposing older, dead parts. **Leaves** Narrow and grey-green, but becoming almost white in dry weather. **Status** Locally common. Found on damp woodland and moors.

Woolly Fringe-moss

Racomitrium lanuginosum
Grimmiaceae Height to 2cm

Locally dominant moss of mountain tops; where it is not trampled it forms deep carpets that cover sizeable areas. **Stems** Long and branched. **Leaves** Narrow, grey-green and tipped with a white, hair-like point. **Spore capsules** seldom produced. **Status** Locally common in upland areas.

Common Tamarisk-moss

Thuidium tamariscinum
Thuidiaceae Spreading

Distinctive moss. **Leaves** Fresh green fronds look feathery or fern-like. **Stems** Main stems are dark. **Status** Common in woodlands among fallen branches and leaf litter.

Wall Screw-moss

Tortula muralis Pottiaceae
Height to 1cm

Forms low, spreading cushions. **Leaves** Oval with rounded tips, ending in a fine point. **Spore capsules** Narrow and held upright on long, slender stalks; yellow when young but ripening brown. **Status** Widespread and common. Grows on old brick walls and rocks.

Haircap mosses

Haircap mosses belong to the genus *Polytrichum* (height to 10cm), several species of which are widespread in the region. All are distinctive, having narrow, spreading leaves, the plants themselves resembling miniature conifer shoots; typically, they form carpets on woodland floors and on decaying wood. **Common Haircap** *P. commune* (**1**) has green leaves while those of **Juniper Haircap** *P. juniperum* (**2**) are tipped reddish. The box-shaped spore capsules are brown when ripe and are held on tall, slender stems.

Bog-mosses

Bog-mosses belong to the genus *Sphagnum* (height to 5cm) and are extremely important wetland plants, many of them growing in the wettest places, often almost floating on water. They act a bit like sponges, absorbing water when there is plenty available and retaining it in times of drought. Those that grow in the wetter parts of heathland bogs can sometimes be picked out at a distance by the fresh green colour of their leaves. A large number of superficially similar species are found in Britain and Ireland; of these, *S. palustre* is found beside lakes and ponds, *S. papillosum* (**1**) is characteristic of peat bogs and *S. recurvum* (**2**) often grows in waterlogged woodland.

Swan's-neck Thyme-moss

Mnium hornum Mniaceae
Height to 8cm

Densely tufted moss that forms cushions. **Stems** Dark. **Leaves** Dark green, narrow and pointed. **Spore capsules** Rather cylindrical and pendulous. **Status** Widespread and common throughout much of the region. Grows on humus and tree roots in woods, mostly on acid soils.

Waved Silk-moss

Plagiothecium undulatum
Plagiotheciaceae Spreading

One of several similar species of prostrate mosses. **Leaves** Ovate, undulate and bright green, tapering to acute tips. **Stems** Up to 15cm long. **Status** Widespread and common. Found on logs and humus on heaths and in open woods, favouring acid soils.

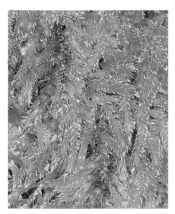

Crescent-cup Liverwort

Lunularia cruciata Lunulariaceae
Spreading

Fresh green liverwort. **Thallus** Lacks a median line and has crescent- or halfmoon-shaped reproductive cups. **Status** Widespread and common. Found on damp ground beside streams, but also often abundant in the well-watered humus of potted plants and beside paths.

Variable-leaved Crestwort

Lophocolea heterophylla
Geocalycaceae Spreading

Moss-like liverwort. **Stems** Branching and trailing, to 2cm long. **Leaves** Come in 2 forms: small ones, and larger, toothed-tipped, overlapping ones that almost hide the former. **Status** Widespread and common in deciduous woods. Grows on the bark of living trees and on fallen branches.

Forked Veilwort

Metzgeria furcata Metzgeriaceae
Spreading

Thallus Long and narrow, barely 2mm wide and only 1 cell thick except for the midrib, which is thickened; the result is a superficial resemblance to a miniature seaweed. **Status** Common throughout. Found on tree trunks, rocks and walls in shady places.

Great Scented Liverwort

Conocephalum conicum
Conocephalaceae Spreading

Carpeting, scented liverwort. **Thallus** Dark green, broad or narrow and fleshy, to 15cm long; the lobes often overlap. On close inspection, its surface is marked with pale dots. **Status** Common throughout on damp rocks and stones, and often found beside woodland streams.

Common Liverwort

Marchantia polymorpha
Marchantiaceae Spreading

Thallus Divided and dark green. The lobes overlap, and the surface bears shallow cups and umbrella-shaped reproductive structures: stalked, rayed females and toadstool-like males. **Status** Widespread and common. Grows on shady river banks and the well-watered compost of potted plants.

Greater Featherwort

Plagiochila asplenioides
Plagiochilaceae Spreading

A delicate, leafy liverwort that resembles a miniature version of Maidenhair Spleenwort (*see* p.176). **Leaves** Oval, overlapping and borne in 2 rows on a thickened stem. **Status** Widespread and fairly common. Grows on damp, shady banks.

Overleaf Pellia

Pellia epiphylla Pelliaceae
Spreading

Familiar, patch-forming liverwort. **Thallus** Broad, flattened and branched, with a thickened midrib. **Spore capsules** Round, black and shiny, borne on green stalks; they appear in spring. **Status** Common and widespread. Found on damp, shady banks, often beside streams.

Spore capsules

Lichens

Colour is mostly lacking in the winter landscape, but take a walk in woodland or along the coast and you will find some ornate and vibrant gems in the form of lichens: some encrust rocks, many grow on the ground while others are attached to tree bark. If you use a hand lens to inspect these organisms, you will discover some fiery colours lurking amongst the subtle greys and greens.

Cladonia floerkeana
Height 3cm

 Encrusting lichen. Forms patches of greyish-white scales from which granular, scale-encrusted stalks arise, topped with bright red, spore-producing bodies. **Status** common on heaths and moors.

Cladonia portentosa
Height 5cm

 Wirewool-like lichen. Forms intricate networks of densely packed blue-grey or whitish strands. **Status** Widespread and common on moors and heaths. Several similar species occur.

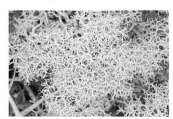

Cladonia pyxidata
Height 3cm

 Cup-shaped, typically greyish green or bluish lichen. Broad, cup-shaped fruiting bodies, borne on short stalks, often have a granular texture. **Status** Common and widespread on heaths or rotting wood. Several similar species occur.

Cladonia uncialis
Height 5cm

 Antler-like lichen. Fruiting bodies are branched and bluish green. **Status** Common and widespread on heaths and moors, sometimes growing in rather waterlogged ground. Several similar species occur.

Graphis scripta
Width 15mm

 Bark-encrusting lichen. Forms an irregularly rounded patch that is blue-grey or green-grey; spore-producing structures are black lines and scribbles that are slit-like. **Status** Common on deciduous trees, notably Hazel and Ash.

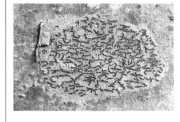

Hypogymnia physodes
Width 25mm

 Much-branched, encrusting lichen; grows on twigs but also on rocks and walls. Forms irregularly rounded patches, smooth and grey on upper surface. **Status** Common and widespread in woods and near coasts. Several similar species occur.

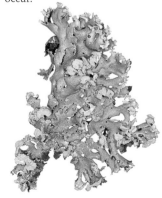

Parmelia caperata
Spreading

 Bark-encrusting lichen. Forms grey-green, rounded patches, with overlapping lobes; surfaces have brown, flat-topped, spore-producing discs. **Status** Fairly common on deciduous tree bark but intolerant of air pollution. Several similar species occur.

Usnea florida
Width 5cm

 Forms tangled masses of long, thin threads, attached to twigs with a holdfast. Cup-shaped spore-producing bodies form at tips of some branches, cups fringed by thin hairs. **Status** Commonest in W. Several similar species occur.

Usnea subfloridana
Length 9cm

Forms dense, tangled masses of thin threads that 'bounce back' when compressed. Attached by a holdfast to twigs but sometimes grows on rocks. Cup-shaped spore-producing bodies are absent. **Status** Common and widespread.

Crab's-eye Lichen
Ochrolechia parella Spreading

Rock-encrusting, patch-forming lichen. Surface is greyish with a pale margin and clusters of raised, rounded and flat-topped spore-producing structures. **Status** Common on walls and rocks, mainly in uplands and western Britain. Several superficially similar species occur.

Map Lichen
Rhizocarpon geographicum Spreading

Aptly-named, encrusting upland lichen. Surface is yellowish and etched with black spore-producing bodies. When neighbouring colonies meet, boundaries are defined by black margins, creating a map-like appearance. **Status** Locally common in uplands.

Yellow Scales
Xanthoria parientina Spreading

Our most familiar and colourful lichen; most spectacular on coasts but also common inland. Forms bright orange-yellow patches on rocks, walls and brickwork. Surface comprises leafy, narrow and wrinkled scales. **Status** Widespread and common thoughout Britain and Ireland.

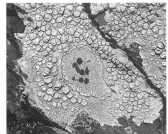

Tree Lungwort
Lobaria pulmonaria Spreading

Attractive lichen. Grows on tree bark. Lung-like in appearance with spreading lobes, pitted with depressions and adorned with bronchiole-like network of veins. **Status** Local, restricted to areas of low pollution and high rainfall, growing on ancient trees.

Oak Moss
Evernia prunastri Width 6cm

Distinctive lichen, usually found on oak branches. Comprises repeatedly divided flattened, partly curled and twisted branches. Upper surface is greenish grey, lower surface is white. **Status** Widespread and common in woodlands throughout Britain and Ireland.

Sea Ivory
Ramalina siliquosa Length 3cm

Tufted, branched lichen of coastal rocks and stone walls. Grows well above high-tide mark but still tolerant of salt spray. Branches are flattened and grey, and bear disc-like spore-producing bodies. **Status** Widespread, often abundant in W Britain.

Caloplaca marina
Width 5cm

Bright orange, encrusting lichen. Forms irregular patches on seashore rocks around high-water mark. Tolerant of salt spray and brief immersion in seawater. **Status** Widespread around coasts of Britain and Ireland; commonest in W.

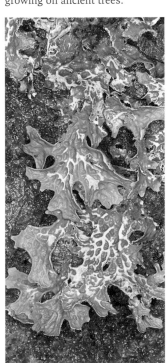

Dogtooth Lichen
Peltigera canina Width 7cm

Mat-forming lichen. Forms dense patches on ground, usually on sandy heathland soils. Attached to ground by root-like structures on lower surface. Upper surface bears tooth-like reproductive structures. **Status** Widespread and common.

Black Shields
Tephromela atra Spreading

Patch-forming lichen. Encrusts seashore rocks, just above high-tide mark; tolerates salt spray. Surface is knobbly and grey; spore-producing structures are rounded and black with pale grey margins. **Status** Commonest on coasts and particularly abundant in W, but also found inland.

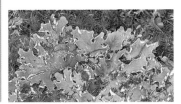

FUNGI

Most people are only aware of fungi for a few weeks in autumn when colourful toadstools decorate the countryside. But fungi are around all year and touch our lives in countless ways. From the yeasts that give us our bread, beer and wine, to the mould that grows on marmalade and is the basis for penicillin and other drugs, they are everyday household companions. In environmental terms, their importance cannot be overstated: they aid the growth of trees and are vital agents of decay and of nutrient recycling. Until comparatively recently, fungi were thought of as plants, albeit rather unusual ones. Today, they are classified as a distinct group, separate from either plants or animals, and in terms of classification they command their own kingdom.

Common Earthball.

The structure of fungi

In most fungi the main body is hidden from view and consists of a mass of thin threads called hyphae, which permeate the species' growing medium, such as rotting wood or leaf litter. The collective name for the hidden body of fungal hyphae is the mycelium, a mass of white threads that you may discover if you break off a piece of bark on a decaying tree stump. Hyphae feed the fungus by producing enzymes to aid the release of nutrients, which are then absorbed along with water from the growing medium.

Sexual reproduction in fungi involves the fusion of hyphae from different fungal colonies. Toadstools are fungal fruiting bodies and are composed of tightly packed hyphae; their role is to produce and disperse spores.

ABOVE: Fungal hyphae growing through rotting wood.
LEFT: Fairy Inkcaps *Coprinus disseminatus* are the visible, external sign of a fungus whose hyphae have a permanent, permeating presence inside the tree on which they appear.

Fungal lifestyles and growing habitats

Fungi are crucial in the recycling of nutrients from dead organic matter back into the environment. But many species also have a complex and positive relationship with living plants, and with trees in particular. In these partnerships, the fungus sheathes the tree roots, its hyphae extending into the soil and also penetrating the tree's root cells; this association is known as a mycorrhiza, which means 'fungus root'. The fungus derives almost all of its energy from the photosynthetic reactions of the tree's leaves. In return, the tree obtains nitrogen and phosphorous, otherwise in short supply, via fungal action in the soil. Many autumn toadstools are the fruiting bodies of these partnerships. A few fungi, such as Honey Fungus *Armillaria mellea*, do attack living trees and elsewhere in the natural world, some parasitise insects, and in a few cases they even attack their own.

A Parasitic Bolete *Pseudoboletus parasiticus*, parasitising a Common Earthball *Scleroderma citrinum*.

Like most boletes, the Scarletina Bolete *Boletus luridiformis* has a myccorhizal association with tree roots.

ABOVE: Collared Parachute *Marasmius rotula*.
BELOW: Russet Toughshank *Collybia dryophila*.

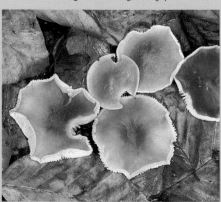

Spore production and dispersal

The part of a fungus that can be seen is called the fruiting body. Its role is to produce spores and to aid their dispersal. In the simplest cases, spores are produced all over the smooth surface of the toadstool. Increasing the surface area available for spore production means that greater numbers can be produced. Gills, pores or teeth on the underside of the toadstool or bracket allow countless millions of spores to be liberated in the life of a single toadstool. Because spores are generally dispersed by the wind, the higher their liberation occurs away from the ground, the more chance they have of being carried far and wide. So, toadstool caps are typically borne on a stalk, and bracket fungi fruiting bodies appear on the trunks or branches of trees, well above the ground.

1 Simple spore-producing surfaces
The fruiting bodies of some fungi, such as the earthtongue *Geoglossum cookeianum* (seen here), are simple structures whose spores are produced over the whole surface.

2 Pored toadstools
Members of the genus *Boletus* (and many bracket fungi) have tubes on the undersides of their caps, inside which spores are produced; the spores are released via external pores.

3 Gilled toadstools
Most familiar common toadstools, including the Cultivated Mushroom *Agaricus bisporus*, have radiating gills on the underside of their cap. Released spores fall from these gills and are carried by the wind; if a cut cap is placed on a sheet of white paper, a detailed spore print is produced.

4 Toothed fungi
A few unusual fungi increase the surface area for spore production with teeth, either on the underside of the cap in hedgehog fungi (genus *Hydnum*), or all over their external surface in species like the Bearded Tooth *Hericium erinaceus* (seen here).

spore print

5 Puffballs and earthstars
These organisms are known as stomach fungi, and their spores are produced inside a rounded sac that opens via a terminal pore. If raindrops hit the sac, the internal pressure is enough to puff out a plume of spores. Some are anchored to the ground, but in the Collared Earthstar *Geastrum triplex* (seen below), the sac's outer casing spreads and forms radiating arms to improve stability.

6 Cup fungi
In cup fungi, spores are usually produced on the inner surface and are washed out by raindrops. With the Common Bird's-nest *Crucibulum laeve* (seen below) and related fungi, the spores are borne in egg-like packets, which are flung out of the cup and dispersed by falling rain.

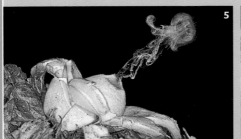

Animal agents

Slugs can wreak havoc on toadstools, especially if the season is dry and other sources of water are in short supply. Though they damage the appearance of otherwise pristine toadstools, they probably don't do any real damage, unless the toadstool fails to mature. Undigested spores are inadvertently transported by the slugs and often taken underground. Research suggests that in some fungi, slugs are the most important agents of spore dispersal. There are parallels here with the Stinkhorn *Phallus impudicus*, and related species, whose spores are produced in stinking slime that attracts feeding flies, which in turn spread the spores when they fly off.

A Dusky Slug *Arion subfuscus* feasting on a toadstool.

Alien invaders

Britain's collection of fungi is constantly changing. Though there are plenty of losses, associated with habitat destruction and degradation, there have also been some gains in recent years, notably species that have been introduced inadvertently by man. The Redlead Roundhead *Stropharia aurantiaca* and the Red Cage Fungus *Clathrus ruber* are spreading thanks to the increased use of woodchip mulch; the former is now locally abundant in some flowerbeds. Perhaps the weirdest of all is the Starfish Fungus *Aseroë rubra*, which comes from the southern hemisphere and now grows in a few woods in southern England; it emits an unspeakable smell that attracts flies, which disperse its spores.

Starfish Fungus.

Fungi

British and Irish fungi are so varied in appearance and structure that a lifetime could be spent studying them. Even species with a typically toadstool-like appearance are tremendously variable, although broadly speaking the undersides of their caps are covered either in pores, in the case of boletes, or gills.

Boletes

Although they differ in size, shape and colour, all boletes have tubes underneath the cap that open and allow the release of spores via pores. Many species grow in Britain, and just a few of the most striking and widespread ones are described here. The **Cep** *Boletus edulis* (**1**) (height to 25cm) is the most sought-after culinary species and grows in deciduous woodland; its cap is brown, the pores are white at first but become creamy or yellow, and the stem is fat and bulbous. The **Red-cracking Bolete** *B. chrysenteron* (**2**) (height to 10cm) grows in deciduous woodland and has a buffish-brown cap that cracks to reveal red flesh; the pores are buffish yellow and the stem is flushed red. The **Bay Bolete** *B. badius* (**3**) (height to 15cm) grows in woodland and its cap colour ranges from tan to buff; the pores are yellow but bruise bluish green, and the white flesh flushes blue when cut. The **Orange Birch Bolete** *Leccinum versipelle* (**4**) (height to 25cm) grows under birches and has a domed, orange-brown cap; the pores are greyish white, the flesh is white but blackens when cut, and the white stem is marked with dark scales. The **Larch Bolete** *Suillus grevillei* (**5**) (height to 20cm) grows under Larch and has a viscid yellowish-orange cap; the pores are yellow and the stem is sticky, bearing a ring.

Milkcaps

Milkcaps belong to the genus *Lactarius* and are so called because their cap and gills produce droplets of 'milk' when cut or scratched. In most species the milk is white, but if the colour changes on contact with air then this can be a useful aid to identification, as can the taste. Many species grow in Britain, three examples of which are described here. The **Woolly Milkcap** *L. torminosus* (**1**) (height to 10cm) grows under birch and has a reddish-orange cap with darker concentric rings and woolly hairs; the stem is buff, the gills are whitish and the milk is white. The **Ugly Milkcap** *L. turpis* (**2**) (height to 8cm) grows under birch and is dirty brown and sticky, often with flakes of leaf sticking to the cap; its milk is greyish. The **Yellowdrop Milkcap** *L. chrysorrheus* (**3**) (height to 8cm) has a smooth orange-buff cap with concentric rings; the gills are pinkish white and the milk is white, soon turning yellow.

Brittlegills

The brittlegills belong to the genus *Russula* and are recognisable as a group by their usually colourful caps, clean-looking white stems and pale (white in many species) gills. They are associated with woodland trees. A large number of species grow in Britain, a small selection of which is described here. The **Charcoal Burner** *R. cyanoxantha* (**1**) (height to 9cm) grows in deciduous woodland and has a greyish-lilac cap that is typically blotched with black and reddish purple; the gills are white and feel slightly greasy. The **Beechwood Sickener** *R. nobilis* (**2**) (height to 10cm) grows under Beech and has a bright red cap and slightly creamy gills; it is a poisonous species. The **Yellow Swamp Brittlegill** *R. claroflava* (**3**) (height to 12cm) has a bright yellow cap with grooved margins and off-white gills; it grows in damp birch woodland. The **Common Yellow Brittlegill** *R. ochroleuca* (**4**) (height to 12cm) grows in drier deciduous woodland; its cap is ochre-yellow and its gills are white. The **Purple Brittlegill** *R. atropurpurea* (**5**) (height to 7cm) grows under oaks and Beech; its cap grades from almost black in the centre to reddish purple around the edge, and its gills are off-white.

The dreaded Honey Fungus

Just a mention of its name is enough to strike fear in the heart of any gardener, such is the reputation of **Honey Fungus** *Armillaria mellea* for attacking and killing prize trees. In the countryside at large, however, the species has a positive role to play and is a vital agent in the processes of death, decay and the recycling of nutrients. Its cap is brown and slightly scaly, the gills are pale buff and the stem has a ring; typically, Honey Fungus appears in sizeable clusters on or near tree stumps.

Parasol

Macrolepiota procera
Height to 30cm

An impressive and edible grassland species. **Cap** Egg-shaped when young; the mature cap is flat and marked with brown scales. **Gills** White and widely spaced. **Stem** Brown with a scale-like pattern. **Status** Locally common and widespread in meadows and pastures.

Stinking Dapperling

Lepiota cristata
Height to 6cm

A dainty species that is poisonous and should not be handled. **Cap** Pale with brown scales, the colour most intense in the centre. **Gills** White and widely spaced. **Stem** Pale and slightly scaly. **Status** Locally common in the S, sometimes in woods and often in flowerbeds.

Fungi

Agarics

Members of the genus *Amanita*, all of which are associated with trees, have the classic toadstool look, with colourful caps and tall, clean-looking stems. All species arise from a basal sac (called a volva) and their caps are often adorned with remains of the veil (which protected the developing cap); a ring is also present on the stem of some species. For most people, the **Fly Agaric** *A. muscaria* (**1**) (height to 15cm) is the archetypal toadstool, with its white-flecked red cap and contrasting white stem; it grows in troops under birch. ***A. excelsa*** (**2**) (height to 12cm) has a grey-buff cap with white flecks and a stem marked with white scales; it grows in deciduous woodland. The **Blusher** *A. rubescens* (**3**) (height to 15cm) is similar, but the veil flecks on the cap are pinkish grey and the stem blushes red. The **Panther Cap** *A. pantherina* (**4**) (height to 15cm) is a striking toadstool with a chocolate-brown cap and white flecks; it grows under Beech and oaks. As its name suggests, the **Deathcap** *A. phalloides* (**5**) (height to 15cm), a beechwood specialist, is poisonous and should not even be touched, let alone eaten; it has a sickly smell, green-tinged cap, white gills and stem, and grows from an obvious volva. The **False Deathcap** *A. citrina* (**6**) (height to 12cm) smells of potato and grows in deciduous woodland on acid soils; its pale greenish-yellow cap has brownish flecks and the base of the stem is bulbous. The **Tawny Grisette** *A. fulva* (**7**) (height to 12cm) often grows with oaks and has a tawny-brown cap that is marked with marginal striations; its basal volva is obvious but it lacks a ring. The **Grisette** *A. vaginata* (**8**) (height to 12cm) has a grey cap with striking marginal striations; its clean-looking stem arises from a large volva.

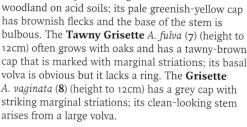

Tricholomas

Members of the genus *Tricholoma* are variable, and the identification of many species is nearly impossible in the field – even some experts give up on them! All they have in common is a typical toadstool-like appearance, and that's about it. Of the numerous species found in our region, two are reasonably distinctive. The **Sulphur Knight** *T. sulphureum* (1) (height to 10cm) is sulphur-yellow and smells strongly of coal gas; it grows in deciduous woodland on acid soils. The **Birch Knight** *T. fulvum* (2) (height to 8cm) has a brown, viscid cap, a buffish-brown, fibrous stem and smells mealy; it grows under birches.

Whitelaced Shank

Megacollybia platyphylla
Tricholomataceae
Height to 15cm

Woodland species with a long, root-like structure. **Cap** Brown, fibrous and often splits around the margins. **Gills** White and widely spaced. **Stem** Buff and fibrous. **Status** Widespread and locally common in deciduous woodland.

Aniseed Funnel

Clitocybe odora
Tricholomataceae
Height to 12cm

Attractive and distinctive toadstool that smells strongly of aniseed. **Cap** Blue and fibrous. **Gills** Whitish. **Stem** Whitish and streaked. **Status** Widespread and locally common in deciduous woodland throughout.

Clouded Agaric

Clitocybe nebularis
Tricholomataceae
Height to 15cm

Woodland toadstool that often grows in troops. **Cap** Blue-grey to lilac, grading to brown in the centre. **Gills** Creamy white. **Stem** Fibrous and swollen at the base. **Status** Locally common in deciduous woodland.

Butter Cap

Collybia butyracea
Tricholomataceae
Height to 15cm

Distinctive woodland toadstool. **Cap** Brown, palest around the margin, and with a greasy, buttery texture. **Gills** White. **Stem** Tough, pale brown, tapering upward and bulbous at the base. **Status** Locally common in woodland.

Waxcaps

Waxcaps are delightful little grassland fungi belonging to the genus *Hygrocybe*. Most are extremely colourful and their caps – and sometimes stalks – have a greasy, waxy or viscid texture. Several dozen species grow in our region, two of the most attractive of which are described here. The **Heath Waxcap** *H. laeta* (1) (height to 5cm) grows in acid grassland and is bright yellow. The **Blackening Waxcap** *H. conica* (2) (height to 8cm) has a conical cap that starts off red or yellow, but soon turns black; it grows in meadows and coastal dunes.

Fungi

Deceiving fungi

Two members of the genus *Laccaria* are popularly known as 'deceivers' because they are so variable in appearance. Try to learn them in all their forms (fortunately, their respective colours are pretty constant), as this will help with the identification of other, similar fungi. They are both woodland species. The **Deceiver** *L. laccata* (**1**) (height to 5cm) has an irregularly rounded cap that is orange-brown; its gills are buff, and the stem is twisted and concolourous with the cap. The **Amethyst Deceiver** *L. amethystina* (**2**) (height to 5cm) is wholly lilac or purple; the cap is domed at first but flattens irregularly with age.

Blewits

Blewits belong to the genus *Lepista* and are tough fungi. The **Field Blewit** *L. saeva* (**1**) (height to 8cm) has a brownish cap, pinkish-brown gills and a lilac-tinged stem; it grows in fields and woodland margins. The **Wood Blewit** *L. nuda* (**2**) (height to 8cm) is a species of deciduous woodland and is often entirely tinged lilac or purple.

Bonnet fungi

Members of the genus *Mycena* are known collectively as bonnet fungi. All are dainty and rather brittle, with slender stems. Apart from a few distinctive species, most are hard to identify to species level with certainty, and even experienced mycologists find it a challenge. The **Burgundydrop Bonnet** *M. haematopus* (**1**) grows on rotting wood and yields red 'milk' when damaged. The **Lilac Bonnet** *M. pura* (**2**) has a cap that is either pink or lilac and a stem that is flushed with the cap colour; it grows on the woodland floor.

Tawny Funnel

Lepista flaccida Tricholomataceae
Height to 6cm

Attractive species. **Cap** Orange-buff and funnel-shaped. **Gills** Decurrent and white. **Stem** White. **Status** Widespread and fairly common. Grows in broadleaved woodland.

Porcelain Fungus

Oudemansiella mucida Tricholomataceae Width to 7cm

Distinctive, rather translucent fungus. **Cap** White and slimy. **Gills** White. **Stem** Slender and variable in length. **Status** Locally common. Grows on the dying and fallen branches of deciduous trees, mainly Beech.

Rooting Shank

Xerula radicata Marasmiaceae
Height to 25cm

Deeply rooting toadstool. **Cap** Wrinkled and buffish grey. **Gills** White. **Stem** Long, slender and palest towards the top. **Status** Locally common. Grows in deciduous woods.

Rooting Shank

Stubble Rosegill

Volvariella gloiocephala
Pluteaceae Height to 15cm

Emerges from a sac-like volva. **Cap** Olive-grey, striated and shiny when dry, viscid when wet. **Gills** Pink. **Stem** White. **Status** Local, growing on manured ground and woodchip piles.

emerging cap

Brown Rollrim

Paxillus involutus Paxillaceae
Height to 12cm

Familiar, rather grubby-looking toadstool. **Cap** Tan to dirty brown in colour; flattened at first and then funnel-shaped but with an inrolled margin. **Gills** Decurrent and brown. **Stem** Brown. **Status** A common woodland species, associated with birch.

Cortinarius fungi

The genus *Cortinarius* is another of those groups that are extremely difficult to assign to a particular species, and many mycologists are content to leave detailed identification to the experts. A common name for the group is webcap – in many species the web-like remains of the veil cling to the margins of the cap and adorn the upper stem. The **Variable Webcap** *C. anomalus* (height to 7cm) is a typical woodland webcap, with a brown cap and gills and a fibrous stem.

Common Rustgill

Gymonpilus penetrans
Cortinariaceae Height to 8cm

Tufted species. **Cap** Yellow-tan and fibrous. **Gills** Yellow at first, becoming rust-spotted as the spores mature. **Stem** Buffish, gradually stained darker by spores. **Status** Widespread and common. Grows on conifer stumps and woodchip mulch.

Sulphur Tuft

Hypholoma fasciculare
Strophariaceae Height to 8cm

Woodland species. **Cap** Sulphur-yellow, dark in the centre. **Gills** Yellow. **Stem** Often curved. **Status** Widespread and extremely common. Grows in clumps on dead stumps and the fallen branches of deciduous trees.

Common Rustgill underside

Common Rustgill

BELOW: Sulphut Tuft

Sheathed Woodtuft

Kuehneromyces mutabilis
Strophariaceae Height to 7cm

Familiar tufted woodland fungus. **Cap** Orange-brown, drying yellow from the centre and looking 2-coloured. **Gills** Buffish yellow. **Stem** Fibrous with a ring. **Status** Common, forming tufted clumps on stumps and dead branches.

Blue Roundhead

Stropharia caerulea
Strophariaceae Height to 9cm

Attractive blue toadstool. **Cap** Frilly veil remains persist around its margin. **Gills** Pinkish buff. **Stem** Has a coating of white scales and a ring. **Status** Locally common. Found in conifer woods and on woodchip mulch.

Mushrooms & odd-looking fungi

While there are some truly delicious species lurking in our woodlands, in reality most either taste unpleasant or inedibly tough; and a few are actually poisonous. So caution is the best approach: do not eat wild fungi unless you are absolutely certain of their identity.

Mushrooms

With a few exceptions, most members of the genus *Agaricus* are edible and delicious, including among their numbers the Cultivated Mushroom *A. bisporus*. All have pink gills that become brown as the spores mature. Identifying wild-growing *Agaricus* mushrooms is often difficult, but two extremes forms are reasonably distinct.

Field Mushroom
Agaricus campestris Agaricaceae
Height to 8cm

 Familiar meadow species. **Cap** Smooth or only slightly shaggy. **Gills** Pink. **Stem** White. **Status** Common and widespread in grassland.

The Prince
Agaricus augustus Agaricaceae
Height to 12cm

 Impressive shaggy species. **Cap** Has scaly surface. **Gills** Pink. **Stem** White and flecked. **Status** Locally common in woodland.

TOP: Field Mushroom; ABOVE: The Prince

Inkcaps

Inkcaps have caps that eventually blacken and liquefy, the spores being washed away by water rather than dispersed by air. Changes in classification are underway, but here our common species are treated as belonging to the genus *Coprinus*.

Common Inkcap
Coprinus atramentarius
Coprinaceae Height to 7cm

 Familiar inkcap. **Cap** Yellow-buff with an orange centre at first, but this soon blackening. **Gills** White then brown. **Stem** Slender. **Status** Common and widespread, often growing beside paths and tracks.

Shaggy Inkcap
Coprinus comatus Coprinaceae
Height to 15cm

 Our most familiar inkcap. **Cap** Shaggy, whitish and bell-shaped; liquefies from the bottom upwards. **Gills** White then pink. **Status** Widespread and common beside tracks and on verges.

Magpie Inkcap
Coprinus picaceus Coprinaceae
Height to 12cm

 Distinctive piebald species. **Cap** Shaggy, and black and white. **Gills** White then pink. **Stem** White. **Status** Locally common under Beech or on woodchips.

Fairy Inkcap
Coprinus disseminatus
Coprinaceae Height to 4cm

 Impressive inkcap that grows in large troops at the base of deciduous trees. **Cap** Silvery white. **Gills** Reddish. **Stem** White and slender. **Status** Widespread and common.

Edible and delicious

Connoisseurs of edible fungi often shun cooking with Cultivated Mushrooms. Instead, following mainly French and Italian traditions, they opt for wild species with sublime flavours and bizarre appearances. Four of the favourites are widely available dried, but they also grow wild in Britain. The **Horn of Plenty** *Craterellus cornucopioides* (**1**) grows under Beech in small troops, while the bright yellow **Chanterelle** *Cantharellus cibarius* (**2**) (height to 10cm), which smells of apricots, favours deciduous woods of all types. The **Wood Hedgehog** *Hydnum repandum* (**3**) (height to 10cm) grows in leaf litter and has teeth rather than gills on the underside of its cap. Unusually among our fungi, the **Morel** *Morchella esculenta* (**4**) (height to 12cm) appears in spring, not autumn, and its cap is a convoluted honeycomb.

Wood Cauliflower

Sparassis crispa Sparassidaceae
Width to 25cm

Bizarre fungus, the size and shape of a cauliflower. **Surface** Much divided into buffish, convoluted, overlapping, leafy-looking lobes. **Status** Widespread and locally common. Grows at the base of mature conifers.

Spindle-like fungi

Look carefully on the woodland floor and among meadow grasses and you may find some strange, spindly little fungi growing in tufts; in some species the lobes simply taper, but in others they are branched and antler-like. The **Meadow Coral** *Clavulinopsis corniculata* (**1**) forms yellow antler- or coral-like tufts and grows in grassland. **Yellow Stagshorn** *Calocera viscosa* (**2**) is its woodland counterpart, while **Upright Coral** *Ramaria stricta* (**3**) grows on part-buried stumps in woodland and is warm buff in colour.

Stinkhorns

Stinkhorns are among the few fungal species that can be detected in the field by smell alone. The spore-laden mucus that coats the tip of the fungal stalk is foul-smelling and attracts not only mycologists but also flies, the latter feeding on the mucus and subsequently disperse the spores. The **Stinkhorn** *Phallus impudicus* (**1**) (height to 15cm) is the largest and most common species, while the **Dog Stinkhorn** *Mutinus caninus* (**2**) (length to 10cm) usually grows on or near rotting stumps and has a more conical, reddish tip. Both species are found in woodland.

Brackets & soil-dwelling fungi

Most people's idea of fungi is mushrooms and toadstools, but in reality fungal fruiting bodies come in a wide variety of other shapes and structures. Bracket-type forms are perhaps the most striking of these, although if you search at ground level in woods, heaths and meadows you may come across some truly weird and wonderful examples.

Brackets

The term 'bracket fungus' is a non-scientific one that applies to a wide range of species, some of which are unrelated. However, for the novice mycologist it is a valid way of describing this eclectic group of species, all of which appear as level brackets on the trunks or fallen branches of trees.

Oyster Mushroom

Pleurotus ostreatus Lentinaceae
Width to 13cm

Edible and delicious species. **Upper surface** Grey-buff. **Lower surface** Has whitish gills. **Status** Widespread and common; forms tiers of brackets on tree trunks, mainly Beech.

Hen of the Woods

Grifola frondosa Coriolaceae
Width to 30cm

Appears as spreading tiers of brackets at the base of deciduous tree stumps. **Upper surface** Grades from buff to grey with a pale margin. **Lower surface** Has pores. **Status** Common and widespread.

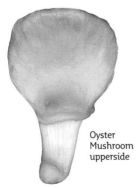

Oyster Mushroom upperside

LEFT: Oyster Mushroom
BELOW: Hen of the Woods

Chicken of the Woods

Laetiporus sulphureus Coriolaceae Width to 20cm

Appears as tiers of thick, knobbly brackets on the trunks of living deciduous trees. **Upper surface** Bright yellow. **Lower surface** Has pale pores. **Status** Common and widespread.

INSET: **Chicken of the Woods upperside**

Birch Polypore

Piptoporus betulinus Coriolaceae
Width to 20cm

Forms semicircular brackets on birch trunks. **Upper surface** Buffish brown. **Lower surface** Has white pores. **Status** Common and widespread.

BELOW: **Birch Polypore**

Southern Bracket

Ganoderma australe
Ganodermataceae Width to 30cm

Grows on living Beech. **Upper surface** Brown, knobbly and hard. **Lower surface** Has white pores that release brown spores. **Status** Common and widespread.

Beefsteak Fungus

Fistulina hepatica Fistulinaceae
Width to 25cm

Grows on deciduous trees, mainly oak. **Upper surface** Blood-red. **Lower surface** Has whitish pores. **Status** Common and widespread.

Variable Oysterling

Crepidotus variabilis
Crepidotaceae Width to 3cm

Grows as kidney-shaped brackets on fallen twigs of deciduous trees. **Upper surface** Pale cream and downy. **Lower surface** Has pinkish-buff gills. **Status** Common and widespread.

Hairy Curtain Crust

Stereum hirsutum Stereaceae
Width to 4cm

Grows on dead wood and forms irregular tiers of tough, rubbery brackets with wavy margins. **Upper surface** Greyish and hairy. **Lower surface** Orange. **Status** Common and widespread.

Bleeding Oak Crust

Stereum gausapatum Stereaceae
Width to 4cm

Appears as brackets or encrustations on dead stumps. **Upper surface** Grey-brown, turning red if cut or bruised. **Lower surface** Buffish. **Status** Common and widespread.

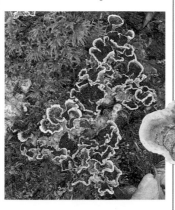

Turkeytail

Trametes versicolor Coriolaceae
Width to 8cm

Aptly named, growing as tiers of thin brackets on dead stumps. **Upper surface** Has concentric zones of colour. **Lower surface** Has white pores. **Status** Common and widespread.

Smoky Bracket

Bjerkandera adusta Coriolaceae
Width to 8cm

Grows as overlapping tiers of pale brackets on dead conifer stumps. **Upper surface** Brownish. **Lower surface** Has smoky-grey pores. **Status** Common and widespread.

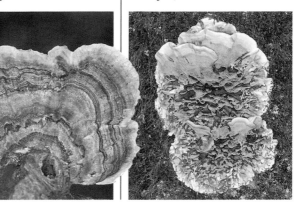

Earthstars

Earthstars are so bizarre in appearance that they attract their own band of dedicated followers. More than a dozen species occur in our region but most are rare; the following three, while enigmatic in occurrence, are still found on a regular basis. The commonest is the **Collared Earthstar** *Geastrum triplex* (**1**) (width to 7cm), small colonies of which are found in woodlands. The **Sessile Earthstar** *G. fimbriatum* (**2**) (width to 3cm) is another woodland species; in maturity its rays curl up and give it the appearance of an ornate button. The **Striate Earthstar** *G. striatum* (**3**) (height to 5cm) is often found in churchyards, typically growing under mature Yews or other conifers; its spore sac is borne on a slender stalk.

Ground-dwelling fungi

Ground level is not only a good place to find typical toadstools, but is also home to more unusual fungal forms. The **Earthfan** *Thelephora terrestris* (**1**) (width to 5cm) forms patches of downy, fan-shaped, bracket-like lobes; it grows in conifer woods and on heaths. The **Field Bird's-nest** *Cyathus olla* (**2**) appears as clusters of small, greyish cups, inside which egg-like spore capsules are borne; these are dispersed when splashed by raindrops.

Puffballs, fleshy & encrusting fungi

Not only do fungal fruiting bodies vary considerably in terms of shape and structure depending on the species, but their texture ranges from soft and fleshy to hard and nodular. Whatever they look and feel like, however, all serve the same dual functions – the production and dispersal of spores.

Earthballs and puffballs

These intriguing fungi produce their spores inside a ball-like sac; when these are mature the surface of the fruiting body splits or opens via a pore (depending on the species), allowing the spores to be expelled. The **Stump Puffball** *Lycoperdon pyriforme* (**1**) (height to 6cm) appears in clusters on the woodland floor, and its spore sac is borne on a stalk. The **Common Earthball** *Scleroderma citrinum* (**2**) (width to 8cm), another woodland species, also often appears in groups and the balls have a rough, warty texture. Most impressive of all is the **Giant Puffball** *Calvatia gigantea* (**3**) (width to 35cm), a grassland species; it splits and disintegrates when the spores are ripe, but before that it can reach the size of a football.

White Saddle

Helvella crispa Helvellaceae
Height to 15cm

An unusual-looking fungus. **Cap** Creamy white; strangely convoluted and distorted, resembling melted plastic. **Stem** Greyish white and deeply furrowed. **Status** Fairly common, growing along rides and verges.

Earpick Fungus

Auriscalpium vulgare
Auriscalpiaceae Height to 4cm

Bizarre fungus that is easily overlooked. **Cap** Borne on a stem, spore-producing and armed with peg-like teeth on the lower surface. **Stem** Slender. **Status** Widespread and fairly common. Grows on conifer cones.

INSET: cap underside

Fleshy fungi

Although many fungi have a tough, dryish feel, some of those that grow on twigs or on the ground are soft and fleshy, often rather unnervingly so. The most convincingly flesh-like species is the **Jelly Ear** *Auricularia auricula-judae* (**1**) (width to 5cm), which appears as ear-like brackets on dead branches of deciduous trees, especially Elder. The **Yellow Brain** *Tremella mesenterica* (**2**) (width to 7cm) is a distinctive winter species that appears as a convoluted, orange-yellow brain-like mass on dead deciduous branches. **Jelly Rot** *Phlebia tremellosa* (**3**) (width to 5cm) appears as a rubbery, gelatinous encrustation on dead deciduous timber, and **Wrinkled Crust** *P. radiata* (**4**) (width to 8cm), while similarly gelatinous, has a surface that is adorned with radially wrinkled patches abutting one another. A search at ground level may reveal the **Orange Peel Fungus** *Aleuria aurantia* (**5**) (width to 8cm), which does indeed look like a picnicker's discarded peel, or **Hare's-ear** *Otidea onotica* (**6**) (width to 6cm), which has a strangely skin-like texture and colour.

Bark-encrusters and outgrowths

In early winter, bare twigs and fallen branches are host to several small but interesting fungi, some of which are extremely colourful. The **Green Elfcup** *Chlorociboria aeruginescens* (**1**) (width to 4mm) stains wood blue-green and has tiny, saucer-shaped fruiting bodies; it favours dead oak. **Coral Spot** *Nectria cinnabarina* (**2**) (width to 3mm) appears as masses of pinkish-orange pinheads on the dead and dying twigs of deciduous trees. **Black Bulgar** *Bulgaria inquinans* (**3**) (width to 4cm) grows as collections of close-pressed, black rubbery discs growing mainly on dead oak branches. **King Alfred's Cakes** *Daldinia concentrica* (**4**) (width to 7cm) is also dark, but its fruiting bodies are extremely hard; it appears on dead deciduous branches. **Candlesnuff Fungus** *Xylaria hypoxylon* (**5**) (height to 5cm) is an intriguing species that appears as flattened, antler-like stems arising from dead wood; it starts off white but blackens as the spores mature.

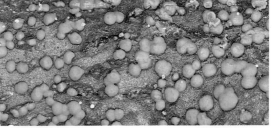

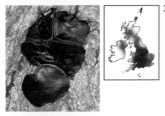

MAMMALS

Fox.

Being members of the group ourselves, Mammals (class Mammalia) are easy for us to identify with. But studying them is a different matter – many are wary or nocturnal, or both. From the smallest shrew to the largest whale, all mammals have the following in common: they generate their own body heat and maintain a more or less constant internal temperature by the circulation of blood. In land-based mammals, hair on the skin is a defining character and instrumental in the regulation of heat loss.

Mammal reproduction

In all mammals, fertilisation takes place internally and, with the exception of marsupials, the developing foetus is nourished via a placenta, in which it is implanted at an early stage. Young placental mammals are born with varying degrees of independence and maturity, from baby mice born blind and naked, to a newborn deer capable of standing on its own four legs within an hour or so of birth.

Young mammals, such as this Grey Seal, suckle milk from their mothers as their main source of food in early life.

Mammal lives

Mammals have a fundamental advantage over most other animals, except birds: they can generate body heat and maintain a fairly constant internal temperature. This enables them to be largely independent of whatever the weather can throw at them and the spectrum of mammal species can exploit almost all terrestrial and aquatic habitats. In land-based mammals, hairy or furry coats act as good insulators and help reduce heat loss; because of our strongly seasonal climate, many species have thicker coats in winter than in summer.

LEFT: Although whales and dolphins have completely abandoned terrestrial life, the nostrils of this Blue Whale are a reminder that cetaceans are still air-breathing mammals and must return to the surface periodically.
BELOW LEFT: Mountain Hares have a thick winter coat that keeps them warm in even the harshest of winter weather.
BELOW: The senses of sight, smell and hearing are acute and honed to perfection in most land mammals; in the case of this Rabbit, they allow it to detect danger and are a great aid to survival.

Meet our mammals

There are nine main groups (biologists call them orders) of mammals that are either native or have a long-established feral ancestry in Britain and Ireland. These are: shrews, Hedgehog and Mole; bats; rodents; hares and Rabbit; whales and dolphins; seals; carnivores; horses; and even-toed ungulates.

Shrews, Hedgehog and Mole

Often referred to as insectivores, shrews, Hedgehogs and Moles actually eat a much wider range of small prey. They have rather poor eyesight but a strong sense of smell. Their teeth are needle-like and arranged in a continuous row, and their fur is velvety. They belong to the order Insectivora.

Common Shrew

Bats

Bats (order Chiroptera) are strictly nocturnal mammals that feed primarily on flying insects, notably moths and beetles. They are the only mammal group to have mastered flight in as complete a manner as birds. They use echolocation rather than sight to locate prey and for orientation.

Brown Long-eared Bat

Rodents

The rodents comprise a varied group that includes squirrels, mice, voles and rats; they are placed in the order Rodentia. Most rodents are seed-eaters and their teeth suit their diet: sharp-edged front teeth (incisors) prepare the food, and ridged grinding teeth (molars) macerate it before it is swallowed.

Wood Mouse

Hares and Rabbit

Despite superficial similarities, hares and Rabbit are unrelated to rodents and are placed in the order Lagomorpha. Like most rodents, they are herbivores, but typically they eat vegetation rather than seeds using sharp front teeth (incisors) and grinding cheek teeth. They are fast runners with powerful back legs, with short tails.

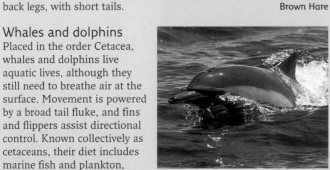

Brown Hare

Whales and dolphins

Placed in the order Cetacea, whales and dolphins live aquatic lives, although they still need to breathe air at the surface. Movement is powered by a broad tail fluke, and fins and flippers assist directional control. Known collectively as cetaceans, their diet includes marine fish and plankton, depending on the species.

Common Dolphin

Seals

Unlike cetaceans, seals give birth on land. While their terrestrial gait is awkward and shuffling, in water they are graceful and speedy, powered by flippers. They have well-developed eyes and prominent whiskers, and our two native species feed mainly on fish. Seals are placed in the order Pinnipedia.

Grey Seal

Carnivores

Placed in the order Carnivora, this group is represented in Britain and Ireland by dogs, cats and mustelids (species such as Badger, Weasel, Otter and their relatives). All carnivores catch, kill and eat live prey (often other mammals), although some are omnivorous to a degree. Carnivore teeth comprise sharp incisors, dagger-like canines and grinding cheek teeth.

Badger

Horses

True wild horses have long been extinct in our region but their domesticated ancestors live on and are placed in the order Perissodactyla. Horses are fast runners with highly evolved limbs and a single hoof on each leg. Their diet is vegetarian and opposing sets of front incisor teeth allow grass and other vegetation to be sheared off.

Horse

Even-toed ungulates

Deer, cattle, goats, pigs and sheep are included in this group. All have limbs that end in two pairs of toes, only one of which bears the animal's weight. They have a complex ruminant digestion whose efficiency allows them to survive on a comparatively meagre vegetarian diet. Males (and, in some cases, females) develop antlers or horns. All are placed in the order Artiodactyla.

Cow

Small mammals

The term 'small mammal' is used collectively to describe insectivores (shrews, Hedgehog and Mole) and the small members of the rodent family, namely voles and mice.

Hedgehog

Erinaceus europaeus Erinaceidae
Length 23–27cm

Has a unique coating of protective spines. Feeds mainly on invertebrates but will take food put out by people. Nocturnal and hibernates Oct–Apr. **Adult** Has spines (modified hairs) on its back; these are erectile and form an effective deterrent when the animal rolls into a defensive ball (below). Head and underparts covered in coarse hairs. Muzzle-shaped head ends in a sensitive nose. **Voice** Utters pig-like squeals in distress, and grunts when courting. **Status** Familiar garden resident and frequently seen as a road casualty. Distribution is more patchy in the countryside at large.

Mole

Talpa europaea Talpidae
Length 14–18cm

Burrowing, tunnel-dwelling mammal whose cylindrical body is covered in black fur. Feeds mainly on earthworms. Presence detected above ground by 'molehills' (spoil heaps of excavated soil). **Adult** Has velvety fur and spade-like front feet, these armed with sharp claws and used for digging. Eyes are tiny and external ears are absent. Head is elongated into a sensitive snout. **Voice** Mostly silent. **Status** Common in meadows and woods with well-drained, invertebrate-rich soil. Beneficial as a soil aerator.

Pygmy Shrew

Sorex minutus Soricidae
Length 7–10cm

Our smallest land mammal. Active throughout the 24-hour period. Hunts invertebrates mainly at ground level, but also climbs well and sometimes found in bird boxes. **Adult** Has dense fur: dark brown on upperparts and flanks, and contrastingly pale greyish on underparts. Note the pointed, whiskered snout and small, beady eyes; ears partly hidden by fur. Tail is long relative to body length. **Voice** High-pitched squeaks. **Status** Widespread and common in woodland margins, hedgerows, meadows and moors.

family of babies

Did you know?

In the garden, the Hedgehog is seen as an ally because it feeds on unwelcome invertebrates such as slugs and snails. However, where it has been introduced to certain Scottish islands, such as North Ronaldsay and North Uist, its presence is not welcome since it kills and eats the eggs and nestlings of ground-nesting birds. Attempts are now underway to eradicate it from particularly sensitive locations.

Common Shrew

Sorex araneus Soricidae
Length 9–14cm

Spends much of its brief life (a year or so) in ground cover or underground, hence easily overlooked. Leads a frenetic life, its search for invertebrate food never ending. **Adult** Has velvety fur, dark brown above, with buffish flanks grading to greyish white on underparts. Head extends to a pointed, whiskered snout. Eyes are tiny and beady and ears are small. Tail is relatively short compared to that of Pygmy Shrew. **Voice** High-pitched squeaks and shrill screams. **Status** Widespread and common in hedgerows, grassland and woodland margins.

Water Shrew

Neomys fodiens Soricidae
Length 12–17cm

Distinctive bicoloured shrew, usually seen near fresh water. Swims well. **Adult** Has dense fur: blackish upperparts and flanks contrast with whitish underparts. In water, the fur traps a layer of air that makes submerged animal look silvery. Fringes of hairs on tail and relatively large hind feet assist swimming. **Voice** High-pitched squeaks. **Status** Widespread except in N Scotland and on islands. Favours slow-flowing and well-vegetated streams and Watercress beds. Pollution and disturbance are agents of its decline.

Hazel Dormouse

Muscardinus avellanarius Gliridae
Length 13–17cm

Iconic nocturnal small mammal and a conservation symbol. Mainly arboreal and hibernates in winter. Nests are usually made from shredded Honeysuckle bark. **Adult** Has predominantly golden-brown coat with paler throat and belly, and tail has coating of golden fur. Note the large, beady eyes and rounded ears. Feet have flexible toes, used when climbing. **Voice** Mainly silent. **Status** Woodland species. Thrives best where mature oaks, coppiced Hazel and Honeysuckle grow together. Local, and threatened by habitat destruction.

Island shrews

In addition to our three widespread shrews, two more species are found on remote islands. Both are extremely similar to one another, are widespread in mainland Europe, and may have been introduced accidentally to our region by Man. The **Lesser White-toothed Shrew** *Crocidura suaveolens* (length 8–12cm, Soricidae) is restricted in our region to the Isles of Scilly, and to Jersey and Sark in the Channel Islands; it has grey-brown fur and white-tipped teeth. The Greater White-toothed Shrew *C. russula* (length 8–13cm, Soricidae) is similar in appearance but in our region occurs only on Guernsey, Alderney and Sark in the Channel Islands.

Lesser White-toothed Shrew

Edible Dormouse

Glis glis Gliridae
Length 28–33cm

Plump rodent that recalls a miniature Grey Squirrel. **Adult** Fat-bodied in autumn but sleek in spring after losing weight during hibernation. Has a mainly grey coat but throat and belly are whitish and hint of a dark stripe is sometimes seen down back. Note the large eyes, rounded ears and long toes. Tail is long and bushy. **Voice** Chattering grunts and squeals. **Status** Introduced and now widespread in the Chilterns. Favours mature deciduous woodlands and mature gardens.

Dormouse hibernation

Both our species of dormice hibernate through the winter months. Little is known about the habits of the **Hazel Dormouse** (**1**), but most animals probably pass the winter in a state of torpor, in an underground nest among tree roots. The **Edible Dormouse's** (**2**) winter nest is often located in a tree hole. Both species fatten up in the autumn and lose up to half their body weight during hibernation. Unsurprisingly, animals that are underweight in the autumn often die during the winter.

Small rodents

The tally of small rodents in Britain and Ireland includes four species of voles and four species of mice. Although their numbers vary considerably throughout the seasons and from year to year, their prodigious breeding capacity ensures that they are the most numerous wild mammals in the region, with populations in the millions.

Bank Vole

Clethrionomys glareolus
Muridae Length 13–17cm

Plump, richly coloured vole. Makes an underground nest and radiating shallow tunnel network; forages for seeds and fruits above ground. **Adult** Has a compact body and mainly reddish-brown fur, paler and greyer on chest and belly. Relative to Field Vole, has large ears and long tail. Island subspecies are larger and heavier than mainland animals. **Voice** Squeaks if alarmed. **Status** Common in deciduous woodland, hedgerows and field margins; introduced to S Ireland. Island subspecies occur on Skomer (below), Jersey, Mull and Raasay.

Skomer Vole

Field Vole

Microtus agrestis Muridae Length 11–16cm

Locally abundant grassland rodent. Makes a network of concealed surface runways and of tunnels just below soil surface or through compacted roots of grasses. Diet comprises mainly grass roots. **Adult** Has a plump body and relatively shorter tail and smaller ears than Bank Vole. Coat colour is mainly grey-brown, palest on chest and belly. **Voice** Utters shrill squeaks in alarm. **Status** Common in grassy habitats, from lowland meadows to upland moors. Widespread, but absent from Ireland and most islands.

Water Vole

Arvicola terrestris Muridae Length 20–32cm

Charming waterside mammal, equally at home in water or on river banks. **Adult** Has a plump body and reddish-brown fur. Head is relatively large and rounded. Front feet grasp vegetation during eating, and hind feet are used for swimming. Tail is bristly and relatively shorter than that of Brown Rat, which has a naked tail. **Voice** Mostly silent. **Status** Likes clean, slow-flowing or still waters with marginal vegetation and steep, muddy banks, into which it burrows. Thrives only where this habitat is maintained and predatory American Mink are controlled.

young animal

Island voles

Microtus arvalis (length 12–17cm, Muridae) is a widespread vole in mainland Europe, where it is known as the Common Vole. Although absent from, and not native to, most of Britain, it occurs as an ancient introduction to two far-flung island chains, where it has evolved into distinct subspecies. The **Orkney Vole** *M. a. orcadensis* (1) has been present on those islands for at least 5,000 years and lives on moors and coastal cliffs. The **Guernsey Vole** *M. a. sarnius* (2) is found only on Guernsey and lives in undisturbed grassy habitats.

Harvest Mouse

Micromys minutus Muridae
Length 10–15cm

Our smallest rodent. In summer, it weaves a tennis ball-sized nest among grass stems. **Adult** Has mainly golden brown coat but throat, chest and belly are white. Feet possess a good grip for climbing; tail prehensile. **Voice** Mostly silent. **Status** Now confined to wildlife-managed meadows, Bramble patches and dry reedbeds.

Predators and prey

Small mammals, particularly Bank and Field voles and the Wood Mouse, are crucially important in the diets of predators. The impact that Weasels and Stoats have on their numbers is difficult to assess because these predatory mammals are so hard to study, but the relationship between small rodent prey and avian predators is far easier to discern. Anyone who holds the mistaken belief that in a natural setting predators control prey animals should consider the cases of species such as the Barn Owl: in years when vole numbers are poor, owl breeding success is correspondingly low, with many young simply starving to death. So, the reality is that in this natural situation prey numbers dictate and control predator numbers. Compare this with the case of domestic cats (*see* p.206).

Wood Mouse

Apodemus sylvaticus Muridae
Length 15–22cm

Our most familiar mouse. Its underground nest and tunnel network serve as a refuge. After dark, forages for seeds, nuts and fruits above ground; climbs well. **Adult** Has a classic mouse shape, with pointed head, compact body and long tail. Coat is mainly yellowish brown above, with dark vertebral band along dorsal surface of head and body. Yellowish flank colour grades to whitish on underparts. **Voice** Utters frantic squeals in distress. **Status** Common in woodland but also found in most other terrestrial habitats, including scrub and gardens.

family of babies

Yellow-necked Mouse

Apodemus flavicollis Muridae Length 18–25cm

Large, mainly nocturnal mouse. Climbs well. **Adult** Similar to Wood Mouse but larger overall, with relatively larger ears, eyes and feet, and longer tail; coat is richer brown on upperparts and shows clearer demarcation between upperparts and clean-looking white underparts. Has a broad, yellow band on throat (in Wood Mouse, yellow on the throat is, at most, a discrete spot). **Voice** Squeals loudly in distress. **Status** Distribution is patchy, and only locally common in deciduous woodland.

House Mouse

Mus domesticus Muridae
Length 14–19cm

Archetypal mouse, ancestor of domesticated pet mice. Diet is eclectic. Presence detected by musky smell. **Adult** Has a compact head and body, roughly the same length as tail. Coat ranges from yellowish brown to grey-brown and is darker above than below. Ears are relatively large. **Voice** High-pitched squeaks. **Status** Probably introduced to Britain during the Iron Age. Formerly abundant but now less so. Favours sites where food is stored (factories and farm barns) and has a truly commensal association with Man.

Rats, squirrels, hares & Rabbit

Rats and squirrels are large rodents. By contrast, although they are superficially similar to look at, hares and Rabbit are unrelated and are classified as lagomorphs. Most species are conspicuous animals in the countryside and, despite differences in their classification, all have sharp front incisor teeth, with which they shear vegetation and other foodstuffs, and grinding teeth at the back of the mouth.

Brown Rat

Rattus norvegicus Muridae
Length 30–50cm

Familiar and highly adaptable rodent. Omnivorous and swims and climbs well. **Adult** Recalls an outsized mouse but with a larger, plumper body, shorter ears, shorter legs (but larger feet) and a thicker tail. Fur is coarse and mainly brown, grading to grey on underparts. Tail looks scaly, with sparse bristles. **Voice** Utters agonising screams in distress. **Status** First recorded here in 1720 as a boat stowaway. Now widespread and abundant, especially in areas where food is discarded.

The demise of the Black Rat

Until the 18th century, the **Black Rat** *Rattus rattus* (length 30–45cm, Muridae) was the only one of its kind in the region, having been introduced by the Romans. Following the arrival of the Brown Rat, its numbers dwindled and today it is restricted to dockland areas in London and Bristol, and on the Shiant Islands off the Isle of Lewis. It is a nocturnal, omnivorous rodent with a darker, shaggier coat than its cousin; the ears are rather large and the tail is relatively long.

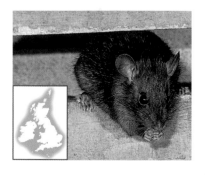

Red Squirrel

Sciurus vulgaris Aplodontidae
Length 35–45cm

Our native squirrel. **Adult** Has a compact body with a large head, tufted ears and bushy tail. In summer, coat is mainly orange-brown with creamy-white underparts; tail bleaches paler with time. In winter, coat is thicker and darker on back and ear tufts are more pronounced. **Voice** Chattering calls. **Status** Formerly widespread prior to introduction of Grey Squirrel; now common only in N Britain in mature conifer forests, and with a few S outposts (e.g. Thetford Forest in Norfolk, the Isle of Wight and Brownsea Island, Dorset).

Grey Squirrel

Sciurus carolinensis Aplodontidae
Length 45–55cm

Abundant alien; our most familiar squirrel. Omnivorous and renowned for its cunning in exploiting food sources. **Adult** Has a plump but elongated body and long, bushy tail. Rounded ears lack ear tufts. Coat is mainly grizzled grey with a whitish chest and belly; some individuals are variably tinged with brown in summer. **Voice** Utters a teeth-smacking 'tchack' when alarmed. **Status** Native to North America, introduced here in 1876. Now widespread and common, its adaptability allowing it to thrive in woods, parks and gardens.

Rabbit

Oryctolagus cuniculus Leporidae
Length 40–55cm

Familiar long-eared social burrowing mammal. Lives in tunnel complexes called *warrens*. Mainly nocturnal or crepuscular; its diet is vegetarian. **Adult** Has mainly greyish-brown fur with rufous nape and pale greyish underparts. Long ears have rounded brown tips and tail is dark above and white below. Legs are long but relatively shorter than those of hares. **Voice** Squeals loudly in alarm. **Status** Introduced but now widespread and common in grassland, scrub and on roadside verges. Population numbers regularly crash, locally, due to outbreaks of myxomatosis.

young Rabbit

Brown Hare

Lepus europaeus Leporidae
Length 50–75cm

Rabbit-like but with longer legs and ears. Other structural and behavioural differences allow separation of Brown Hare and Rabbit. Brown Hare is a fast runner and does not burrow. Performs 'boxing' displays while courting. **Adult** Has a brown coat grizzled with grey and black, especially on back. Coat is thicker, darker and redder in winter than summer. Ears are black-tipped (brown-tipped in Rabbit); tail is dark above with a pale fringe (uniformly dark above in Rabbit), and whitish below. Has 'wild'-looking eyes located high on sides of head. **Voice** Mostly silent. **Status** A native that has declined owing to farming practices and persecution; still locally common on farmland and grassland.

BELOW: **Brown Hares boxing**

Mountain Hare

Lepus timidus Leporidae
Length 45–60cm

Has shorter ears and more compact body than Brown Hare. Tail is uniformly white and ears are tipped black. **Adult Scottish race** Has a greyish-brown coat in summer, palest on underparts and most rufous on head. Underfur is blue-grey. In winter, acquires a thick whitish coat except for buffish nose. **Adult Irish race** Has a reddish-brown summer coat and variably buffish-brown and white winter coat. **Voice** Mostly silent. **Status** Scottish animals favour heather moors and mountains, while Irish Mountain Hares favour more grassy habitats.

ABOVE: **winter coat**; BELOW: **summer coat**

Carnivores

Although, broadly speaking, all our carnivorous mammals have similar ways of feeding – to a greater or lesser degree they kill and eat other animals – and are classified in the order Carnivora, they are divided into three different families. The Wildcat is a member of the cat family (Felidae), the Fox belongs to the dog family (Canidae) and remaining species (Stoat, Weasel, Polecat, American Mink, Pine Marten, Otter and Badger) are placed in the family Mustelidae.

Wildcat

Felis silvestris Felidae
Length 75–100cm

Resembles a large domestic tabby cat, but subtle differences allow separation. Mainly nocturnal; rarely seen. **Adult** Has greyish-brown fur with vertical stripes along flanks and dark vertebral line on back. Tail is thick, bushy and blunt-tipped, marked with 3–5 dark and discrete bands ('wild-type' domestics have ill-defined flank stripes and tapering tail). **Voice** Mews, purrs and spits like a domestic cat. **Status** Now confined to Scotland; favours moors, forest margins and lower mountain slopes.

Fox

Vulpes vulpes Canidae
Length 95–130cm

Adaptable dog-like carnivore, but with a catholic diet that includes fruits and berries. Mainly nocturnal; its daytime shelter is called an *earth*. **Adult** Has a thick, mainly orange-brown coat with whitish jaws, white underparts and white tip to tail. Feet and backs of ears are blackish. **Voice** A yelping scream is uttered mainly by females for a brief period in winter, to mark the breeding season. **Status** Common, widespread and adaptable, found in towns and cities as well as the countryside.

Domestic cats

DNA testing has recently shown the majority of Wildcats in the United Kingdom to be Wildcat/domestic cat hybrids. The number of domestic cats - currently estimated at over 6 million in the United Kingdom - is far higher than could be supported naturally because they are fed by humans. This means they have a major impact on the population of their prey - native wildlife - as well as posing a serious threat to the purity of the Wildcat genes.

Stoat

Mustela erminea Mustelidae
Length 25–45cm

Voracious predator with a long, sinuous body. Catches mainly rodents and Rabbits. Males are larger than females. **Adult** Has a summer coat comprising orange-brown upperparts and white underparts; tail is black-tipped. In winter, some animals in the N acquire a white 'ermine' coat; tip of tail remains black. **Voice** Utters high-pitched calls in alarm. **Status** Widespread but seldom common; still widely persecuted. In Ireland, sometimes confusingly referred to as a *weasel*. Favoured habitats include farmland, woodlands, marshes and moors.

Weasel

Mustela nivalis Mustelidae
Length 20–40cm

Voracious predator of voles and mice. Has a long, sinuous body and recalls a small Stoat, except for shorter tail, which is a uniform colour, not black-tipped. **Adult** Has orange-brown upperparts and sides, and white underparts, including throat; note clear demarcation between the two colours. Male is larger than female. **Voice** Utters high-pitched hisses in alarm. **Status** Widespread and fairly common in a range of habitats, including woods, hedgerows and scrub; absent from most islands and Ireland.

Polecat

Mustela putorius Mustelidae
Length 40–60cm

Nocturnal predator, mainly of Rabbits. Likely ancestor of the ferret. **Adult** Has coarse fur, dark brown overall but flanks and belly are a paler buff. Face has white marks on either side of nose and on forehead, and a dark 'mask' through eyes. **Voice** Screams and hisses when alarmed. **Status** Widespread a century ago but eradicated in much of the region by the mid-20th century by human persecution and the disappearance of Rabbits through myxomatosis. Currently staging a comeback, but beware confusion with feral ferrets.

Pine Marten

Martes martes Mustelidae
Length 55–80cm

Secretive and mainly nocturnal mammal. Diet includes small mammals, Rabbits, birds and frogs. Agile climber, capable of catching Red Squirrels. **Adult** Has a slender body and long, bushy tail. Head is pointed and ears are relatively large. Fur is thick, sleek and mainly dark orange-brown with a creamy-yellow throat patch. **Voice** Mostly silent. **Status** Formerly widespread but often persecuted and eradicated from many areas; now restricted mainly to remote parts of Scotland and Ireland. Favours broken ground, often with tree cover.

Otter

Lutra lutra Mustelidae
Length 95–135cm

Sinuous swimmer with a bounding gait on land. Feeds mainly on fish. **Adult** Has a long, cylindrical body, with short legs and long, thickset tail. Blunt head has sensitive bristles and toes are webbed. Coat is mainly brown but chin, throat and belly are whitish. Fur has water-repellent properties, appearing sleek in water but 'spiky' when dry. **Voice** Mostly silent. **Status** Persecuted and poisoned (by agricultural pesticides) until mostly extinct in lowland Britain by the 1960s. Now recovering and recolonising its former haunts.

American Mink

Mustela vison Mustelidae
Length 45–65cm

Introduction from North America. Active predator of aquatic life; instrumental in the demise of Water Vole. Hunts in the afternoon and after dark. Swims well and buoyantly. **Adult** Has a slender body and bushy tail. Fur is soft, silky and typically dark brown. Male is larger than female. **Voice** Utters high-pitched calls when alarmed. **Status** Originated from fur farms ('liberated' animals and escapees) and now very widespread along waterways throughout. Often the subject of eradication programmes.

Badger

Meles meles Mustelidae
Length 65–80cm

Distinctive nocturnal mammal, spending daytime in its tunnel complex (*sett*). Omnivorous and opportunistic feeder. Its facial markings are unmistakable. **Adult** Has coarse fur, greyish on back and flanks, and blackish on underside and legs. Head is elongated into a snout that is marked with longitudinal black and white stripes. Legs are short, and blunt tail has a white tip. **Voice** Mostly silent. **Status** Locally common where farmland, meadows and woods occur side by side; also occurs on the fringes of suburbia.

Towns & urban wildlife

For the majority of people in Britain and Ireland, the urban environment is the one with which they are most familiar. Although towns and cities are conspicuously lacking in many species, it would be a mistake to assume that they are without wildlife interest altogether. Many of our plants and animals are extremely adaptable and have successfully colonised this seemingly unpromising environment. Town parks and gardens can play host to an extraordinary variety of flowers and wildlife.

These days, **Hedgehogs** (*see also* p.200) probably reach their highest densities in suburban gardens. Foxes, too, are increasingly seen on the fringes of towns and cities, but in mammal terms it is probably the ubiquitous Brown Rat that thrives best alongside Man, benefiting from our wasteful lifestyles.

Garden ponds are important breeding sites for amphibians, such as **Common Frogs** (*see also* p.331), as their natural habitats in the countryside disappear. Amphibians are tied to water when breeding but spend the rest of the year on land. So the terrestrial environment in the garden is important too.

Floral colour is important to most gardeners and many flowers in the gardening palette are also good sources of pollen for bees and nectar for butterflies, such as the **Small Tortoiseshell** (*see also* p.338), and other insects. Enlightened gardeners encourage flowers, such as the Butterflybush, mainly for insects.

Just sit in your local park for a while and you will notice a surprising variety of birds, animals and invertebrates that make their home there. Bold Grey Squirrels are common, but in many areas, deer too have begun to venture into surprisingly urban environments. Sympathetically managed graveyards are refuges for wildflowers, lichens and mosses. And even in the heart of the busiest cities, a lake or pond will attract gulls, ducks, Coots and other waterbirds.

If you have a garden, you can create your own private haven for wildlife. Try to think of the bigger picture and rather than focusing on a particular group of animals (butterflies or birds, for example), include plants, invertebrates and other forms of life in your plans. Plant native shrubs and trees because they attract more insects than alien species. Favour lawns and flower borders over decking and paving, and leave plenty of wild areas where native plants and small animals such as

Common Toads and Slow-worms can live undisturbed.

Gardening for wildlife always involves selectively encouraging or removing certain things. But to achieve this end, don't use chemicals that kill invertebrate 'pests' or plant 'weeds'. They are seldom specific and many of the target 'pests' and 'weeds' are beneficial in the grand scheme of things: aphids are food for predators; slugs and snails are eaten by shrews, Hedgehogs and amphibians; and many so-called 'weeds' are food for moth larvae.

A dozen or more species of birds nest in most large gardens but also breed in the countryside at large. But the **Swift** (*see also* p.286) is exceptional – these days it seldom nests away from urban sites, using the loft spaces of houses and churches. House Martins and Starlings also favour urban areas.

Many people tolerate wild plants in their garden or actually encourage them, if their flowers are colourful. Green gardeners go further, growing less appealing plants for their wildlife value. **Common Nettle** (*see also* p.34) is particularly important as it is food for the caterpillars of Small Tortoiseshell and Peacock butterflies.

Trees and shrubs are popular garden features. Soil type and drainage influence what can be grown but the specific choice is largely a matter for the gardener. **Rowan** (*see also* p.25), which produce a valuable crop of fruits that are food for migrant birds, are a good option for those seeking to encourage wildlife.

Mammal skulls

Being robust and well constructed, mammal skulls usually remain intact long after the death of an animal.

Small skulls APPROX. SCALE: 80% lifesize

Despite their diminutive sizes and rather delicate construction, small mammal skulls are occasionally found intact. As with the skulls of larger animals, their dentition and overall structure provide clues about the diet and lifestyle of the previous owner. In many cases, their appearance is unique enough to allow certain identification.

Lesser White-toothed Shrew LENGTH 1.6cm

Common Shrew LENGTH 2cm

Greater Horseshoe Bat LENGTH 2cm

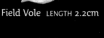
Field Vole LENGTH 2.2cm

Mole LENGTH 3.2cm

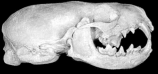
Weasel LENGTH 3.2cm

Brown Rat LENGTH 5cm

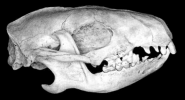
Stoat LENGTH 5cm

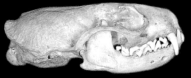
Hedgehog LENGTH 5.6cm

Mink LENGTH 6cm

Polecat LENGTH 6.5cm

Grey Squirrel LENGTH 6cm

Rodent skulls APPROX. SCALE: lifesize

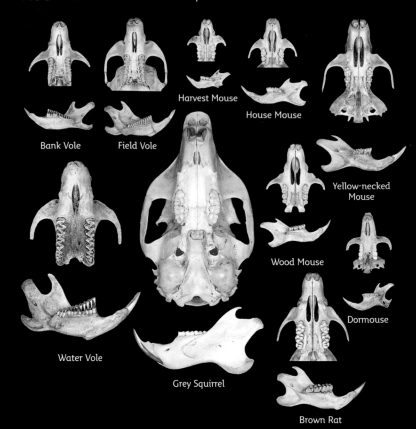

Harvest Mouse

House Mouse

Bank Vole

Field Vole

Yellow-necked Mouse

Wood Mouse

Water Vole

Dormouse

Grey Squirrel

Brown Rat

Rodent teeth APPROX. SCALE: 3.5× lifesize

For the sake of clarity, the grinding teeth isolated from the upper right jaws of Britain's native small rodents are shown here side by side.

Harvest Mouse LENGTH 3mm

Wood Mouse LENGTH 4mm

Yellow-necked Mouse LENGTH 5mm

House Mouse LENGTH 4mm

Dormouse LENGTH 4mm

Field Vole LENGTH 6mm

Bank Vole LENGTH 6mm

Water Vole LENGTH 10mm

Brown Rat LENGTH 7mm

PREMOLAR
MOLAR
Grey Squirrel LENGTH 10mm

NOTE: in the Grey Squirrel and Dormouse, the above grinding teeth comprise 1 premolar (the top tooth in each illustration) plus 3 molars. In all the other teeth shown here, premolars are absent and the grinding teeth comprise just 3 molars. The teeth are shown to scale and come from adult animals. Bear in mind that the age of any rodent will have a bearing on its size and hence the size of its teeth.

RIGHT: A rodent's incisor teeth grow throughout its life. The exposed cutting edge may appear quite short (*below*) but by removing one of this Grey Squirrel's incisors (easily done in most rodents) its true extent is revealed (*above*).

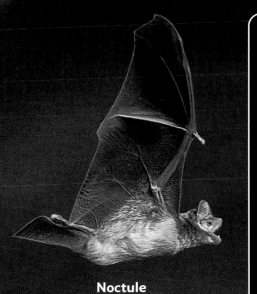

Noctule

BAT IDENTIFICATION IN THE FIELD

SPECIES	FLIGHT TIME	FLIGHT	FOOD	HABITAT
Greater Horseshoe	1hr after sunset onwards	slow, fluttering	picked from ground	wooded pastures
Lesser Horseshoe	Evening onwards	fast, rapid wingbeats	Gleaned from foliage	wooded habitats
Daubenton's	30min after sunset onwards	fast, low over water	caught over water	open woodland; near water
Natterer's	strictly nocturnal	rapid; may hover	caught in flight	woodland, farmland
Brandt's	late evening onwards	aerobatic, mid-level	caught in flight	woodland; near water
Whiskered	strictly nocturnal	direct, low level	caught in flight	meadows, gardens; near water
Bechstein's	20min after sunset onwards	fluttering, aerobatic	caught in flight	undisturbed woodland
Serotine	20min after sunset onwards	fluttering, low level	caught in flight	parkland, woodland
Noctule	late evening onwards	powerful, direct	caught in flight	woodland and parks
Leisler's	sunset onwards	fast, treetop level	caught in flight	woodland
Barbastelle	dusk onwards	slow, deliberate	sometimes over water	parks, wooded gardens
Long-eareds	strictly nocturnal	fluttering, hovers	Gleaned from foliage	woodland, gardens
Pipistrelles	20min after sunset onwards	fast, erratic	caught in flight	woodland margins, gardens

Brandt's Bat

Serotine

Pipistrelles (Common and Soprano)

Barbastelle

Whiskered Bat

Bat boxes

Bat boxes, made from untreated wood, provide excellent roosting sites for bats. They are most effective when used in woodland and best sited on a tree trunk, placed well out of harm's way with the aid of a ladder. The greatest chances of occupancy come if three are sited close to one another on the east, north and west sides of the trunk. Because of the law, once in place, bat boxes can only be inspected by licensed bat workers.

Bats need friends

Sadly, many people have their first close encounter with a bat when they find an injured one - cats and cars take a heavy toll. Fortunately, however, a network of bat hospitals around the country help these unfortunate animals and many recover to the point where they can be released into the wild again. Bat rescuers are highly trained and licensed, which is vital because bats are so delicate. These organisations are voluntary, so the best way to assist them is by donating money.

Deer

Deer (family Cervidae) have a vegetarian diet, browsing or grazing foliage according to the availability of food. As mammals go, they are large animals, ranging from the Muntjac, which is roughly the size of a Labrador dog, to the pony-sized Red Deer.

Red Deer

Cervus elaphus Cervidae
Shoulder height 100–130cm

The male (*stag*) is our heaviest deer. Size varies with region – S animals are larger than N ones. Lives in separate-sex herds for much of year. **Adult** Reddish in summer but dark brown in winter. Has whitish rump patch and buffish-brown tail. Only males have antlers: these appear in spring, mature in autumn and are shed in winter. Number of antler points increases with age. **Juvenile (*calf*)** Reddish brown with white spots. **Voice** Female (*hind*) bleats and male utters bellowing roars during autumn rut. **Status** Common and native in Scotland. Also occurs locally in Lake District, Exmoor, New Forest and Norfolk.

ABOVE: stag
LEFT: calf
BELOW: hind

Sika Deer

Cervus nippon Cervidae Shoulder height 70–90cm

Has body proportions of Fallow Deer but pointed antlers like Red. **Adult** Reddish brown with whitish spots in summer, dark grey-brown in winter. Rump is whitish with black margin; tail is white with dark median line above. Male's antlers appear in spring, mature in autumn and are shed in winter. Number of antler points increases with age. **Juvenile (*calf*)** Reddish brown with whitish spots. **Voice** Male utters blood-curdling screams during autumn rut. **Status** Introduced from Far East. Feral populations exist in several parts of the region, favouring wooded country. Mainly nocturnal.

ABOVE: stag; INSET: winter stag; BELOW: hind and calf

Fallow Deer

Dama dama Cervidae
Shoulder height 80–100cm

Medium-sized deer. Mainly nocturnal. Lives in separate-sex herds for much of year. **Adult** Reddish brown with whitish spots in summer. Usually dark grey-brown in winter, but some are black or creamy white. All have whitish rump with dark margin and blackish tail with white margin. Male (*buck*) grows broad, palmate antlers in spring and early summer; these are shed by late winter. Antler size and complexity increase with age. Female (*doe*) does not have antlers. **Juvenile (*fawn*)** Reddish brown with whitish spots. **Voice** Female has a barking alarm call; male utters a belching groan in autumn rut. **Status** Introduced. Now widespread but local in woodland, farmland and scrub.

ABOVE: dark buck
RIGHT: buck in velvet
BELOW: doe
BOTTOM: fawn

Roe Deer
Capreolus capreolus Cervidae
Shoulder height 65-70cm

Charming native species. Rather territorial and mostly solitary. **Adult** Has white on its muzzle, and black nose and 'moustache'. Coat is reddish brown in summer and greyish brown in winter. Male (*buck*) has short, branch-like antlers from spring to early winter; their size and complexity increases with age. Note whitish oval mark on rump. Female (*doe*) is less stocky than male and lacks antlers; its whitish rump marking is like an inverted heart. **Juvenile (*fawn*)** Reddish brown with white spots. **Voice** Utters a barking call in alarm. **Status** Widespread in Scotland and locally common in England in wooded farmland.

ABOVE: **summer buck**; BELOW LEFT: **winter buck**
BELOW RIGHT: **fawn**

Chinese Water Deer
Hydropotes inermis Cervidae
Shoulder height 55-60cm

Small, secretive deer. **Adult** Reddish buff in summer, greyish brown in winter. Its black nose contrasts with the otherwise white muzzle, and its beady black eyes have a white surround. Ears are large and antlers are absent in both sexes. With age, the upper canines develop into projecting tusks; these are longer in males than females. **Juvenile (*fawn*)** Reddish brown with white spots. **Voice** Barks and screams in alarm. Males have a whistling call during the rut. **Status** Escaped from Whipsnade Zoo early in the 20th century. Feral populations now found from Buckinghamshire to East Anglia. Favours marsh habitats, including fens and reedbeds.

Muntjac
Muntiacus reevesi Cervidae
Shoulder height 38-45cm

Tiny, unobtrusive deer. Territorial and mainly solitary. Browses low vegetation. **Adult** Mainly reddish brown with a whitish chest and belly. Has a large head (with converging dark stripes on forehead) and short legs. Tail is long, reddish brown above but whitish below and conspicuous when raised in alarm. Male (*buck*) develops tusk-like upper canine teeth; antlers appear in autumn and are shed the following summer. Female (*doe*) does not grow antlers. **Juvenile (*fawn*)** Tiny and reddish brown with white spots. **Voice** A piercing bark. **Status** Introduced from Far East, now locally common in woods and gardens.

Reindeer
Rangifer tarandus Cervidae
Shoulder height 0.9-1.2m

Long-legged deer. Feeds on low-growing plants, including mosses and lichens. Both sexes have antlers - used to clear snow in winter for feeding. **Adult** Grey-brown; coat is thickest in winter. Male (*bull*) is thickset, with asymmetrical, palmate antlers from early spring to midwinter. Female (*cow*) has shorter antlers that lack palmations; they are shed in May. **Juvenile (*calf*)** Greyish brown. **Voice** Grunting sounds. **Status** Formerly native in the region but driven to extinction by the 12th century. Domesticated animals introduced from Scandinavia now roam the Cairngorms.

ABOVE: **bull**; INSET: **cow and calf**

Deer-watching

Most deer are wary of people and hide in cover in the daytime, making them difficult to spot - they blend in with their surroundings to a surprising degree. If you want to get more than a fleeting glimpse of a fleeing animal, then knowledge of deer habits and behaviour can greatly improve your chances. Deer have acute senses of smell, hearing and sight, the latter particularly honed when it comes to movement. So don't expect just to be able to walk up to them, and always approach from a downwind direction.

Top tips for deer watching
- Sit, or park your car, in a concealed location (having taken into consideration wind direction to ensure your scent is not being blown towards where you hope to see the deer) before dawn, or in the late afternoon, overlooking a woodland margin from where deer are likely to emerge.
- Look for Roe Deer in late winter when the leaves are off the trees and the animals are more concerned with each other than with being observed.
- Never take a dog with you.
- Red and Fallow deer are easiest to find and observe during the autumn.
- Richmond Park, Surrey, has feral Red and Fallow deer; Fallow are easy to see in the New Forest, Hampshire; Sika are best seen at Arne, Dorset; Roe are widespread in Hampshire and Dorset woods.

Domesticated animals

Since humans arrived in Britain in the wake of the last Ice Age, several larger mammals, for example Wolf, Brown Bear, Lynx and Beaver, have been driven to extinction. But there have been additions to the list, too, in the form of introduced domesticated farm animals. Several of these also had wild ancestors that once lived here but are now extinct.

Goat

Capra hircus Bovidae
Shoulder height 60–90cm

Familiar domesticated animal. Sure-footed on steep, broken terrain. **Adult** Feral Goat is shorter and stockier than domesticated forms. Coat is long, shaggy and variably coloured, often a piebald mixture of grey, black and whitish. Male (*billy*) is larger and bulkier than female (*nanny*) and has recurved, ringed horns that increase in size with age. Many have a 'beard' and tassels on the chin. Female is smaller than male, with shorter horns. **Juvenile (*kid*)** Lacks horns. **Voice** Utters a warning whistle. Females summon their kids by bleating. **Status** Feral Goat is found in a number of locations. **Domesticated goats** Domesticated for more than 10,000 years, prized for its hair, milk, hide and meat. Probably brought to area by the first Neolithic human settlers. Popular breeds include British Saanen Goat (white), British Toggenberg (brown and white), Anglo-Nubian (Roman-nosed, lop-eared) and British Alpine (black and white).

ABOVE: Wild Goat
RIGHT: British Toggenberg

Sheep

Ovis aries Bovidae
Shoulder height 50–70cm

The most ancient sheep breed are Soays, which live in a few remote regions, living in mixed-sex flocks. **Adult** Coat is reddish brown in summer, darker and thicker in winter; comprises thick, rigid hair and thin, curly wool. Male (*ram*) has curved, ribbed horns with a slight spiral twist; they increase in length and curvature with age. Female (*ewe*) is smaller with shorter horns. **Juvenile (*lamb*)** Resembles female, but with shorter, cleaner coat. **Voice** In lambing season, mothers and lambs utter familiar 'baaing' and bleating calls respectively. **Status** Feral populations remain on St Kilda and on a few other islands such as Lundy. Soays were probably brought to Soay in the St Kilda islands by Viking settlers. **Domesticated sheep** Domesticated for millennia, kept for milk, wool, meat and hide. Many regional domesticated breeds of sheep occur, including Dorset Horn, South Down, Suffolk, Cheviot and Hampshire Down.

Soay ram

TOP: Hampshire Down
ABOVE: Suffolk

TOP: South Down
ABOVE: Dorset Horn

Cattle

Bos taurus Bovidae
Shoulder height 1–1.5m

Familiar domesticated animal, present as a number of distinct breeds. Modern breeds are extremely variable in appearance. **Adult** Has variably coloured coat according to breed. **Male (*bull*)** is larger and stockier than female (*cow*). Horns are present in some breeds but not others. **Juvenile (*calf*)** Resembles small, hornless adult. **Voice** Bulls of all breeds bellow loudly. **Status** In the past, many breeds were regionally distinct, adapted to the needs and climate of particular parts of Britain; today, little regional separation exists. Chillingham Cattle probably resemble the ancestors of our domesticated breeds and a feral herd has been present in Chillingham Park, Northumberland, for 700 years. **Domesticated cows** Dairy breeds include Friesian, Guernsey, Jersey and Dairy Shorthorn. Breeds of beef cattle include Aberdeen Angus, Highland, Devon, Sussex, Hereford and Beef Shorthorn. Dual-purpose breeds include Belted Galloway and Red Devon.

Many non-pedigree cattle are difficult to assign to a specific breed.

Highland

Friesian

Jersey

Pig

Sus scrofa Suidae
Length 1–1.5m

Stocky, well-built animal and a familiar domesticated animal. **Adult** Has a laterally flattened body. Head tapers to a blunt snout. Note the small eyes and relatively long ears. Male (boar) is more powerfully built than female (sow). **Juvenile (*piglet*)** Recalls a miniature adult. **Voice** Foraging animals grunt while feeding; a barking call is uttered in alarm. **Status** Domesticated Pig breeds include Large White, Berkshire, Middle White, Tamworth, Gloucester Old Spot and Saddleback.

Saddleback

Berkshire

Middle White

Wild Boar

The ancestor of the domesticated pig, the **Wild Boar** *Sus scrofa* (length 1–1.5m) has a grizzled grey-brown coat comprising just bristle like guard hairs in summer but dense, with underfur, in winter; it is often obscured by mud. The male (boar) is more powerfully built than the female (sow) and has protruding, upwards-pointing, tusk-like, lower canine teeth. The juvenile (piglet) is reddish brown with longitudinal white stripes. Wild Boar became extinct here in the 17th century. Reintroduced animals and escapees have formed feral populations, mainly in Sussex and Kent.

Horse

Equus caballus Equidae
Shoulder height 100–150cm

In layman's terms, 'pony' is used imprecisely to describe a small horse. However, several regional breeds do qualify more formally for the description (*see* box). These are hardy and live semi-feral lives, usually in small social groups. **Status** Until 9,000 years ago, Wild Horses still roamed Britain; the first Neolithic human settlers caused their extinction, replacing them with domesticated animals. Numerous horse breeds now exist; Ponies are the smallest and least developed of these.

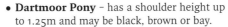

Ponies of Britain and Ireland

All pony breeds are named after the region in which they roam free; familiar examples include the following:

- **Dartmoor Pony** – has a shoulder height up to 1.25m and may be black, brown or bay.
- **Exmoor Pony** – usually dark reddish brown and up to 1.25m at the shoulder. Note the dark nose, pale muzzle and pale eye surround.
- **New Forest Pony** – stands 1.2–1.4m at the shoulder. Its colour is variable but reddish brown is typical.
- **Shetland Pony** – tiny, usually 1m at the shoulder.
- **Welsh Mountain Pony** – stands 1.1–1.2m at the shoulder. It is usually brown or grey in colour.
- **Fell Pony** – stands up to 1.3m at the shoulder and lives on the fells of Westmorland and Cumberland. It is usually brown or grey in colour.
- **Dales Pony** – found east of the Pennines, and usually dark brown or blackish in colour and 1.4m at the shoulder.
- **Highland Pony** – usually black or brown in colour and stands 1.25–1.45m at the shoulder.
- **Connemara Pony** – tolerates wet and wild weather, is usually bay or grey in colour and stands 1.3–1.4m at the shoulder.

Przewalski's Horse

Equus ferus przewalskii, the last remaining race of the otherwise extinct Wild Horse, is the ancestor of modern domesticated horses and ponies. It no longer occurs in the wild in its native Mongolia, but small herds are kept in captivity in our region, occasionally being used as grazers for nature conservation purposes. It is a stocky, short-legged, pony-sized animal, and is grey-brown or dun overall, often with a dark vertebral stripe and a mealy nose.

Dartmoor Pony

Exmoor Pony

New Forest Pony

Shetland Pony

Connemara Pony

Mammal tracks, trails & signs

You will be amazed at how much you can learn about mammals in the countryside simply by looking for tracks, trails and signs. With larger tracks, trails and droppings, it is usually possible to determine the identity of the mammal that made them.

Footprints, tracks and trails

Soft soil, mud and sand that is damp but not waterlogged is usually the best medium for the formation and preservation of good-quality tracks. The autumn and winter months usually offer the best opportunities for finding tracks.

Heavy falls of snow give an excellent opportunity to gain an insight into the activities of shy or nocturnal species. The tracks here were made by a Rabbit.

Dog Commonly found on land to which the public and their pets have access. Track size depends upon the breed of dog. Note the four toes and visible claw marks in clean prints.

Cat Prints are rounded in overall appearance, with four toes. As the claws are retractable, no marks are visible.

Fox Tracks are more oval than those of a dog. Note the four toes and the presence of claw marks. Typically, fox tracks follow a straight line.

Badger Prints are broad, with five toes and conspicuous claw marks.

Rabbit Tracks have long prints left by the hind feet and smaller, rounded ones formed by front feet; these often appear on top of one another in slow-moving animals.

Fallow Deer Print has relatively narrow slots that are usually more or less parallel to one another. Sika Deer make similar prints but these are typically broader.

Red Deer Slots are relatively long and often splay outwards towards the tip.

So-called rutting rings, here formed by a Roe Deer chasing prospective mates around a fixed point, are a sign of the species' presence in woodland.

Roe Deer Rather dainty tracks, which comprise slots that are broad-based but distinctly pointed at the tip.

Muntjac Tiny prints with slots that are proportionately narrower than those of Roe Deer.

Horse The imprint of an unshod Horse or Pony is almost unmistakable given the unique shape of the single toe on which the animals stand.

Sheep Tracks may be confused with those of a medium-sized deer. The tips may be slightly splayed depending on the speed of movement.

Goat Tracks are relatively narrow, the slots often touching at the base but cloven at the tip.

Cow Usually unmistakable on account of the size. Note that the two slots generally abut one another.

Underground homes

Although some terrestrial mammals do not use homes or refuges, many do and recognition of these can help in identification. Some species live underground in an excavated tunnel perhaps although all you are likely to see is the entrance.

RIGHT: A Badger's underground home is known as a 'sett'. A typical site would be in well-drained woodland soil or in a hedgerow bank. BELOW: The refuge used by an Otter is called a 'holt'. The entrance to its riverside home is often made more obvious by the presence of a mud-slide running from the mouth of the burrow to the water's edge.

Tell-tale marks left by rodents

The sizes of their mouths, and the structure of their skulls and teeth, influence how different rodent species nibble and gnaw their food. In many instances, their teeth leave conspicuous – and quite different – marks on the remains of their food, and these are sometimes easy to discern with the naked eye.

Many trees and shrubs serve as food sources for rodents, but the Hazel stands out because of the importance of its nut in the diet of many woodland species. Five rodent species gnaw open the hazelnut's hard case to extract its nutritious contents, each leaving telltale marks on the empty shell. These marks can be seen with comparative ease using a hand lens. It is most significant in the detection of the rare and endangered Hazel Dormouse, which leaves the neatest and most perfect of holes in a nibbled nut.

When hazelnuts turn a rich brown, they are fully ripe and feasted upon by woodland rodents.

Bank Voles enlarge the initial hole they have gnawed in the nut with their noses inside the shell. Their lower incisors leave a series of tooth marks on the sloping inner cut; the outside of the shell usually remains relatively unscathed.

A **Wood Mouse** will gnaw at the hole it has made in the nut with its nose outside. The lower incisors are inside the hole as it gnaws and they leave teeth marks on the inner slope; the upper incisors create scratch marks on the outside of the shell.

A **Yellow-necked Mouse** opens hazelnuts in a similar fashion to a Wood Mouse, and leaves similar marks.

A **Dormouse** creates a hole in the nut and uses its incisors to chisel around the hole, in a circular fashion, to enlarge it. The result is an extremely smooth inner edge that slopes, not at an acute angle, but usually at 90 degrees or more.

A **Grey Squirrel** will usually crack the nut neatly in half, leaving just a small nick at the top of one of the shell halves where it prised the nut open with its incisors (*left*). Inexperienced young squirrels often shatter the hazelnut shell in an attempt to gain entry, leaving a jagged cut margin (*right*).

Mammal droppings   APPROX. SCALE: lifesize

Mammal droppings are useful for identifying the presence of secretive animals; many species use them to define territories.

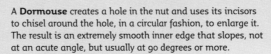

Rabbit   Brown Hare   Mountain Hare   Brown Rat   Grey Squirrel

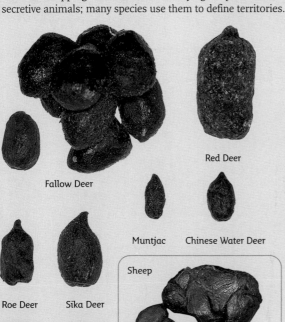

Fallow Deer

Red Deer

Muntjac   Chinese Water Deer

Roe Deer   Sika Deer

Goat   Sheep

Sheep

APPROX. SCALE: 1.6× lifesize

Pine Marten

Mink

Stoat

APPROX SCALE: 50% lifesize

Weasel

Fox

Dog

House Mouse   3×

Wood Mouse

Yellow-necked Mouse   3×

Field Vole   3×

Bank Vole   3×

Water Vole

3×

3×

Hedgehog   3×

Common Pipistrelle

Large cetaceans

Being essentially oceanic animals, whales are among the hardest of our mammals to observe, although even the sight of a blow or fin is enough to set any pulse racing. The largest species are mostly filter-feeders but Sperm and Long-finned Pilot whales have impressive teeth and a carnivorous diet to match.

Blue Whale

Balaenoptera musculus Balaenopteridae
Length 25–30m

The world's largest living animal. Feeds on tiny swarming planktonic crustaceans (krill). **Adult** Has a huge streamlined body; bluish grey overall but mottled with greyish white. Seen just below the water surface, body looks very blue. A single ridge extends from nostrils to tip of rostrum. Throat has 70–90 pleats that allow a huge expansion when feeding. Flippers are relatively small, tailstock is thick and dorsal fin is small and set far back (*see also* p.229). **Status** Decimated by 20th-century whaling and possibly recovering. A small NE Atlantic population lives in deep waters with upwellings, and individuals are seen regularly in the Bay of Biscay and Western Approaches.

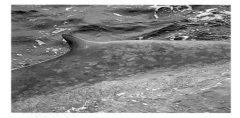

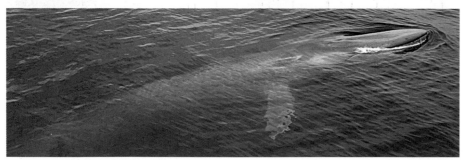

Fin Whale

Balaenoptera physalus Balaenopteridae
Length 18–22m

Second only in terms of size to Blue Whale. Uniquely, has asymmetrical markings on its head. A fast swimmer, catching shoaling fish. **Adult** Has a slender body with a narrow, pointed head. Body is mainly dark grey, palest on underparts and throat. Right side of lower jaw is whitish (as are front half of right side of baleen plates and tongue), whereas left side is same colour as rest of body. A single median ridge extends from nostril to tip of rostrum. Flippers are relatively long. Dorsal fin is curved, rather large and set a long way back. Tailstock is thick and tail fin is large (*see also* p.229). **Status** Mainly oceanic but sometimes seen in relatively shallow inshore waters. The W coast of Ireland is a hotspot.

Sei Whale

Balaenoptera borealis Balaenopteridae
Length 12–15m

(Pronounced 'say'.) Similar to Fin Whale but smaller and with a characteristic blow and dive sequence. Rises almost horizontally to the surface, so its blow, head, back and dorsal fin can be seen together (Fin and Minke whales rise in a gliding arch). A fast swimmer that feeds on krill. **Adult** Has a streamlined body; blue-grey, palest on underside. Head is broader and less pointed than that of Fin Whale and colour is symmetrical. Has a single median ridge on rostrum, flippers are uniformly coloured and dorsal fin is tall and strongly recurved (*see also* p.229). **Status** Enigmatic and comparatively little known. Occurs off NW Scotland, Shetland and the Orkney Isles in summer.

BELOW: **Sei Whale**

Humpback Whale

Megaptera novaeangliae Balaenopteridae
Length 12–15m

Large cetacean that is often active at the surface, where it breaches and engages in flipper- and tail-slapping. Feeds on shoaling fish and krill. **Adult** Has a stream-lined, bulky body; broad head is covered in lumpy tubercles. Pectoral fins are long and mainly white; dorsal fin is short and stubby. Tail is typically dark above and white with black markings below. Before making a deep dive, it arches its back strongly and the tailstock and fluke come clear of the water (*see also* p.229). **Status** Tolerates comparatively shallow seas if the feeding is good, and hence sometimes seen from land. Occurs regularly off W Ireland, mainly in winter.

Minke Whale

Balaenoptera acutorostrata Balaenopteridae
Length 8–10m

(Pronounced 'minky'.) The smallest baleen whale but still large and impressive. Tail is not revealed when animal dives. Feeds on shoaling fish. **Adult** Streamlined with a narrow, pointed snout. Upperparts are dark grey; underparts are whitish and broad bands of paler coloration extend up flanks and are sometimes visible on animals at the surface. Curved dorsal fin is set far back on body. A single ridge runs from nostrils to tip of rostrum. Has a diagnostic broad white spot or band on upper surface of flipper, easily seen in swimming animals (*see also* p.229). **Status** Favours relatively shallow waters of the continental shelf and is regularly seen from land; Mull and W Ireland are hotspots.

Sperm Whale

Physeter macrocephalus Physeteridae
Length 16–20m

Huge, distinctive whale with large, bulbous head. Tail is raised clear of water prior to a deep dive. Dives to 2,000m or more in search of Giant Squid and sharks. Females live in social groups.

Adult Blackish grey with whitish scars from fights and violent encounters with prey. Head is roughly one-third of body length and mass. Lower jaw is slender and has sharp teeth. Lacks a dorsal fin, but note the dorsal 'hump' and series of knobbly lumps. Flippers are small and tail is large. Has a single blowhole (baleen whales have 2), sited at front of head and angled slightly forward and to the left (*see also* p.229). **Status** Favours deep oceanic waters and is sometimes seen from ferries. Healthy individuals are seldom found in inshore waters.

Long-finned Pilot Whale

Globiocephala melaena Delphinidae
Length 4–6m

Medium-sized cetacean with distinctive head and dorsal fin shapes. Lives in groups (*pods*). Head, back and dorsal fin are visible simultaneously when cruising at surface. **Adult** Mainly blackish with a greyish saddle-shaped mark behind dorsal fin. Has a white thighbone-shaped mark from throat to vent, seen when breaching. Head is blunt-ended, forehead is domed and flippers are long and sickle-shaped. Dorsal fin is broad-based and curved. **Status** Regular in British and Irish waters.

Smaller cetaceans & seals

Although unrelated, cetaceans and seals have evolved to live in the same environment and have broadly similar diets comprising mainly fish. Unlike whales and dolphins, seals have retained many of their ties to land – they haul out onto rocks on a daily basis and give birth above the high-tide mark.

Killer Whale

Orcinus orca Delphinidae
Length 4–9m

Distinctive, well-marked cetacean. The largest dolphin. Social, living in 'pods' of 5–20 animals. Feeds on fish, squid, seals and other cetaceans. **Adult male** Has mainly blackish upperparts with a grey saddle-like patch behind dorsal fin. Underparts are white and a band of white extends onto flanks. Also has a white patch behind eye. Dorsal fin is up to 1.8m tall, triangular and upright, sometimes even forward-leaning. Flippers are broad and paddle-shaped. **Adult female** Smaller, with a much shorter, shark-like dorsal fin (*see also* p.229). **Status** Widespread in the NE Atlantic and occasional in British and Irish waters. Seen close to land if the feeding is good.

Risso's Dolphin

Grampus griseus Delphinidae
Length 3–3.5m

Large, distinctive blunt-nosed dolphin. Lives in 'pods' of 3–15 animals. **Adult** Greyish brown overall, darkest on dorsal fin, flippers and tail, and palest on face, throat and belly. Older animals become very pale and their upper surface is heavily criss-crossed with white scars. Head is blunt-ended and forehead is split down middle – from upper lip to blowhole – by a deep crease. Dorsal fin is tall, pointed and slightly recurved. Flippers are long and narrow, and tail fin is broad. **Status** Regularly seen in relatively shallow coastal water, mainly off W Britain.

ABOVE: **Risso's Dolphin**; BELOW: **Killer Whale**

Short-beaked Common Dolphin

Delphinus delphis Delphinidae
Length 1.8–2.3m

Our most regularly encountered dolphin. Gregarious, living in schools of tens or hundreds of animals. **Adult** Streamlined, with a pattern of overlapping stripes and bands of pigmentation. Dark grey above and whitish below, with a broad, tapering yellow band running along flanks from eye and mouth to just behind dorsal fin; a grey band continues along flanks towards tail. Overall, the yellow and grey patches resemble an hourglass. Flippers are narrow and black, with a black line running forward from their base to the throat. Dorsal fin is broadly triangular and curves backwards slightly. **Status** Mostly oceanic, typically seen from boats out of sight of land.

Bottlenose Dolphin

Tursiops truncatus Delphinidae
Length 2.5–4m

Bulky, muscular dolphin. Social, found in schools of 3–4 animals. Its diet includes fish, crabs and shrimps. Playful at the surface. **Adult** Greyish brown overall, darkest above and palest on throat and belly; a mid-grey band is sometimes seen on flanks. Beak is rather short and blunt (fancifully bottle-like), with lower jaw extending beyond upper one. Flippers are rather long and pointed; dorsal fin is tall, curved backwards and almost shark-like (*see also* p.229). **Status** Some are pelagic while other groups favour coastal waters. Moray Firth in Scotland and the Shannon Estuary in Ireland are hotspots.

Deep-sea dolphins

Our smaller inshore cetaceans have oceanic relatives that spend most of their lives far out to sea. Fortunately, ferry crossings often allow sightings of these enigmatic creatures. The **White-beaked Dolphin** *Lagenorhynchus albirostris* (**1**) (length 2.4-2.8m, Delphinidae), which lives off Orkney and Shetland, and in the northern North Sea generally, occurs in schools of 3-15 and has a black back and dorsal fin, a broad, pale band behind the eye and an oblique greyish-white stripe on the flanks; the snubnosed beak is pale, as are the throat and belly. Seen above the water, the beak usually looks pale pinkish grey but, when submerged, the filtering effects of seawater make it appear gleaming white. The dorsal fin is sickle-shaped. The **Atlantic White-sided Dolphin** *Lagenorhynchus acutus* (**2**) (length 2-2.5m, Delphinidae), which is easiest to see in the Irish Sea and on crossings to Shetland, lives in groups of 5-200 and has a dark back and dorsal fin and a grey band on the flanks, above which are white and yellowish stripes. The belly is white, this colour extending as a band to the eye. The beak is short and stubby, the flippers are small and the dorsal fin is tall and curved.

Harbour Porpoise

Phocoena phocoena Phocoenidae
Length 1.4-1.9m

Our smallest cetacean. Lives in groups of 3-15 animals. Playful at the surface. **Adult** Has a stout, streamlined body, blunt head and no beak. Flippers are small and oval, and dorsal fin is triangular with a concave trailing edge. Tailstock is thick and tail fin is broad. Upperparts are mainly dark grey while underparts are whitish. Has a bluish-grey patch on flanks, roughly between eye and start of dorsal fin. **Status** Favours inshore waters, including estuaries, river mouths and sheltered coastal bays, frequently within sight of land. Moray Firth and the Inner Hebrides are hotspots. Large numbers are killed each year by the fishing industry.

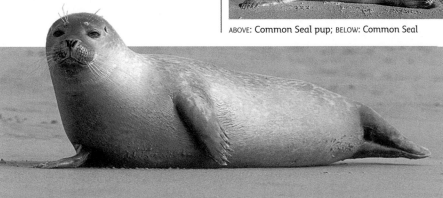

Common Seal

Phoca vitulina Phocidae
Length 1.2-1.9m

Has a 'friendly'-looking face. Hauled-out seals are easy to observe. **Adult** Greyish brown but variably mottled with darker spots, underside paler than upperside. Dry coat looks shiny, especially if coated in sand. Bridge of nose has a concave outline (convex in Grey Seal) and muzzle is blunt, creating a dog-like appearance. Seen from the front, nostrils are close together at base and splayed in a V-shape (separated from, and more parallel to, one another in Grey Seal). Front flippers have claws, and powerful hind flippers effect propulsion when swimming. Males are larger than females. **Juvenile (*pup*)** Born with a marbled grey-brown coat. **Voice** Mainly silent. **Status** Favours sheltered seas on E coast of England, and Scotland and Ireland generally.

ABOVE: **Common Seal pup**; BELOW: **Common Seal**

Grey Seal

Halichoerus grypus Phocidae
Length 2.2-3.2m

Bulky seal with a proportionately large head. Inquisitive in water. Hauls out for long periods. **Adult** Greyish overall with dark blotchy spots; fewer, larger spots than on Common Seal. Males are larger and darker than females. In profile, it looks 'Roman nosed': bridge of nose is convex, more pronounced in males than females. From the front, nostrils are distinctly separated and more or less parallel to one another, not V-shaped. Fore flippers have sharp claws, and hind flippers propel the animal through water. **Juvenile (*pup*)** Born with white fur; moults after a few weeks. **Voice** Utters low, moaning calls. **Status** Often found on rocky shores and tolerates rough seas. Widespread on W coasts and locally in the North Sea.

ABOVE: **bull**
LEFT: **bull in profile**
BELOW: **pup**

Watching seals & cetaceans

Because seals, whales and dolphins are marine mammals and humans are not, getting good views of them may seem an unlikely prospect. But if you know where to visit and the best time to go, seal sightings are guaranteed and good views of cetaceans are entirely possible.

In some locations, Grey Seals in particular are often actively curious about human intruders who venture into their domains.

Seal hotspots

Although they are true marine mammals, seals are tied to the land in a manner not seen in whales and dolphins. Each day they bask out of water, giving terrestrial observers the chance to get close to them. Where they are persecuted by those with commercial fishing interests, seals are understandably wary, but in places where they are actively protected and encouraged they can be positively indifferent to humans.

One of the best places to watch Common and Grey seals is **Blakeney Point on the north Norfolk coast** (1). Although encounters with seals are brief, views can be spectacularly close, with boats passing within a couple of metres of hauled-out animals. Good views of Grey Seals can also be seen on the **Farne Islands in Northumberland** (2). Probably the best land-based views of seals are at **Donna Nook** (3) on the Lincolnshire coast.

Common or Grey seal?

Adult Grey Seals are on average larger and more bulky than Common Seals, but in the field this is not especially useful for identification. A good way of separating the two is to look at the shape of the nostrils: front on, those of a **Common Seal** (1) converge towards the bottom and look rather heart-shaped, whereas **Grey Seal** (2) nostrils are more parallel to one another.

1

2

Cetacean hotspots

British and Irish seas and the northeast Atlantic were heavily exploited by the whaling industry during the 19th and 20th centuries and whale numbers were decimated. On top of these limited numbers, a combination of regular gales and the Atlantic swell mean that sea conditions are rarely ideal for observation.

So, until comparatively recently, many people dismissed the prospect of seeing whales and dolphins on a regular basis in British and Irish waters. But today there is a very real prospect of being able to encounter these magnificent creatures for yourself. In part, this is due to overall numbers of larger cetaceans creeping up. But thanks are due to a band of British and Irish enthusiasts whose perseverance in the field has produced a better understanding of cetacean occurrence, distribution and habits in our region.

A breaching Humpback Whale is a spectacular sight that will stay with the observer for the rest of his or her life.

Although land-based cetacean-watching can be somewhat hit-and-miss, a few regions in Britain and Ireland are fairly reliable hotspots. They include **west Cornwall and the Isles of Scilly** (1) (Common, Bottlenose, Atlantic White-sided and Risso's dolphins, and Minke Whale); **Hartland Point on the north Devon coast** (2) (Harbour Porpoise); the **Moray Firth** (3) (Bottlenose Dolphin and Harbour Porpoise); **Shetland and the Orkney Isles** (4) (Common Dolphin, Killer Whale and Long-finned Pilot Whale); the **Hebrides** (5) (Minke Whale and Harbour Porpoise); and the **west and south coasts of Ireland** (6) (Bottlenose, Common and Risso's dolphins, and Fin and Minke whales).

What kind of cetacean?

As most views of whales and dolphins are unlikely to be close enough to recognise them through their body shape, it is worth knowing other forms of identification. The height of the 'blow' made when the animal exhales is useful for larger cetaceans – and at a distance. The shape of the fin is another although you will need to be closer to make a definite identification.

Comparing spouts When a cetacean exhales, moisture in its breath condenses and appears as a spout. Dolphin and smaller whale spouts are impossible to see except on calm days, but those of larger whales can be seen from a distance. Spout height is hard to judge at a distance, it varies throughout the blow, and is affected by both wind speed and direction.

10M

0M

| Blue Whale = 10m max | Fin Whale = 6m max | Sperm Whale = 5m max | Humpback Whale = 3m max | Sei Whale = 3m max | Minke Whale = 1.5m max |

Comparing fins Whales and dolphins spend much of the time underwater and are at the surface for only comparatively brief periods. As the animal appears, the dorsal fin becomes visible and its size, particularly relative to the body as a whole, and its shape and angle, can be useful in identification.

Blue Whale   Fin Whale   Sei Whale   Humpback Whale   Sperm Whale   Minke Whale   Killer Whale   Bottlenosed Dolphin

Whale-watching experiences

LEFT: As a whale swims underwater, the force from its tail leaves a telltale 'fluke print' of calm sea at the surface.
BELOW LEFT: Land-based observations of Minke Whales are the highlight of a visit to the Hebrides, although typically views are rather distant.
BELOW: Fortunate beachcombers may sometimes discover washed-up whale bones such as this Minke Whale vertebra.

RIGHT: Sadly, dead animals such as this Sperm Whale are sometimes washed up on our beaches. They provide an unrivalled – if rather smelly – chance to examine cetaceans close up.

Dolphin deaths

Tens of thousands of dolphins are killed each year in British and Irish waters by the fishing industry. Most simply drown, trapped in nets from which they are unable to escape to the surface to breathe. Of these mortalities, several hundred are washed up on our shores. Sadly, this is the only way that most people will ever see oceanic species such as the Striped Dolphin *Stenella coeruleoalba* in our region. Think of this image the next time you eat fish and chips.

BIRDS

The ancestors of birds took to the air some 150 million years ago and since then their ability to fly has allowed them to occupy most terrestrial habitats and many aquatic ones too. In Britain and Ireland, birds are the most diverse and numerous vertebrates. By and large, they are colourful or well marked and many are easy to see in urban, rural and coastal habitats, making birding the most popular branch of natural history.

Puffin.

Nests and eggs

In common with their reptilian ancestors, birds lay eggs, inside which their young develop. Eggs are laid in a nest, which varies from a rudimentary scrape to an intricately woven basket depending on the species. The eggs are protected by a hard, chalky outer casing and the developing chick gains its nutrition from the egg's yolk.

Although it is a bulky bird, the Woodpigeon builds a surprisingly rudimentary nest of twigs.

A Song Thrush's nest is a beautifully woven cup of grasses, lined with a smooth coating of mud.

A Lapwing's egg – medium sized and pointed at one end.

A Coal Tit's egg – small and rather rounded.

Social behaviour

Although some birds lead rather solitary lives except during the breeding season, many species are gregarious and either breed in colonies or spend the winter months in sizeable groups. Complex behaviour patterns allow individuals to rub along with one another.

Rooks spend most of their lives in the company of others of the same species, nesting communally and feeding in flocks. Members of a breeding pair will sometimes forage together and maintain their bond with touchingly gentle interactions.

Like most duck species, Wigeon pairs are essentially solitary during the breeding season. In winter, however, they form huge flocks that feed and roost together. With so many eyes on the look out for danger, there really is safety in numbers.

Territoriality and attracting a mate

Most bird species are territorial to a degree, at least during the breeding season. The territory may be small - the jabbing distance of a beak in the case of many seabirds - but in others the range they defend includes good feeding grounds as well as nesting habitat. Many species advertise their presence using sound, both to attract a mate and to announce ownership to prospective rivals.

ABOVE: The males of many gamebird species, such as this Capercaillie, gather at communal display sites (called leks) to attract females.
RIGHT: Like other songbirds, the male Dartford Warbler sings throughout the breeding season, and is aggressive towards intruders in his territory.

Diet

Not only are there birds in all habitats in Britain and Ireland, but almost all sources of food are exploited by one species or another. Some birds are purely vegetarian, feeding perhaps on seeds, fruits or shoots, while many more include invertebrates - particularly insects - in their diet during the summer months. A few are strict predators, taking prey that includes other birds.

ABOVE: There are clues to a Greenshank's diet in its appearance: its long legs and bill allow it to feed in deep water on aquatic invertebrates.

RIGHT: Although Buzzards are partial to the flesh of Rabbits, they will also eat the humble Common Earthworm.

Plumage and moult

Day-to-day life causes wear and tear to feathers, so birds moult their old ones for a new set at regular intervals; the timing of this event (annual in some birds and twice a year in others) varies according to the species. Many birds have significantly different plumages in summer and winter, and males, females and immature birds can look like entirely different species to the uninitiated.

BELOW: An adult male Black Redstart looks striking in breeding plumage but dowdy when moulting.

ABOVE: Immature and female Black Redstarts are drab by comparison with the male.

Parental care

Newly hatched birds are vulnerable and chick mortality is high. The degree to which newly hatched birds can fend for themselves, and the effort invested by their parents in looking after them, varies considerably. With songbirds and many other bird families, adults incubate and feed their young (which are defenceless at first) at least until they fledge, and often for several weeks afterwards. Among gamebirds, the young are active almost from the moment they hatch and leave the nest straight away; even so, they still depend on their parents to a degree for shelter and protection, and to be guided to areas where food is plentiful.

A Tawny Owl chick will often leave the nest before it can fly properly; it will be fed by its parents for several weeks.

Aquatic lives

Most bird species live terrestrial lives, their encounters with water limited to drinking and bathing. However, a few specialised groups have evolved to take advantage of the food available in water. Ducks and geese are obvious examples, but they are quite at home on land too. Divers and grebes, however, can barely waddle more than a few feet on dry ground.

Although adept at swimming at the surface, dabbling ducks such as this Mallard cannot dive and instead have to up-end in order to feed on submerged plants.

Red Kite.

primaries

tail feathers

carpal patch

axillaries (armpit)

underwing coverts

undertail coverts

tertials

secondaries

Flight, feathers and feather care

Arguably the most significant role played by feathers is in aiding flight. They are, however, also vital for thermal insulation, and patterns and colours confer species and gender identity on their owners; for some species, feathers are important as camouflage. Different feathers fulfil different functions; those associated with flight are structurally quite different from those that insulate.

A flight feather has a rigid shaft and a zip-like arrangement that locks the barbs together.

Body feathers vary, but this one has downy, insulating properties towards its base while the outer barbs contribute to the contour of the bird.

Because feathers are so important, birds such as this immature Herring Gull spend a lot of time preening, to keep them clean and to ensure their structure remains intact.

Divers & grebes

Divers and grebes are consummate waterbirds. Their entire lives, apart from when they are nesting, are spent in water. They swim both above and below the water using their modified feet (lobed toes in grebes, webbed feet in divers). In flight, the head and neck are held outstretched. Divers belong to the family Gaviidae and grebes to Podicipedidae.

Typical diver silhouette in flight. This is a Red-throated Diver, summer adult

Red-throated Diver

Gavia stellata Gaviidae
Length 55–65cm

Swims low in the water with head and dagger-like bill tilted upwards. Dives frequently. Sexes are similar. **Adult** In summer, has blue-grey on face and sides of neck, red throat and black and white lines on back and lower sides of neck. Upperparts are otherwise brownish grey and underparts are whitish. In winter, grey upperparts are spangled with small white spots; underparts are white. **Juvenile** Similar to winter adult but grubby-looking. **Voice** Mostly silent. **Status** Scarce breeding species; nests beside freshwater pools. Outside the breeding season found in shallow coastal seas; locally common in winter.

Easiest to see in winter.

Black-throated Diver

Gavia arctica Gaviidae
Length 60–70cm

Swims buoyantly with bill held level. Dives frequently. Sexes are similar. **Adult** In summer, has blue-grey nape and head, and black throat; sides of neck have black and white lines. Back is dark with white spots; underparts are white. In winter, upperparts are mainly grey-black and underparts whitish; note white patch on flanks at water level in swimming birds. **Juvenile** Similar to winter adult but grubby-looking. **Voice** Mostly silent. **Status** Rare breeding species on large Scottish lochs. Scarce in winter, mainly in coastal waters.

Easiest to see in winter on S and E coasts of England. Do not try to approach closely when nesting.

Great Northern Diver

Gavia immer Gaviidae
Length 75–85cm

Buoyant waterbird. Large bill is held level or very slightly elevated. Sexes are similar. **Adult** In summer, has a black neck with two rows of white stripes. Upperparts are blackish with white patches on mantle and spots elsewhere, and underparts are gleaming white. Bill is dark. In winter, has dark grey upperparts and whitish underparts with a dark half-collar on neck. Bill is greyish. **Juvenile** Similar to winter adult but back feathers have pale margins. **Voice** Silent in our region. **Status** Non-breeding visitor to coastal seas, favouring both rocky shores and large bays; occasional on inland reservoirs.

Easiest to see in winter.

TOP: **summer adult** (*see also* **top right of page**)
ABOVE: **winter adult**

TOP: **summer adult**
ABOVE: **winter adult**

TOP AND BELOW: **summer adult**
ABOVE: **juvenile**

Did you know?

Divers and grebes are so adapted to an aquatic life that they seldom abandon water willingly. In the breeding season, divers nest on land but within a metre or so of the water's edge. Grebes build floating nests of vegetation, located amongst emergent vegetation. Both divers and grebes are superb swimmers and, depending on the species, feed on fish or aquatic invertebrates.

Little Grebe

Tachybaptus ruficollis Podicipedidae
Length 25–29cm

Dumpy, buoyant waterbird with a powderpuff of feathers at its rear end. Dives frequently for fish and aquatic invertebrates. Wings are rounded and uniform grey-brown. Sexes are similar. **Adult** In summer, is mainly brownish but neck and cheeks are chestnut. Pale-tipped dark bill has lime-green spot at base. In winter, has mainly brown upperparts and buffish underparts. **Juvenile** Recalls winter adult but with pale throat and black stripes on face. **Voice** A whinnying trill. **Status** Fairly common resident of freshwater ponds and slow-flowing rivers; in winter, also seen on sheltered coasts and estuaries.

Best located by its very distinctive call.

summer adult; INSET: winter adult

Great Crested Grebe

Podiceps cristatus Podicipedidae
Length 46–51cm

Graceful waterbird with a slender neck and dagger-like bill. White wing panels are revealed in flight. Dives frequently. Sexes are similar. **Adult** In summer, has grey-brown upperparts and mainly whitish underparts; head has a black cap and crest, and an orange-buff ruff bordering paler cheeks. Bill is pink and eye is red. In winter, has drab grey-brown and white plumage. **Juvenile** Recalls winter adult but has dark stripes on cheeks. **Voice** Wails and croaks in the breeding season. **Status** Locally common breeding species on lakes and reservoirs. Widespread in the winter, when it is also found in inshore seas.

Look for displaying pairs in spring.

ABOVE: **summer adult**; INSET: **winter adult**; BELOW: **adults displaying**

Red-necked Grebe

Podiceps grisegena Podicipedidae
Length 40–45cm

Has striking summer plumage and yellow-based bill. White wing panels are seen in flight. Sexes are similar. **Adult** In summer, has a red neck and upper breast; head has white-bordered pale grey cheeks and black cap. Upperparts are otherwise grey-brown and underparts are whitish with grey streaks on flanks. In winter, loses neck colours but often retains hint of reddish collar. Cheek pattern is less well defined and ear coverts are grubby. **Juvenile** Similar to winter adult but with more extensive red on neck. **Voice** Mostly silent. **Status** Scarce winter visitor to sheltered coasts and lakes.

Easiest to see in winter.

summer adult; INSET: winter adult

Black-necked Grebe

Podiceps nigricollis Podicipedidae
Length 28–34cm

Buoyant waterbird with uptilted bill, steep forehead and beady red eye. A white patch on trailing edge of wing is seen in flight. Sexes are similar. **Adult** In summer, has a blackish head, neck and back, with golden-yellow tufts on face. Flanks are chestnut. In winter, has mainly blackish upperparts and white underparts; can be told from similar Slavonian by head shape and greater extent of black on cheeks. **Juvenile** Similar to winter adult but grubby-looking. **Voice** Mostly silent. **Status** Scarce winter visitor to sheltered coasts; occasional on inland reservoirs. A few pairs nest on shallow, well-vegetated lakes.

Easiest to see in winter.

summer adult; INSET: winter adult

Slavonian Grebe

Podiceps auritus Podicipedidae
Length 31–38cm

Buoyant waterbird with beady red eye. Flattish crown and white-tipped bill (both mandibles are curved) allow separation from Black-necked. White patches on both leading and trailing edges of wings are seen in flight. Sexes are similar. **Adult** Has a reddish-orange neck and flanks. Back is black and black head has golden-yellow plumes. In winter, has black upperparts and white underparts, with a clear demarcation between black cap and white cheeks. **Juvenile** Similar to winter adult. **Voice** Trills and squeals at nest. **Status** Scarce winter visitor to sheltered coasts. Rare breeder in Scotland, on shallow lochs.

Visit RSPB's Scottish reserves in spring.

summer adult; INSET: winter adult

Storm-petrel to herons

Fulmar and shearwaters, storm-petrels and Gannet are all seabirds that lead oceanic lives outside the breeding season. Cormorant and Shag favour inshore waters, while herons are found mainly in shallow waters.

British Storm-petrel

Hydrobates pelagicus Hydrobatidae
Length 14–16cm

Our smallest seabird. Flutters low over water with dangling feet when feeding. Note the square-ended tail. Sexes are similar. **Adult** Dark sooty brown except for white rump. A close view reveals white bar on underwing. **Juvenile** Similar but with pale wingbar on upperwing. **Voice** Silent at sea; gurgles and purrs at nest. **Status** Very locally common breeder. Nests in burrows on islands that are visited after dark. Otherwise, seldom seen near land. Seen from ferries.

Leach's Storm-petrel

Oceanodroma leucorhoa Hydrobatidae
Length 16–18cm

Longer-winged than British Storm-petrel; flight is ever-changing, with powerful wing-beats and glides. Sexes are similar. **Adult** Dark sooty grey except for pale panel on upper-wing coverts. Tail fork and grey central line on rump can be hard to see. Underwings are all dark. **Juvenile** Similar. **Voice** Silent at sea; weird gurgling rattles are heard at nest. **Status** Oceanic. Very locally common but hard to see. Visits colonies after dark. Autumn gales may produce sightings.

Manx Shearwater

Puffinus puffinus Procellariidae
Wingspan 70–85cm

Skims low over the sea on stiffly held wings. Contrasting dark upperparts and mainly white underparts are seen as birds bank and glide. Gregarious when the feeding is good. Sexes are similar. **Adult** Has blackish upperparts and mainly white underparts with dark wing margins. **Juvenile** Similar. **Voice** Silent at sea; after dark, nesting birds utter strangled coughing calls. **Status** Fairly common summer visitor. Seen mostly at sea; visits land only to breed, after dark. Nests in burrows on remote islands. Visit Skomer or Bardsey for close-up views.

Did you know?

Manx Shearwaters (*below*) and our storm-petrels are masterful seabirds but they are vulnerable to attack by predators while nesting on land. So, they nest in burrows and return to their colonies only after dark.

Manx Shearwater at burrow entrance.

Fulmar

Fulmarus glacialis Procellariidae
Wingspan 105–110cm

Has tube nostrils, stiffly held wings and gliding flight. Swims buoyantly. Sexes are similar. **Adult** Typically has blue-grey upperwings and back. Head, underparts and tail are white. Has a dark smudge around eye. Arctic birds are blue-grey and are seen occasionally. **Juvenile** Similar to adult. **Voice** Gurgling cackles and grunts. **Status** Locally common. Nests on sea-cliffs. Otherwise, at sea. Easy to see in N Britain.

Gannet

Morus bassanus Sulidae
Wingspan 165–180cm

Large seabird with powerful wing-beats; glides well. Bill is dagger-like. Plunge-dives to catch fish. Sexes are similar. **Adult** Has mainly white plumage with black wingtips; head has a buffish wash. **Juvenile** Has dark brown plumage speckled with white dots in first year; adult plumage is acquired over next 4 years. **Voice** Silent at sea; nesting birds utter grating calls. **Status** Very locally common. Nests colonially but otherwise marine. Bempton Cliffs or Bass Rock offer spectacular views.

TOP: adult
ABOVE: juvenile

Cormorant

Phalacrocorax carbo
Phalacrocoracidae
Length 80–100cm

Dark waterbird with heavy, hook-tipped bill. Swims low in the water, propelled by large webbed feet. Wings are often held outstretched when perched. Sexes are similar. **Adult** In summer, is mainly dark with an oily sheen and black-bordered brownish feathers on wings, and white on thigh and head. Eye is green and skin at base of bill is yellowish. In winter, white feathering is absent. **Juvenile** Has brown upperparts and whitish underparts. **Voice** Silent except at nest, when it utters nasal and guttural calls. **Status** Common, favouring sheltered seas and large freshwater lakes. Breeds colonially. Easy to see on many flooded gravel pits.

Shag

Phalacrocorax aristotelis Phalacrocoracidae
Length 65–80cm

Smaller than Cormorant, with more slender bill. Leaps in order to submerge. Often perches with wings held outstretched. Sexes are similar. **Adult** All dark but with an oily green sheen. Has a yellow patch at base of bill and a distinct crest. In winter, loses crest; colours at base of bill are subdued. **Juvenile** Has dark brown upperparts, buffish underparts and pale throat. Crest peaks on forehead (peaks on rear of crown in juvenile Cormorant). **Voice** Silent except at nest, when it makes harsh grunting calls. **Status** Locally common on rocky coasts. Nests colonially on sea cliffs. Easy to see at most seabird colonies.

Spoonbill

Platalea leucorodia
Threskiornithidae
Length 70–80cm

Unmistakable. Spoon-shaped bill is swept from side to side in shallows to catch crustaceans and fish. Sleeps with bill tucked under wings – cf. Little Egret. Sexes are similar. **Adult** Has whitish plumage and black bill with yellow tip; in the breeding season, has a crest and base of bill and breast are flushed yellow. **Juvenile** Similar but legs and bill are dull pink. **Voice** Silent. **Status** Rare nesting species and non-breeding visitor from Europe. Most records are coastal. Visit N Norfolk in late summer.

Bittern

Botaurus stellaris Ardeidae
Length 70–80cm

Shy and superbly camouflaged, hence hard to see in reedbeds. Posture is usually hunched, but 'skypoints' and sways if alarmed. Bill is dagger-like, and legs and feet are long and powerful. Superficially owl-like in flight. Sexes are similar. **Adult** Has brown plumage with intricate dark markings. **Juvenile** Similar but crown and 'moustache' are paler. **Voice** Territorial males 'boom' in spring. **Status** Favours large reedbeds with shallow water for feeding. Sometimes seen in smaller wetlands outside the breeding season. Visit N Norfolk reedbed reserves in spring.

Little Egret

Egretta garzetta Ardeidae
Length 55–65cm

Unmistakable pure white heron-like bird. The long black legs have bright yellow toes. Feeds actively in water, often chasing small fish. Has a hunched posture when resting. In flight, neck is held in an 'S' shape and legs are trailing. Sexes are similar. **Adult** Has pure white plumage. Note the yellow eye. Nape plumes are seen in breeding plumage. **Juvenile** Similar. **Voice** Mostly silent. **Status** Recent arrival to Britain, now locally common on coasts and, increasingly, inland wetlands. Easy to see on S England estuaries.

Grey Heron

Ardea cinerea Ardeidae
Length 90–98cm

Familiar wetland bird. Stands motionless for long periods. Flies on broad wings with slow, deep wingbeats; neck is held hunched. Sexes are similar. **Adult** Has a whitish-grey head, neck and underparts, with dark streaks on front of neck and breast; note the white forecrown and black sides to crown leading to black nape feathers. Back and upperwings are blue-grey, and flight feathers are black. Dagger-like bill is yellowish. **Juvenile** Similar but crown and forehead are dark grey. **Voice** Utters a harsh *krrarnk* in flight. **Status** Common resident. Favours freshwater wetlands but seen on coasts in winter. Easy to find.

LEFT: juvenile
BELOW: adult

Swans & geese

Swans and geese are the largest wildfowl (family Anatidae). Their long necks are held outstretched in flight and they have vegetarian diets. Adult swan plumage is white, while geese are grey or grey-brown.

Mute Swan
Cygnus olor Anatidae
Length 150–160cm

cygnet

Large, distinctive and familiar waterbird. Swimming birds hold their long neck in an elegant curve. Family groups are a feature of lowland lakes in spring. Typically tolerant of people. In flight, shallow, powerful wingbeats produce a characteristic throbbing whine. Sexes are similar but bill's basal knob is largest in males. **Adult** Has white plumage, although crown may have an orange-buff suffusion. Bill is orange-red with black base. **Juvenile** Has grubby grey-brown plumage and dull pinkish-grey bill. **Voice** Mostly silent. **Status** Our commonest swan; the only resident species. Found on freshwater habitats, besides which it nests; in winter, also occurs on sheltered coasts.
Easy to see on most wetlands and a familiar sight even in the heart of London.

TOP: adult; ABOVE: adult; BELOW: juvenile

Whooper Swan
Cygnus cygnus Anatidae
Length 150–160cm

Similar to Mute Swan but separable using bill shape and colour. Typically holds its neck straight, not curved. Seen in medium-sized flocks comprising several family groups. Sexes are similar. **Adult** Has mainly pure white plumage, although head and upper neck are sometimes stained orange. Bill is triangular and rather long, with a yellow patch that extends beyond nostril. **Juvenile** Has grubby buffish-grey plumage and dark-tipped pale pink bill. **Voice** Loud bugling calls. **Status** A handful of pairs breeds here each year, but best known as a winter visitor; several thousand are present Oct–Mar.
Visit Welney Wildfowl & Wetlands Trust (WWT) reserve in Cambridgeshire in winter.

TOP: adult; ABOVE: adult; BELOW: juvenile

Bewick's Swan
Cygnus columbianus Anatidae
Length 115–125cm

Our smallest swan. Usually seen in medium-sized flocks comprising family groups that remain as such throughout their winter stay here. Similar to larger Whooper Swan but separated by its relatively shorter neck and different bill pattern. Sexes are similar. **Adult** Has mainly pure white plumage. Bill is wedge-shaped but proportionately shorter than that of Whooper; its yellow colour typically does not extend beyond start of nostrils and yellow patch is usually rounded, not triangular. **Juvenile** Has grubby buffish-grey plumage and dark-tipped pink bill. **Voice** Various honking and bugling calls. **Status** Winter visitor, with 10,000+ birds found at traditional sites: flooded grassland, marshy meadows and, occasionally, arable farmland.
Visit the Ouse Washes in Cambridgeshire, or Slimbridge Wildfowl and Wetland Trust centre in Gloucestershire in winter for spectacular views and large numbers of birds.

family group of Whooper Swans

ABOVE: adult; BELOW: juvenile;

Brent Goose

Dark-bellied Brent

Branta bernicla Anatidae
Length 56–61cm

Our smallest goose and similar in size to a Shelduck. Subtle plumage differences allow separation of the two subspecies that winter here: Pale-bellied Brent *B.b. hrota* (which breeds on Svalbard and Greenland) and Dark-bellied Brent *B.b. bernicla* (which breeds in Russia). Healthy birds are invariably seen in sizeable, noisy flocks. In flight, looks dark except for white rear end. All birds have black bill and black legs. Sexes are similar. **Adult Pale-bellied** Has blackish head, neck and breast; side of neck has narrow band of white feathers. Note the neat division between dark breast and pale grey-buff belly. Back is a uniform dark brownish grey. **Adult Dark-bellied** Similar but belly is darker and flanks paler. **Juveniles** Similar to respective adults, but note pale feather margins on back and absence of white markings on side of neck; white on neck is acquired in New Year. **Voice** Very vocal, uttering a nasal *krrrut*. This is the sound of winter on many of our estuaries. **Status** Winter visitor to coasts and found on most estuaries in south and east England, and around much of the Irish coast.

Spectacular numbers can be found on the Solent, in Hampshire, and along the north Norfolk coast. In places where they are not disturbed, the birds are often remarkably indifferent to human observers.

ABOVE: **Pale-bellied Brent adult**
BELOW: **Dark-bellied Brent adult**

Flocks

Outside the breeding season, geese and swans live together in large, typically single-species groups called flocks. They feed and roost together, relying on the principle of safety in numbers – there are always plenty of eyes on the look out for danger. Swans and migratory goose species even migrate to and from their Arctic breeding grounds in flocks. In winter, huge flocks of geese, sometimes comprising tens of thousands of birds, can be seen at the following locations: Brent Geese at Farlington Marshes, Hampshire, and Holkham, Norfolk; White-fronted and Pink-footed geese at Holkham, Norfolk; and Barnacle Geese at Loch Gruinart, Islay.

Dark-bellied Brents

Barnacle Goose

Branta leucopsis Anatidae
Length 58–69cm

Small, well-marked goose, seen in large, noisy flocks. All birds have a black bill and legs. Looks strikingly black and white in flight. Sexes are similar. **Adult** Has a mainly white face with a black line from bill to eye; black crown and nape merge with black neck and breast. Belly is whitish grey with faint dark barring on flanks; back is grey with well-defined black and white barring. Stern is white while tail is black. **Juvenile** Similar but white elements of plumage are often tinged yellow and barring on back is less well defined. **Voice** Loud barking calls. **Status** Winter visitor to coastal farmland and saltmarshes.

Islay and Solway Firth are hotspots

Canada Goose

Branta canadensis Anatidae
Length 95–105cm

Large, familiar goose with long neck and upright stance. All birds have a blackish bill and dark legs. In flight, wings appear uniformly grey-brown while stern is white. Sexes are similar. **Adult** Has white cheeks on an otherwise black head and neck. Body is mainly grey-brown, darkest on back (pale feather margins create barring) and palest on breast. Stern is white and tail is dark. **Juvenile** Similar but barring on back is less distinct. **Voice** Utters loud, disyllabic trumpeting calls in flight. **Status** Introduced but now our most widespread goose; commonest in lowland England, usually near fresh water, often on nearby grassland.

Easy to see.

Goose bills

Although goose bills differ in size, their basic plan is common to all species: they are rather broad and flattened, and are used to crop vegetation. Geese often graze in grassland but their long neck also allows them to feed on submerged aquatic plants.

Geese & ducks

Geese belonging to the genus *Anser* are often referred to collectively as 'grey geese'; subtle differences in plumage and more obvious differences in bill structure and colour allow separation of the four species that occur regularly in Britain and Ireland. The geese and ducks described here are large, plump-bodied birds with long necks and robust bills, adapted for grazing or filter-feeding depending on their diet. Their feet are webbed and propel them when swimming. The Shelduck and naturalised Egyptian Goose are both appreciably smaller than grey geese.

Greylag Goose

Anser anser Anatidae
Length 75–90cm

Largest *Anser* goose and the only one that breeds in Britain. Feral populations confuse the species' wild status. Compared to other 'grey' geese, is bulky and more uniformly grey-brown. Pink legs and heavy pinkish-orange bill help with identification. In flight, pale forewings, rump and tail contrast with darker flight feathers. Sexes are similar. **Adult** Greyish with dark lines on side of neck, barring on flanks and pale margins to back feathers. Bill is pale-tipped. **Juvenile** More uniformly grey-brown than adult and bill lacks pale tip. **Voice** Loud honking calls. **Status** Locally common resident, mainly in the N. Wild migrants boost numbers in winter. Favours wetlands and reservoirs.

 Easiest to see in N Britain in winter.

Bean Goose

Anser fabilis Anatidae
Length 65–85cm

Heavy-looking 'grey' goose with bulky bill and orange legs. Neck is long compared to that of Pink-footed. In flight, wings are mainly dark, but note the pale bars on upper surface. Dark rump and tail contrast with narrow white band that defines tail-base. Typically forms single-species flocks. Sexes are similar but 2 sub-species occur. **Adult** Has dark brown on head and neck, grading to paler brown on breast and belly. Stern is white and back is dark brown, the feathers having pale margins. Ssp. *fabilis* (from N Europe) is long-necked; its large bill is well marked with orange. Ssp. *rossicus* (from N Siberia) is smaller with a shorter neck; its smaller bill has a small extent of orange. Some adults of both subspecies show white at base of bill. **Juvenile** Similar to respective adults but bill and leg colours dull. **Voice** A nasal, trumpeting cackle. **Status** Scarce winter visitor; favours marshes and wet grassland.

 Small numbers winter in N Norfolk.

Pink-footed Goose

Anser brachyrhynchus Anatidae
Length 60–75cm

Similar to Bean Goose, but smaller and more compact; smaller bill is marked with pink. Pink leg colour is diagnostic. In flight, note the pale blue-grey back, rump and upperwing coverts, and extent of white on tail. Forms single-species flocks. Sexes similar. **Adult** Dark chocolate-brown head and upper neck, grading to buffish brown on breast and belly. Back is blue-grey with pale feather margins. **Juvenile** Similar but back is buffish and feathers lack clear pale margins; leg and bill colours are dull. **Voice** A nasal, trumpeting cackle, higher in pitch than that of Bean Goose. **Status** Locally common winter visitor, mainly from Iceland; favours stubble fields and grassland.

East Anglia and Lancashire north to E Scotland are the main wintering areas; Holkham in Norfolk is a hotspot.

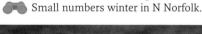
adult *fabilis*

adult *rossicus*

White-fronted Goose

Anser albifrons Anatidae
Length 65–75cm

Adults have a striking white patch on forehead. Two subspecies occur: Greenland White-front *A.a. flavirostris*, with an orange bill and overall darker plumage; and smaller, pink-billed European White-front *A.a. albifrons*. All birds have orange legs and mainly dark wings with faint pale wingbars. Sexes are similar. **Adult Greenland** Has a dark brown head grading to paler brown on neck and underparts; note the black patches on belly and large white forehead patch. Back is dark grey-brown and stern is white. Bill tip is white. **Adult European** Shorter-necked and paler, especially on head, belly and back. Bill tip is white. **Juveniles** Similar to respective adults, but lack white forehead patch and black belly markings; bill tip is dark. **Voice** Barking, musical calls. **Status** Locally common winter visitor; Greenlands visit Ireland and NW Scotland, Europeans visit England and S Wales. Favours wet grassland. Slimbridge WWT is a hotspot.

Shelduck

Tadorna tadorna Anatidae
Length 55–65cm

Goose-sized duck with bold markings. In poor light, looks black and white. In flight, note the contrast between white wing coverts and black flight feathers. Sifts mud for small invertebrates. Nests in burrows. Sexes separable with care. **Adult male** Mainly white but with dark green head and upper neck (looks black in poor light), chestnut breast band, black belly stripe and flush of orange-buff under tail. Legs are pink and bill is bright red with a knob at base. **Adult female** Similar but bill's basal knob is much smaller. **Juvenile** Has mainly buffish-grey upperparts and white underparts. **Voice** Courting males whistle while female's call is a cackling *gagaga*... **Status** Common on most estuaries and mudflats; local at inland freshwater sites. Migrates to favoured sites like Bridgwater Bay in Somerset for its summer moult. Easy to see on most estuaries.

male

female

Egyptian Goose

Alopochen aegyptiacus Anatidae
Length 65–72cm

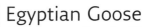

Striking Shelduck-sized bird. In flight, bold white patch on innerwing is a useful identification aid. Bill and legs are pink. Sexes are similar. **Adult** Has a grubby-white head and neck; pale eye is surrounded by dark patch. Orange-buff breast is clearly separated from paler neck and darker, grey-buff belly; note the small dark patch on centre of breast. Back is dark grey-brown; note the white and chestnut on wing, seen in resting birds. **Juvenile** Similar but colours are duller; dark breast spot and patch around eye are absent. **Voice** Mostly silent. **Status** Introduced from Africa but now well established and increasing. Usually seen near water, often on nearby grassland. N Norfolk and the Thames Valley/M4 corridor are hotspots.

Flight and 'V' formations

If you live in Scotland, one of the most evocative sounds of autumn is the call of grey geese flying high overhead as they migrate south from their Arctic breeding grounds. Typically, long-distance flights of geese take on a 'V' formation and are known as 'skeins'. Such formations are not just the consequence of following the leader, but are also thought to improve aerodynamic efficiency. When flying to and from roosts and feeding grounds, geese tend to fly in loose flocks.

Dabbling ducks

Although they occasionally up-end in shallow water to feed, or make a half-hearted attempt to submerge, dabbling ducks are so buoyant that much of their diet is consumed while dabbling in the shallows or, indeed, by grazing on land. They belong to the genus *Anas*.

Mallard

Anas platyrhynchos
Anatidae
Length 50–65cm

male

Our most familiar duck. In flight, both sexes show white-bordered blue speculum. Sexes are dissimilar. **Adult male** Has a yellow bill and shiny green head and upper neck, this separated from chestnut breast by striking white collar. Underparts are grey-brown except for black stern and white tail. Back is grey-brown grading to reddish brown. Legs and feet are orange. In eclipse, the male resembles an adult female but note the yellow bill and well-defined reddish-brown breast. **Adult female** Has an orange-brown bill and mottled brown plumage. Legs and feet are dull orange-yellow. **Juvenile** Similar to adult female. **Voice** Males utter whistles and nasal calls. Females utter familiar quack. **Status** Widespread, commonest on lowland lakes and rivers.

👓 Found on almost every wetland habitat.

Gadwall

Anas strepera
Anatidae
Length 46–55cm

Rather nondescript dabbling duck. Distant males look grey and brown but a close view reveals intricate feather patterns. In flight, both sexes show white in speculum; male also has chestnut on innerwing. Sexes are dissimilar. **Adult male** Has a buffish head and neck, clearly separated from finely patterned grey breast and flanks. Centre of belly is white; black stern is a useful identification feature. Has a dark bill and yellow legs. In eclipse, male resembles adult female. **Adult female** Has mottled brown plumage with a greyish head and yellow bill. **Juvenile** Resembles adult female. **Voice** Males utter a croaking call and females a Mallard-like quack. **Status** Found on shallow fresh water; dabbles for water plants.

👓 Easiest to see in winter on flooded gravel pits.

Wigeon

Anas penelope Anatidae
Length 45–47cm

Male is colourful and attractive. Forms large flocks outside the breeding season. Sexes are dissimilar. **Adult male** Has a mainly orange-red head with yellow forehead. Breast is pinkish; rest of plumage is mainly finely marked grey except for white belly and black and white stern. In flight, has a white patch on wing. Bill is pale grey and dark-tipped. In eclipse, resembles adult female although white wing patch is still evident. **Adult female** Mainly reddish brown, darkest on head and back. Note, however, the white belly and stern. In flight, lacks male's white wing patch. Bill is grey and dark-tipped. **Juvenile** Resembles adult female. **Voice** Males utter an evocative *wheeeoo* whistle. **Status** Very scarce breeder but locally common winter visitor to estuaries and wet grassland.

👓 Easy to see on coasts in winter.

TOP: male; ABOVE: female

TOP: male; ABOVE: female

TOP: male
ABOVE: female

male

Plumages and moult

Male and female ducks of the same species are typically very different to look at. Males usually have a gaudy plumage that is used in display. In contrast, female plumage is usually subdued and marbled brown, making them camouflaged while sitting on eggs. Ducks moult their feathers mostly in the summer months, after the breeding season has finished. For a brief period, males acquire a female-like plumage called 'eclipse', but are restored to their usual splendour by autumn.

Teal

Anas crecca Anatidae
Length 34–38cm

male

Our smallest duck. Forms flocks outside the breeding season. Often nervous and flighty. In flight, both sexes show a white-bordered green speculum. Sexes are otherwise dissimilar. **Adult male** Has a chestnut-orange head with yellow-bordered green patch through eye. Plumage is otherwise finely marked grey except for black-bordered yellow stern and horizontal white line along flanks. Bill is dark grey. In eclipse, resembles adult female. **Adult female** Has mottled grey-brown plumage. Bill is grey with hint of yellow at base. **Juvenile** Similar to adult female but a warmer buff. **Voice** Males utter a ringing whistle, females a soft quack. **Status** Associated with water. Nests in small numbers beside pools and bogs, mainly in N. Locally common outside breeding season on freshwater marshes, estuaries and mudflats.

 Easy to see on coasts in winter.

Garganey

Anas querquedula Anatidae
Length 37–41cm

male

Teal-sized summer visitor to the region. Favours emergent wetland vegetation, hence unobtrusive. In flight, male shows a pale blue-grey forewing and white-bordered greenish speculum; female's speculum is brown. Sexes are dissimilar. **Adult male** Has a reddish-brown head and broad white stripe above and behind eye. Breast is brown but otherwise plumage is greyish, except for mottled buffish-brown stern. In eclipse, resembles adult female but retains wing patterns. **Adult female** Has mottled brown plumage; similar to female Teal but bill is uniform grey and has a pale spot at base. **Juvenile** Resembles adult female. **Voice** Males utter a diagnostic rattle. **Status** Scarce breeding bird but fairly widespread on migration.

 Easiest to see in spring in SE England.

Shoveler

Anas clypeata Anatidae
Length 44–52cm

male

Unmistakable because of bill shape. Usually unobtrusive. In flight, male shows a blue forewing panel and white-bordered green speculum; in the female, the blue is replaced by grey. Sexes are dissimilar overall. **Adult male** Has a shiny green head, white breast, and chestnut on flanks and belly. Stern is black and white and back is mainly dark. Has a yellow eye and dark bill. In eclipse, resembles adult female although body is more rufous and head is greyer. **Adult female** Has mottled buffish-brown plumage and yellowish bill. **Juvenile** Similar to adult female. **Voice** Males utter a sharp *tuk-tuk* and females a soft quack. **Status** Scarce breeding species on freshwater wetlands. More common at other times of year.

 Easiest to see in winter.

TOP: male; ABOVE: female

TOP: male; ABOVE: female

TOP: male; ABOVE: female

BELOW & RIGHT: male; ABOVE: female

Pintail

Anas acuta Anatidae
Length 51–66cm

Recognised by its elongated appearance; the male is unmistakable. In flight, male's grey wings and green speculum are striking; in the female, white trailing edge on innerwing is obvious. Sexes are dissimilar in other regards. **Adult male** Has a chocolate-brown head and nape, with white breast extending as stripe up side of head. Plumage is otherwise grey and finely marked, but note the cream and black stern, and the long, pointed tail, often held at an angle. In eclipse, resembles adult female but retains wing pattern. **Adult female** Has mottled buffish-brown plumage. **Juvenile** Similar to adult female. **Voice** Males utter a whistle, females a grating call. **Status** Rare breeding species (on freshwater marshes) but fairly common in winter, often on estuaries.

 Easiest to see in winter.

Diving ducks

Members of the genus *Aythya* are all ducks that dive routinely in order to feed, and are found on freshwater lakes and sheltered coasts, depending on the species. Due to migration from elsewhere in Europe, they are most widespread and numerous in winter, the majority forming sizeable flocks at this time of year.

Greater Scaup

Aythya marila Anatidae
Length 42–51cm

Bulky diving duck. Recalls Tufted Duck but has rounded head and lacks tufted crown. Gregarious outside the breeding season. In flight, has striking white wingbar. Sexes are dissimilar in other respects. **Adult male** Has a green-glossed head and dark breast (these look black in poor light). Belly and flanks are white, back is grey and stern is black. Has a yellow eye and dark-tipped grey bill. In eclipse, dark elements of the plumage are buffish brown. **Adult female** Has mainly brown plumage, palest and greyest on flanks and back. Note the white patch at base of bill. **Juvenile** Similar to adult female but white on face is less striking. **Voice** Mostly silent. **Status** A few pairs breed here but best known as a local winter visitor, mostly to sheltered coasts.

Easiest to see in winter.

TOP: male; ABOVE: female

Tufted Duck

Aythya fuligula Anatidae
Length 40–47cm

male

Familiar diving duck. The tufted crown is useful in identification. Gregarious in winter. In flight, note the white wingbar. Sexes are dissimilar in other regards. **Adult male** Has mainly black and white plumage; a purplish sheen on the head is seen in good light. Has a yellow eye and black-tipped blue-grey bill. In eclipse, white elements of the plumage are buffish brown. **Adult female** Has mainly brown plumage, palest on flanks and belly. Has white at base of bill (less than in female Scaup), yellow eye and black-tipped blue-grey bill. **Juvenile** Similar to adult female but colours duller. **Voice** Males utter a soft peep. **Status** Common on lakes and gravel pits. Several thousand pairs breed; winter numbers are boosted by migrants from Europe.

Easy to find.

TOP: male; ABOVE: female

Pochard

Aythya ferina Anatidae
Length 42–49cm

male

Diving duck with long bill, curving forehead and peaked crown. Gregarious in winter, often with Tufted Ducks. Both sexes have a dark bill with pale grey band. In flight, all birds have grey wings with a dark trailing edge to outer flight feathers. Sexes are dissimilar in other regards. **Adult male** Has a reddish-orange head, black breast, finely marked grey flanks and back, and black stern. In eclipse, black elements of plumage are sooty brown. **Adult female** Has a brown head and breast, grey-brown back and flanks, and pale 'spectacle'. **Juvenile** Resembles adult female but browner. **Voice** Mostly silent. **Status** Scarce breeder but locally common in winter. Favours flooded gravel pits and lakes.

Easiest to find in winter.

MAIN PIC: male; INSET: female

Ferruginous Duck

Aythya nyroca Anatidae
Length 38–42cm

Diving duck with white wingbar on upperwing, white underwings and white belly. In all birds, cap is peaked and bill is mainly grey; a pale band separates the grey from a dark tip. Sexes are separable with care. **Adult male** Has rich reddish-brown plumage, darkest on back and almost black on rump and tail. Has a white stern and belly (the latter visible only in flight) and white eye. **Adult female** Similar but duller and eye is dark. **Juvenile** Similar to adult female but duller. **Voice** Mostly silent. **Status** Scarce non-breeding visitor. Status is confused by presence of escapees.

Feeds among vegetation, so easily overlooked.

male; INSET: female

Diving ducks

Diving ducks swim underwater to feed. To overcome their buoyancy and submerge they have to make a little leap (*right*) at the surface.

Introduced ducks

Non-native ducks are widely kept in captivity but most retain a wild streak and individuals regularly escape to freedom. With Ferruginous Duck and Red-crested Pochard, the presence of escapees casts doubt on the origins of potentially vagrant birds seen in the wild. The presence of Ruddy, Mandarin and Wood ducks in the region is entirely due to escapes from captivity.

Red-crested Pochard

Netta rufina Anatidae
Length 54–57cm

 Large, distinctive diving duck; the male is unmistakable. Associates with other diving ducks. In flight, both sexes show striking white wingbars. Sexes are dissimilar in other regards. **Adult male** Has a rounded, bright orange head, black neck, breast, belly and stern, and white flanks. Back is grey-buff and bill is bright red. In eclipse, resembles adult female but retains red bill. **Adult female** Has mainly grey-buff plumage, darkest on back and above eye, and pale cheeks. Bill is mainly dark with a pink tip. **Juvenile** Resembles adult female but bill is uniformly dark. **Voice** Mostly silent. **Status** Occurs in mainland Europe; some records may be genuine vagrants but most sightings are certainly escapees and feral populations are now established. Favours lakes and flooded gravel pits.

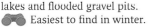 Easiest to find in winter.

TOP: male; ABOVE: female

Ruddy Duck

Oxyura jamaicensis Anatidae
Length 35–43cm

 North American species with established feral populations. Tail is regularly cocked up. Male is unmistakable but female could be confused with Little Grebe in winter plumage. **Adult male** Has orange-chestnut body plumage, white cheeks and stern, black cap and nape, and bright blue bill; colours are brightest in spring. **Adult female** Has grey-brown plumage with paler cheeks and a dark line from base of bill; bill is dull blue-grey. **Juvenile** Similar to adult female or eclipse male. **Voice** Mostly silent. **Status** Favours lakes and flooded gravel pits with vegetated margins. Escaped from Slimbridge several decades ago; now locally common in lowland Britain.

 Easiest to find in winter.

TOP: male; ABOVE: female

Wood Duck

Aix sponsa Anatidae
Length 43–50cm

 Distinctive duck; the male is particularly striking. Sexes are dissimilar. **Adult male** Has a shiny greenish-blue crown and mane, with white lines. Chin and throat are white, extending as white lines on face. Breast is maroon, flanks are buff and back is greenish; white lines separate these areas. **Adult female** Has mainly brown plumage, darkest on back and head. Breast and flanks have buffish streak-like spots. Has a white 'spectacle', small amount of white on throat and dark bill. **Juvenile** Resembles adult female. **Voice** Mostly silent. **Status** Introduced from North America. Found on ornamental lakes but a tiny feral population also exists.

Most obvious in spring.

Mandarin Duck

Aix galericulata Anatidae
Length 41–49cm

The male is stunningly colourful, with elaborate tufts and plumes. Sexes are strikingly dissimilar. **Adult male** Has a mane of orange, white, greenish and brown feathers, with white above eye and elongated orange plumes arising from cheek. Note the orange sail-like feathers on back, dark breast with vertical white stripes, brown flanks and white stern. Bill is bright red with a pale tip. **Adult female** Grey-brown overall, darkest on back, with pale buffish spots on flanks. Has a white belly, white 'spectacle', and white at base of bill and on throat. Bill is dull pink with a pale tip. **Juvenile** Resembles adult female. **Voice** Mostly silent. **Status** Introduced from China and now local on wooded lakes in the S.

Most visible in spring.

TOP: female
ABOVE: male

male

female

Diving ducks & sawbills

While many ducks have a vegetarian diet, several diving species are carnivorous: coastal species such as Eiders feed primarily on molluscs, while so-called 'sawbills' (members of the genus *Mergus*) specialise in catching fish. With a few exceptions, most of the species described here are gregarious to a greater or lesser degree outside the breeding season.

Long-tailed Duck

Clangula hyemalis Anatidae
Length 40–47cm

Elegant diving duck, at home in the roughest of seas. Dives for bottom-dwelling invertebrates. In flight, has dark wings and mainly white underparts. Sexes are dissimilar in other respects. **Adult male** In winter and spring, looks mainly black, grey and white, with buffish patch around eye and pink band on bill. In summer and in eclipse (both seldom seen here), has mainly brown and black plumage, with white on belly and flanks and pale buff eye-patch; bill is dark. **Adult female** In winter, is mainly brown and white; face is white except for dark cheek patch and crown. In summer, similar but face is mainly brown with pale eye-patch. **Juvenile** Similar to adult female in summer. **Voice** Males utter a nasal *ow-owlee*. **Status** Mainly a winter visitor, commonest in the N. Favours shallow coastal seas.

 Easiest to find in winter in NE Scotland.

winter male displaying; INSET: winter female

summer male

Eider

Somateria mollissima
Anatidae Length 50–70cm

male

Bulky, flock-forming sea-duck. Wedge-shaped bill forms a continuous line with slope of forehead. In summer, female flocks supervise a 'crèche' of youngsters. Sexes are dissimilar. **Adult male** Has mainly black underparts and white upperparts, except for black cap, lime-green nape and pinkish flush on breast. In eclipse, plumage is brown and black, with some white feathers on back and pale stripe above eye. **Adult female** Brown with darker barring. **Juvenile** Similar to adult female but with pale stripe above eye. **Voice** Males utter an endearing, cooing *ah-whooo*. **Status** Almost exclusively coastal. Nests close to seashore and feeds in inshore waters, diving for prey such as mussels. Easiest to find in N Britain.

male; INSET: female

Eider down

In order to protect their eggs from the elements, female Eiders line their nests with downy feathers plucked from their own breast. The feathers have fantastic thermal insulation properties, something that has not gone unnoticed by man. In parts of the world such as Iceland, Eiders are encouraged to nest in artificially high densities on man-made islands and a proportion of the down is then removed to make duvets and padded jackets. If done sensitively, the female Eider replaces the missing down and breeding success is not affected.

Common Scoter

Melanitta nigra Anatidae
Length 44–54cm

The male is our only all-black duck. Rather long tail is sometimes raised when swimming. Gregarious outside the breeding season. In flight, looks mainly dark but paler flight feathers can sometimes be seen. Sexes are dissimilar. **Adult male** Has uniformly black plumage. Head sheen is visible only at close range. Bill is mostly dark but with a yellow ridge; its base is bulbous. **First-winter male** Has browner plumage and all-dark bill. **Adult female** Has mainly dark brown plumage with pale buff cheeks. **Juvenile** Resembles adult female. **Voice** Mostly silent. **Status** Rare breeding bird, found on vegetated N lakes and lochs. Locally fairly common in winter, on coasts with sandy seabeds.

 Easiest to find in winter in N Britain.

male; INSET: female

Velvet Scoter

Melanitta fusca Anatidae
Length 51–58cm

Bulky diving duck. Larger than Common Scoter, with which it associates. All birds have white inner flight feathers, most obvious in flight. Head markings are useful in identification. Sexes are dissimilar. **Adult male** Has mainly black plumage, with striking white patch below pale eye; white on the closed wings is sometimes visible when swimming. Bill is two-toned: yellow and blackish. **First-winter male** Lacks white under eye. **Adult female** Mainly dark sooty brown, with pale patches on cheek and at base of bill; the bill is dark. **Juvenile** Similar to adult female. **Voice** Mostly silent. **Status** Scarce non-breeding visitor from Scandinavia, present mainly Oct–Mar. Favours coasts with sandy seabeds. Commonest in the N. Search among flocks of Common Scoter.

LEFT: female
BELOW: male

Goldeneye

Bucephala clangula
Anatidae
Length 42–50cm

male

Compact diving duck. Both sexes are easily recognised. In flight, all birds show white on innerwings (extent is greatest in males). Sexes are dissimilar in other respects. **Adult male** Has mainly black and white plumage. Rounded, peaked, green-glossed head has a yellow eye and striking white patch at base of bill. In eclipse, resembles adult female but retains more striking white wing pattern. **Adult female** Has a mainly grey-brown body, pale neck, dark brown head and yellow eye. **Juvenile** Similar to adult female but with a dark eye. **Voice** Displaying males utter squeaky calls and rattles. **Status** Scarce breeding species, mainly in the N. Locally common in winter, mostly on estuaries but also on inland lakes and flooded gravel pits.

 Easiest to find in winter.

TOP: male; ABOVE: female

Smew

Mergus albellus Anatidae
Length 38–44cm

male

Elegant little diving duck. Male is stunning and unmistakable. Female might be confused with a grebe in winter plumage. Sexes are dissimilar. **Adult male** Looks pure white at a distance but a close view reveals black patch through eye and black lines on breast and back. In eclipse (not seen here), resembles adult female but retains more extensive white on wing. **Adult female, juvenile and 1st-winter birds** Have an orange-red cap and nape (hence their name 'Redhead' Smews), white on cheek and throat, and grey-brown body. **Voice** Silent. **Status** Occasional and unpredictable winter visitor. Turns up on flooded gravel pits, reservoirs and lakes in variable numbers.

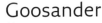 Usually appears at the onset of severe weather at traditional wintering grounds in the Netherlands and elsewhere.

TOP: male; ABOVE: female

Goosander

Mergus merganser
Anatidae
Length 58–66cm

male and female

Large, elegant diving duck. The narrow mandibles have serrated edges. In flight, upper surface of male's innerwing is white; in the female, white is restricted to trailing edge. Sexes are dissimilar in other regards. **Adult male** Has a bright red bill, green-glossed head (looks dark in poor light), mainly pink-flushed white body and black back. In eclipse, resembles adult female but retains white wing pattern. **Adult female** Has a reddish bill, orange-red head with a shaggy crest, and greyish body that is palest on breast; chin is white. **Juvenile** Resembles a dull adult female. **Voice** Displaying males utter ringing calls. **Status** Fairly common freshwater species beside wooded upland rivers; nests in tree holes. An influx from mainland Europe boosts winter numbers; occurs on reservoirs, lochs and flooded gravel pits.

 Easiest to find in winter.

TOP: male; ABOVE: female

Red-breasted Merganser

Mergus serrator Anatidae
Length 52–58cm

Slim duck with a shaggy, spiky, crest. Dives frequently in search of fish. In flight, all birds show white on upper surface of inner-wing; extent is greatest in males. Sexes are dissimilar in other regards. **Adult male** Has a narrow red bill, green head, white neck and orange-red breast. Flanks are grey and back is black. In eclipse, similar to adult female but retains extensive white on wing. **Adult female** Has a red bill, dirty-orange head and nape, and paler throat; body is otherwise greyish buff. **Juvenile** Resembles adult female. **Voice** Mostly silent. **Status** Nests beside fish-rich lakes and rivers, mainly in the N; locally common. European migrants boosts numbers in winter, then common on estuaries and sheltered coasts.

Easiest to find in winter.

ABOVE: male
FAR LEFT: female
LEFT: male

Osprey, hawks, eagles & kites

Birds of prey have sharply hooked bills, powerful feet and sharp talons. They comprise the families: Pandionidae (Osprey); Accipitridae (including hawks, kites, eagles and buzzards); and Falconidae (falcons).

Osprey

Pandion haliaetus Pandionidae
Wingspan 145–160cm

Classic fish-eating raptor. Can look gull-like in flight but its fishing technique is unmistakable: it hovers, then plunges talons first into water. Sexes are similar. **Adult** Has mainly brown upperparts, except for pale crown; underparts are mainly whitish with darker chest band. In flight from below, looks pale overall with dark carpal patches, a dark band along base of flight feathers and a dark terminal band on barred tail. **Juvenile** Similar but darker markings are less distinct. **Voice** Various whistling calls. **Status** Migrant visitor, almost always seen near water. Nests near lakes and lochs, mainly in the N. Breeding range is gradually extending S; also introduced in places (notably Rutland Water). Passage migrants sometimes linger at fish-rich lakes.

Easy to see at RSPB's Loch Garten reserve.

Goshawk

Accipiter gentilis Accipitridae
Wingspan 100–115cm

Impressive Buzzard-sized raptor. In flight, note the broad, rounded wings and relatively long but thickset, barred tail. Soaring birds fan their tails and splay their fluffy white undertail. A close view reveals an orange eye, yellow legs and feet, and a striking pale supercilium. Sexes are similar but male is smaller than female. **Adult** Has mainly grey-brown upperparts; pale underparts are marked with fine dark barring. **Juvenile** Has brown upperparts; buffish underparts are marked with dark, teardrop-shaped spots. **Voice** Utters a harsh *kie-kie-kie* in the breeding season. **Status** Scarce but also easily overlooked. Favours wooded habitats with adjacent open country.

Easiest to see late Feb–early Mar, when territorial birds are displaying. At that time, typically observed soaring over suitable woodland. Seen by chance at other times.

Sparrowhawk

Accipiter nisus Accipitridae
Wingspan 60–75cm

Widespread but secretive raptor that catches small birds in flight in surprise low-level attacks. Has relatively short, rounded wings, a long barred tail, long legs and staring yellow eyes. Male is much smaller than female and also separable on plumage details. **Adult male** Has blue-grey upperparts; pale underparts are strongly barred and reddish brown on body and wing coverts. **Adult female** Has grey-brown upperparts and pale underparts with fine, dark barring. **Juvenile** Has brownish upperparts, and pale underparts with broad, brown barring. **Voice** Utters a shrill *kew-kew-kew* in alarm. **Status** Common, associated mainly with wooded habitats, both rural and suburban.

Easiest to see in spring; males perform aerial displays.

BELOW: male
BOTTOM: female

Golden Eagle

Aquila chrysaetos Accipitridae
Wingspan 190–225cm

 Majestic raptor. Seen in flight from a distance, could be confused with a soaring Buzzard, but note the proportionately longer wings (narrowing appreciably towards base) and relatively long tail. Catches Mountain Hares and Red Grouse but also feeds on carrion in winter. Sexes are similar. **Adult** Has mainly dark brown plumage with paler margins to feathers on back and golden-brown feathers on head and neck. Tail is dark-tipped and barred but can look uniformly dark in flight silhouette. **Juvenile** Similar but has white patches at base of outer flight feathers; tail is mainly white but with a broad, dark tip. **Sub-adult** Gradually loses white elements of juvenile plumage by successive moults over several years. **Voice** Mainly silent. **Status** Resident of remote upland regions, mainly in Scotland. Favours open moorland and mountains.

 Easiest to find in winter.

ABOVE: **adult**; BELOW: **juvenile**

Success stories

Over the centuries, birds of prey have had their ups and downs. They have been – and to a degree still are – persecuted by gamekeepers and those with shooting interests, and in the 1960s were poisoned by agricultural pesticides, notably DDT. So it is good to learn that there are some success stories, too. Under its own steam, the Buzzard has spread eastwards in recent decades and is now our commonest raptor. Two other species – the White-tailed Eagle and Red Kite – have had some help from conservationists and are now also doing rather well.

White-tailed Eagle

Haliaeetus albicilla Accipitridae
Wingspan 190–240cm

 Immense raptor with long, broad, parallel-sided wings and a relatively short, wedge-shaped tail. Surprisingly manoeuvrable, despite its size, catching fish and waterbirds while hunting low over water. Sexes are similar. **Adult** Has mainly brown plumage, palest on head and neck. At rest, white tail is often obscured by wings. Bill and legs are yellow. In flight from below, looks mainly dark except for paler head and neck and white tail. **Juvenile** Similar but looks darker overall and tail is uniformly dark. **Sub-adult** Acquires adult plumage over successive moults; the last immature feature to disappear is a dark terminal band on tail. **Voice** Mournful whistling calls. **Status** Formerly just a rare visitor from mainland Europe but now reintroduced successfully to certain Scottish islands. Mainly coastal but sometimes seen soaring over open moorland.

 Mull and Islay are hotspots.

Red Kite

Milvus milvus Accipitridae
Wingspan 145–165cm

Graceful raptor, identified in flight by its deeply forked tail (twisted to aid flight control) and long, bowed wings. Seldom spends much time on the ground but sometimes perches in trees. Sexes are similar. **Adult** Has a pale grey head but otherwise mainly reddish-brown plumage. Eye, base of bill and legs are yellow. Seen in flight from below, note the reddish-brown body and underwing coverts, silvery-grey tail and patch on primaries, and otherwise dark wings. From above, tail appears red while reddish-brown back and wing coverts contrast with dark flight feathers. **Juvenile** Resembles dull adult with pale margins to wing-covert feathers. **Voice** Utters shrill calls in flight, like somebody whistling for their dog. **Status** Confined to central Wales until the late 1980s; reintroduction programmes mean it is now also locally common in England and Scotland.

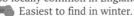 Easiest to find in winter.

Harriers & buzzards

Birds of prey (raptors) are daytime predators. All are expert flyers but they have evolved different flight styles to help them capture their particular prey (see 'Birds of prey in flight', pp.252–3). At one extreme, harriers have slow, buoyant flight, and search the ground for prey. At the other, falcons are mostly fast-flying and catch targeted prey in rapid flight.

Montagu's Harrier

Circus pygargus Accipitridae
Wingspan 100–115cm

Graceful, narrow-winged raptor with a slow, buoyant flight. Sexes are dissimilar. **Adult male** Has mainly blue-grey plumage with a small white rump, black wingtips, single dark bar on upperwing and two dark bars on underwings; note the chestnut barring on underwing coverts and streaking on belly. **Adult female** Has pale brown plumage with darker barring on wings and tail, streaking on body underparts and narrow white rump. **Juvenile** Recalls adult female, but underparts and underwing coverts are orange-red and unstreaked. **Voice** Mainly silent. **Status** Rare summer breeding visitor (perhaps 5–10 pairs), mainly in S England and favouring arable farmland or heathland. Passage migrants also seen in spring and autumn.

Nests are invariably guarded and only occasionally publicised. To observe the species, either visit a coastal birdwatching site in spring or autumn, or search suitable habitats during the breeding season.

ABOVE: female; BELOW: male

Hen Harrier

Circus cyaneus Accipitridae
Wingspan 100–120cm

Distinctive raptor that hunts slowly and effortlessly, low over the ground. Sexes are dissimilar but confusion with corresponding plumages of Montagu's Harrier is possible. Time of year and habitat aid identification: of the two, a harrier seen in open country in winter will be a Hen. **Adult male** Has pale blue-grey plumage except for white belly, white rump and black wingtips. **Adult female** Brown with darker barring on wings and tail, streaking on body underparts and narrow white rump. **Juvenile** Similar to adult female, but breast and underwing coverts are more reddish and upperwing coverts are brighter. **Voice** Mainly silent. **Status** Breeds on upland and N moorland (500–1,000 pairs). During winter, birds spread to more low-lying areas throughout the region.

Easiest to see in winter; try the N Norfolk coast and New Forest in Hampshire. In May–Jul, sightings are common on the Isle of Man and Orkney.

TOP: female; ABOVE: male

Marsh Harrier

Circus aeruginosus Accipitridae
Wingspan 110–125cm

Graceful wetland raptor, usually seen in flight quartering the ground slowly. Sexes are dissimilar. **Adult male** Reddish brown except for blue-grey head, grey, unbarred tail and black wingtips. **Adult female** Mainly dark brown, except for pale leading edge to wings, pale cap and chin, and reddish-brown tail. **Juvenile** Similar to adult female but tail is dark brown. **Voice** Mainly silent. **Status** Scarce breeder (100 or so pairs nest), favouring extensive reedbeds and marshes. On migration, also seen in other open country habitats.

Visit an East Anglian wetland and reedbed reserve in spring, such as RSPB Titchwell or Minsmere. Pairs perform aerobatic displays at the start of the breeding season, and aerial food passes when feeding their young.

TOP: male; ABOVE: female

Buzzard

Buteo buteo Accipitridae
Wingspan 115–130cm

Our commonest medium-sized raptor. Soars effortlessly, with broad wings often held in a shallow 'V' and tail fanned out. Plumage is variable but most birds are brown overall. Sexes are broadly similar, although females are larger than males. **Adult** Has rather uniformly brown upperparts. Seen perched, breast is finely barred and typically paler than throat or belly. In flight and from below, flight feathers and tail are grey and barred; wings have a dark trailing edge and tail has a dark terminal band. Body and underwing coverts are contrastingly dark, and pale bands on breast and underwing coverts can sometimes be seen. **Juvenile** Similar but lacks dark terminal tail band and dark trailing edge to wings. **Voice** A distinctive mewing *pee-ay*. **Status** Has staged a comeback in recent years, from near-extinction in many areas: there may now be 20,000 pairs and rising. Typically, adult Buzzards are territorial residents, favouring open country with scattered woodland for nesting.

Usually most vocal and easiest to observe at the start of the breeding season.

Rough-legged Buzzard

Buteo lagopus Accipitridae
Wingspan 125–140cm

Medium-sized, rather long-winged raptor. Similar to Buzzard but plumage details and behaviour allow separation. Pale tail on upperside is diagnostic. Hovers more frequently than Buzzard, often low over the ground. Sexes are separable with care. **Adult male** Has brown upperparts, except for white tail, which has a dark terminal band and smaller second bar. From below, underparts appear rather pale except for dark head, carpal patches, wingtips and trailing edge to wing. **Adult female** Similar but shows a dark belly and single dark terminal band on tail. **Juvenile** Similar to adult female but dark markings are indistinct. **Voice** Mainly silent. **Status** Scarce winter visitor from mainland Europe (20 or fewer in most years), mainly to E England and Scotland. Harsh weather and/or low prey density in Europe can lead to larger influxes. Favours open habitats, including coastal marshes and grassland.

Visit coastal areas in East Anglia or E Kent in winter.

Honey-buzzard

Pernis apivorus Accipitridae
Wingspan 135–150cm

Specialised raptor that feeds on larvae and adults of bees and wasps by raiding their nests. Similar to Buzzard, but with proportionately longer tail and long wings that are held slightly downcurved when soaring. Closer views reveal a relatively small, almost Cuckoo-like head. Sexes are similar. **Adult** Typically has brownish upperparts, pale underparts, grey head and evenly barred grey tail. Eye is yellow and bill is rather long and narrow. In flight and from below, note the evenly barred tail, dark carpal patch and conspicuous barring on underwing. **Juvenile** Similar but usually browner overall and with less distinct barring on underwing coverts. **Voice** Mainly silent. **Status** Scarce summer visitor from Africa. Associated with undisturbed woodland where its insect prey is numerous. Breeding numbers vary but are small. Also seen on migration.

To see passage migrants, visit the E coast of England in early Sep when E or SE winds are blowing. During the breeding season, visit the New Forest in Hampshire.

Buzzard

Did you know?

Honey-buzzards are the only British raptors that regularly eat bees and wasps, while Buzzards and Red Kites will eat most prey. In time of hardship, particularly in winter, the latter two will scavenge already-dead animals, such as road-kill Rabbits, which is why you often see them close to motorways.

Falcons

Falcons are small to medium-sized raptors with rather pointed wings and agile, rapid flight. Most species catch their prey on the wing and some are capable of extraordinary bursts of speed.

Kestrel

Falco tinnunculus Falconidae
Wingspan 65–80cm

Familiar small falcon that habitually hovers where look-out perches are not available. Feeds on small mammals, ground-dwelling birds and insects. Sexes are dissimilar. **Adult male** Has a spotted orange-brown back, blue-grey head and blue-grey tail with black terminal band. Underparts are creamy buff with bold black spots. In flight from above, dark outerwing contrasts with orange-brown innerwing and back. **Adult female** Has barred brown upperparts and pale creamy-buff underparts with dark spots. In flight from above, the contrast between brown innerwing and dark outerwing is less distinct than with male and tail is barred. **Juvenile** Resembles adult female but upperparts are more reddish brown. **Voice** A shrill *kee-kee-kee...* **Status** Common and widespread in open, grassy places. Breeding success is dependent upon prey populations, notably those of Field Voles and Wood Mice. Easy to see hovering over roadside verges.

TOP RIGHT & ABOVE: **male**; RIGHT: **female**

Merlin

Falco columbarius Falconidae
Wingspan 60–65cm

Our smallest raptor. Typically seen in low, dashing flight in pursuit of prey such as Meadow Pipits. Also perches on fenceposts or rocky outcrops. Sexes are dissimilar. **Adult male** Has blue-grey upperparts and buffish, streaked and spotted underparts. In flight from above, note the contrast between blue-grey back, innerwings and tail, and dark wingtips and dark terminal band on tail. **Adult female** Has brown upperparts and pale underparts with large brown spots. In flight from above, upperparts look rather uniformly brown with numerous bars on wings and tail. **Juvenile** Resembles adult female. **Voice** Mostly silent, but a shrill *kee-kee-kee...* is uttered in alarm near the nest. **Status** During the breeding season in spring and summer, found on upland moorlands. In winter, moves S and to lowland areas, with numbers boosted by migrants from Iceland. Easiest to find near coasts in winter.

LEFT: **male**; ABOVE: **female**

Hobby

Falco subbuteo Falconidae
Wingspan 70–85cm

Elegant falcon. Aerial mastery allows it to catch agile prey, including Swifts, hirundines and even dragonflies. In silhouette, it has proportionately longer and narrower wings than a Peregrine and a longer tail. Generally unobtrusive. Sexes are similar. **Adult** Has blue-grey upperparts and pale, dark-streaked underparts. Has a dark 'moustache', white cheeks and reddish-orange 'trousers'. **Juvenile** Similar but lacks reddish 'trousers' and underparts look buffish overall. **Voice** Utters a shrill *kiu-kiu-kiu...* in alarm. **Status** Scarce summer visitor; breeds mainly in S and SE England. Favours heathland and farmland with scattered woods. On migration, a Hobby could turn up almost anywhere. Between 500 and 1,000 pairs are probably present in the region in the summer months. The New Forest in Hampshire is a hotspot.

Peregrine

Falco peregrinus Falconidae
Wingspan 95–115cm

One of our most impressive raptors. Soars on broad, bowed wings but stoops with wings swept back at speeds in excess of 320km/h on prey, such as pigeons. Sexes are similar but male is smaller than female. **Adult** Has dark blue-grey upperparts and pale, barred underparts. Note the dark mask on face and powerful yellow legs and feet. In flight from above, looks uniformly dark grey, although rump may appear paler; from below, pale underparts are barred and the contrast between pale cheeks and throat, and dark moustache, is striking. **Juvenile** Similar but upperparts are brownish while paler underparts are suffused with buffish orange. **Voice** Utters a loud *kek-kek-kek...* in alarm. **Status** Widespread resident in N and W Britain and Ireland. The population is recovering following a crash caused by pesticide poisoning in the 1960s. Favours mountains and coastal cliffs but increasingly nests in towns and cities. Easiest to see on rugged coasts in W Britain; family parties are vocal in summer.

adult attacking Lapwing

Hobby-watching

Hobbies typically arrive back in England in late April, sometimes before migrant Swallows and martins – important prey species – have returned in good numbers. Consequently, many Hobbies turn to insects as a food source, with emerging mayflies and dragonflies being extremely important in this respect. Where the feeding is good at this time of year – such as on the flooded gravel pits in the Thames Valley/M4 corridor – a dozen or more Hobbies may gather to feed.

Birds of prey in flight

Most observations of birds of prey are made of birds in flight. Use these pages to familiarise yourself with their relative shapes, colours and distinctive diagnostic features. Judging the size of a bird in flight is notoriously difficult, as it is often hard to assess how far away the bird is. The images are roughly all in proportion to one another.

dark trailing edge

dark carpal patch

slender neck and head

dark terminal band

barred underwing

all adults

Honey-buzzard

pale form

reddish-brown unbarred plumage

adult

dark trailing edge

dark terminal band

dark chest band

Buzzard

adult male

grey upperwing

barred underwing

reddish-brown plumage

adult female

barred underwing

owl-like face

Montagu's Harrier

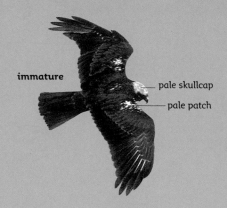

immature

pale skullcap

pale patch

adult female

pale underwings with dark wingtip

adult male

Marsh Harrier

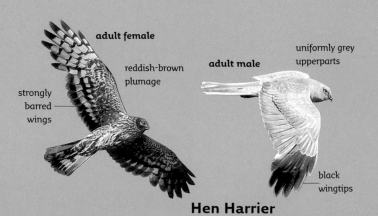

adult female

reddish-brown plumage

strongly barred wings

adult male

uniformly grey upperparts

black wingtips

Hen Harrier

broad-based wings

rounded wingtips

strongly barred underparts

adults

Sparrowhawk

immature

pale patches on wings

pale base to tail

adult

proportionately long tail and long wings

Golden Eagle

adult

white tail

immature

dark tail

White-tailed Eagle

adult

dark carpal patch

dark belly

Rough-legged Buzzard

pale body and wing coverts

adult

Osprey

forked tail

white patch on wings

adults

reddish plumage

Red Kite

barred underwing

adult male

Merlin

pointed wingtips

adults

Peregrine

dark hood and 'moustache'

adult

proportionately long tail and narrow wings

dark hood

Hobby

proportionately long wings compared to Sparrowhawk

barred tail, often fanned when soaring

adult

Goshawk

adult male

blue-grey head

orange-brown underparts

adult female

brown upperparts

barred tail

Kestrel

Moorlands & mountains

Apart from parts of the coast, moorlands and mountains are the only habitats in Britain and Ireland that still retain a feeling of 'wilderness'. Inhospitable terrain and often inclement weather mean they can be challenging places to visit, but those who make the effort will be rewarded with the discovery of many remarkable plants and animals, many found nowhere else.

Harsh winter weather and limited food availability mean moors and mountains are too inhospitable for most British mammals. But the **Mountain Hare** (*see also* p.205) typically shuns lowland areas and has a coat for all seasons: blue-grey and comparatively short in summer, but thick and mainly white in winter.

Heathers thrive on acid soils and do well on upland moors. **Heather** (*see also* p.104) is the most widespread species and it turns entire moorland slopes purple when flowering in late summer. Young shoots are eaten by Red Grouse and so moors are sometimes managed to favour this gamebird.

Even in summer, uplands can often be cold, windy and wet – hardly the ideal place for delicate butterflies. But the tiny **Mountain Ringlet** (*see also* p.344), along with a number of other specialist invertebrates, is found nowhere else in Britain and Ireland. The upland invertebrate season is brief – mainly May to August.

In centuries gone by, all but the highest British peaks were wooded, but clearance and heavy grazing by sheep ensured that the natural woodland was removed and could not regenerate. In general terms, moorland is the dominant habitat in upland areas, although the characteristic plants and appearance varies considerably from region to region. In a few parts of Wales, northern England and Scotland, mountains dominate the landscape. Sometimes rising to altitudes above the level at which trees would grow if they were allowed to do so, these uplands harbour unique communities of plants and animals.

Upland vegetation is influenced by factors such as underlying soil type, rainfall and altitude. Areas of neutral soils, where grasses and rushes thrive, are often fairly uniform in appearance. Here, the landscape is often maintained by the nibbling attentions of sheep. In places where the underlying soil is acid, heather moors, reminiscent in appearance of lowland heaths, often develop; they are dominated by Heather and Bell Heather. Where the underlying soil is lime-rich, a far more diverse flora develops and the invertebrate communities tend to be more varied and interesting.

Uplands are blessed with generous rainfall and relatively unpolluted air. These factors, combined with a comparative lack of competing vegetation, mean that lichens and non-flowering plants often thrive on moors and mountains. Species that encrust rocks, such as the **Map Lichen** (*see also* p.183), are common.

The **Ptarmigan** (*see also* p.256) is one of only a handful of British birds that make the mountains their home. Its seasonally varying plumage matches its surroundings and its tameness lends it an endearing quality. A Ptarmigan's hardy qualities allow it to endure winter cold and cope with snow covering its feeding grounds.

Moorlands & mountains hotspots
1. **Cairngorms National Park (NP)**, Aberdeenshire; 2. **Ben Lawers National Nature Reserve**, Perth and Kinross; 3. **Lake District NP**, Cumbria; 4. **North York Moors NP**, North Yorkshire; 5. **Snowdonia NP**, north Wales; 6. **Dartmoor NP**, Devon

Gamebirds

Robust and hardy birds with stout bodies and relatively short, rounded wings, gamebirds have a largely vegetarian diet. Some species perform elaborate displays at communal leks, at the start of the breeding season. As their name suggests, gamebirds have been hunted for centuries. Sadly, some of our native species are still hunted.

Red Grouse

Lagopus lagopus scoticus Tetraonidae
Length 37–42cm

Familiar moorland gamebird. If alarmed, takes to the air explosively; flight comprises bouts of rapid wingbeats and long glides on bowed wings. Both sexes have uniformly dark wings but are separable with care. **Adult male** Has chestnut-brown plumage overall; fine feather markings are visible at close range. Note the red wattle above eye. **Adult female** Paler, more buffish-grey and marbled plumage than male: well camouflaged when sitting on nest. **Juvenile** Resembles adult female but with less well-marked plumage. **Voice** A distinctive *go-back, go-back, go-back*. **Status** Confined to heather moorland, feeding primarily on shoots of Heather and related plants. Moors are sometimes managed by selective burning to benefit Red Grouse (this encourages young plant growth). Easy to see on heather moors in N Britain; males are most conspicuous in spring.

TOP: male; ABOVE: female

Ptarmigan

Lagopus mutus Tetraonidae
Length 34–36cm

Hardy mountain gamebird. Indifferent to observers but easily overlooked: unobtrusive and blends in well with surroundings. In flight, both sexes reveal white wings and black tail. Forms small flocks outside the breeding season. Sexes are separable with care. **Adult male** In winter, white except for dark eye, lores and bill. In spring and summer, has mottled and marbled greyish-buff upperparts, the amount of white on back decreasing with time; belly and legs are white, while striking red wattle fades by midsummer. **Adult female** In winter, is white except for black eye and bill. In spring and summer, has finely barred buffish-grey upperparts; the extent of white on back diminishes with time. **Juvenile** Resembles a uniformly brown female. **Voice** A rattling *kur-kurrrr* call. **Status** Confined to the Scottish Highlands, favouring rocky ground with plentiful mountain vegetation. Commonest above 1,000m but at lower altitudes further N. Visit the Cairngorms.

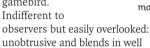

male

BELOW: summer male
INSET: winter male

female on nest

Capercaillie

Tetrao urogallus Tetraonidae Length 60–90cm

Huge gamebird. Male is half as big again as female with different plumage. **Adult male** Blackish with greenish sheen on breast. Has brownish wings, red wattle above eye and white spot at base of wing. **Adult female** Has barred grey-brown plumage with orange-brown breast. **Juvenile** Resembles a small, dull female. **Voice** Displaying males utter clicks and loud pops. **Status** Favours mature Scots Pine forest. Current population results from reintroductions. Easiest to see at at RSPB's Loch Garten reserve.

INSET: female
BELOW: male

Did you know?

Male Black Grouse and Capercaillies (*below*) gather at communal display grounds, known as 'leks', in the spring. Here, they puff themselves up, call and strut their stuff in order to vie for the attentions of onlooking females. Leks are traditional sites that are used year after year.

Black Grouse

Tetrao tetrix Tetraonidae
Length 40–45cm

Moorland gamebird. Sexes are dissimilar. **Adult male** Has mainly dark plumage with red wattle above eye. Displaying bird elevates and spreads lyre-shaped tail, revealing white undertail coverts. **Adult female** Has barred orange-brown plumage. **Juvenile** Resembles small, dull female. **Voice** Displaying males utter bubbling calls. **Status** Scarce. Prefers mixed grass and heather moors. Easiest to see from a car.

ABOVE: male
INSET: female

The pheasant-shooting industry

Pheasants are big business. Although around 3 million of these alien birds form a stable breeding population, the species' exact British status is hard to assess accurately at any given time because of the Pheasant-rearing industry. An estimated 15 million captive-bred birds are released each autumn, of which some 7 million are shot in the winter months and a further quota die of natural causes. Many naturalists find it hard to reconcile the Pheasant's undeniable beauty with the impact upon native wildlife of human activities undertaken in its name.

Pheasant

Phasianus colchicus Phasianidae
Length 55–90cm

Male is colourful and unmistakable. Birds take to the wing noisily and explosively when flushed. Sexes are dissimilar. **Adult male** Typically has orange-brown body plumage, a blue-green sheen on head, a large and striking red wattle, and a long, orange, barred tail; some birds have a white collar. Violet-blue birds are occasionally released. **Adult female** Mottled buffish brown with a shorter tail than male. **Juvenile** Resembles a small, short-tailed female. **Voice** Territorial males utter a loud, shrieking call, followed by a bout of vigorous wing-beating. In alarm, a loud *ke-tuk, ke-tuk, ke-tuk* is uttered as the bird flies away. **Status** Introduced and established here since the 11th century. Widespread and commonest in wooded farmland. Easy to find.

TOP: male; ABOVE: male, purple form

female

Golden Pheasant

Chrysolophus pictus Phasianidae
Length 60–100cm

Secretive. Male is gaudy. Sexes are dissimilar. **Adult male** Is mainly red with yellow crown and barred 'cape', yellow rump, and blue on wings and back; tail is buffish with intricate markings. **Adult female** Buffish brown and barred. **Juvenile** Recalls a short-tailed female. **Voice** Male utters a shrill call. **Status** Native to China. Naturalised locally in dense woodland. Birds at RBG Kew are tame.

TOP: male; ABOVE: female

Lady Amherst's Pheasant

Chrysolophus amherstiae Phasianidae
Length 60–120cm

Male is unmistakable; female is similar to a female Golden Pheasant but separable. Sexes are dissimilar. **Adult male** Mainly whitish below and black above; some dark feathers have a blue sheen. Black-edged white feathers form a cape; note red on rear of crown, and red and yellow on rump. Tail is long and grey with intricate black markings. **Adult female** Reddish brown with greyish cheeks and a paler, unbarred belly. **Juvenile** Resembles a short-tailed female. **Voice** Male utters shrieks, typically at night. **Status** Native to China. Naturalised in Bedfordshire and nearby counties. Favours dense conifer plantations. Shy and hard to observe.

Smaller gamebirds, crakes, rails & Crane

Partridge and Quail are the smallest gamebirds, and like the unrelated Corncrake are associated with arable farmland and grassland. The Rail and Crake family (Rallidae) is associated with wetlands. Cranes (family Gruidae) also favour wetlands, ideally with wide-open vistas.

Quail

Coturnix coturnix Phasianidae
Length 16–18cm

 Tiny, secretive gamebird. Sexes are separable with care. **Adult male** Has mainly brown, streaked plumage, palest and unmarked on belly. Head has dark stripes; otherwise pale throat has a black centre and is defined by dark lines. **Adult female** Similar but has pale throat. **Juvenile** Similar to adult female. **Voice** Song is a diagnostic, trisyllabic phrase, often rendered as 'wet-my-lips'. **Status** Scarce migrant visitor whose numbers vary from year to year (100–1,000 calling birds). Arable farmland (especially Barley) is favoured.

 Easily detected by its call. Patience is required to see one.

male; INSET: female

Grey Partridge

Perdix perdix Phasianidae
Length 29–31cm

Well-marked gamebird. Usually seen in small parties. Hunted and consequently wary; prefers to run from danger. Sexes are separable with care. **Adult male** Has mainly grey, finely marked plumage with an orange-buff face, large chestnut mark on belly, maroon stripes on flanks and streaked back. **Adult female** Similar but mark on belly is small. **Juvenile** Grey-buff with a hint of adult's dark markings. **Voice** A choked, harsh *kierr-ikk* call. **Status** Native of grassland and arable farmland with mature hedgerows. Once abundant, now scarce owing to modern farming methods.

 Easiest to see in winter.

male; INSET: female

Red-legged Partridge

Alectoris rufa Phasianidae
Length 32–34cm

 Dumpy, well-marked gamebird. Forms small parties (called 'coveys') outside the breeding season. Hunted and often wary. Prefers to run from danger but flies low on stiffly held wings. Sexes are similar. **Adult** Has a red bill and legs, and white throat bordered with gorget of black spots. Plumage is otherwise mainly blue-grey and warm buff except for black and white barring on flanks. **Juvenile** Has grey-buff plumage with a hint of adult's dark markings. **Voice** A loud *ke che-che, ke che-che...* call. **Status** Introduced but well established, mainly on arable farmland with mature hedgerows and scattered woods.

 Most vocal in spring; coveys are easiest to see in winter.

Water Rail

Rallus aquaticus Rallidae
Length 23–28cm

 Secretive wetland bird. Distinctive call is heard far more frequently than bird is seen. In profile, note dumpy body, short tail and long bill. Seen head on, body is laterally compressed. Sexes are similar. **Adult** Has mainly blue-grey underparts, reddish-brown upperparts, and black and white barring on flanks. Bill and legs are red. **Juvenile** Similar but duller. **Voice** A pig-like squeal and various choking calls. **Status** Favours reedbeds and marshes; migrants are sometimes found on streams and Watercress beds.

 Least tricky to see in winter; sometimes ventures out of cover briefly.

Corncrake

Crex crex Rallidae
Length 27–30cm

A secretive bird that is easier to hear than see. Sexes are similar. **Adult** Has sandy-brown upperparts; dark feather-centres create a 'scaly' look. Face, throat, breast and belly are blue-grey while flanks are barred chestnut and white. In flight, note chestnut wing patch and dangling legs. **Juvenile** Greyer than adult. **Voice** Territorial males utter a ceaseless *crek-crek, crek-crek...*, mainly at night. **Status** Migrant visitor that has declined due to modern farming practices. Now restricted to traditionally managed hay meadows.

Easiest to hear on Outer Hebrides and W Ireland.

Moorhen

Gallinula chloropus Rallidae
Length 32–35cm

Familiar wetland bird. Swims with jerky movements and constantly flicks its tail. Often tame on urban lakes. Sexes are similar. **Adult** Can look blackish but has a dark blue-grey head, neck and underparts, and a brownish back, wings and tail. Has a yellow-tipped red bill and frontal shield, and yellow legs and long toes. Note the white feathers on sides of undertail and white line along flanks. **Juvenile** Greyish brown with white on throat, sides of undertail coverts and along flanks. **Voice** A loud *kurrrk*. **Status** Common resident on all sorts of wetland habitats, from village ponds to flooded gravel pits and lakes.

Easy to see.

adult; INSET: juvenile

Coot

Fulica atra Rallidae
Length 36–38cm

Robust waterbird, often found with Moorhens. Has lobed toes. Feeds by up-ending, making shallow dives or grazing waterside vegetation. Gregarious outside the breeding season. Sexes are similar. **Adult** Has blackish plumage, darkest on head and neck. Note the white bill and frontal shield on head, and beady red eye. Legs are pale yellowish. In flight, shows a white trailing edge on otherwise dark, rounded wings. **Juvenile** Has dark greyish-brown upperparts and white on throat and front of neck. **Voice** A loud *kwoot* call. **Status** Common resident, found on freshwater wetland habitats; numbers are boosted in winter by migrants.

Easy to see.

adult; INSET: juvenile

Spotted Crake

Porzana porzana Rallidae
Length 19–22cm

Secretive wetland bird. Sexes are similar. **Adult** Has mainly brown upperparts and blue-grey underparts, all adorned with white spots; note the dark-centred feathers on back and striking barring on flanks. Bill is yellow with red base; face is marked with black and undertail coverts are pale buff. Legs and feet are greenish. **Juvenile** Lacks adult's dark face and throat, and blue-grey elements of the plumage are buffish grey instead. **Voice** Male's territorial call is a repetitive whiplash-like whistle, uttered after dark. **Status** Migrant visitor. Favours impenetrable wetlands.

Easiest to find in autumn.

Crane

Grus grus Gruidae
Length 95–115cm

Stately, long-legged, long-necked bird with a bushy tail-end. In flight, wings are broad and long; flies with neck and legs outstretched. Typically wary. Sexes are similar. **Adult** Has mainly blue-grey plumage with black and white on head and neck; back sometimes appears rather brown. Note the patch of red on forecrown. **Juvenile** Similar but head is pale buffish grey and it lacks adult's black and white markings. **Voice** A loud, trumpeted, rolling *krrruu*. **Status** Rare breeding resident in Norfolk; also a scarce passage migrant and occasional winter visitor.

Visit the Norfolk Broads in winter and you might be lucky.

Introducing waders

Waders are the favourite bird group of many ornithologists: most species are attractive to look at, with elegant proportions and often beautifully patterned and colourful plumage. A wide range of bill sizes and shapes among the various species is reflected in the functional elegance of their feeding habits. And to add to their allure, many waders are associated with wild, untamed habitats.

Bills, food and feeding

Most, but not all, waders are associated with wetland habitats and the majority of British species are found in coastal habitats outside the breeding season. To avoid competition for food, different species have evolved bills and feeding habits to fill discrete and separate niches.

Like all plovers, the Ringed Plover has a rather short, stubby bill, which is used to pick up small items of food – typically marine invertebrates – that are stranded or live on the seashore or estuary.

Turnstones live up to their name, turning over stones and strandline seaweeds in search of sandhoppers and other invertebrates.

The long, downcurved bill of a Curlew is typically used to probe soft mud for worms. But this individual is showing its versatility and has caught a crab.

Wader families

Waders, or birds in the classification order Charadriiformes, comprise several family groups: **Stone-curlew** (Burhinidae); **Avocet** (Recurvirostridae); **Oystercatcher** (Haematopodidae); plovers and **Lapwing** (Charadriidae); and sandpipers, stints, 'shanks', curlews, godwits and allies (Scolopacidae).

Phalaropes

Red-necked Phalarope female

Phalaropes are extraordinary waders. They are consummate swimmers that feed on surface plankton, and apart from a brief time spent on land while nesting, they live out their lives at sea – they even feed on water during the breeding season. The roles of the sexes is reversed, so that although the female lays the eggs, it is the male that incubates them and has the duller plumage. As an added bonus, typically phalaropes are incredibly indifferent to human observers, and if you don't mind getting your feet wet they can sometimes be viewed at a range of just a metre or two (*see* descriptions on pp.270-1).

Avocets feed by sweeping their incredibly fine, upturned bill from side to side in shallow water and mud, picking up tiny invertebrates.

Wader bills are very sensitive near the tip and flexible enough to allow buried worms to be captured; this Black-tailed Godwit is flexing the tips of its mandibles.

Nesting and breeding behaviour

Waders often have strikingly different summer and winter plumages: that seen in the breeding season is typically more gaudy and colourful, especially in males. Their nests are usually rudimentary, at best a small cup tucked away inside a tussock of grass, or perhaps just a bare scrape in gravel or sand. At the start of the breeding season, many wader species perform vocal displays to define their territories and attract and retain mates; some species also perform aerial displays.

ABOVE: Accompanied by distinctive calls, a territorial Lapwing performs skydiving displays in early spring.
LEFT: If you visit a relatively undisturbed stretch of coast in spring, you will soon know if you have entered an Oystercatcher's territory by the loud alarm calls. This particular bird is directing its attentions at a potential rival.

ABOVE: The Golden Plover's evocative song captures the essence of the lonely upland moors it inhabits during the breeding season.
RIGHT: Most of the waders we see on the coast in winter migrate north in order to breed, many nesting well inside the Arctic. Just a few Purple Sandpipers find remote Scottish mountain tops a suitable substitute for tundra.

Flight and migration

Waders are strong flyers and most have proportionately long wings. While their powers of flight are put to good use on a daily basis in the winter months, they are essential for spring and autumn migration. Some of our wintering species breed in Greenland, Siberia or Arctic Canada, and it is thought that many individuals make, at most, just a single stop on their journey.

Safety in numbers

Typically, most waders are territorial and nest independently of others of the same species. However, outside the breeding season many form single-species flocks that feed and roost together. Working on the principle that there is safety in numbers, individual birds benefit from having an army of watchful eyes on the look out for danger.

LEFT: Most of the Black-tailed Godwits we see in Britain and Ireland in winter breed in Iceland and make the journey to and from their nesting grounds in one go.

BELOW: The most spectacular wader flock formations seen are of Knots, which gather at hotspots like the Wash in vast flocks. At high tide, as the birds are pushed off the mudflats, they provide birdwatchers with an unrivalled spectacle.

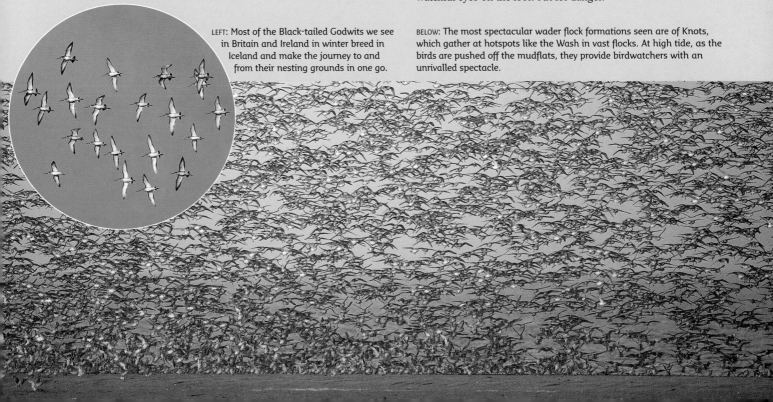

Plovers & other waders

Although many wader species habitually feed in water, affiliations to aquatic habitats vary throughout the group. Plovers are among the most terrestrial, while the superficially plover-like Stone-curlew thrives in relatively dry terrain, gaining the moisture it needs from its food.

Avocet

Recurvirostra avosetta Recurvirostridae
Length 43cm

Elegant wader with distinctive black and white plumage. Feeds by sweeping its upcurved bill (unique amongst British waders) from side to side through water. Gregarious outside the breeding season. Sexes are similar. **Adult** Has mainly white plumage with black on crown, nape and wings. Legs are blue and bill is black. **Juvenile** Similar but black elements of plumage are dark brown. **Voice** A ringing *klueet-klueet...* call. **Status** Favours shallow, brackish coastal lagoons in the breeding season. In winter, is found on estuaries, mainly in SW England.

In the breeding season, visit East Anglian RSPB reserves (especially Minsmere and Titchwell). In winter, easy to see on the Exe and Tamar estuaries in Devon.

Stone-curlew

Burhinus oedicnemus Burhinidae
Length 38–45cm

Secretive, dry-country wader best known for its eerie calls. Well camouflaged and hard to spot. In flight, looks long-winged and gull-like, with a striking black and white pattern on upperwing. Sexes are similar. **Adult** Streaked sandy-brown plumage; black and white wingbars can be seen in standing birds. Has yellow legs, a black-tipped yellow bill and large yellow eyes. **Juvenile** Similar but markings are less distinctive. **Voice** Utters strange Curlew-like wails at dusk and throughout the night. **Status** Nests on downland, Breckland heaths and arable fields.

Easiest to see at Weeting Heath reserve in Breckland.

Dotterel

Charadrius morinellus Charadriidae
Length 22cm

Tame, mountain wader. Sexes are separable. **Summer adult** Reddish-orange breast and belly, black-bordered white collar and blue-grey throat. Face is whitish; note white supercilium and dark cap; upperparts are mostly grey-brown. Legs are yellow. Male is duller than female. **Winter adult** Grey-buff with a broad, pale buff supercilium and pale breast band. **Juvenile** Similar to winter adult but with scaly-looking back. **Voice** A soft *pierrr* call. **Status** Rare migrant visitor, mainly to the Scottish Highlands.

Sometimes seen near the summit of Cairngorm, but please avoid disturbance.

Oystercatcher

Haematopus ostralegus Haematopodidae
Length 43cm

Distinctive wader with striking black and white plumage and a loud alarm call. Powerful bill is used to hammer molluscs off rocks. Sexes are similar. **Adult** In summer, has black upperparts and white underparts with clear demarcation between the two on breast. Note the red bill, pinkish legs and beady red eye. In winter, similar but with white half-collar. **Juvenile** Similar to summer adult but black elements of plumage are brownish and the bill and leg colours are subdued. **Voice** A loud, piping *peep* call. **Status** Breeds commonly on the coast and beside inland lakes and rivers in the N. Mainly coastal in winter, favouring estuaries and mudflats.

Widespread in winter. Human disturbance excludes it from most suitable nest sites in the S.

adult Dotterel; INSET: *juvenile*

Lapwing

Vanellus vanellus Charadriidae
Length 30cm

Pied-looking wader with spiky crest. Has rounded black and white wings and a distinctive call. Sexes are separable in summer. **Summer adult male** Has green- and purple-sheened dark upperparts; underparts are white except for orange vent and black foreneck. Note the black and white markings on face. **Summer adult female** Similar but has less distinct black neck markings and shorter crest. **Winter adult** Similar to summer female but throat and foreneck are white, and back feathers have buffish fringes. **Juvenile** Similar to winter adult but crest is short and back looks scaly. **Voice** A choked *pee-wit* call. **Status** Fairly common nesting species of undisturbed grazed grassland, moors and arable farmland; numbers have declined seriously. Migrants from Europe boost numbers in winter. Easiest to find in winter; often coastal.

summer adult

The sound of the wilderness

For many people, the mournful 'song' of the Golden Plover is the most enchanting sound associated with our rolling upland moors, evoking a sense of untamed wilderness. Territorial birds utter an almost pleading *pu-peeoo* call to advertise ownership of nesting grounds, and in particularly good locations a hill walker will encounter adjacent territories within earshot of one another, every few hundred metres.

Golden Plover

Pluvialis apricaria Charadriidae
Length 28cm

Beautifully marked wader whose call is evocative of desolate uplands in summer. Gregarious outside the breeding season; often associates with Lapwings. In flight, note the white underwings. Sexes are sometimes separable in summer. **Adult** In summer, has spangled golden upperparts bordered by a white band. In most males, belly is black, grading to grey on neck and face. Most females have less distinct dark underparts and face is often whitish. Breeders from N Europe (seen on migration) have darker underparts than British birds. In winter, underparts are pale, and head, neck and back are streaked golden. **Juvenile** Similar to winter adult. **Voice** Utters a *peeoo* flight call and a plaintive *pu-peeoo* in summer. **Status** Locally common breeding species on N upland moors and mountains. Widespread in winter on grassland and arable fields. Fairly easy to see in winter but most rewarding in summer.

LEFT: **summer male**
BELOW: **summer female**

LEFT: **winter adult**
BELOW: **winter**

Grey Plover

Pluvialis squatarola Charadriidae
Length 28cm

Plump-bodied coastal wader. Best known in winter plumage, although breeding plumage is sometimes seen in newly arrived migrants or those soon to depart. In flight, note the black 'armpits' on otherwise white underwings. Typically solitary. Sexes are similar. **Adult** In winter, looks grey overall but upperparts are spangled with black and white and underparts are whitish. Legs and bill are dark. In summer plumage, has striking black underparts (sometimes rather mottled in females), separated from spangled grey upperparts by a broad white band. **Juvenile** Resembles winter adult but plumage has a buff wash. **Voice** A diagnostic trisyllabic *pee-oo-ee* call, like a human wolf-whistle. **Status** Nests in the high Arctic; a coastal non-breeding visitor to Britain and Ireland. Easy to find on most estuaries and mudflats.

ABOVE: **winter adult**; INSET: **winter adults**

summer adult

Plovers & sandpipers

Smaller plovers, sandpipers and stints are small but interesting waders, and as they usually feed in the open they can be easy to watch. Some species are confusingly similar but each has a diagnostic combination of plumage, behaviour and habitat preferences to aid identification.

Kentish Plover

Charadrius alexandrinus Charadriidae
Length 15–17cm

Dumpy little coastal plover. Looks much paler overall than other similarly sized species. Note the white wingbar in flight. Sexes are dissimilar. **Summer adult male** Has sandy-brown upperparts and white underparts. Sandy crown has black at front and rufous at back. Has black through eye and a dark patch on side of breast. Legs and bill are black. **Summer adult female, juvenile and winter adult** Similar but black plumage elements are pale sandy brown (the same colour as upperparts); legs are dull brown. **Voice** A soft *bruip* call. **Status** Scarce, short-staying passage migrant, found on sandy estuaries.

👓 Visit a S coast estuary in spring to find this species; fairly regular at the Fleet, behind Chesil Beach in Dorset.

male

female

Little Ringed Plover

Charadrius dubius Charadriidae
Length 15–17cm

Slim-bodied little plover that lacks a white wingbar. Sexes are similar. **Summer adult** Has sandy-brown upperparts and white underparts with a black collar and breast band, and black and white markings on head. Has a black bill, yellow legs and yellow eye-ring. Female has duller black head plumage elements than male. **Juvenile** (Also winter adult – not seen in the region.) Black plumage elements are replaced by sandy brown. Breast band is usually incomplete, leg and eye-ring colours are dull, and head lacks pale supercilium seen in juvenile Ringed Plovers. **Voice** A *pee-oo* call. **Status** Locally fairly common, nesting around margins of flooded gravel pits and other man-made sites. Migrants turn up at freshwater sites outside the breeding range and sometimes on the coast.

👓 Relatively easy to find near gravel pits along the Thames Valley/M4 corridor.

adult

juvenile

Ringed Plover

Charadrius hiaticula Charadriidae
Length 17–19cm

Small, dumpy coastal wader. Runs at speed before pausing to pick food items from the ground. Note the white wingbar in flight. Sexes are separable. **Summer adult male** Has sandy-brown upperparts and white underparts with a black breast band and collar. Has black and white markings on face, and white throat and nape. Legs are orange-yellow and bill is orange with a dark tip. **Summer adult female** Similar but black elements of plumage are duller. **Winter adult** Similar to summer adult but most black elements of plumage are sandy brown, and has a pale supercilium. Leg and bill colours are dull. **Juvenile** Similar to winter adult but breast band is often incomplete. **Voice** A soft *tuu-eep* call. **Status** Locally common. Nests mainly on sandy or shingle beaches, sometimes inland. Coastal outside the breeding season; numbers are boosted by migrants from Europe.

👓 Easy to find on mudflats in winter.

adult on nest

adult male summer

juvenile

Dunlin

Calidris alpina Scolopacidae
Length 17–21cm

winter

Several different races, with differing bill lengths, occur here outside the breeding season. Forms large flocks in winter. **Summer adult** Has a reddish-brown back and cap, and whitish underparts with a bold black belly and streaking on neck. Male is usually more boldly marked than female. **Winter adult** Has uniform grey upperparts and white underparts. **Juvenile** Has reddish-brown and black feathers on back; pale feather margins align to form 'V' patterns. Underparts are whitish but with black streak-like spots on flanks and breast; head and neck are brown and streaked. **Voice** A *preeit* call; breeding 'song' comprises a series of whistling calls. **Status** Local breeding species on damp moorland and mountain habitats. Locally abundant outside the breeding season owing to an influx of migrants from the Arctic.

Easy to find on most estuaries.

ABOVE: adult breeding plumage; BELOW: winter adult

Dunlin – the wader 'yardstick'

Being common but having such variable plumage and structure, the Dunlin is the yardstick by which to judge other small waders. Get to know it in its different plumages and you will have overcome a big hurdle in wader identification. Start with juvenile Dunlins in autumn – watch their feeding habits and behaviour as much as their appearance. Spots on the flanks are a guarantee that you are looking at a juvenile Dunlin (*below*).

Curlew Sandpiper

Calidris ferruginea Scolopacidae
Length 19–21cm

Small, elegant wader. Similar to Dunlin but with a longer, obviously downcurved bill and white rump. Sexes are hard to separate. **Adult** In summer, has a spangled reddish-brown, black and white back, and (briefly) brick-red on face, neck and underparts (appears mottled in moulting migrants). Males are brighter than females. In winter (seldom seen here), has greyish upperparts and white underparts. **Juvenile** (Commonest plumage encountered here.) Has a scaly-looking back, white belly and buffish breast; note the pale supercilium. **Voice** A soft *prrrp* call. **Status** Breeds in the high Arctic and seen here as a scarce passage migrant, usually on estuaries and coastal pools, and often with Dunlins.

Easiest to find in autumn, when juveniles predominate.

ABOVE: adult breeding plumage
LEFT: juvenile

Little Stint

Calidris minuta Scolopacidae
Length 13–14cm

Recalls a tiny short-billed Dunlin with frantic feeding activity. Sexes are similar. **Adult** In summer, has white underparts, a reddish-brown back, and variably rufous on head and neck. Has yellow 'V' on mantle, and pale supercilium that forks above the eye. In winter, mainly grey above and white below. **Juvenile** Has white underparts. Dark back and wing feathers have pale fringes that align to form white 'V' markings. Note buffish flush on side of breast, pale supercilium that forks above eye, pale forecrown and dark centre to crown. **Voice** A shrill *stip* call. **Status** Regular passage migrant, mostly in autumn.

Favours coastal pools in autumn.

TOP: adult breeding plumage
ABOVE: winter adult
LEFT: juvenile

Temminck's Stint

Calidris temminckii Scolopacidae
Length 14–15cm

Tiny, slim-bodied wader with a deliberate, almost creeping, feeding action. Compared to Little Stint, has short yellow legs, a longer tail and wings, and a slightly downcurved bill. Note the clear demarcation between dark breast and white underparts and white outer-tail feathers. Sexes are similar. **Adult** In summer, has grey-brown upperparts and a streaked grey head, neck and breast; many of the back feathers have dark centres. Underparts are white. In winter (unlikely to be seen here), has uniform grey-brown upperparts and white underparts. **Juvenile** Has white underparts and brownish upperparts with a scaly-looking back. **Voice** A trilling call. **Status** Scarce passage migrant; favours margins of shallow freshwater pools. A few pairs breed in Scotland.

Regular on East Anglian coasts.

ABOVE: adult
LEFT: juvenile

Sandpipers & other waders

If you visit the coast during the winter months you are certain to find several species of medium-sized waders without too much trouble. A few have very specific habitat requirements, such as sandy beaches or rocky shores, but the greatest diversity is found on estuaries and mudflats. Although several wader species seemingly share the same habitat, they each exploit different feeding niches to avoid competing with one another.

Sanderling

Calidris alba Scolopacidae
Length 20cm

 winter

Small, robust wader. Seen in flocks running at speed along edges of breaking waves on sandy beaches. Has a white wingbar and black bill and legs. Sexes are similar. **Adult** In winter, has grey upperparts and white underparts. In summer (sometimes seen in late spring or early autumn), plumage is flushed with red on head and neck and has dark-centred feathers on back; underparts are white. **Juvenile** Similar to winter adult but many back feathers have dark centres. **Voice** A sharp *plit* call. **Status** Locally common non-breeding visitor, mainly to sandy beaches; occasional on shingle or mudflats.

Easy to find on almost any undisturbed sandy beach in winter.

Knot

winter adults

Calidris canutus
Scolopacidae
Length 25cm

Dumpy, robust wader. Forms large flocks in winter. Has a white wingbar but otherwise lacks distinctive features in non-breeding plumage. Sexes are similar. **Adult** In winter, has uniform grey upperparts and white underparts. Bill is dark and legs are dull yellowish green. In summer plumage (sometimes seen in late spring or early autumn), has an orange-red face, neck and underparts; back is marked with black, red and grey. Legs and bill are dark. **Juvenile** Resembles winter adult but has a scaly-looking back and peachy flush to breast. **Voice** A sharp *kwet* call. **Status** Non-breeding visitor to Britain and Ireland. Locally common in winter on estuaries and mudflats.

Visit the RSPB's Snettisham reserve, adjacent to the Wash; given suitable tide conditions, flock formations here are spectacular.

Purple Sandpiper

winter adult

Calidris maritima
Scolopacidae
Length 21cm

Plump-bodied wader. Unobtrusive but typically confiding. Legs are yellowish in all birds and bill has a yellowish base. A white wingbar is seen in flight. Sexes are similar. **Adult** In winter, is uniform blue-grey on head, breast and upperparts, darkest on back; belly is white and flanks are streaked. In summer plumage (sometimes seen in late spring), has reddish-brown and black feathers on back and dark ear coverts on otherwise streaked grey-brown face. **Juvenile** Recalls winter adult but feathers on back have pale margins, creating a scaly look; neck, breast and flanks are streaked. **Voice** Utters a sharp *kwit* call in flight. **Status** A few breed in Scotland. Best known as a local non-breeding visitor to rocky shores and headlands.

Look for feeding birds just where waves are breaking on rocky shores.

TOP: winter adult; ABOVE: adult breeding plumage

ABOVE: juvenile; LEFT: adult breeding plumage

TOP: adult breeding plumage; ABOVE: winter adult

Ruff

Philomachus pugnax Scolopacidae
Length 23–29cm

Small-headed wader with variable plumage and size. Has a slightly downcurved bill, orange-yellow legs and, in flight, a narrow white wingbar and white sides to rump. Male is smaller than female and, in the breeding season, has unique head decorations. **Summer adult male** Has brownish upperparts, many feathers with black tips and bars. On breeding grounds, briefly has facial warts and a variably coloured ruff and crest feathers. **Summer adult female** Has grey-brown upperparts, many feathers with dark tips and bars; underparts are pale. **Winter adult** Has rather uniform grey-brown upperparts and pale underparts. **Juvenile** Recalls winter adult but has a buff suffusion and scaly-looking back. **Voice** Mostly silent. **Status** Rare breeding species on freshwater wetlands. Passage migrant on coastal freshwater pools; scarce in winter.

👓 Easiest to find in autumn.

ABOVE LEFT: **juvenile**
LEFT: **male breeding plumage**; BELOW: **summer adult female**

Redshank

winter adult

Tringa totanus Scolopacidae
Length 28cm

Medium-sized wader with a shrill alarm call. In flight, note white trailing edge to wings, white back and rump, and trailing red legs. Sexes are similar. **Adult** In summer, is mainly grey-brown above and pale below, but back is marked with dark spots and neck, breast and flanks are streaked. Note the faint, pale supercilium and eye-ring; base of bill is reddish. In winter, has uniform grey-brown upperparts, head, neck and breast, with paler, mottled underparts. Bill and leg colours are dull. **Juvenile** Recalls winter adult but plumage is browner overall, back feathers have pale marginal spots, and legs and base of bill are dull yellow. **Voice** Utters a yelping *tiu-uu* alarm call. Its song is musical and yodelling. **Status** Locally common nesting species in damp grassland, moors and marshes. Migrants boost numbers outside the breeding season and are common on coasts in winter.

👓 Easy to find on estuaries or mudflats in winter.

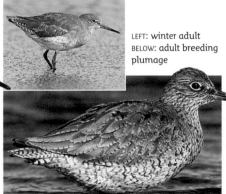

LEFT: **winter adult**
BELOW: **adult breeding plumage**

Spotted Redshank

Tringa erythropus Scolopacidae
Length 30cm

Similar to Redshank but with a longer bill and legs. In flight, note the uniform upperwings (no white trailing edge) and distinctive flight call. Often feeds in deep water and sometimes swims. Sexes are similar. **Adult** In breeding plumage (seen in late spring and summer), is mainly black with a white eye-ring and dotted white fringes to back feathers; incomplete breeding plumage is more typically observed. In winter, has pale grey upperparts and clean, whitish underparts. Legs are reddish and note the pale supercilium. **Juvenile** Recalls winter adult but plumage is darker overall and underparts are barred; legs are orange-yellow. **Voice** A diagnostic *tchewit* call. **Status** Regular but scarce passage migrant; winters in small numbers on estuaries, mainly in the S.

👓 Learning its distinctive call will greatly improve chances of detection.

TOP: **adult breeding plumage**; ABOVE LEFT: **adult incomplete breeding plumage**; ABOVE RIGHT: **adult winter**

Greenshank

Tringa nebularia Scolopacidae
Length 30–31cm

Elegant, long-legged wader. Beautifully patterned but can look very white at a distance. Has yellowish-green legs and a long, slightly upturned, grey-based bill. Sexes are similar. **Adult** In summer, has grey-brown upperparts with black centres to many back feathers. Head, neck and breast are streaked but underparts are white. In winter, upperparts are pale grey and underparts are white. **Juvenile** Recalls winter adult but upperparts are darker and browner. **Voice** A distinctive *tchu-tchu-tchu* call. **Status** Scarce breeding species of Scottish blanket bogs. Passage migrant, mainly to coasts, and local in winter.

👓 Easiest to find on estuaries in autumn.

BELOW: **juvenile**
INSET RIGHT: **adult breeding plumage**
INSET BELOW: **winter adult flight**

Sandpipers & larger waders

Novice wader-watchers sometimes find sandpipers hard to identify. Concentrate on plumage features and calls to separate them. This is also the key with the larger Curlew and Whimbrel, but the dumpy Turnstone's distinctive appearance and habits make identification straightforward.

adult Common Sandpipers

Wood Sandpiper adult breeding plumage; INSET: adult showing rump

Green Sandpiper; INSET: adult showing rump

Common Sandpiper

Actitis hypoleucos Scolopacidae
Length 18–20cm

Active little wader with a bobbing gait and elongated tail-end. Flies low over water on bowed, fluttering wings: note the white wingbar and absence of a white rump. Sexes are similar. **Adult** Has warm brown upperparts with faint dark centres and barring to feathers of back and wings. Head and neck are grey-brown with a clear demarcation between dark breast and white underparts, the white extending up sides of breast. **Juvenile** Similar but wing covert feathers are barred. **Voice** A whistling *tswee-wee-wee* call. **Status** Fairly common summer visitor, nesting beside upland and northern rivers and lakes. Widespread and fairly common passage migrant, found at inland sites on coasts and most in evidence in Apr and Aug–Sep. A handful of individuals overwinter.
Look for birds bobbing among waterside stones on upland and northern streams in spring. At other times, often seen beside coastal pools.

Wood Sandpiper

Tringa glareola Scolopacidae
Length 19–21cm

Elegant wader. Legs are yellowish and relatively long. Has a pale supercilium and, in flight, note white rump and barred tail. Sexes are similar. **Adult** Has brownish, spangled upperparts. Head and neck are streaked and has faint streaks and spots on otherwise pale underparts. **Juvenile** Similar but upperparts are browner and marked with pale buff spots. Voice A *chiff-chiff-chiff* flight call. **Status** Fairly common passage migrant, visiting coastal freshwater pools; rarer inland. A few pairs breed in Scottish Highlands.
Easiest to find on coastal pools in autumn.

Green Sandpiper

Tringa ochropus Scolopacidae
Length 21–23cm

White-rumped wader with a bobbing gait. Sexes are similar. **Adult** Has dark brown upperparts with small pale spots. Head and neck are streaked; has clear demarcation between dark, streaked breast and clean white underparts. Legs are greenish yellow. **Juvenile** Similar but pale spotting on upperparts is more noticeable. **Voice** A trisyllabic *chlueet-wit-wit* flight call. **Status** Passage migrant and winter visitor to fresh water. A few overwinter.
Visit a Watercress bed in winter.

Turnstone

Arenaria interpres Scolopacidae
Length 23cm

Pugnacious wader with a stout, triangular bill, used to turn stones in search of invertebrates. Feeds unobtrusively. All birds have reddish-orange legs and a black and white wing pattern in flight. Sexes are similar. **Adult** In summer, has orange-red on back, white underparts and bold black and white markings on head. Males have brighter back colours than females and more distinct black head markings. In winter, has grey-brown upperparts, including head and neck. Breast is marked with a blackish band showing a clear demarcation from white underparts. **Juvenile** Similar to winter adult but upperparts are paler and back feathers have pale fringes. **Voice** A rolling *tuk-ut-ut* in flight. **Status** Non-breeding visitor to coasts. Widespread and common. Easy to find on almost every beach or estuary.

ABOVE: adult breeding plumage; INSET: adult breeding plumage flight; BELOW: winter adult

Curlew; INSET: adults

Curlew

Numenius arquata Scolopacidae
Length 53–58cm

Large, distinctive wader with a long, downcurved bill. Its call is evocative of lonely, windswept uplands during spring and summer, and coasts in winter. Sexes are similar, although male has a shorter bill than female. **Adult** Has mainly grey-brown plumage, streaked and spotted on neck and underparts; belly is rather pale. **Juvenile** Similar but looks more buffish brown overall, with fine streaks on neck and breast, and an appreciably shorter bill. **Voice** Utters a characteristic *curlew* call and a bubbling song on its breeding grounds. **Status** Locally common breeding species on N and upland habitats. Almost exclusively coastal outside the breeding season. Common on most estuaries and mudflats. Listen for the diagnostic call.

Whimbrel

Numenius phaeopus Scolopacidae
Length 40–45cm

Smaller than Curlew, with a shorter bill, diagnostic head markings and distinctive call. Sexes are similar. **Adult** Has grey-brown to buffish-brown plumage with fine, dark streaking on neck and breast. Head pattern comprises two broad, dark lateral stripes on an otherwise pale crown, and a pale supercilium. **Juvenile** Similar but warmer buff overall. **Voice** Bubbling call comprises 7 notes that descend slightly in pitch from start to finish. Its song is confusingly similar to that of Curlew. **Status** Rare breeding species on boggy moorland. Coastal passage migrant; overwinters in small numbers. Learn the distinctive call by listening to recordings. Search coasts in Apr and Sep, or visit the Shetland Isles in May–Jun.

Whimbrel; INSET: adult

Godwits & smaller waders

Godwits are long-billed waders with a showy summer plumage. By contrast, snipe and woodcock, with similarly long bills, are camouflaged. Phalaropes are everyone's favourites; these atypical swimming waders are stunning in summer, striking in winter and easy to get close to.

Black-tailed Godwit

Limosa limosa Scolopacidae
Length 38–42cm

Long-legged wader with a long, straight bill. In flight, has a black tail, white rump and white wingbars on upperwing. Sexes are dissimilar in summer. **Breeding plumage adult male** Has a reddish-orange face, neck and breast. Greyish back is spangled with reddish brown and belly is whitish with barring on flanks. **Breeding plumage adult female** Similar but reddish elements of plumage are less intense. **Winter adult** Grey-brown, palest on belly; undertail is white. **Juvenile** Recalls winter adult but has an orange suffusion on neck and breast, and pale fringes and dark spotting on back feathers. **Voice** A *kwe-we-we* call in flight. **Status** Rare British breeder; the Ouse and Nene washes are strongholds and it favours wet grassland. Icelandic migrants boost numbers outside the breeding season; locally common on muddy estuaries.

Easiest to see in winter.

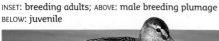

INSET: breeding adults; ABOVE: male breeding plumage
BELOW: juvenile

Bar-tailed Godwit

Limosa lapponica Scolopacidae
Length 35–40cm

Large wader with a long, slightly upturned bill. Looks shorter-legged than Black-tailed. In flight, note the absence of a wingbar on upperwing; white rump extends as a wedge to lower back and tail is barred. Sexes are dissimilar in summer. **Breeding plumage adult male** Has a reddish-orange head, neck and underparts. Back is spangled grey, black and pale buff. **Breeding plumage adult female** Has a buffish-orange wash on head, neck and breast, a pale belly and a greyish back. **Winter adult** Has a grey-brown head, neck and upperparts; underparts are pale. **Juvenile** Recalls winter adult but has a buffish wash to head, neck and upperparts. **Voice** A sharp *kve-wee* call in flight. **Status** Nests in the Arctic; non-breeding visitor to coastal Britain and Ireland.

Easiest to find in winter on estuaries and beaches.

INSET: breeding adults
ABOVE: adult non-breeding plumage
LEFT: adult breeding plumage

Red-necked Phalarope

Phalaropus lobatus Scolopacidae
Length 18cm

Confiding wader that habitually swims, picking food from the water's surface. Non-breeding life is spent at sea. **Summer adult female** Has brown upperparts, with buff feather margins on back. Note the white throat, dark cap and reddish-orange neck; grey breast and mottled flanks grade to white underparts. **Summer adult male** Similar but duller. **Winter adult** Has grey upperparts, nape and hindcrown, white underparts and a black eye-patch. **Juvenile** Recalls winter adult but grey elements of plumage are mainly brown or buff; grey feathers are acquired gradually. **Voice** A sharp *kip* call. **Status** Mainly an Arctic breeding species. Britain and Ireland are at the S limit of its range and hence rare here and found mainly in the N. Nests beside freshwater pools. Best known as scarce passage migrant, seen mainly in wake of severe gales.

Visit the Shetland Isles in spring or check coasts for storm-driven birds in autumn.

ABOVE: adult summer female
BELOW: breeding plumage females fighting

juvenile

Grey Phalarope

Phalaropus fulicarius Scolopacidae
Length 20–21cm

Confiding wader that habitually swims and spends its non-breeding life at sea. Bill is shorter and stouter than that of similar Red-necked and has a yellow base. Sexes are dissimilar in breeding plumage (seen here occasionally); female is much more colourful than male. **Winter adult** Has grey upperparts, white underparts, dark cap and nape, and black 'panda' mark around eye. **Summer adult female** Has an orange-red neck and underparts, dark crown, white face patch and buff-fringed dark back feathers. **Summer adult male** Similar but duller. **Juvenile** Recalls winter adult but plumage is tinged buff and back feathers are dark with buff fringes. **Voice** A sharp *pit* flight call. **Status** Nests in the high Arctic, winters in tropical seas and seen here on migration, mostly in autumn but sometimes in spring. Mainly coastal but occasionally occurs on reservoirs. Look for storm-driven birds during and after autumn gales.

ABOVE: adult winter
LEFT: adult winter
BELOW: female summer

Woodcock

Scolopax rusticola Scolopacidae
Length 35–38cm

Dumpy, long-billed wader with short legs and cryptic (camouflaged) plumage. Mainly nocturnal. Sexes and ages are similar. **Adult and juvenile** Have marbled chestnut, black and white plumage, palest and more extensively barred on underparts. Note the large eyes, located high on head, giving birds almost complete all-round vision. **Voice** Males utter soft duck-like calls and explosive squeaks at dusk. **Status** Associated with wooded habitats; both mixed and deciduous woods favoured. Needs a mosaic of open areas and dense canopy cover in breeding season. Migrants from as far as Russia boost winter numbers, when it is very locally common. Many are shot. Easiest to detect in spring: visit likely looking woodland at dusk and watch and listen for roding birds (*see* box right). Prolonged close views are tricky because the bird is so difficult to spot.

Jack Snipe

Lymnocryptes minimus Scolopacidae
Length 18–20cm

Dumpy wader. Much smaller than Snipe, with a shorter bill and legs, and more striking head and back markings. Pumps body up and down as it walks. Easy to overlook: plumage is cryptic (camouflaged) and bird is very reluctant to fly. Sexes and ages are similar. **Adult and juvenile** Have mainly brown upperparts with intricate, cryptic, dark feather markings. Note the striking yellow stripes on back; a greenish sheen is sometimes discerned. Head has dark and pale buff stripes, including a forked, pale supercilium. Neck and breast are streaked and underparts are white. **Voice** Mostly silent. **Status** Non-breeding visitor in small numbers. Favours muddy margins of pools and marshes, where tangled dead rush and grass stems match its cryptic plumage. Most reliably seen on the Isles of Scilly in autumn. Otherwise, observations are a matter of luck.

Snipe

Gallinago gallinago Scolopacidae
Length 25–28cm

Distinctive, even in silhouette: has a dumpy body, rather short legs and a very long, straight bill. Feeds by probing its bill in a sewing machine-like manner. Sexes and ages are similar. **Adult and juvenile** Have mainly buffish-brown upperparts, beautifully patterned with black and white lines and bars. Note the distinctive stripes on head, streaked and barred breast and flanks, and white underparts. **Voice** Utters one or two sneeze-like *kreech* calls when flushed. Performs a 'drumming' display in the breeding season, the sound caused by vibrating tail feathers. **Status** Locally common and invariably associated with boggy ground. In the breeding season, favours marshes, meadows and moorland bogs. In winter, numbers are boosted by migrants, found on a wide range of wetland habitats. Sadly, tens of thousands of birds are shot each year. Easiest to see in spring, when displaying.

adult; INSET: adult 'drumming'

Roding

Roding is the name given to the territorial displays performed by male Woodcocks in spring. Typically, they fly along a regular route, usually a woodland ride, uttering soft duck-like quacks and strange squeaking sounds. Birds will normally repeat their 'circuit' every 15 minutes or so: if you sit quietly, you will probably be treated to more than one performance.

Introducing skuas, gulls & terns

For many observers, gulls, skuas and terns are the archetypal seabirds, although of course auks, petrels and shearwaters also have claims to this title. Outside the breeding season, skuas and some terns live entirely marine lives, well away from British shores. The majority of gull species are coastal at best, and comparatively few species actually merit being called 'seagulls'.

The wing flashes of Great Skuas are used in display: the birds are extremely territorial on their breeding grounds.

Skuas

Skuas (family Stercorariidae) are powerfully built seabirds with mainly rather dark plumage in most species. All are summer visitors to Scotland, wintering at sea in southern oceans. Skuas seldom willingly come close to land except during the breeding season. They have predatory and scavenging habits, either killing smaller seabirds or harassing larger ones into regurgitating their last meal, a tactic referred to as food parasitism.

Gulls

These medium to large birds (family Laridae) have mainly white adult plumage and an affiliation, for at least part of the year, with water. They have a scavenging or predatory diet and robust bills, the size of which matches their feeding predilections. Gulls have webbed feet, which are used for swimming, and their wings are comparatively long and narrow, affording them powerful flight.

The Kittiwake is among the most marine of the gulls, shunning land outside the breeding season.

Terns

Terns (family Sternidae) are elegant and slim-bodied when compared to gulls. However, their delicate appearance is deceptive: they are robust birds whose long, pointed wings give them a buoyant and powerful flight. Terns are fish-eaters that plunge-dive in search of prey. Outside the breeding season, most spend their lives at sea in southern oceans. Some species migrate extraordinary distances each year.

BELOW: With its short legs – characteristic of the group as a whole – the Arctic Tern is not at home on land, but it is a master of the air.

Moult, plumage and ageing in gulls

Depending on the species, gulls take between two and four years from fledging to acquire full adult plumage. This is done in a series of twice-annual moults, and with each successive change of feathers the bird in question loses juvenile characters. But the changes do not stop there. Gulls look different in summer and winter. The differences may be subtle – streaking on the head in winter, for example – or they may be more striking – several species acquire a dark hood in the breeding season. As a result, experienced birdwatchers can tell at a glance the age of any given gull.

A Black-headed Gull acquires full adult plumage in the second winter after it hatches; thereafter, for the rest of its life it has a dark hood in summer months but loses this each time it moults into winter plumage.

1st winter · adult winter · adult summer

Identifying gulls

The identification of gulls may seem a daunting prospect to the novice birdwatcher, but perseverance for a year or so will certainly lead to a mastery of the basics. On the plus side, gulls are large birds and most are indifferent enough to people to allow close scrutiny of them with binoculars, and their plumage can be examined in minute detail if you use a telescope. Black-headed and Herring gulls are the commonest species, so if you get to know these two in all their varying plumages you will have a couple of yardsticks by which to judge more unusual birds.

If you search among flocks of Black-headed Gulls in winter, with luck you may come across a Mediterranean Gull. Features to look for, and to compare with the Black-headed, include wing colour (mainly white in the Mediterranean, but with a white leading edge and dark trailing edge in the Black-headed), bill size (more robust in the Mediterranean) and menacing-looking dark smudges on the face.

Gull feeding

Black-headed and Herring gulls are among the most numerous of our larger birds and part of their success is due to their diet. Like most gull species, they are opportunistic and omnivorous, and will take a free meal courtesy of man's activities: visit a rubbish dump or watch farmland being ploughed if you need any confirmation of this.

Although Great Black-backed Gulls are arch scavengers, they are capable predators too: this juvenile is in the act of swallowing a Bass. Small seabirds also feature in the diet of this species.

Black-headed Gulls are a familiar sight on farmland, and flocks appear – seemingly out of nowhere – when activities such as ploughing or root-crop harvesting begin: the birds are looking for soil invertebrates.

Gull nesting

Most gull species nest in colonies. This is partly because suitable breeding habitat tends to be local, obliging the birds to nest in close proximity to one another. There is also safety in numbers: an army of pecking beaks is more likely to deter a would-be predator than a lone bird. Most large gull species nest on sea cliffs, while smaller gulls such as the Black-headed favour coastal marshes and the margins of freshwater wetlands.

Herring Gull colonies are found on isolated sea cliffs and islands, typically where no ground-living predatory mammals occur. In a successful colony, nests are almost within pecking distance of one another.

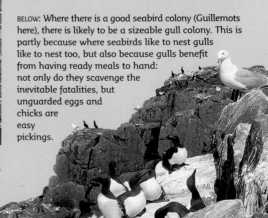

BELOW: Where there is a good seabird colony (Guillemots here), there is likely to be a sizeable gull colony. This is partly because where seabirds like to nest gulls like to nest too, but also because gulls benefit from having ready meals to hand: not only do they scavenge the inevitable fatalities, but unguarded eggs and chicks are easy pickings.

Smaller gulls

With their mainly white plumage, smaller gulls are conspicuous and familiar, both on the coast and inland, especially in winter. The Black-headed Gull is by far the commonest, but groups of gulls are always worth checking for more unusual species.

Little Gull

Larus minutus Laridae
Length 25–28cm

Our smallest gull. Flight is buoyant. Sexes are similar. **Adult** In summer is pale grey above with white wingtips, a dark hood, dark bill and short, reddish legs. In flight, upperwings have a white trailing edge; underwings are dark with a white trailing edge. In winter, similar but loses dark hood; the otherwise white head has dark smudges on crown and ear coverts. **Juvenile** Has a striking black bar (forming a 'W') on upperwings and back, and dark markings on mantle, nape and ear coverts, and dark tail band; plumage is otherwise white. **First-winter bird** Similar to juvenile but back is pale grey, hence dark bar is seen only on wings. Adult plumage is acquired over next 2 years. **Voice** A sharp *kyeck* call. **Status** Scarce passage migrant and winter visitor; mainly coastal.

Search sheltered coves during gales.

TOP: adult breeding plumage; INSET: 1st winter
ABOVE: adult winter

Mediterranean Gull

Larus melanocephalus Laridae Length 36–38cm

adult summer

Similar to Black-headed but has a stouter bill; adult has uniformly pale wings. Sexes are similar. **Adult** In summer, has a pale grey back and wing coverts, and white flight feathers. Note the black hood and white 'eyelids'; bill is mainly red, with a yellow tip and black sub-terminal band. Legs are deep red. In winter, loses dark hood; whitish head has a menacing look created by dark smudges. **Juvenile** Has grey-brown upperparts with pale margins to back feathers. Note the darkish flush on breast. Bill and legs are dark; tail has a dark terminal band. **First-winter bird** Similar to juvenile but with a plain grey back and dark smudges on head. Adult plumage is acquired by 3rd winter. **Second-year bird** Resembles adult (at respective times of year) but with variable black in wing-tips. **Voice** *Cow-cow-cow* call. **Status** Very locally common, usually with Black-headeds. A few nest in S England. More widespread outside the breeding season.

Easiest to find in winter. The Solent, Folkestone and Portland Harbour are hotspots. Often comes if food is offered.

1st winter

ABOVE: adult winter
RIGHT: adult breeding plumage

Black-headed Gull

Larus ridibundus Laridae
Length 35–38cm

Our most numerous medium-sized gull. Plumage is variable but a white leading edge to the outerwings is a consistent feature. Forms single-species flocks. Sexes are similar. **Adult** In summer, has a grey back and upperwings, white underparts and a chocolate-brown hood. Legs and bill are red. In flight, trailing edge of outerwing is black. In winter, loses dark hood; white head has dark smudges above and behind eye. **Juvenile** Has an orange-brown flush to upperparts, dark feathers on back, dark smudges on head, and dark tip to tail. Acquires adult plumage by second winter through successive moults. **First-winter bird** Retains many juvenile plumage details but loses rufous elements and gains a grey back. **First-summer bird** Still has juvenile-type wing pattern but gains a dark hood. **Voice** Raucous calls include a nasal *kaurrr*. **Status** Widespread and numerous. Commonest on coasts and inland freshwater sites, but also in towns and on farmland; often follows the plough. Nests colonially beside water. Migrants from Europe boost winter numbers.

Hard to miss.

TOP: adult summer
ABOVE: adult winter
LEFT: adult summer
BELOW: 1st winter

The perfect storm

Seabird enthusiasts dream about the perfect storm – weather conditions that drive oceanic seabirds close enough to the coast for land-based observation. September is the ideal month, as this is when Arctic-breeding species are moving south on migration. The best British seabird watching sites are on the Cornish coast, which needs to be battered by southwesterly force 8-9 winds for at least 24 hours to blow birds into the Irish Sea and Bristol Channel from the Atlantic. Winds then need to veer northwest overnight and into the morning: migrating birds continue to battle their way south, but the coastline is an obstacle in their path and the onshore winds ensure that they hug the shore.

Juvenile Kittiwake (*left*) and Sabine's Gull.

Common Gull

Larus canus Laridae
Length 40-42cm

Medium-sized gull. Sexes are similar. **Adult** In summer, has a grey back and upperwings with a white trailing margin; black wingtips have white spots. Plumage is otherwise white. Bill is yellowish and legs are yellowish green. In winter, similar but with dark streaks on head and neck; bill is duller with a dark sub-terminal band. **Juvenile** Has pale-margined brown back feathers and upperwings. Head and underparts are pale with dark streaks, while neck and breast look grubby. Adult plumage is acquired over 2 years. **First-winter bird** Similar to juvenile but has a grey back; bill is pink with a dark tip. **Second-winter bird** Similar to adult but with more black on outerwing and a broader band on bill. **Voice** A mewing *keeow*. **Status** Locally common. Nests near water. Widespread in winter on coasts and farmland.

1st winter

adult winter

 Easy to find in winter.

adult winter

Kittiwake

Rissa tridactyla Laridae
Length 38-42cm

adult summer

A true 'sea gull': its non-breeding life is spent entirely at sea. Sexes are similar. **Adult** Has a blue-grey back and upperwings, and black wingtips; plumage is otherwise white. Bill is yellow, and eye and legs are dark. In flight, wingtips look dipped in black ink. In winter, similar but head has grubby patches behind eye. **Juvenile** Has black 'V' markings on upperwing; back and upperwing coverts are grey. Note the triangle of white on flight feathers, dark tip to tail, black half-collar and dark markings on head; bill is dark. **First-winter bird** Similar to juvenile but gradually loses dark half-collar and black tail-tip. **Voice** Utters a diagnostic *kittee-wake* call when nesting. **Status** Nests colonially on coastal cliff ledges. Non-breeding period is spent at sea.

 A visit to a Kittiwake colony is a 'must' for all birdwatchers. Dunbar Harbour (in S Scotland) and the Farne Islands are superb.

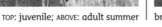

TOP: **juvenile**; ABOVE: **adult summer**

Sabine's Gull

Larus sabini Laridae
Length 30-35cm

Distinctive seabird. Can be confused only with a juvenile Kittiwake but upperwing patterns are separable with care. Sexes are similar. **Adult** In summer, has a blue-grey back and upperwings, dark hood, dark wingtips with white spots, and dark bill with yellow tip. In flight, upperwing pattern is diagnostic: triangular patches of black, white and grey. Tail is forked. In winter, similar but dark smudges on nape replace dark hood. **Juvenile** Has an upperwing pattern similar to that of adult but triangle of grey is replaced by scaly grey-brown. Forked tail is dark-tipped. **Voice** Silent. **Status** Nests in the high Arctic and winters at sea in southern oceans. Seen here mainly as an offshore passage migrant in autumn. Does not willingly come close to land.

 Late-summer pelagic boat trips from Cornwall usually see small numbers. Severe W gales sometimes drive birds close to W coast headlands.

ABOVE: **adult breeding plumage**; BELOW RIGHT: **juvenile**

breeding plumage

SMALLER GULLS **275**

Large gulls

Our larger gulls are familiar to most people, and tend to be bold and relatively unafraid of people, so good views can usually be had. That just leaves the small matter of sorting out their varied and occasionally confusing plumages to make a positive identification.

Herring Gull

Larus argentatus Laridae
Length 56–62cm

Noisy, familiar bird and our most numerous large gull species. Often follows boats. Bold when fed regularly. Sexes are similar. **Adult** In summer, has a blue-grey black and upperwings, with white-spotted, black wingtips; plumage is otherwise white. Legs are pink, bill is yellow with an orange spot, and eye is yellow with an orange-yellow ring. In winter, similar but with dark streaks on head and nape. **Juvenile and first-winter birds** Mottled grey-brown with streaked underparts. Legs are dull pink, bill is dark, and spotted pale tail has a dark tip. Adult plumage is acquired over 3 years. **Second-winter bird** Similar but has a grey back and grey areas on upperwing. Tail is white with a dark tip. **Third-winter bird** Resembles winter adult but has more black on wingtips and hint of a dark tail band. **Voice** A distinctive *kyaoo* and anxious *ga-ka-ka*. **Status** Common, mainly coastal in summer, nesting on sea cliffs and in seaside towns. Widespread and more numerous in winter owing to a migrant influx. Easy to find and identify.

Yellow-legged Gull

Larus cachinnans Laridae
Length 52–60cm

Similar plumage to that of Herring Gull but adult has yellow legs. Typically, consorts with other large gull species. Sexes are similar. **Adult** In summer, has a grey back and upperwings (darker than on Herring Gull), with more black and less white in wingtips. In winter, similar, sometimes with small dark streaks on head. **Juvenile and first-winter birds** Have a grey-brown back and wing coverts, and otherwise dark wings. Head, neck and underparts are streaked, and are paler than on similarly aged Herring Gull. Adult plumage is acquired over 3 years. **Voice** Similar to that of Herring Gull but more nasal. **Status** S European counterpart of the Herring Gull and rather scarce non-breeding visitor to Britain and Ireland. Most frequent in winter. Found among flocks of Herring Gulls. Could turn up almost anywhere, but rubbish tips and reservoir roosting sites are prime locations.

Lesser Black-backed Gull

Larus fuscus Laridae
Length 53–56cm

Similar to Herring Gull but adult has a dark grey back and upper-wings and bright yellow legs. Confusion is possible with adult Yellow-legged Gull (upperparts are paler). Sexes are similar. **Adult** In summer, has a dark grey back and upperwings. Black wingtips are darker than rest of upperwing, which has a white trailing edge; rest of plumage is white. Bill is yellow with an orange spot. Iris is yellow and orbital ring is red. In winter, similar but with streaks on head and neck, and duller leg and bill colours. **Juvenile and first-winter birds** Have streaked and mottled grey-brown plumage, palest on head. Upperwings are dark brown and whitish tail is dark-tipped. Eye and bill are dark. Adult plumage is acquired over 3 years. **Second-winter bird** Similar to first-winter but with a grey back, pinkish legs and dark-tipped pink bill. **Third-winter bird** Resembles a heavily streaked winter adult. **Voice** A distinctive *kyaoo* and an anxious *ga-ka-ka*. **Status** Locally common in summer, nesting colonially on sea cliffs and islands. Most migrate S to the Mediterranean outside the breeding season; small numbers remain, often roosting on reservoirs. Easiest to see at breeding colonies.

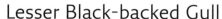

ABOVE AND BELOW: adult

ABOVE: juvenile; BELOW: adult

TOP: 1st winter
ABOVE: adult
INSET: adult

Great Black-backed Gull

Larus marinus Laridae
Length 64–79cm

Our largest gull. Compared to Lesser Black-backed adult has darker back, massive bill and pink legs. Sexes are similar. **Adult** Has an almost uniformly dark back and upperwings; wingtips are only marginally darker than rest of wings. Note white patch at tip of wings and broad white trailing edge. Plumage is otherwise white. Bill is yellow with an orange spot. **Juvenile and first-winter birds** Mottled and streaked grey-brown. In flight, brown upperwings have pale panels and inner primaries. Bill is dark, legs are dull pink and whitish tail is dark-tipped. Adult plumage is acquired over 3 years. **Voice** A deep *kaa-gaga* call. **Status** Local and coastal in the breeding season. Often nests near seabird colonies and pairs are territorial. Outside the breeding season, more widespread inland.

Present at most seabird cliffs.

adult; INSET TOP: **1st winter**; INSET RIGHT: **adult**

Glaucous Gull

adult

Larus hyperboreus Laridae
Length 62–68cm

Bulky, pale gull with white wingtips. Bill is massive and legs are pinkish in all birds. Sexes are similar. **Adult** Has a pale grey back and upperwings with white wingtips and trailing margin. Plumage is otherwise white with variable streaking on head and neck in winter plumage. Eye has pale iris and yellow orbital ring. **Juvenile** Mainly pale buffish grey with pale primaries. Pink bill is dark-tipped. **Immatures** Become paler at first and acquire adult plumage over 3 years. **Voice** A *kyaoo* and an anxious *ga-ka-ka*. **Status** Scarce winter visitor, commonest on N coasts in late winter.

Try busy fishing harbours in winter.

adult; INSET: **2nd winter**

Iceland Gull

Larus glaucoides Laridae
Length 52–60cm

Smaller and longer-winged than Glaucous, with smaller bill. Legs are pink at all times. Sexes are similar. **Adult** Has a pale grey back and upperwings, with white primaries and trailing edge. Plumage is otherwise white with variable dark streaks on head and neck in winter. Bill is yellowish with an orange spot. Eye has pale iris and red orbital ring. **Juvenile** Pale grey-buff with white primaries. Bill is dark with a dull pink base. **Immatures** Become paler at first; acquire adult plumage over 3 years. **Voice** A *kyaoo* call and an anxious *ga-ka-ka*. **Status** Scarce winter visitor.

N Scotland and W Ireland are hotspots.

adult; LEFT INSET: **adult**; RIGHT INSET: **1st winter**

Skuas & terns

Terns and skuas are masters of the air, even when faced with gales at sea. Small tern species buoyantly forge ahead against raging winds, while skuas treat even storm-force blows with indifference. They visit land to breed but otherwise most lead oceanic lives.

Great Skua

Stercorarius skua Stercorariidae
Length 48–52cm

Bulky seabird. Gull-like but note the large head, dark legs and dark bill. In flight, shows a striking white wing patch. Part scavenger, part predator and food parasite of Gannet. Sexes are similar. **Adult** Brown with buff and golden-brown streaks. **Juvenile** Uniformly dark brown and rufous. **Voice** Mostly silent. **Status** Locally common summer visitor and passage migrant. Nests near seabird colonies in Scotland; Orkney and Shetland are strongholds. Passage birds invariably seen at sea flying past exposed headlands.

Visit Hermaness, on the Shetland island of Unst, for guaranteed sightings. Good seawatching sites may yield sightings in spring and autumn during onshore gales.

Arctic Skua

Stercorarius parasiticus Stercorariidae
Length 46cm

Aerobatic, graceful seabird with deep wingbeats and narrow, pointed wings. Food parasite of Arctic Tern and Kittiwake. Adults have a wedge-shaped tail and pointed streamers. Sexes are similar but adults occur in two morphs. **Adult pale phase** Has a white neck, breast and belly, dark cap and otherwise grey-brown plumage. Note the faint yellowish flush on cheeks. **Adult dark phase** Uniformly dark grey-brown. **Juvenile** Dark rufous brown. **Voice** Utters nasal calls near nest. **Status** Local summer visitor to Scottish coasts; coastal passage migrant elsewhere.

Orkney and the Shetlands are breeding hotspots. Autumn seawatching produces regular sightings.

Arctic Tern

Sterna paradisaea Sternidae
Length 35cm

Graceful seabird with a buoyant flight. Plunge-dives for fish. Sexes are similar. **Adult** Has grey upperparts, black cap and pale underparts, palest on cheeks and darkest on belly. Has a uniformly red bill, short, red legs and long tail streamers. In flight from below, flight feathers look translucent, with a narrow, dark trailing edge to primaries. **Juvenile** Has white underparts, an incomplete dark cap and scaly grey upperparts. In flight from above, has a dark leading edge and white trailing edge to innerwing. Legs and bill are dull. **Voice** Utters a harsh *krt-krt-krt* call near nest. **Status** Locally common summer visitor and passage migrant. Colonial nester, always near coasts; commonest in the N.

Easy to see on the Orkney, Shetland and Farne islands.

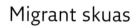

Migrant skuas

It is always a thrill to see Great and Arctic skuas on passage. But part of the fun of seawatching comes from seeing the unusual and two Arctic-breeding skua species fit the bill. Both are seen mainly on exposed western coasts in spring and autumn. The **Pomarine Skua** *Stercorarius pomarinus* (**1**) (length 42–50cm) is larger than an Arctic Skua; adults (shown) have long, spoon-shaped tail streamers. Pale forms predominate and juveniles are barred dark grey-brown. The **Long-tailed Skua** *S. longicaudus* (**2**) (length 36–42cm) resembles a pale phase Arctic Skua but adults (shown) have much longer tail streamers. Juveniles are barred grey-brown.

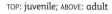

TOP: juvenile; ABOVE: adult

TOP: adult dark phase; ABOVE: adult pale phase

Common Tern

Sterna hirundo Sternidae
Length 35cm

Similar to Arctic Tern but separable with care. Sexes are similar. **Adult** In summer, has grey upperparts, black cap and whitish underparts. Compared to Arctic, note the black-tipped orange-red bill, longer red legs and paler underparts. In flight from below, only inner primaries look translucent and wings have a diffuse dark tip. Non-breeding plumage (sometimes seen in late summer) is similar, but with white on forehead and dark shoulder bar; bill and legs are dark. **Juvenile** Has white underparts, incomplete dark cap and scaly grey upperparts; in flight from above, leading and trailing edges of innerwing are dark. **Voice** A harsh *kreeear* call. **Status** Widespread summer visitor, commonest on coasts; also nests on flooded gravel pits and reservoirs. Widespread coastal passage migrant.

👓 Easy to see in spring on S and E coasts.

Freshwater terns

Terns of the genus *Sterna* are mostly coastal birds, although Common Terns also breed beside inland wetlands. In contrast, members of the genus *Chlidonias*, or 'marsh terns', are mostly associated with fresh water and are only occasionally seen over the sea. The only species to occur here regularly is the **Black Tern** *C. niger* (length 24cm); it has bred on occasions but its true status is that of a scarce passage migrant. In breeding plumage, it is mainly grey and black, with a white undertail. From July onwards, the plumage appears increasingly mottled as the mainly grey and white nonbreeding plumage is acquired. Juveniles recall a winter adult with a brownish grey and scaly back.

Roseate Tern

Sterna dougallii Sternidae
Length 38cm

Our rarest breeding tern. Sexes are similar. **Adult** In summer, pale grey above, with a dark cap and whitish underparts, flushed pink. Has a red-based dark bill, red legs and long tail streamers. In flight, looks very pale. By late summer acquires white on forehead and loses tail streamers. **Juvenile** Has white underparts, partial dark cap and scaly upperparts; upperwings are uniform except for dark leading edge to innerwing. **Voice** A disyllabic *chew-vik* call. **Status** Rare breeding bird and scarce passage migrant.

👓 A few pairs breed on the Farne Islands.

Sandwich Tern

Sterna sandvicensis Sternidae
Length 41cm

Striking seabird with a buoyant flight and distinctive call. Sexes are similar. **Adult** In summer, is pale grey above, with a dark, crested cap, and otherwise white plumage. Legs are black and long, black bill is yellow-tipped. Looks very white in flight. Non-breeding plumage (seen from late summer onwards) is similar but forehead is white. **Juvenile** Similar to winter adult but back is barred and scaly. **Voice** A harsh *chee-urrick* call. **Status** Locally common summer visitor and early returning migrant. Coastal, nesting on shingle beaches and islands.

👓 Easy to find on coasts of S England and East Anglia.

adult summer; INSET: adult non-breeding

Little Tern

Sterna albifrons Sternidae
Length 24cm

Tiny, pale tern. Hovers and plunge-dives after fish and shrimps. Sexes are similar. **Adult** In summer, has a grey back and upperwings, mainly black cap and otherwise white plumage. Has black-tipped yellow bill and yellow-orange legs. In flight, wingtips are dark. By late summer, has acquired white forehead and dull leg and bill colours. **Juvenile** Similar to winter adult but back looks scaly. **Voice** A raucous *cree-ick* call. **Status** Local summer visitor on shingle and sandy islands and beaches.

👓 Easy to see on coasts of the Solent and N Norfolk in late spring.

Auks

Auks belong to the family Alcidae and are true seabirds, coming to land only during the breeding season and then to precarious cliffs at the very edge of the marine environment. In many ways they are the northern hemisphere equivalent of penguins, one obvious difference being that they can fly.

Guillemot

Uria aalge Alcidae
Length 42cm

adult winter

Nests in dense breeding colonies. Swims well and flies on whirring wingbeats. Sexes are similar. **Adult** In summer, is dark brown above (darkest in N birds) and white below. So-called 'Bridled Guillemot' has a white 'spectacle' around eye. In winter, all birds have white cheeks and throat with black line behind eye. **Juvenile** Recalls winter adult but with a smaller bill. **Voice** Growling calls at breeding colonies. **Status** Locally numerous at colonies. Moves offshore in winter.
👓 Easy to see in May-Jul at breeding colonies such as the Farne Islands.

adult summer
Bridled Guillemot flying

Puffin

Fratercula arctica Alcidae
Length 30cm

Endearing seabird. Flies on narrow wings with whirring wingbeats. Swims well and dives frequently for fish. Sexes are similar. **Adult** In summer, has mainly dark upperparts with a dusky face; underparts are white. Legs are orange-red and bill is huge, flattened and marked with red, blue and yellow. In winter, similar but with a dark grey face and smaller, duller bill. **Juvenile** Similar to winter adult but with a small, dark, dull bill. **Voice** Utters groaning calls at the nest. **Status** Locally common. Comes ashore only in the breeding season. Colonial nester, excavating burrows in grassy cliffs. Only storm-driven, sick or oiled birds are seen near land in winter.
👓 Easy to see at large colonies in spring and early summer.

Razorbill

Alca torda Alcidae
Length 41cm

Bulky seabird with a distinctive bill and essentially black and white plumage. Swims well and flies on whirring wingbeats. Sexes are similar. **Adult** In summer, has a black head, neck and upperparts, and white underparts; note the white wingbar. Bill is large and flattened, with vertical ridges and white lines. In winter, similar but throat and cheeks are white and bill is smaller. **Juvenile** Recalls winter adult but with smaller bill. **Voice** Mostly silent. **Status** Locally common in seabird colonies on rocky coasts in the W and N. Nests under boulders and in crevices. Pelagic at other times; healthy birds seldom come near land in winter. Off-duty nesting birds usually conspicuous.
👓 Easiest to see on Shetland and Orkney in Jun-Jul.

Black Guillemot

Cepphus grylle Alcidae Length 34cm

Charming coastal auk. Sexes are similar. **Adult** In summer, mainly sooty-brown except for white wing patch. Has red legs. In winter, has scaly grey upperparts and white underparts; black wings and contrasting white wing patch are retained. **Voice** High-pitched whistling calls. **Status** Local year-round resident in inshore waters of Ireland and N and W Scotland.
👓 Visit Shetland and Orkney in summer for guaranteed sightings. Only ever locally common near suitable breeding areas, such as raised boulder beaches.

ABOVE: **adult summer**
INSET: **adult winter**

adult summer; INSET: adult winter

adult summer

adult winter

Puffin
adult
summer

Razorbill
adult
summer

Guillemot
adult
summer

Black Guillemot
adult summer

Little Auk

Alle alle Alcidae
Length 20cm

Our smallest auk, with a dumpy body, short neck and tiny, stubby bill. Flies on whirring wingbeats and can look almost Starling-like in the air. Swims well and dives frequently. Sexes are similar. **Adult** In winter, has a black cap, nape and back, and white underparts; at close range, note the white lines on wings and tiny white crescent above eye. Breeding plumage adults and juveniles not seen in our region. **Voice** Silent at sea. **Status** Winter visitor from Arctic breeding grounds, where locally abundant. Probably numerous in N North Sea in winter but seldom comes close to land by choice.

Sometimes seen flying offshore during severe gales; storm-driven birds occasionally turn up on coastal pools or even inland. Visit East Anglia and NE Scotland in Jan or Feb.

LEFT: adults winter
BELOW: adult winter

Sandeels

Puffins and, to a lesser degree, Razorbills and Guillemots rely on a plentiful supply of Sandeels to feed their young. However, industrial-scale fishing continues to devastate stocks of this once-abundant fish, so that the breeding success of these auks, as well as birds such as Arctic Terns, suffers as a consequence.

Puffin with beak full of Sandeels.

Auk nesting

A visit to an auk colony is a memorable experience and in a few locations, mainly in Scotland, it is possible to see all four of Britain's breeding auk species on the same general area of cliff. In terms of nesting options, however, they have very different preferences. Each is adapted to occupy a very different niche and hence avoid undue competition for prime nesting sites, which are much in demand.

1 Puffins nest in burrows, excavated into soft, clifftop soil.
2 Guillemots nest on rock ledges, supporting their single egg on their feet.
3 Razorbills lay their single egg in a rock crevice or gully.
4 Black Guillemots nest among boulders in loose colonies.

Doves, pigeons & Cuckoo

Pigeons and doves belong to the family Columbidae and are familiar medium-sized birds; Feral Pigeons and Collared Doves are an everyday sight even in towns and cities. Although superficially similar, the Cuckoo (family Cuculidae) is unrelated and wary, and has an intriguing lifestyle.

Woodpigeon

Columba palumbus Columbidae
Length 41cm

Plump, familiar bird whose 'song' is a common countryside sound, as is the loud clatter of wings heard as the bird flies off. Forms flocks outside the breeding season. Sexes are similar. **Adult** Has mainly blue-grey plumage with pinkish maroon on breast. Note the white patch on side of neck and, in flight, the prominent, transverse white wingbars, dark wingtips and dark terminal band on tail. **Juvenile** Similar but white mark on neck is missing. **Voice** Sings a series of *oo-OO-oo, oo-oo* phrases. **Status** Abundant on farmland and lightly wooded countryside. Increasingly common in towns. Easy to find in lowland regions.

Stock Dove

Columba oenas
Columbidae
Length 33cm

Similar to Woodpigeon but slimmer and separable using plumage details. Rather solitary but forms flocks outside the breeding season. In flight, wings are flicked. Sexes are similar. **Adult** Has blue-grey upperparts and paler grey underparts. Note the pinkish-maroon flush to breast and iridescent green patch on side of neck; wings have two narrow black bars on upper surface and broad, dark trailing edge. **Juvenile** Similar but wingbars are faint. **Voice** During the breeding season, utters repetitive *oo-u-look* call. **Status** Locally common in lowland and wooded arable farmland. Unobtrusive in the breeding season; the 'song' is the best way of determining its presence.

Feral Pigeon or Rock Dove?

Known by the scientific name *Columba livia*, these birds are essentially one and the same. The **Rock Dove** (**1**) (length 33cm) is a shy, scarce bird of wild cliffs and coasts in northern and western parts of the region, while the **Feral Pigeon** (**2**) is its domesticated descendant and urban counterpart, abundant in towns and cities but also seen on farmland. The Feral Pigeon occurs as a variety of colour forms but true Rock Doves show little variation. Both form flocks and utter a range of cooing calls, and the sexes in both are similar.

Adult and juvenile Rock Doves have blue-grey plumage, palest on the upperwings and back, and flushed pinkish maroon on the breast. They have two dark wingbars and a dark-tipped tail. In flight, note the small white rump; the upperwings have a dark trailing edge and narrow wingbar. Feral Pigeons occur in a spectrum of colour forms, from almost black to pure white, some appearing very similar to the ancestral Rock Dove.

INSET: white form

ABOVE & INSET: Woodpigeon

ABOVE & RIGHT: Stock Dove

Collared Dove

Streptopelia decaocto
Columbidae Length 32cm

Recent arrival to Britain and Ireland but now commonly seen and heard. Often in pairs. Sexes are similar. **Adult** Has mainly sandy-brown plumage with a pinkish flush to head and underparts, and a dark half-collar on nape. Black wingtips and white outer-tail feathers are striking in flight. Bill is dark and legs are reddish. **Juvenile** Similar but colours are duller and lacks black half-collar. **Voice** Repetitive song comprising a repeated *oo-oo-oo* phrase. **Status** First seen here in the 1950s. Now common and widespread.

 Easy to find.

Cuckoo

Cuculus canorus Cuculidae
Length 33–35cm

Secretive summer visitor with an intriguing lifestyle (*see* box). Heard more than seen. Recalls a Sparrowhawk in flight. Feeds mainly on hairy caterpillars. Sexes are sometimes separable. **Adult male and most females** Have a blue-grey head, neck and upperparts; underparts are white and barred. **Some adult females** Brown and barred on head, neck and upperparts; underparts are white with dark bars. **Juvenile** Similar to brown adult female but with white nape. **Voice** Males utter a distinctive *cuck-oo* song; females have bubbling call. **Status** Local summer visitor. Range is dictated by occurrence of the songbirds it uses for nest parasitism; host species include Meadow Pipit, Dunnock and Reed Warbler.

 Easiest to hear and see from late April until the end of May near reedbeds; thereafter silent and secretive.

ABOVE: juvenile
BELOW: adults

Turtle Dove

Streptopelia turtur Columbidae
Length 27cm

Small, well-marked dove with a distinctive song. Flight is fast and direct, with flicking wingbeats. Sexes are similar. **Adult** Has a blue-grey head, neck and under-parts, with a pinkish-buff flush on breast and white barring on neck. Back and wing coverts are chestnut: dark feather centres and pale margins create a scaly appearance. Long, mainly black tail looks wedge-shaped in flight owing to white corners. **Juvenile** Similar but duller and lacks neck markings. **Voice** Song is a diagnostic purring *coo*. **Status** Local summer visitor. Declining due to modern farming. Found on farmland and downland.

 Listen for its distinctive call.

The secret life of a Cuckoo

The Cuckoo is renowned for its unusual parasitic breeding behaviour, whereby the female lays an egg in a songbird's nest. Having evicted its companion eggs or chicks, the young Cuckoo is then fed by its hosts until it fledges. Beyond the selection of a suitable host nest in which to lay, the parent Cuckoo plays no part in the upbringing of its offspring. Typically, adult birds begin migrating back to Africa before the juveniles are fully fledged.

Cuckoo chick and Reed Warbler 'parent'.

Owls & Nightjar

Owls are predators, recognised by their dumpy bodies and fixed, forward-pointing eyes. Although some hunt by day, all can feed in low light (some in pitch darkness) aided by their keen hearing and eyesight. The Barn Owl belongs to the family Tytonidae; all other species are in the family Strigidae. The unrelated Nightjar (family Caprimulgidae) is also nocturnal.

Barn Owl

Tyto alba Tytonidae
Length 34–38cm

Mainly nocturnal but sometimes hunts from late afternoon onwards. Feeds mostly on small grassland mammals. Flight is leisurely and slow on rounded wings. Responds well to nestbox schemes. Sexes are similar. **Adult and juvenile** Have orange-buff upperparts speckled with tiny black and white dots. Facial disc is heart-shaped and white. In flight, underwings appear pure white. **Voice** Utters a blood-curdling call at night. **Status** Vulnerable and generally scarce resident.

Sometimes caught in car headlights, feeding over verges beside country roads. Usually easily found in large wetland reserves in East Anglia in winter.

Tawny Owl

Strix aluco Strigidae
Length 38–40cm

Our most familiar owl. Strictly nocturnal; roosts in tree foliage during the day. Flight is leisurely on broad, rounded wings. Sexes are similar. **Adult and juvenile** Have streaked, variably chestnut-brown or grey-brown plumage, palest on underparts. Eyes are dark. In flight, underwings look pale. Young birds typically leave nest while still downy and white. **Voice** A sharp *kew-wick* and well-known hooting calls; most vocal in late winter and early spring. **Status** Fairly common resident of woodland habitats where small mammals are common; also occurs in gardens and suburban parks.

Presence detected by hearing calls in late winter. In daytime, listen for telltale alarm calls of mobbing songbirds.

Arctic wanderer

A **Snowy Owl** *Bubo scandiacus* (length 55–65cm) is an always welcome visitor and a prized sighting for birdwatchers. As befits its snowy-white plumage, this huge and unmistakable bird hails from the Arctic and typically favours tundra-like northern or upland habitats. In the past, the species has tried to nest on Shetland, but these days its status is that of an occasional visitor. Snowy Owls are often active during the day, but even when perched their sheer size and mainly white plumage make them conspicuous. Males have essentially pure white plumage but females and immature birds are marked with grey barring. Visit the Shetland Isles in late winter for a chance of finding this species.

BELOW: **female**; RIGHT: **male**

Short-eared Owl
Asio flammeus Strigidae
Length 35–40cm

Well-marked owl that often hunts in daylight. Flight is leisurely, often with stiffly held wings. Perches on fenceposts. Sexes are similar. **Adult and juvenile** Have buffish-brown plumage, heavily spotted and streaked on upperparts; underparts are streaked but paler. Facial disc is rounded; note the yellow eyes and short 'ear' tufts. **Voice** Displaying birds sometimes utter deep hoots. **Status** Local and rather scarce. Nests on upland moors but outside the breeding season favours lowland marshes, grassland and heaths, particularly near coasts. Winter numbers boosted by European birds.

Uplands of S Scotland and Mainland Orkney are good prospects in the breeding season. In winter, visit coastal S and E England.

Long-eared Owl
Asio otus Strigidae
Length 32–35cm

Strictly nocturnal; sometimes caught in car headlights after dark or glimpsed at its daytime winter roost. In flight, told from a Short-eared by the orange-buff patch that contrasts with otherwise dark upperwing. When alarmed, sometimes adopts an upright posture with 'ear' tufts raised. Sexes are similar. **Adult and juvenile** Have streaked dark brown upperparts and paler underparts. Orange-buff facial disc is rounded; note the orange eyes and long 'ear' tufts. **Voice** Deep hoots are sometimes uttered in spring. **Status** Nests in isolated conifer plantations and scrub thickets with adjacent open country. Disperses outside the breeding season: winter roost sites include coastal and wetland scrub and hedgerows. An influx of European birds boosts winter numbers.

Hard to see, except by chance.

Little Owl
Athene noctua Strigidae
Length 22cm

Our smallest owl. Has a large-headed, short-tailed and overall dumpy silhouette. Partly diurnal and seen perched on fenceposts and dead branches. Sometimes bobs head and body when agitated. Sexes are similar. **Adult** Has brown upperparts with whitish spots; pale underparts have dark streaks. Note the yellow eyes. **Juvenile** Duller and lacks spots on head. **Voice** Calls include a cat-like *kiu*, uttered repeatedly in early evening. **Status** Introduced from mainland Europe in the 19th century. Now widespread and fairly common in S Britain. Nests in tree holes and cavities in stone walls.

To detect its presence, listen for distinctive calls at dusk. At the start of the breeding season members of a breeding pair will often perform vocal 'duets'.

Nightjar
Caprimulgus europaeus Caprimulgidae
Length 24–27cm

Nocturnal; camouflaged plumage makes it hard to find in daytime. Often seen as a silhouette in flight: looks long-winged and narrow-tailed. Has a huge gape used to catch flying moths. Sexes are separable. **Adult male** Has intricate brown, grey and black markings that resemble tree bark. Has white patches near wingtips and corners of tail. **Adult female and juvenile** Similar but lack white on wings and tail. **Voice** Territorial males utter a distinctive churring song after dark. **Status** Migrant visitor to lowland heaths and heather moors.

To hear male Nightjars, visit a S heathland at dusk in May.

Swift to woodpeckers

The birds described here are all active and distinctive in their own ways. The Swift is a consummate flyer that spends most of its life on the wing, Kingfisher, Hoopoe, Bee-eater and Ring-necked Parakeet are dazzlingly colourful, and most woodpecker species are instantly recognisable as such.

An exotic touch

As with things culinary, spice is added to the birdwatching diet by sightings of exotic and colourful species from far afield. More at home in the Mediterranean, the **Hoopoe** *Upupa epops* (**1**) (Length 25–28cm) has a long, downcurved bill and pale pinkish-brown plumage with white barring on the wings and back; its barred pink crest is raised in excitement. In flight it is striking, but its habit of creeping along the ground can make it hard to spot when feeding. The Hoopoe utters a diagnostic *hoo-poo-poo* call and turns up occasionally on the S coast in spring and autumn. With similar southerly origins, the **Bee-eater** *Merops apiaster* (**2**) (Length 26–29cm) is a stunningly colourful bird that catches insects in flight. All colours of the rainbow are found in its plumage and its bill is long and dagger-like. The Bee-eater utters a bubbling *pruuupp* call and is mainly a rare migrant visitor in spring and autumn, although it has bred here on occasions. Although it looks perfectly at home here, the **Ring-necked Parakeet** *Psittacula krameri* (**3**) (Length 40–42cm) is an introduced species from Asia. This mainly green, long-tailed alien has a powerful bill and males have a dark-bordered pinkish neck ring. All birds utter loud, squawking calls and feral populations are established in many leafy suburbs; W London is a stronghold.

1

male

2

Swift

Apus apus Apodidae
Length 16–17cm

Invariably seen in flight: has an anchor-shaped outline and mainly dark plumage. Catches insects on the wing. Sexes are similar. **Adult** Has mainly blackish-brown plumage with a pale throat. Tail is forked but often held closed in active flight. **Juvenile** Darker overall but throat and forehead are paler. **Voice** Loud screaming calls are uttered in flight. **Status** Locally common summer visitor. Nests in churches and loft spaces. Feeding birds gather where insects are numerous.
👓 Easy to see in late spring in villages and towns. Most adults leave Britain in Aug.

Kingfisher

Alcedo atthis
Alcedubudae
Length 16–17cm

Dazzling bird with dagger-like bill. Perches on branches that overhang water and plunge-dives after small fish. Flies low over water on whirring wings. Excavates nest burrows in river banks. Sexes are separable. **Adult male** Has orange-red underparts and mainly blue upperparts with an electric-blue back. Legs and feet are red and bill is all dark. **Adult female** Similar but base of lower mandible is flushed red. **Juvenile** Similar to adult but bill tip is pale. **Voice** Call is high-pitched. **Status** Widespread resident of lowland streams and lakes.
👓 Often sits unobtrusively in shade.

adult female

Woodpeckers

Britain has three species of true woodpeckers (family Picidae): the tiny Lesser Spotted and its much larger cousin the Great Spotted, and the plump and robust Green Woodpecker. All have dagger-like bills and a skull structure and constitution that allows them to hammer at dead wood in search of food and to excavate nesting chambers.

Lesser Spotted Woodpecker

Dendrocopus minor Picidae
Length 14–15cm

Unobtrusive and easily overlooked. Sexes are separable with care. **Adult male** Has a black back and wings, with white barring. Underparts are grubby white with dark streaking. Face is white and nape is black; a black stripe runs from bill, around ear coverts to sides of breast. Note the white-flecked red crown. **Adult female and juvenile** Similar but with a black crown. **Voice** Territorial males utter a raptor-like piping call in spring. Drumming is rapid but rather faint. **Status** Local resident of deciduous woodland and parkland; often associated with Alders.

Easiest to find in spring, when males are most vocal.

ABOVE & BELOW: **male**

Great Spotted Woodpecker

Dendrocopus major
Picidae Length 23–24cm

female

Our commonest pied woodpecker. Flight is undulating. Sexes are separable. **Adult male** Mainly black on back, wings and tail, with white 'shoulder' patches and narrow white barring; underparts are grubby white. Face and throat are white, while cap and nape are black and connect via a black line to a stripe running from base of bill. Has a red nape patch and vent. **Adult female** Similar but red nape patch is absent. **Juvenile** Recalls adult male but has a red crown and subdued red vent colour. **Voice** Utters a loud *tchick* alarm call. Males 'drum' loudly in spring. **Status** Widespread, commonest in S.

Visits peanut feeders.

LEFT: **juvenile**; BELOW: **male**

Green Woodpecker

Picus viridis Picidae
Length 32–34cm

Unobtrusive and often shy. Climbs trees but also feeds on ground, mainly on ants. Flight is undulating. Sexes are separable. **Adult male** Has greenish-olive upperparts and whitish underparts. Has a red crown, black 'mask' and red-centred black 'moustache'. In flight, yellowish rump is striking. **Adult female** Similar but 'moustache' is all black. **Juvenile** Recalls adult male but is heavily spotted. **Voice** Distinctive *yaffling* and yelping calls. **Status** Fairly common in open woodland, parks and gardens.

Easily overlooked in summer when it tends to be secretive. Presence is easiest to detect by learning its distinctive call.

ABOVE & RIGHT: **male**
BELOW: **juvenile**

A woodpecker with a difference

The **Wryneck** *Jynx torquilla* (length 16–17cm) is a woodpecker relative whose intricate plumage markings look like tree bark and afford it superb camouflage. It feeds mainly on the ground and is fond of ants. Sadly, the days when the Wryneck was a widespread breeding bird in the region are long gone. A handful of pairs nests in the Scottish Highlands, but it is now best known as a scarce passage migrant, found mainly in autumn on southern and eastern coasts of England.

Introducing passerines

More commonly known as perching birds, passerines are our most varied group of birds both in terms of numbers of species and diversity of appearance and habitat preferences. Included among them are 22 families of birds that occur regularly in Britain and Ireland and some of our most familiar garden and countryside species. They vary in size from the tiny Goldcrest (our smallest bird) to the Buzzard-sized Raven.

Perching birds

Their common name refers to their ability to perch and all passerines have feet that allow them to grip with ease. Three toes point forwards and one faces back, which not only provides support and allows the bird to stand upright on level ground, but also means it can grasp comparatively slender twigs and branches with a sure grip.

LEFT: The Blue Tit is one of our most familiar passerines.
ABOVE: A sure-footed grip allows this Brambling to move with confidence among branches and twigs.

Vocalisation

Passerines are generally extremely vocal birds and males of some species are among the finest songsters in the bird world. Many territorial male passerines advertise ownership of breeding grounds and attract and retain mates by loud and characteristic songs. All species also have a repertoire of calls that serve a variety of behavioural functions, including alarm (for example, at the presence of a predator) or contact (with other members of the species in feeding flocks or on migration).

ABOVE: Shortly after his arrival in spring, and once he has found a suitable territory, a male Whitethroat advertises ownership by singing his scratchy song.

The Robin's song can be heard intermittently at most times of the year, although territorial males are most vocal in spring.

Safety in numbers

Most passerine birds lead rather solitary lives during the breeding season and nest in relative isolation from pairs of the same species. But outside the breeding season, some species form sizeable flocks that migrate, feed and roost together. Congregating in flocks means there are always plenty of eyes watching for danger, making it harder for predators to launch surprise attacks; living in a flock also lessens the probability of an individual becoming a victim.

The Starling is the most familiar flock-forming passerine bird in the winter months. Flocks are particularly spectacular as they go to roost, with group formations performing swirling, aerobatic displays that appear, to our eyes, to be synchronised.

Food and feeding

The diet of passerines is as varied as the appearance of the birds themselves, but for many species small invertebrates are important for at least part of the year – typically in spring and summer, when they are nesting. Some, such as warblers, feed almost exclusively on insects, spiders and other invertebrates, while sparrows and buntings rely primarily on seeds as a source of nutrition. Many crow family members are arch scavengers that, to a certain extent, have predatory habits too. But in shrikes, the predilection for live prey reaches its apogee; the birds behaving like miniature raptors and even having hook-tipped bills to aid in the dismemberment of victims.

Despite its small size, the Red-backed Shrike is a formidable predator, tackling prey the size of small lizards and large insects. Some shrikes even keep 'larders' of prey on hand to eat later.

The diet of many passerines changes with the seasons. Redwings, for example, will feed on worms when they are available, but happily switch to fruits and berries in autumn and winter.

Many passerines adapt quickly to new food sources – Blue Tits (*right*) and Siskins (*above*) are avid consumers of peanuts at man-made feeders.

A clue to the Pied Flycatcher's diet lies in its name. Insects are scarce here in winter, so to avoid starvation the birds migrate to better feeding grounds in Africa after nesting has finished.

The Goldfinch is a specialised feeder whose slender bill is carefully adapted to extract Teasel seeds from the spiky seedhead with maximum efficiency.

During the summer months, a Waxwing's diet includes plenty of insects, but in winter it will consume berries with relish.

Plumage

With many passerines, visual differences between the sexes are subtle (to our eyes at least): think of the Wren, Dunnock or Rook. Behavioural differences play an important part for the birds themselves, but in certain species it is only when a male is heard singing, or nesting behaviour is seen, that human observers can be certain of the bird's gender. But among a select group of passerines (certain finches, buntings and chats, for example) there are striking differences in plumage, although as a general rule these are much more apparent when the bird is in breeding plumage than during the winter months.

male

female

The differences between the sexes in the Chaffinch are striking, particularly during the breeding season.

Martins, Swallow & larks

Despite their comparatively small size, Swallows and martins (family Hirundinidae) are powerful flyers, able to catch small insects on the wing and undertake annual migrations to Africa for the winter months. Larks belong to the family Alaudidae and, although capable of sustained flight, feed on the ground, including both insects and seeds in their diet.

Sand Martin

Riparia riparia Hirundinidae
Length 12cm

Typically seen hawking for insects over water, sometimes even picking them off the surface. Sexes are similar. **Adult** Has sandy-brown upperparts and mainly white underparts with a brown breast band. Tail is short and forked. **Juvenile** Similar but has pale margins to back feathers. **Voice** A range of rasping twitters. **Status** Widespread summer visitor. Nests colonially, excavating burrows in sandy banks beside rivers and sand and gravel quarries.

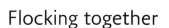 Easy to find in summer near colonies.

In spring and autumn, sometimes gathers in large numbers over lakes in the S.

House Martin

Delichon urbica Hirundinidae
Length 12–13cm

Recognised by overall black and white appearance and striking white rump. Sexes are similar. **Adult** Has mainly blue-black upperparts with a white rump; underparts are white. **Juvenile** Similar but underparts are grubby and upperparts are duller. **Voice** Utters a distinctive *prrrt* call in flight. A twittering song is often delivered from overhead wires near nest. **Status** Locally common summer visitor. Typically constructs a hemispherical mud nest under house eaves and overhangs, in loose colonies. On migration, often feeds on insects and congregates over fresh water.

Easy to find in most towns and villages during the breeding season.

Swallow

Hirundo rustica Hirundinidae
Length 19cm

Recognised in flight by pointed wings and long tail streamers. Sexes are similar but male has longer tail streamers than female. **Adult** Has blue-black upperparts and white underparts except for a dark chest band and brick-red throat and forecrown. **Juvenile** Similar but has shorter tail streamers and a buff throat. **Voice** Utters a sharp *vit* call in flight; male sings a twittering song, often from overhead wires near nest. **Status** Common and widespread in summer. Usually nests in villages and farmyards, its half-cup-shaped mud nest typically sited under eaves or in a barn. Migrants congregate over fresh water and roost in reedbeds.

Easy to find in the breeding season.

LEFT AND BELOW: male

Flocking together

Our three common hirundine species are often seen feeding together, especially shortly after their arrival in spring and before their departure for Africa in winter. Typically, they are spotted hawking for insects over lakes and marshes at these times. Prior to migration, they sometimes perch alongside one another on overhead wires, allowing easy comparison of the species.

Skylark

Alauda arvensis Alaudidae
Length 18cm

Has nondescript plumage; best known for its incessant song, delivered in flight. Sexes are similar. **Adult** Has streaked sandy-brown upperparts and paler underparts; breast is streaked and flushed buff. Short crest is sometimes raised. In flight, note the whitish trailing edge to wings and white outer-tail feathers. **Juvenile** Similar but with a scaly-looking back. **Voice** Rapid song comprises trills, whistles and elements of mimicry. Call is a rolling *chrrrp*. **Status** Favours grassy habitats, including meadows, heaths and arable farmland. Has declined alarmingly due to changes in farming practices, but still common. Forms flocks outside breeding season; upland breeding birds move to lowlands in winter. Easy to see and hear in areas of undisturbed grassland.

Woodlark

Lullula arborea Alaudidae
Length 15cm

Unobtrusive and easily overlooked but for its wonderful yodelling song. Sexes are similar. **Adult** Has streaked sandy-brown upperparts and mainly pale underparts; breast is streaked and flushed buff. Note the chestnut ear coverts, pale supercilium and black and white markings on wing. **Juvenile** Similar but with a scaly-looking back. **Voice** Song comprises fluty, yodelling notes. Call is a yodelling *deet-luee*. **Status** Local, restricted to heathy habitats with short turf (for feeding), longer grassland (for nesting) and scattered trees. Nomadic outside the breeding season. Best looked and listened for on Hampshire and Surrey heaths in Apr.

Shorelark

Eremophila alpestris Alaudidae
Length 16–17cm

Well marked but unobtrusive when feeding among saltmarsh plants. Sexes are similar but female is duller than male. **Summer adult** (Sometimes seen on migration.) Has grey-brown or sandy-brown upperparts, streaked on back. Underparts are mainly white but with a black breast band and faint buff streaks on flanks. Head is yellow with a black band through eye and ear coverts; black forecrown extends to 2 projecting 'horns'. **Winter adult** Similar but duller and 'horns' are absent. **Juvenile** Similar to winter adult but upperparts have pale spots. **Voice** Flight call is a thin *see-seer*. **Status** Best known as scarce winter visitor to coastal saltmarshes. Usually found in small flocks. Visit coastal East Anglia or N Kent in winter.

TOP: adult summer; ABOVE: adult winter

Pipits & wagtails

Pipits and wagtails (family Motacillidae) are slim, long-tailed passerines with needle-like bills. They feed actively on tiny insects and other invertebrates. Some are sedentary but others are long-distance migrants.

Meadow Pipit

Anthus pratensis Motacillidae
Length 14–15cm

Rather nondescript, streaked brown bird. Forms loose flocks outside the breeding season. Sexes are similar. **Adult** Has streaked brown upperparts and pale underparts with dark streaks; has a buffish-yellow flush to flanks and breast, most noticeable in autumn. Has a pale, unmarked throat, pale eye-ring and hint of a short, pale supercilium. Legs are pinkish and outer-tail feathers are white. **Juvenile** Similar but with less extensive streaking. **Voice** A *pseet-pseet-pseet* call. A descending song is delivered in flight but starts and ends on ground. **Status** Common and widespread resident. Favours rough, grassy habitats; upland birds move to lowlands outside the breeding season and European migrants boost winter numbers. Easy to find in suitable habitats.

Tree Pipit

Anthus trivialis Motacillidae
Length 15cm

Similar to Meadow Pipit but separable with care using plumage details, voice and habitat preferences. Sexes are similar. **Adult** Has streaked sandy-brown upperparts. Underparts are pale, whitish and unmarked on throat and belly, but boldly streaked and flushed with yellow-buff on breast and flanks. Note the striking pale supercilium and dark sub-moustachial stripe. Legs are pinkish and outer-tail feathers are white. **Juvenile** Similar. **Voice** Flight call is a buzzing *spzzzt*. Song (often delivered in flight but starts and ends from different tree perches) is an accelerating trill ending with thin notes. **Status** Widespread migrant summer visitor, commonest in the W and N. Favours open woodland and heaths with scattered trees. Learn the call and song by listening to recordings.

Rock Pipit

Anthus petrosus Motacillidae
Length 16–17cm

Bulky, dark pipit invariably found within sight of the sea. Sexes are similar. **Adult and juvenile** Have streaked dark grey-brown upperparts and rather grubby yellowish underparts, heavily streaked on breast and flanks. Throat is pale, and note the indistinct pale supercilium and eye-ring, and dark sub-moustachial stripe. Legs and bill are dark and outer-tail feathers are grey. **Voice** Utters a single *pseet* call. A Meadow Pipit-like song is delivered in flight; starts and ends on a cliff-side rocky outcrop. **Status** Locally common resident, found on rocky coasts and cliffs in summer; commonest in the N and W. More widespread, but still coastal outside the breeding season. Easy to find in suitable habitats.

TOP: adult spring; ABOVE: adult displaying

Continental pipit

Formerly treated as a Continental race of the Rock Pipit, the **Water Pipit** *Anthus spinoletta* (length 16–17cm) is superficially similar but is a winter visitor to Britain and Ireland and is invariably found near fresh water. During the winter months it has streaked dark buffish-brown upperparts and pale underparts, streaked and flushed buffish brown on the breast and flanks; the throat is white and unmarked, and the whitish supercilium contrasts with the dark eye-stripe. Before birds depart in April, they usually acquire breeding plumage, where the underparts become unmarked and flushed pinkish on the breast; the back is brown and the head and neck are grey. Look for Water Pipits at Watercress beds and sewage works in SE England in winter.

adult summer

Pied Wagtail
Motacilla alba yarrellii Motacillidae
Length 18cm

Familiar black, grey and white bird that pumps its tail up and down and has a distinctive call. Sexes are dissimilar. **Adult male** In summer, has mainly white underparts and a black breast and upperparts; note the white face, white wingbars, dark legs and bill, and white outer-tail feathers. In winter, similar but throat is white and black on breast is less extensive. **Adult female** Recalls adult male in various seasons but back is dark grey. **Juvenile and first-winter birds** Have greyish upperparts, a black rump and whitish underparts; note the whitish wingbars and yellowish wash to face. **Voice** A loud *chissick* call. **Status** Favours bare ground and short grassland, often near farms, on playing fields or in car parks. Easy to see.

Grey Wagtail
Motacilla cinerea Motacillidae
Length 18cm

Elegant waterside bird whose strikingly long tail is continually pumped up and down. Sexes are dissimilar. **Summer adult male** Has blue-grey upperparts and lemon-yellow underparts. Note the black bib, white sub-moustachial stripe and white supercilium. Bill is dark, legs are reddish and outer-tail feathers are white. **Summer adult female** Similar but bib is whitish and variably marked with grey, while underparts are paler with yellow colour confined mainly to vent. **Winter adult and juvenile** Similar to respective summer plumages but with white throats. **Voice** Uttcrs a sharp *chsee-tsit* call in flight. **Status** Favours fast-flowing stony streams and rivers; commonest in the N and W. Easy to find in suitable habitats.

Yellow Wagtail
Motacilla flava flavissima Motacillidae
Length 16–17cm

Long-tailed wetland bird. Sometimes feeds at feet of grazing animals. Sexes are dissimilar. **Adult male** Has greenish-yellow upperparts, yellow underparts, white outer-tail feathers and whitish wingbars. **Adult female** Similar but less colourful. **Juvenile** Has olive-buff upperparts and pale underparts, with a whitish throat, yellow flush to undertail, white outer-tail feathers and pale wingbars. Blue-headed Wagtail *M.f. flava* (race from mainland Europe) sometimes turns up; male has bluish cap and ear coverts, and white supercilium. **Voice** A distinctive *tsree-ee* call (all birds). **Status** Local summer visitor to damp, grazed grassland. Learn the distinctive call to detect the presence of hidden feeding birds.

TOP: male winter; ABOVE: juvenile

TOP: male summer; INSET: male; ABOVE: male winter

TOP: male; INSET: male; ABOVE: female

Continental wagtail

The **White Wagtail** *Motacilla alba alba* (length 18cm) is the mainland European counterpart of the Pied Wagtail and is easily overlooked because the two are so similar. All birds have a grey (not black) rump, although this is not always easy to discern. The adult male also has a grey (not black) back, with a clear demarcation from the black nape and hindcrown in summer. In terms of behaviour and voice, White and Pied wagtails are identical. The White Wagtail replaces the Pied in the Channel Islands and a few pairs also breed in Scotland, particularly on the Shetland Isles. Elsewhere, it is a fairly common passage migrant, with most sightings on the coast in autumn.

male summer

Small perching birds

Most habitats in Britain, from windswept moors to town centres, have their complement of small perching birds as long as food is available. The species illustrated here are placed in a range of families. With the exception of the Waxwing, which eats mainly berries during its winter visits to Britain, all feed primarily on invertebrates.

Wren

Troglodytes troglodytes Troglodytidae
Length 9–10cm

Tiny, dumpy bird that cocks its tail upright. Unobtrusive and often creeps through low vegetation, although call is distinctive. Sexes are similar. **Adult and juvenile** Have dark reddish-brown upperparts with barring on wings and tail. Underparts are greyish white with a buff wash to flanks; note the striking pale supercilium. Bill is needle-like and legs are reddish. **Voice** Utters a loud, rattling alarm call; its warbling song ends in a trill. **Status** Widespread resident of all sorts of habitats with dense undergrowth.

Easy to find; heard more often than seen.

Dipper

Cinclus cinclus Cinclidae
Length 18cm

Dumpy waterside bird that perches on river boulders. Flies low over water. Submerges readily in search of invertebrates. Sexes are similar. **Adult** Has dark grey-brown wings, back and tail. Head is reddish brown and throat and breast (bib) are white. Belly grades from reddish chestnut at front to blackish brown at rear. Legs and feet are stout and powerful. **Juvenile** Has greyish upperparts and barred, pale underparts. **Voice** A shrill *striitz* call. **Status** Fairly common but local on fast-flowing streams and rivers.

Easy to find on suitable rivers in Wales and N England.

ABOVE: **juvenile**; RIGHT: **semi-submerged adult**

Dunnock

Prunella modularis Prunellidae
Length 13–14cm

House Sparrow-like bird with a thin, warbler-like bill. Mostly rather skulking. Sexes are similar. **Adult** Has a streaked chestnut-brown back. Underparts are mostly bluish grey but flanks are streaked with brown and chestnut. Face is bluish grey with brown streaking on ear coverts and crown. Bill is dark and legs are reddish pink. **Juvenile** Similar but has bolder streaking. **Voice** Song is warbler-like; usually delivered from a prominent perch. Alarm call is a thin *tseer*. **Status** Common resident of woodlands, hedgerows and gardens with plenty of cover.

 Easiest to see in spring, when males are singing.

Robin

Erithacus rubecula Turdidae
Length 13–14cm

Distinctive bird. Garden-dwellers are bold and inquisitive. Sexes similar. **Adult** Has an orange-red face, throat and breast, bordered by blue-grey on sides but with a sharp demarcation from white belly. Upperparts are buffish brown with a faint buff wingbar. **Juvenile** Has brown upperparts, marked with buff spots and teardrop-shaped streaks; pale buff underparts have darker spots and crescent-shaped marks. **Voice** Song is plaintive and melancholy. Alarm call is a sharp *tic*. **Status** Widespread resident, commonest in the S.

 Easy to find in gardens and parks.

Wheatear

male

Oenanthe oenanthe Turdidae
Length 14–16cm

Open-country bird with white rump and tail. Sexes are dissimilar. **Adult male** Has a blue-grey crown and back, black mask and wings, and pale underparts flushed orange-buff on breast. **Adult female** Has mainly grey-brown upperparts, darkest on wings. Face, throat and breast are pale orange-buff and underparts are otherwise whitish. **First-winter birds** Have grey- to buffish-brown upperparts and buffish underparts. **Voice** A sharp *chak* alarm call, like two pebbles knocked together. Song is warbling. **Status** Locally common summer visitor to moors and open grassland.

 Easy to find.

adult; INSET: juvenile

male

male; INSET TOP: female
INSET RIGHT: juvenile

Urban winter visitor

The **Waxwing** *Bombycilla garrulus* (length 18cm) is a much-admired bird, so called because the adults have red, wax-like projections on the wings. The plumage is mainly pinkish buff but note the crest, black throat and black mask. The rump is grey, the undertail is chestnut and the dark tail has a broad yellow tip (narrower in females than males). Waxwings breed in northern mainland Europe and are winter visitors to Britain. In most years there are just a few records, but once every decade or so they appear in large numbers. They are often indifferent to people and turn up in town parks and gardens.

Chats & relatives

Chats and related perching birds (family Turdidae) are related to thrushes. They feed primarily on insects and other invertebrates, some are fine songsters and most are migratory to a greater or lesser extent.

Stonechat

Saxicola torquata Turdidae
Length 12–13cm

Small, compact bird. When perched, flicks its short, dark tail and utters a harsh alarm call. Sexes are dissimilar. **Adult male** Has a blackish head, white on side of neck, and a dark back. Breast is orange-red, grading into pale underparts. In autumn, pale feather fringes make head appear paler. **Adult female** Similar but colours are muted and plumage is more streaked. **First-winter bird** Has streaked sandy-brown upperparts and head, and buffish-orange underparts. **Voice** A harsh *tchak* call, like the sound of two pebbles knocked together. Song is rapid and warbling. **Status** Locally common on heaths, commons and gorse-covered slopes near coasts. Some dispersal, mainly to coasts, occurs in winter.

 Easy to find on S England heaths in spring.

LEFT: female
BELOW: male

Whinchat

Saxicola rubetra Turdidae
Length 12–14cm

Colourful, Stonechat-like bird with whitish sides to tail base. Sexes are dissimilar. **Adult male** Has brown, streaked upperparts with a white stripe above eye. Margins of throat and ear coverts are defined by a pale stripe; throat and breast are orange; underparts are otherwise whitish. **Adult female** Similar but colours and contrast are less intense. **First-winter bird** Similar to adult female but upperparts are more spotted. **Voice** Utters a whistling *tic-tic* alarm call. Song is rapid and warbling. **Status** Local summer visitor, favouring rough grassy slopes with scattered scrub.

Its habit of perching makes detection easy in suitable habitats.

LEFT: female
BELOW: male

Redstart

Phoenicurus phoenicurus Turdidae
Length 14cm

Colourful Robin-sized bird. Its dark-centred red tail is pumped up and down when perched. Sexes are dissimilar. **Adult male** Has a grey back, nape and crown, black face and throat, and orange-red underparts (most colourful on breast). Note the white forehead and supercilium. **Adult female** Has grey-brown upperparts and head, and an orange wash to pale underparts. **First-winter birds** Recall respective adult plumages but feathers have pale margins. **Voice** A soft *huiit* call. Song is tuneful but melancholy. **Status** Locally common summer visitor to open woodland with plenty of tree holes for nesting and limited ground cover.

Easiest to find in N and W England and Wales.

TOP: male
ABOVE: female
BELOW: male

An urban success story

Once extremely rare as a breeding species, the **Black Redstart** *Phoenicurus ochruros* (length 14cm) has benefited from urban sprawl and indeed often thrives in areas where industrial dereliction prevails. It is a bold bird that perches conspicuously, quivering its striking red tail in an obvious manner. Adult males are particularly striking, with slate-grey body plumage that is darkest on the face and breast. By comparison female and immature birds are rather drab, with mainly grey-brown body plumage. In a strange way, the Black Redstart's song sometimes matches its surroundings: it includes curious crackling, static-like phrases. Between 50 and 100 pairs nest here each year, but the species is more numerous as a passage migrant and occasional winter visitor to south coasts.

immature

male

Bluethroat

Luscinia svecica Turdidae
Length 13–14cm

Robin-sized bird that feeds on the ground. Unobtrusive but obvious when seen well. Red sides to base of tail are diagnostic. Sexes are dissimilar. **Adult male** Has mainly grey-brown upperparts and whitish underparts, with a white super-cilium and iridescent blue throat and breast, bordered below by bands of black, white and red; typically, blue 'throat' has white or red central spot depending on race. Blue colour is masked by pale feather fringes in autumn.
Other plumages Blue on throat is obscured by pale feather margins, or replaced by cream or white, depending on individual's sex and age.
Voice A sharp *tchick* call. **Status** Scarce passage migrant.

 Most are recorded near coasts in autumn, during E winds; North Sea coasts are most reliable.

Nightingale

Luscinia megarhynchos Turdidae
Length 16–17cm

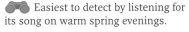

Secretive bird, best known for its powerful musical song, sometimes sung at night. Silent birds are easily overlooked. Sexes are similar. **Adult and juvenile** Have rich brown upperparts overall; tail and rump are warmer reddish chestnut than back and note the hint of grey on face and sides of neck. Underparts are greyish white, suffused pale buffish brown on breast.
Voice Song is rich and varied, and includes fluty whistles and clicking sounds; typically the bird starts with a rich, whistling *tu-tu-tu-tu*.
Status Local and declining summer visitor, favouring coppiced woodland and scrub.

Easiest to detect by listening for its song on warm spring evenings.

Nightingale

TOP: **male autumn**; ABOVE: **juvenile**

Thrushes

Although the plumages of the different thrush species (Turdidae) vary considerably, family similarities are relatively easy to discern. All have plump bodies, pointed bills and relatively long tails, and they typically move with a hopping gait on the ground. Most species feed primarily on the ground and, in addition to invertebrates, include both fruit and berries in their diet.

Nomadic lifestyles

Fieldfares and Redwings are locally abundant in much of Britain and Ireland during the winter months, having flown here from breeding grounds in northern Europe. They form large flocks that often feed on farmland, the two species regularly mixing together. However, because winter food supplies are unpredictable and the weather fickle, they seldom stay in the same place for very long. Instead, they are nomadic and fly considerable distances, often every few days, in search of new feeding grounds. This lifestyle is reflected in their powerful flight.

Blackbird

Turdus merula Turdidae
Length 25–28cm

Familiar ground-dwelling bird. Sexes are dissimilar. **Adult male** Has uniformly blackish plumage. Legs are dark but bill and eye-ring are yellow. **First-winter male** Similar but bill is dark and eye-ring is dull. **Adult and first-winter females** Brown, darkest on wings and tail, and palest on throat and streaked breast. **Juvenile** Similar to adult female but marked with pale spots. **Voice** Utters a harsh and repeated *tchak* alarm call, often at dusk. Male has a rich, fluty, varied song. **Status** Common and widespread in gardens and woodland, and on farmland and coasts. In winter, upland birds move to lower levels and migrants arrive from Europe. Easy to find.

female

BELOW: Blackbird male

Ring Ouzel

Turdus torquatus Turdidae
Length 25–26cm

Upland counterpart of Blackbird. Typically alert and wary. Sexes are dissimilar. **Adult male** Has mainly black plumage with a striking white crescent on breast and pale fringes to wing feathers. Legs are dark, bill is yellowish and feathers on underparts have pale fringes. **Adult female** Similar but dark elements of plumage are browner and pale crescent on breast is grubby white. **First-winter bird** Looks rather dark, with pale feather fringes all over and a hint of adult's pale crescent on breast. **Voice** Utters a harsh *tchuck* alarm call. Song is short bursts of fluty phrases. **Status** Local summer visitor to moorland and lower mountain slopes. Easiest to find in N and W Britain; views are typically distant and brief.

LEFT: Ring Ouzel male

female

Fieldfare

Turdus pilaris
Turdidae
Length 24–26cm

Large, plump thrush. Associates with Redwing in winter flocks. Sexes are similar. **Adult** Has a blue-grey head, chestnut back and pale supercilium. Breast and flanks are flushed orange-yellow and heavily spotted; underparts are otherwise whitish. In flight, note the pale grey rump and white underwings. **Juvenile** Similar but note the pale spots on wing coverts. **Voice** A harsh *chack-chack-chack* call; night-migrating flocks can sometimes be detected by these calls. Song (seldom heard here) comprises short bursts of fluty phrases. **Status** Common winter visitor to farmland and open country. A few pairs breed each year, mainly in the N. Usually easy to find in winter in open countryside.

Redwing

Turdus iliacus Turdidae
Length 20–22cm

Small, well-marked thrush. Forms flocks in winter and mixes with Fieldfare. Sexes are similar. **Adult** Has grey-brown upperparts; pale underparts are dark-spotted and flushed with orange-red on flanks and underwings. Has white stripes above eye and below cheeks. **Juvenile** Similar but has pale spots on upperparts and subdued colours on flanks. **Voice** Utters a thin, high-pitched *tseerp* in flight; often heard on autumn nights from migrating flocks. Song (seldom heard here) comprises short bursts of whistling, fluty phrases. **Status** Common winter visitor to farmland and open, lightly wooded countryside. A few pairs breed here each year, mainly in the NW.

Easy to find in winter in open countryside.

Song Thrush

Turdus philomelos Turdidae
Length 23cm

Dainty, well-marked thrush with a beautiful, distinctive song. Sexes are similar. **Adult** Has warm brown upperparts with hint of an orange-buff wingbar. Underparts are pale but well marked with dark spots; note the yellowish-buff wash to breast. In flight, reveals orange-buff underwing coverts. **Juvenile** Similar but markings and colours are less intense. **Voice** Utters a thin *tik* call in flight. Song is loud and musical; phrases are repeated 2 or 3 times. **Status** Common but declining resident of woods and gardens. Migrants arrive from mainland Europe in autumn.

Easy to see and hear, but for stunning views visit the Isles of Scilly.

ABOVE: **looking for food**; BELOW: **juvenile**

Mistle Thrush

Turdus viscivorus Turdidae
Length 27cm

Appreciably larger than Song Thrush. Unobtrusive but has a distinctive call and song. Sexes are similar. **Adult** Has grey-brown upperparts with hint of a white wingbar. Underparts are pale with large dark spots and flanks are washed orange-buff. In flight, note the white underwings and white tips to outer-tail feathers. **Juvenile** Similar but back has white, teardrop-shaped spots. **Voice** Utters a loud, rattling alarm call. Song contains brief phrases and long pauses; often sung in dull weather. **Status** Fairly common resident of open woodland, parks and mature gardens.

Listen for its song in spring. Individuals defend berry-laden Holly bushes in winter.

INSET TOP: **juvenile**; INSET RIGHT: **adult showing white underwings**

Warblers

Most warblers (family Sylviidae) are seasonal breeding visitors to Britain: long-distance migrant species spend much of their non-breeding lives in Africa. Territorial male warblers sing distinctive songs in spring.

A heathland emblem

The perky little **Dartford Warbler** *Sylvia undata* (length 12–13cm), often seen perched on a Gorse spray with its tail cocked up, is emblematic of heathland conservation. Adults have blue-grey upperparts, reddish underparts with a white belly, a beady red eye and reddish eye-ring, and pinkish-yellow legs; males are brighter than females. The species is often first detected by sound: it utters a *tchrr-tche* alarm call and has a rapid, scratchy warbling song. Dartford Warblers are restricted to Gorse-covered heathland areas in southern England and are mainly resident.

male
INSET: **female**

Reed Warbler

Acrocephalus scirpaceus Sylviidae
Length 13–14cm

Wetland warbler with nondescript brown plumage but a distinctive song. Sexes are similar. **Adult** Has sandy-brown upperparts with a reddish-brown flush to rump. Underparts are pale with a buffish flush to flanks. Legs are dark and bill is needle-like. Note hint of a pale supercilium and eye-ring. **Juvenile** Similar but upperparts are warmer brown and underparts are more intensely flushed buff. **Voice** A sharp *tche* call. Song contains grating and chattering phrases (some are repeated 2 or 3 times), plus elements of mimicry. **Status** Locally common summer visitor to reedbeds.
👓 Easy to see and hear in suitable habitats.

Reed Warbler or Marsh Warbler?

To look at, a **Marsh Warbler** *Acrocephalus palustris* (length 13–14cm) is very similar to a Reed Warbler. The best way to separate the two is by song: a Marsh's is rich and varied, including amazing mimicry of both other European songsters and species found in its African wintering grounds. It also favours subtly different habitats, namely rank waterside vegetation (including nettles and Brambles) rather than reedbeds. Subtle plumage and structural differences also exist between the two species. An adult Marsh has grey-brown upperparts (not 'warm'), including the rump, and pale underparts suffused yellow-buff. The legs are pinkish (not dark) and the soles of the feet look yellowish. The species is a rare summer visitor and just a handful of pairs breed here.

Sedge Warbler

Acrocephalus schoenobaenus Sylviidae
Length 12–13cm

Well-marked wetland warbler with a distinctive song. Sexes are similar. **Adult** Has dark-streaked sandy-brown upperparts and pale underparts, flushed orange-buff on breast and flanks. Head has a dark-streaked crown, striking pale supercilium and dark eye-stripe. **Juvenile** Similar but breast is faintly streaked. **Voice** Utters a sharp *chek* alarm call. Song comprises rasping and grating phrases interspersed with trills and whistles. **Status** Widespread summer visitor to rank marshy vegetation, scrub patches on the fringes of reedbeds and overgrown ditches.
👓 Easiest to detect by sound; songsters are sometimes conspicuous.

Lesser Whitethroat

Sylvia curruca Sylviidae
Length 12–13cm

Small, rather short-tailed warbler with retiring habits but a distinctive song. Sexes are similar. **Adult and juvenile** Have a blue-grey crown, dark mask and grey-brown back and wings. Underparts are pale, whitish on throat but washed pale buff on flanks. Have dark legs, a dark-tipped grey bill and a pale iris. **Voice** Utters a harsh *chek* alarm call. Song is a tuneless rattle, sung on one note, usually preceded by a short warbling phrase. **Status** Fairly common summer visitor, mostly to S and SE England. Favours areas of scrub and hedgerows with dense Hawthorn and Blackthorn.

 Its song is the best clue to the species' presence.

Garden Warbler

Sylvia borin Sylviidae
Length 14–15cm

Warbler with a nondescript plumage but an attractive song. Sexes are similar. **Adult and juvenile** Have uniform grey-brown upperparts and pale underparts with a buffish wash to breast and flanks. Legs are grey, grey bill is short and stubby, and note the subtle grey patch on side of neck (not always easy to see). **Voice** Call is a sharp *chek-chek*. Song is rich and warbling; could be confused with Blackcap's song but is even more musical and almost thrush-like. **Status** Fairly common summer visitor to deciduous woodland and areas of mature scrub.

 Usually sings from dense cover and hence it is usually difficult to see well.

Blackcap

Sylvia atricapilla Sylviidae
Length 14–15cm

Distinctive warbler with a musical song. Sexes are dissimilar. **Adult male** Has grey-brown upperparts, dusky-grey underparts (palest on throat and undertail), a pale eye-ring and a diagnostic black cap. **Adult female and juvenile** Have grey-brown upperparts, pale buffish-grey underparts (palest on throat and undertail) and a reddish-chestnut cap. **Voice** Utters a sharp *tchek* alarm call. Song is a rich, musical warble; similar to Garden Warbler's but contains jaunty phrases.

Status Common summer visitor to deciduous woodland and scrub. Winters in small numbers.

 Easy to see and hear in spring. Wintering birds sometimes visit garden feeders.

male; INSET: female

Whitethroat

Sylvia communis Sylviidae
Length 13–15cm

Familiar warbler of open country. Males often perch openly. Sexes are dissimilar. **Adult male** Has a blue-grey cap and face, grey-brown back and rufous edges to wing feathers. Throat is white; underparts are otherwise pale, suffused with pinkish buff on breast. Legs are yellowish brown and yellowish bill is dark-tipped. Dark tail has white outer feathers. **Adult female and juvenile** Similar but cap and face are brownish and pale underparts (apart from white throat) are suffused pale buff. **Voice** Utters a harsh *check* alarm call. Song is a rapid, scratchy warble. **Status** Common summer visitor to scrub patches, hedgerows and heaths.

 Easy to see and hear in suitable habitats.

male; INSET: juvenile

Warblers & crests

Most warblers (family Sylviidae) feed primarily on insects and other invertebrates. These range in size from aphids (a favourite among *Phylloscopus* warblers) to more sizeable prey such as moth larvae and bush-crickets (preferred by Cetti's and Grasshopper warblers). Technically speaking, our two 'crest' species are also warblers. Although they are our smallest birds, both undertake surprisingly lengthy migrations and autumn sees an influx of these avian gems from mainland Europe.

Grasshopper Warbler

Locustella naevia Sylviidae
Length 13cm

Skulking warbler, heard more than seen. Sexes are similar. **Adult** Has streaked olive-brown upperparts; underparts are paler but flushed buffish brown on breast. Has long, streaked undertail coverts. **Juvenile** Similar but underparts are usually tinged yellow-buff. **Voice** A sharp *tssvet* call. Song is reeling and insect-like; sung mainly at dusk and at night. **Status** Local summer visitor to rank grassland.

Visit a likely looking area at dusk in May and listen for the distinctive song.

Cetti's Warbler

Cettia cetti Sylviidae
Length 14cm

Unobtrusive wetland warbler whose loud song is heard more than the bird is seen. Sexes are similar. **Adult and juvenile** Have dark reddish-brown upperparts, including tail. Underparts are pale: have a whitish throat, grey face and breast, and grey-buff belly. Legs are reddish and bill is dark-tipped. **Voice** A loud *pluut* call. Song is an explosive *chee, chippi-chippi-chippi*. Most vocal in spring but snatches of song are heard at other times. **Status** Recent colonist, now a local resident of scrubby margins alongside marshes and clumps of bushes in extensive reedbeds.

Easy to hear at many large wetland reserves in S England. Tricky to see well.

Savi's Warbler

Locustella luscinioides Sylviidae
Length 14–15cm

Reedbed specialist that is hard to see. Sexes are similar. **Adult and juvenile** Have uniformly warm brown upperparts. Underparts are paler but flushed buffish brown on breast and flanks; undertail coverts are warm buffish brown. **Voice** Call is a sharp *tviit*. Song is reeling, endless and insect-like; sung mainly at night and virtually inaudible to people with poor hearing. **Status** Restricted to extensive wet reedbeds, mainly in East Anglia and Kent.

A tricky species to see, given the inaccessible nature of its favoured habitats. Visit boardwalked East Anglian sites in spring for the best chances.

Savi's Warbler

Chiffchaff

Phylloscopus collybita
Sylviidae Length 11cm

Tiny warbler, best known for its onomatopoeic song. Sexes are similar. **Adult and juvenile** Have grey-brown upperparts and pale greyish underparts suffused with yellow-buff on throat and breast. Bill is needle-like and legs are black; latter feature helps separate silent individuals from similar Willow Warbler. **Voice** Call is a soft *hueet*. Song is a continually repeated *chiff-chaff* or *tsip-tsap*. **Status** Common summer visitor to mature deciduous woodland with a dense understorey of shrubs. Most migrate S to the Mediterranean in autumn but several hundred overwinter in S Britain. Easy to see and hear.

Chiffchaff

Willow Warbler

Phylloscopus trochilus Sylviidae
Length 11cm

Similar to Chiffchaff but separable using subtle plumage details, colour and voice. Sexes are similar. **Adult** Has olive-green upperparts, a yellow throat, whitish underparts and a pale supercilium. Overall, plumage is brighter than that of Chiffchaff and primary feathers project further. Note the pale supercilium and pinkish-yellow legs. **Juvenile** Similar but paler and more yellow, particularly on underparts. **Voice** *Hueet* call is similar to that of Chiffchaff. Song is a tinkling, descending phrase that ends in a flourish. **Status** Widespread and common summer visitor to wooded habitats, including birch woodland and willow scrub.

Easy to see and hear in spring. Autumn migrants need close scrutiny for certain identification.

ABOVE: adult; BELOW: juvenile

Did you know?

Experienced birdwatchers are able to identify almost all territorial warblers they encounter in spring simply by hearing their song – they don't actually have to see the bird in question. Although this may sound an impressive feat, it actually takes most beginners just a season to learn the songs of our commoner warblers, especially if recordings are played and studied during the process.

Wood Warbler

Phylloscopus sibilatrix Sylviidae
Length 11–12cm

Colourful warbler with a distinctive song and precise habitat requirements. Sexes are similar. **Adult and juvenile** Have olive-green upperparts, a bright yellow throat and supercilium, and clean white underparts. Note the dark eye-stripe and pale pink legs. **Voice** A sharp *tsip* call. Song (likened to the sound of a coin spinning on a metal plate) starts with ringing notes and accelerates to a silvery trill. **Status** Locally common summer visitor to mature woodlands with tall trees, limited ground cover and closed canopy; Sessile Oak woods in the W and N are favoured, and Beech woods elsewhere.

Easiest to find by visiting suitable woodland in spring and listening for its song.

Goldcrest

Regulus regulus Sylviidae
Length 9cm

Our smallest bird. Recalls a *Phylloscopus* warbler but note the large head, white-ringed dark eye and colourful crown stripe. Sexes are dissimilar. **Adult** Has greenish upperparts with 2 pale wingbars, and yellow-buff underparts. Note the black-bordered crown stripe (orange in male, yellow in female). **Juvenile** Similar but crown stripe is absent. **Voice** A thin, high-pitched *tsee-tsee-tsee* call. Song is a series of high-pitched phrases and ends in a flourish. **Status** Common woodland resident; favours conifers but also found in deciduous woods, especially in winter. Migrants from N Europe boost numbers outside the breeding season.

Best located by its high-pitched calls or song. Associates with tit flocks in winter.

TOP & ABOVE LEFT: male; ABOVE RIGHT: female

Firecrest

Regulus ignicapillus Sylviidae
Length 9–10cm

Only marginally larger than Goldcrest. Sexes are dissimilar. **Adult** Has yellow-green upperparts with 2 pale wingbars. Underparts are buffish white but flushed golden yellow on sides of neck. Has a dark eye-stripe, broad white supercilium and black-bordered crown stripe (orange in male, yellow in female). **Juvenile** Similar but crown stripe is absent. **Voice** A thin *tsuu-tsee-tsee* call. Song is a series of thin, high-pitched notes. **Status** Rare breeding bird. In summer, often found in mixed woodland with mature conifers and Holly understorey. Migrants and wintering birds favour coastal woodland and scrub.

Visit Hampshire's New Forest in spring and summer.

TOP & ABOVE: male; INSET: juvenile

Flycatchers & rare warblers

Like warblers, flycatchers (family Muscicapidae) also feed on insects, catching flies and butterflies on the wing. Unusual species of warblers sometimes appear, mainly in autumn, hailing from as far away as Asia. Unsurprisingly, their arrival sets birdwatchers' pulses racing.

Flycatchers

With their bold, active, insect-catching habits, flycatchers are favourites with most birdwatchers. Of the two species that breed in our region, the Spotted is the most familiar, while the Pied is a bird that requires a bit of effort to find. The Red-breasted Flycatcher, a scarce migrant visitor, is a star bird on autumn forays to the coast.

Spotted Flycatcher

Muscicapa striata Muscicapidae
Length 14cm

Charming, perky bird with unremarkable plumage. Recognised by its upright posture and habit of making insect-catching aerial sorties from regular perches. Sexes are similar. **Adult** Has grey-brown upperparts, streaked on crown, and pale greyish-white underparts, heavily streaked on breast. **Juvenile** Similar but has pale spots on back and dark spots on throat and breast. **Voice** A thin *tsee* call. Song is simple and includes thin, call-like notes. **Status** Widespread summer visitor to open, sunny woodland, parks and gardens; often nests around habitation.

Look for fly-catching behaviour. Otherwise easily overlooked.

Pied Flycatcher

Ficedula hypoleuca Musciapidae
Length 12–13cm

Well-marked bird with precise habitat requirements. Forages in tree canopy. Sexes are dissimilar. **Adult male** In summer, has black upperparts, white underparts and a bold white band on otherwise black wings; note small white patch at base of bill. **All other birds (including autumn adult male)** Similarly patterned but black elements of plumage are replaced by brown. **Voice** Utters a sharp *tik* alarm call. Song is sweet and ringing. **Status** Locally fairly common summer visitor, mainly to Sessile Oak woodland; most numerous in Devon, Wales and the Lake District.

Easiest to locate in spring, when males are singing. Hard to locate after that time.

BELOW: male
INSET: female

Red-breasted Flycatcher

Ficedula parva Muscicapidae
Length 11–12cm

Charming little bird. Often unobtrusive but more obvious when flycatching from an exposed branch. All birds have diagnostic white sides to otherwise black tail. Sexes are dissimilar. **Adult male** Has a brown back, blue-grey face, orange-red throat and upper breast, and whitish underparts; note the whitish eye-ring. **Adult female** Has brown upperparts and whitish underparts, smudged buffish on sides of breast. **First-winter bird** Similar to adult female but throat and breast are buffish and wing coverts have pale tips. **Voice** Utters a rattling, Wren-like call. **Status** Scarce passage migrant, seen mostly in autumn, and in first-winter plumage.

Look for it in coastal woodland after E winds in late Sep–early Oct.

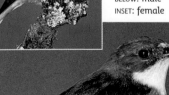

TOP: male; ABOVE: 1st winter

Yellow-browed Warbler

Phylloscopus inornatus Sylviidae
Length 9–10cm

Small, well-marked warbler. Its appearance and frenetic behaviour give it a passing resemblance to Goldcrest. Sexes are similar. **All birds** Have bright olive-green upperparts and whitish underparts. Note the narrow, dark eye-stripe, broad, long, yellow supercilium and 2 pale yellow wingbars. Legs are pinkish. **Voice** A distinctive, drawn-out *tsu-eet* (can sound disyllabic). **Status** Scarce autumn passage migrant; the E coast of England and Isles of Scilly are hotspots. Also occasionally overwinters in S England.

Hard to see because it feeds in dense foliage and is constantly active. Its presence is easiest to detect by learning and listening for its call.

Pallas's Warbler

Phylloscopus proregulus Sylviidae
Length 9–10cm

Tiny, energetic warbler whose active behaviour recalls that of Firecrest or Yellow-browed. **First-winter bird** (The plumage seen in Britain.) Has mainly olive-green upperparts and whitish underparts. Note the striking head pattern: dark eye-stripe, bright yellowish supercilium, and pale median stripe on other-wise dark olive crown. Has 2 pale wingbars; pale rump can be hard to discern. **Voice** Calls include a soft *tchuee*. **Status** Scarce autumn vagrant from Asia.

Look for it in coastal woodland and scrub. The Isles of Scilly and East Anglian coast are hotspots.

Barred Warbler

Sylvia nisoria Sylviidae
Length 16–17cm

Bulky, rather thick-billed warbler. Adult is easy to recognise but is seldom seen here. Juvenile could be confused with Garden Warbler. Sexes are dissimilar. **Adult male** Has blue-grey upperparts and pale underparts marked with dark bars. Note the yellow eye, pale wingbars and pale tip to tail. **Adult female** Similar but duller. **Juvenile** Has grey-brown upperparts and pale underparts washed buff on breast and flanks. Compared to Garden Warbler, note the pale wingbars, subtle crescent-shaped barring on flanks and undertail coverts, pale brown eye and proportionately larger bill. **Voice** Utters a rattling *tchrrrr* alarm call. **Status** Rare passage migrant, mostly on the E coast and in autumn.

Typically found in coastal scrub in late Aug–early Sep after SE winds; Shetland, Orkney and East Anglia are hotspots.

TOP & RIGHT: juvenile

Hippolais warblers

Members of this interesting genus of warblers are summer visitors to mainland Europe. Unfortunately, none breeds in Britain, although we do get occasional visits from passage migrants; look for them at coastal migration hotspots in autumn after southeasterly winds. As a group, *Hippolais* warblers are superficially similar to *Phylloscopus* warblers but have proportionately large heads, a peaked rather than rounded crown, a relatively large, broad-based bill and pale lores. The **Melodious Warbler** *H. polyglotta* (**1**) (length 12–13cm) occurs most regularly and recalls an outsized Willow Warbler. It has uniform olive-green upperparts with a brownish hue to the wings; the underparts (paler in juveniles than adults) are pale yellow, the colour most intense on the throat and breast. The legs are greyish brown and there is a pale eye-ring. The **Icterine Warbler** *H. icterina* (**2**) (length 12–13cm) is similar but is separable with care. It has greyish-green upperparts and pale yellow underparts, which are paler in juveniles than adults. Compared to a Melodious, note the pale panel on the wings, blue-grey legs and long primary projection.

Nestboxes

The breeding season is a critical period for all birds, because adults, eggs and nestlings are particularly vulnerable to predation at this time. They go to great lengths to hide their nests, or to use sites that are inaccessible to predators and protected from the elements. You can do your bit to help them breed in safety by making nestboxes for them to use: with a bit of ingenuity and forethought, a wide variety of species can be simply catered for.

Why use nestboxes?

In the wild, many woodland birds nest in natural crevices and holes in tree branches and trunks. If they are not available, wooden boxes with discreet entrance holes make an ideal substitute. Sadly, natural nest sites are in short supply in many woodlands because of the way the trees are managed. So artificial nestboxes are as invaluable an aid to conservation here as they are in the garden.

The standard nestbox

The standard nestbox is simply a stoutly constructed box with a hole through which the birds gain access; a sloping roof allows rainwater to run off, and you will also need some means of attaching it firmly to a tree or wall. A variety of materials can be used, but treated plywood (particularly marine ply) and native hardwoods such as oak are popular.

Standard nestboxes are designed to accommodate small songbirds and to a degree, the species that occupies any given box can be controlled or at least influenced, by the diameter of the entrance hole. The following hole sizes are typical:

- 2.5cm for Blue Tits
- 2.8cm for Great Tits
- 3.2cm for House Sparrows

Some birds will enlarge the entrance hole to suit their preferences, while a Nuthatch will often plaster the hole, and much of the box too, with mud.

ABOVE: Life inside a standard nestbox is often cramped and somewhat unsanitary by the time the chicks have reached a fair size. This is particularly true of Blue Tits, which have broods of 10 or more.
INSET: There is no more satisfying experience for the garden bird enthusiast than to see a family of newly fledged Blue Tits emerge from a nestbox installed for their benefit.

LEFT: Pied Flycatchers can sometimes be tempted to nest in woodlands where no natural nest holes exist by the provision of nestboxes with a suitably sized entrance hole.

Open-fronted nestboxes

Open-fronted nestboxes attract bolder species of birds that are less prone to attack by predators or that are more able to defend themselves. Robins and Spotted Flycatchers sometimes use such boxes, so long as they are well hidden. Feral Pigeons and, occasionally, Stock Doves will also make use of larger installations.

Tawny Owls will occupy sizeable nestboxes. Typically, these are mounted to the underside of a branch of a large tree.

Communal nestboxes

Most bird species nest in isolation from others of their own kind, and indeed typically from other bird species. In a few cases, however, the presence of pairs in the vicinity is viewed favourably, and such loosely colonial nesting can be accommodated by nestboxes with multiple openings (*see* right). The classic species to benefit from this is the House Sparrow.

You can help halt the decline in House Sparrow numbers by installing a communal nestbox in your garden – a plentiful supply of food will also keep these chirpy visitors happy.

Board and lodging

Some species need little special provision when it comes to nesting. For example, all that Swifts require is a reasonably clear loft space and unimpeded access via the eaves of the house to the outside world.

Tits

Members of the tit family (Paridae) include some of our most familiar and best-loved garden birds. Two other bird families are honorary members of the group: the Bearded Reedling (sometimes called Bearded Tit), which belongs to the family Timaliidae; and the Long-tailed Tit, which belongs to the family Aegithalidae. All the birds featured here feed on seeds and small insects in varying proportions at different times of the year.

Great Tit

Parus major Paridae
Length 14–15cm

Bold, well-marked bird with a distinctive song. Sexes are separable with care. **Adult male** Has white cheeks on an otherwise black head; black throat continues as a black line down centre of breast on otherwise yellow underparts. Upperparts are greenish and blue but note the white wingbar. **Adult female** Similar but with a narrower black line on breast. **Juvenile** Duller than adult. **Voice** Utters a harsh *tche-tche-tche* alarm call. Song is a variation on *teecha-teecha-teecha* theme. **Status** Common resident of lowland woodlands and gardens. Easy to see and hear. Often visits garden feeders.

ABOVE & BELOW: male

Blue Tit

Cyanistes caerulescens Paridae
Length 11–12cm

Familiar garden and woodland bird. Sexes are similar. **Adult** Has a greenish back, blue wings and yellow underparts. Mainly white head is demarcated by a dark blue collar, connecting to a dark eye-stripe and dark bib; cap is blue. Bill is short and stubby and legs are bluish. Male is brighter than female. **Juvenile** Similar but colours are subdued. **Voice** Call is a chattering *tser err-err-err*. Song contains whistling and trilling elements. **Status** Common resident of woods and gardens. One of the easiest birds to see. Often visits garden feeders.

Willow Tit

Poecile montanus Paridae
Length 12–13cm

Separable from Marsh Tit using voice and plumage differences. Sexes are similar. **Adult and juvenile** Have a black cap and bib; cap is dull and bib is relatively large. Cheeks are whitish, back is grey-brown and underparts are pale grey-buff. Compared to Marsh, neck is thicker and note pale panel on otherwise grey-brown wings. Bill is short and legs are bluish. **Voice** A nasal *si-si tchay-thcay-tchay* call. Song is musical and warbling. **Status** Very local resident of damp woodland. Has very specific habitat requirements.

Marsh Tit

Poecile palustris Paridae
Length 12–13cm

Pugnacious woodland bird. Similar to Willow Tit but separable using subtle differences in plumage and voice. Sexes are similar. **Adult and juvenile** Have a black cap and bib; compared to Willow, cap is glossy, not dull, and bib is relatively small. Cheeks are whitish, upperparts are grey-brown and underparts are pale grey-buff. Bill is short and legs are bluish. **Voice** A loud *pitchoo* call. Song is a loud and repeated *chip-chip-chip...* **Status** Locally common resident of deciduous woodland and mature gardens; commonest in the S. Best identified by learning its call. Often visits garden feeders, which Willow seldom does.

Coal Tit
Periparus ater Paridae
Length 10–11cm

Tiny, well-marked and warbler-like bird. Sexes are similar. **Adult** Has white cheeks and a white nape patch on otherwise black head. Back and wings are bluish grey and underparts are pale pinkish buff. Note the 2 white wingbars and dark, needle-like bill. **Juvenile** Similar but colours and markings are less striking. **Voice** A thin call. Song is a repeated *teechu-teechu-teechu...*, higher-pitched and more rapid than that of Great Tit. **Status** Fairly common resident of conifer forests and mixed and deciduous woodland.

Singing males are easy to locate in the breeding season. In winter, look for them among roving mixed-species flocks.

Did you know?
Long-tailed Tits construct beautiful and intricate ball-shaped nests, made from feathers and spiders' silk, and camouflaged with a coating of lichens. They are often built in the cover of dense, spiny bushes such as Gorse, as a protection from predators.

Crested Tit
Lophophanes cristatus Paridae
Length 11–12cm

Easily recognised by its conspicuous crest. Sexes are similar. **Adult** Has a striking black and white barred crest. Note the black line through eye and bordering ear coverts on otherwise mainly whitish head. Black throat and collar demarcate head. Upperparts are otherwise brown and underparts are buffish white. Bill is narrow and warbler-like. **Juvenile** Similar but duller. **Voice** A high trilling call. Song is a rapid series of call-like notes and whistles. **Status** Very local resident, restricted to ancient Caledonian pine forests and mature Scots Pine plantations in the Scottish Highlands.

Visit forests in the Cairngorms.

Bearded Reedling
Panurus biarmicus Timaliidae
Length 16–17cm

Reedbed specialist with a rounded body and long tail. Its distinctive call leads to the affectionate nickname of 'pinger'. Forms flocks outside the breeding season. Sexes are dissimilar. **Adult male** Has a sandy-brown body and tail, with black and white markings on wings. Head is blue-grey with a black 'moustache', beady yellow eye and yellow bill. **Adult female** Similar but head is sandy brown. **Juvenile** Similar to adult female but back is blackish, throat is whiter and eye colour is darker. **Voice** A diagnostic high-pitched *ping* call. Song is seldom heard. **Status** Local, restricted to extensive reedbeds.

Easiest to find in winter, on calm days in East Anglian reedbeds.

male; INSET: female

Long-tailed Tit
Aegithalos caudatus Aegithalidae
Length 14cm

Delightful, long-tailed bird with an almost spherical body. Seen in acrobatic flocks. Sexes are similar. **Adult** Looks black and white overall, but note the pinkish-chestnut patch on shoulders and whitish feather fringes on otherwise black back and wings. Head is mainly whitish with a black band above eye; underparts are whitish, suffused pink on flanks and belly. Bill is dark, short and stubby. **Juvenile** Similar but duller and darker. **Voice** Utters a rattling *tsrrr* contact call and thin *tsee-tsee-tsee*. Soft, twittering song is easily missed. **Status** Common resident of woods, scrub and heathland fringes.

Easy to find; listen for the call.

Nuthatch to Golden Oriole

Treecreepers (family Certhiidae) and Nuthatch (family Sittidae) are woodland birds. Shrikes (family Laniidae) are predators of prey to the size of small mammals. The Starling (family Sturnidae) is a farmland and urban bird while the Golden Oriole (family Oriolidae) favours wet woodland.

A Channel Islands speciality

Visit the Channel Islands and you may see a local speciality: a **Short-toed Treecreeper** *Certhia brachydactyla* (length 12–13cm). Although the species is widespread in mainland Europe, this is the only place in Britain where it occurs. Subtle plumage and structural differences exist between it and its close relative, the Treecreeper, but its voice is very different – it utters a piercing *zeeht* call. Like the Treecreeper, the Short-toed has streaked brown upperparts, but its grubby-white underparts are darker, with a strong buffish wash to the flanks, its bill is longer and the hind claw is shorter. It is found in wooded valleys.

Treecreeper

Certhia familiaris Certhiidae
Length 12–13cm

Easily overlooked as it creeps up tree trunks, using spiky tail for support. Probes bark for insects using needle-like bill. Sexes are similar. **Adult and juvenile** Have streaked brown upperparts and silvery-white underparts, subtly suffused buff towards rear of flanks. Note the grubby-whitish supercilium and broad buffish zigzag barring on wings. **Voice** A thin, high-pitched *tseert* call. Song comprises a short series of high-pitched notes and ends in a trill. **Status** Fairly common woodland resident.

 Its presence is easiest to detect by listening for its call.

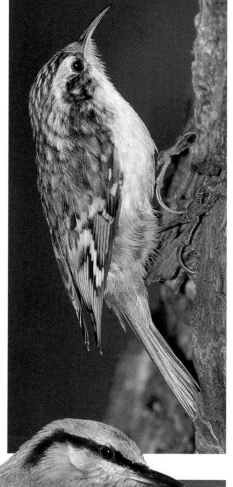

Nuthatch

Sitta europaea Sittidae
Length 14cm

Dumpy, short-tailed woodland bird that often descends tree trunks head first. Sexes are similar. **Adult** Has blue-grey upperparts, a black eye-stripe, white cheeks and orange-buff underparts; on average, males are more reddish buff on flanks than females. **Juvenile** Similar but duller. **Voice** Utters a loud *zwiit*, repeated if bird is agitated. Song is a series of whistling notes. **Status** Fairly common resident of gardens and deciduous and mixed woodland, mainly in England and Wales.

 Extremely vocal in spring. Sometimes visits garden feeders in winter.

Shrikes

Shrikes are bold little predators whose allure derives from their smart appearance and comparative rarity; two species occur regularly. The **Great Grey Shrike** *L. excubitor* (**1**) (length 22–26cm), a scarce winter visitor, has a grey cap and back, white underparts, a broad black mask, and a white patch on the otherwise dark wings. Look for it on heathland, particularly in the New Forest in Hampshire. The **Red-backed Shrike** *Lanius collurio* (**2**) (length 16–18cm), once a locally common summer visitor, now no longer breeds but is still a scarce passage migrant. Adult males have a reddish-brown back, blue-grey cap and nape (with a dark band through the eye), and pink-flushed pale underparts. Adult females are similar but with muted colours, while juveniles are barred brown.

1

2

ABOVE: male
LEFT: juvenile

Starling

Sturnus vulgaris Sturnidae
Length 20–22cm

Familiar urban and rural bird with a swaggering walk. Forms large flocks outside the breeding season. Sexes are separable in summer. **Summer adult male** Has dark plumage with an iridescence seen in good light. Legs are reddish and bill is yellow with a blue base to lower mandible. **Summer adult female** Similar but has some pale spots on underparts and a pale yellow base to lower mandible. **Winter adult** Has numerous white spots adorning dark plumage and a dark bill. **Juvenile** Grey-brown, palest on throat; bill is dark; spotted body plumage is acquired in winter. **Voice** Varied repertoire of clicks and whistles, including mimicry. **Status** Widespread and common but declining. Found in all kinds of open habitats in winter. Often nests in house roofs.

👓 Easy to see.

juvenile   winter

ABOVE: winter
RIGHT: male summer

Golden Oriole

Oriolus oriolus Oriolidae
Length 22–24cm

Stunning and unmistakable, but heard more often than seen. Sexes are dissimilar. **Adult male** Has mainly bright yellow plumage with black on wings and tail. Bill is red. **Adult female** Similar but duller and paler below, with some streaking. **Juvenile** Similar to adult female but upperparts are green and underparts are more heavily streaked. **Voice** Song is a fluty, tropical-sounding *wee-lo-weeow*. Utters harsh cat-like calls in alarm. **Status** Regular passage migrant and scarce breeder; nests in poplar plantations in East Anglia.

👓 Its presence is easiest to detect by hearing its call. Patience is required to get more than a glimpse.

TOP: male; ABOVE: female

Crows

Crows and related birds (family Corvidae) are popularly known as corvids. Intelligent, opportunistic birds, they have a predatory and scavenging diet and the fact that several species are nest-robbers has given them a bad press. Their adaptability is one of the keys to their success.

Magpie

Pica pica Corvidae
Length 45–50cm

Unmistakable black and white, long-tailed bird. Seen in small groups outside the breeding season. Its varied diet includes fruit, insects, animal road-kills, and eggs and young of birds. Sexes are similar. **Adult and juvenile** Mainly black with a white belly and white patch on closed wing. A bluish-green sheen on wings and tail is seen in good light. In flight, outer half of rounded wings is white. **Voice** Utters a loud, rattling alarm call. **Status** Widespread resident of open countryside and gardens.
Easy to find and identify.

Chough

Pyrrhocorax pyrrhocorax Corvidae
Length 38–40cm

Jackdaw-sized corvid with a downcurved red bill, used to probe ground for invertebrates. Forms sociable, noisy flocks outside the breeding season. Superb aeronaut with broad, 'fingered' wingtips. Sexes are similar. **Adult** Has glossy black plumage and reddish-pink legs. **Juvenile** Has duller legs and a dull yellow bill. **Voice** Call is a distinctive *chyah*, uttered while wings are flexed and flicked. **Status** Scarce resident, mainly of coastal sea cliffs in W Britain and Ireland. Has recently recolonised the Lizard in Cornwall.
Easiest to find outside breeding season.

Jay

Garrulus glandarius Corvidae
Length 33–35cm

Colourful, wary bird, identified in flight by its white rump. Buries thousands of acorns each autumn. Sexes are similar. **Adult and juvenile** Have a mainly pinkish-buff body except for a white rump, undertail and lower belly. Wings are black and white with a chequerboard patch of blue, black and white. Note the black 'moustache', streaked pale forecrown and pale eye. **Voice** A loud, harsh scream. **Status** Fairly common woodland resident, commonest where oaks are plentiful.
Easiest to see in autumn.

Chough

Jay

Magpie

Carrion and Hooded crows

Formerly treated as subspecies, two crow species are now recognised in the region. Structurally identical (length 43–50cm), they have different plumages and are separated geographically. The **Carrion Crow** *Corvus corone* (**1**) has glossy black plumage and is found throughout England, Wales and SE Scotland. The **Hooded Crow** *C. cornix* (**2**) is grey and black and is found in NW Scotland, throughout Ireland and on the Isle of Man. Both species are wary and utter a harsh *creeaa-creeaa-creeaa* call.

Rook

Corvus frugilegus Corvidae
Length 43–48cm

Familiar farmland bird. Feeds in large flocks (mainly on soil invertebrates) and occupies noisy, colonial, tree nest sites. Sexes are similar. **Adult** Has black plumage with a reddish-purple sheen. Bill is long, narrow and rather pointed; note the bare patch of whitish skin at base. **Juvenile** Similar but skin at base of bill is feathered. **Voice** A grating *craah-craah-craah...* call. **Status** Common on farmland and grassland. Builds large twig nests in clumps of tall trees.

👓 Easy to find in lowland farmland. Most memorable (and noisy) at the start of the breeding season.

Jackdaw

Corvus monedula
Corvidae
Length 31–34cm

Our most familiar small corvid. Has a swaggering walk and is aerobatic in flight. Forms large flocks outside the breeding season. Sexes are similar. **Adult** Has smoky-grey plumage, darkest on wings and crown, a pale blue-grey eye and a grey nape. **Juvenile** Similar but plumage is tinged brownish and eye is duller. **Voice** A characteristic *chack* call. **Status** Widespread and common resident of farmland, sea cliffs, towns and villages.

👓 Easy to find and identify. Often bold and inquisitive.

Raven

Corvus corax Corvidae
Length 55–65cm

Our largest passerine. Appreciably bigger than Carrion Crow, with a massive bill and shaggy throat. Wary and mostly seen in aerobatic flight; note the thick neck and wedge-shaped tail. Typically seen in pairs. Sexes are similar. **Adult and juvenile** Have black plumage with an oily sheen. **Voice** A loud *cronk* call. **Status** Fairly common resident. Its distribution has a W bias but signs indicate it may be returning to its former haunts in central England. Favours rolling, wooded countryside, desolate upland areas and rugged coasts.

👓 Relatively easy to find on rugged W coasts. Its flight silhouette is diagnostic.

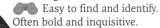

Hedgerows & roadsides

Often taken for granted, hedgerows and roadside verges are important habitats in our increasingly fragmented countryside. They take on the qualities and species of the surrounding landscape, and are vital wildlife corridors, connecting widely separate areas of similar habitat.

Hedgerows

When first created, hedgerows would have been planted with local shrubs and trees – branching and spiny ones in particular. Hence Blackthorn, Hawthorn and willow species predominate. Various elms were popular in many areas, particularly East Anglia, while Beech prevailed in parts of the West Country.

Over the years, natural additions to most hedgerows have occurred as seeds have been transported by the wind or by roosting birds. Given time, hedgerows take on the species composition of neighbouring woodland edge wildlife.

Research has found that there is a reasonably well-defined correlation between the number of woody species in a hedgerow and its age. The exact formula for calculating a hedgerow's age is complicated, but as a rule of thumb, every species of woody shrub in a 30m stretch of hedgerow is likely to correspond to a century of its existence. Of course, this applies only in areas where replanting schemes do not operate. Although some of our hedgerows are truly ancient, most were planted between the 16th and 18the centuries, during the process of land enclosure.

Roadsides

As a rather sad sign of the times, many roadside verges are now important nature reserves, some of them even with official status. Rabbits and Kestrels are often conspicuous, oblivious of the cars whizzing by, and if you brave the traffic noise and make a close inspection, you are likely to find meadow grasses and wild flowers, along with butterflies such as Meadow Brown and Common Blue.

Mature hedgerows are extremely good for songbirds and many species nest there including the **Dunnock** or Hedge-sparrow (*see also* p.295). In addition to providing a safe haven for nesting and roosting, hedgerows provide food in the form of insects during the summer months and fruits and nuts in autumn and winter.

Properly maintained hedgerows are a rich source of berries and fruits. Chief among the edible delights in many areas are **blackberries** (*see also* p.70), the fruits of the Bramble – there can hardly be a stretch of decent hedgerow in lowland Britain and Ireland that cannot supply a succulent meal for free.

Woody shrubs form the backbone of native hedgerows and one common component species is the **Hawthorn** (*see also* p.27). In May, it turns long stretches of hedgerow white with its blossom, while in autumn and winter its red berries provide food for birds and small mammals.

Hedgerow decline

Once such a dominant feature of the British countryside, hedgerows have suffered a dramatic decline in recent decades. In most instances, hedgerows are man-made creations and people control their fate to this day. Regular management – proper laying rather than indiscriminate flailing – is required to keep a hedge in good order. Sadly, this seldom happens and, worse still, vast lengths of hedgerows have already vanished, grubbed up by farmers in the mid-20th century who saw them as a threat to farming 'efficiency'. Today, grants for hedgerow planting schemes are, to some extent, helping to redress the balance.

Moths are usually abundant in mature and well-maintained hedgerows, their larvae feeding on the leaves of the woody shrubs or on the herbaceous plants that grow beneath them. Some butterflies are also found here; perhaps the most characteristic is the **Gatekeeper** or Hedge Brown (*see also* p.345).

Hedgerows have always been important sites for wildlife. Many include the word 'hedge' in their common names. **Upright Hedge-parsley** (*see also* p.99) is a typical hedgerow and roadside verge flower in the spring; other examples include Hedge Bindweed, Hedge Bedstraw and Hedge Woundwort.

Insects are the most conspicuous invertebrates found in hedgerows. If you use a beating tray or sweep net, or even just your eyes in summer, you will discover a wealth of bugs, beetles and flies, as well as sizeable insects such as the **Dark Bush-cricket** (*see also* p.378), which often sits on brambles.

Sparrows & finches

Although underrated by some birdwatchers, sparrows and finches lead interesting and complex lives, some are colourful or well marked, and a few have distinctive songs and calls. Sparrows belong to the family Passeridae while finches are members of the family Fringillidae.

female Chaffinch

House Sparrow

Passer domesticus Passeridae
Length 14–15cm

Familiar because of its affinity for human habitation. Often dust-bathes and sits on roofs, uttering familiar sparrow chirps. Sexes are dissimilar. **Adult male** Has a grey crown, cheeks and rump. Nape, sides of crown, back and wings are chestnut-brown, underparts are pale grey, and throat and breast are black. Bill is dark and legs are reddish. **Adult female** Mainly brown above with buff streaks on back; underparts are pale grey and note the pale buff supercilium behind eye. **Juvenile** Similar to adult female but duller. **Voice** A range of chirping calls. **Status** Fairly common but declining resident, usually found in the vicinity of houses and farms. Easiest to see in urban parks where birds are fed.

TOP: male; ABOVE: female

Tree Sparrow

Passer montanus Passeridae
Length 13–14cm

Well-marked bird, rural counterpart of House Sparrow. Forms flocks outside the breeding season and may feed with finches and buntings in fields. Sexes are similar. **Adult** Has a chestnut cap and striking black patch on otherwise whitish cheeks and side of head; note the black bib. Underparts are otherwise greyish white. Back and wings are streaked brown; note the white wingbars. **Juvenile** Similar but duller. **Voice** Utters chirps and a sharp *tik-tik* in flight. **Status** Scarce and declining resident of untidy arable farms and rural villages. A tree-hole nester that benefits from nestbox schemes. East Midlands and parts of East Anglia are still strongholds. Visits nature reserve feeders in winter.

Chaffinch

Fringilla coelebs Fringillidae
Length 15cm

One of our most familiar birds. Forms flocks outside the breeding season. Sexes are dissimilar. **Adult male** Has reddish pink on face and underparts, a blue crown and nape, and a chestnut back. Note the dark wings and whitish wingbars, and white undertail and vent. White outer-tail feathers are seen in flight. **Adult female and juvenile** Mainly buffish brown, palest on face and underparts; note the pale wingbars (pattern similar to that on male) on otherwise dark wings. **Voice** A *pink pink* call. Song comprises a descending trill with a final flourish. **Status** Common breeding species in a wide range of habitats. Numbers are boosted in winter by migrants from N Europe. Easy to find.

ABOVE: male: BELOW: female

Where have all the sparrows gone?

Once abundant in Britain, sparrows have declined in number catastrophically in recent times. The House Sparrow population is down by 50 per cent compared to a decade ago, and our Tree Sparrow numbers are now a pitiful 3 per cent of what they were in the 1970s. Modern agricultural practices are to blame for the fate of the latter, while tidier gardens and an obsession with decking and paving, as opposed to weedy vegetable patches, deny food to the former.

Brambling

Fringilla montifringilla Fringillidae
Length 14–15cm

Distinctive finch with a white rump. Sexes are dissimilar. **Adult male** In winter, has an orange throat, breast and lesser wing coverts; underparts are white with small dark spots on flanks. Head and back are darkish but with pale feather fringes. Dark wings have pale feather margins and whitish-orange wingbars. Bill is yellow. In breeding plumage (sometimes acquired here), head, back and bill are black. **Adult female** Has similar markings to winter male on wings, breast and underparts, but head is grey-brown with dark lines on sides. **Juveniles** Similar to respective winter adults but duller. **Voice** Calls include a harsh *eeerrp*. Song is a series of buzzing notes. **Status** Common winter visitor, found mainly in mature Beech woodland. A few pairs breed.
 Flocks are thinly scattered.

male winter; INSET: female winter

Siskin

Carduelis spinus Fringillidae
Length 11–12cm

Recognised by a broad yellowish bar on otherwise dark wings, a yellow rump, and yellow sides to otherwise dark tail. Forms flocks outside the breeding season. Sexes are separable. **Adult male** Has yellowish-green upperparts, streaked on back, and a black cap and bib. Breast is flushed yellow-green but underparts are otherwise whitish with streaks on flanks. Note the dark wings and yellow wingbars. **Adult female** Similar but duller, and lacks male's black cap and bib. **Juvenile** Has similar wing and tail patterns to adults but plumage is otherwise streaked grey-brown. **Voice** Whistling or twittering disyllabic calls. Song is a series of twittering phrases. **Status** In the breeding season, locally common in conifer woods. In winter, favours Alder and birch.
 Easiest to find in winter.

Goldfinch

Carduelis carduelis Fringillidae
Length 12cm

Delightful, colourful bird. Yellow wingbars and white rump seen in flight are unique. Sexes are similar. **Adult** Has a striking black and white pattern on head, and a red face. Back is buffish brown and underparts are mainly whitish, suffused pale buff on flanks and sides of breast. Wings are black with a yellow wingbar and white tips to flight feathers; black tail feathers are white-tipped. Bill is narrow, conical and pale pinkish buff. **Juvenile** Mainly pale buffish white, streaked brown on flanks and back. Wings are black with a yellow wingbar. **Voice** A tinkling, trisyllabic call. Song is twittering and rapid. **Status** Common in the breeding season in scrub, deciduous woodland and mature gardens. At other times, forms roving flocks that feed on thistle and Teasel seeds; many birds migrate to mainland Europe in winter.
 Grow Teasels in your garden to be sure of visits from these avian jewels in late winter.

INSET: juvenile

ABOVE: juvenile; BELOW: male

Greenfinch

Carduelis chloris Fringillidae
Length 14–15cm

Familiar greenish finch with a conical pinkish bill, a yellowish patch on wings and yellow sides to base of tail. Sexes are dissimilar. **Adult male** Mainly yellowish green, darkest on back, with grey on face, sides of neck and wings. Intensity of colour increases through winter as pale feather fringes are worn. **Adult female** Similar but duller and faintly streaked. **Juvenile** Recalls adult female but back and pale underparts are obviously streaked. **Voice** Utters a sharp *jrrrup* call in flight. Song comprises well-spaced wheezy *weeeish* phrases or rapid, trilling whistles. **Status** Fairly common. In the breeding season, favours parks, gardens and hedgerows. In winter, forms flocks that visit gardens and arable fields.
 Easiest to find in spring, when males are vocal.

TOP: male; ABOVE: female; BELOW: juvenile

Finches

Finches are small birds whose diet comprises mainly seeds, especially in winter. Bill size and shape varies considerably among our species, each allowing its owner to feed on subtly different sources of food.

Bullfinch

Pyrrhula pyrrhula Fringillidae
Length 16–17cm

Unobtrusive finch whose call and white rump are distinctive. Bill is stubby and dark. Sexes are separable. **Adult male** Has a rosy-pink face, breast and belly. Back and nape are blue-grey and cap and tail are black. Note the white wingbar on otherwise black wings. **Adult female** Similar but duller. **Juvenile** Similar to adult female but head is uniformly buffish brown. **Voice** A soft piping call; pairs sometimes perform duets. Song is quiet and seldom heard. **Status** Fairly common resident of woodlands, hedgerows and mature gardens. Its presence is easiest to detect by learning its call.

LEFT: female
BELOW: male

Hawfinch

Coccothraustes coccothraustes
Fringillidae Length 17–18cm

Giant among finches. Its massive conical bill is used to crack Hornbeam and Cherry seeds. Sexes are separable. **Adult male** Has mainly pinkish-buff plumage with grey on neck and brown on back. Note the broad whitish wingbar, blue-black flight feathers and broad white tip to tail. **Adult female** Similar but duller. **Juvenile** Similar to adult female but plumage patterns are less distinct. **Voice** A sharp, Robin-like *tsic* call. Its quiet song is seldom heard. **Status** Local and rather scarce in mature deciduous woodland, orchards and large gardens.

Hard to see well; best looked for in winter, at fruiting Hornbeam trees.

LEFT: female
ABOVE & BELOW: male

Linnet

Carduelis cannabina Fringillidae
Length 13–14cm

Breeding male is colourful, but at other times is rather nondescript. Forms flocks outside breeding season. Sexes are dissimilar. **Adult male** In summer, has a grey head, rosy forecrown and chestnut back. Pale underparts are flushed rosy pink on breast. Note the whitish patch on wings, pale sides to tail and streaked throat. In winter, rosy elements of plumage are dull or absent. **Adult female** Has a brown back, grey-brown head and streaked pale underparts. Note the whitish patch on wings. **Juvenile** Similar to adult female but more streaked. **Voice** A *tetter-tett* call. Male has a twittering, warbling song. **Status** Common and widespread. Favours heaths and scrub in summer, and waysides and farmland in winter. Easiest to find in spring, when males perch prominently.

ABOVE: male; BELOW: juvenile

Twite

Carduelis flavirostris Fringillidae
Length 13–14cm

Upland counterpart of Linnet. Bill is grey in spring and summer but yellow in autumn and winter (Linnet's is grey at all times). Throat is unstreaked. Sexes are very similar. **Adult male** In summer, has streaked brown upperparts, a pinkish rump and white margins to flight and tail feathers. Pale underparts are heavily streaked. In winter, head and breast are warm buffish brown. **Adult female and juvenile** Similar to winter male but rump is brown, not pink. **Voice** A sharp *tveeht* call. Song is a series of trilling notes. **Status** Local breeder on N heather moors and coasts. In winter, favours saltmarshes and coastal fields, and migrants from N Europe boost numbers.

 Common in spring in the Northern Isles. Norfolk saltmarshes are good in winter.

male summer; INSET: juvenile

Common Crossbill

Loxia curvirostra Fringillidae
Length 15–17cm

Uses crossed-tipped mandibles to extract seeds from conifer cones. Sexes are dissimilar. **Adult male** Mainly red but with brownish wings. **Adult female** Mainly greenish but with brownish wings. **Juvenile** Grey-brown and heavily streaked. **Voice** Utters a sharp *kip-kip-kip* flight call. **Status** Nomadic outside breeding season.

 Listen for its call. Birds often drink at woodland puddles. **Similar species** The **Parrot Crossbill** *L. pytyopsittacus* is an occasional visitor from Scandinavia and is very similar but with a larger bill. Has bred here in the past.

Lesser Redpoll

Carduelis cabaret Fringillidae
Length 12–14cm

Well-marked finch. Forms flocks outside the breeding season and mixes with Siskins. Bill is yellow and conical. Sexes are separable. **Adult male** Has streaked grey-brown upperparts, darkest on back. Underparts are pale but dark-streaked. Note the red forecrown, black bib and lores, white wingbar, pale, streaked rump and often pinkish flush to breast. **Adult female and juvenile** Similar but lack pinkish flush to breast. **Voice** Utters a rattling *chek-chek-chek* call in flight. Song is wheezing and rattling. **Status** Widespread and fairly common. Breeds in birch woodland and favours birches and Alders in winter.

 Easiest to find in winter, in feeding flocks.

BELOW: **Lesser Redpoll**
RIGHT: **Mealy Redpoll**

ABOVE RIGHT: **female**; RIGHT: **male**

Scottish Crossbill

Loxia scotica Fringillidae
Length 15–17cm

Very similar to Common Crossbill; extreme caution is needed with identification. Scottish has a more robust and stout bill than Common, suited to extracting seeds from Scots Pine cones. Sexes are dissimilar. **Adult male** Mainly red with brownish wings. **Adult female** Mainly greenish with brownish wings. **Immature bird** Similar to adults of respective sexes but duller. **Juvenile** Grey-brown and streaked. **Voice** Utters a sharp *kip-kip-kip* flight call, deeper than that of Common. **Status** Found nowhere else in the world other than native Scots Pine forests in Scotland.

 Visit Abernethy and Rothiemurchus forests to be sure of seeing this species.

male

Buntings

Buntings (family Emberizidae) are mainly seed-eaters, especially in winter when they often gather in flocks. They usually feed on the ground, and have stubby bills, rounded bodies and rather long tails.

Yellowhammer

Emberiza citrinella Emberizidae
Length 15–17cm

Colourful bunting with a diagnostic song. Forms flocks outside the breeding season. Sexes are dissimilar. **Adult male** Has a mainly yellow head and underparts, and a reddish-brown back and wings. Note the faint dark lines on head, chestnut flush to breast and streaking on flanks; rump is reddish brown and bill is greyish. In winter, similar but duller and more streaked. **Adult female** Has a streaked greenish-grey head and breast, streaked pale yellow underparts and a brown back; note the reddish-brown rump. **Juvenile** Similar to adult female but more streaked. **Voice** A rasping call. Song is rendered as 'a little bit of bread and no cheese'. **Status** Fairly common resident of farmland and open country with scrub and hedges. Winter flocks often feed on arable fields.

Easiest to detect by its song.

Cirl Bunting

Emberiza cirlus Emberizidae
Length 16–17cm

Well-marked bunting. Its olive-grey rump allows separation from Yellowhammer (has a red-brown rump) at all times. Sexes are dissimilar. **Adult male** Has black and yellow on head. Breast, nape and crown are greenish grey and underparts are yellow, flushed and streaked chestnut on flanks; back is reddish brown. In winter, colours are duller. **Adult female** Has dark and yellowish stripes on head, a streaked greenish-grey crown, nape and breast, and streaked yellowish underparts. Back is reddish brown. **Juvenile** Similar to adult female but paler. **Voice** A sharp *tziip* call. Song is a tuneless rattle, recalling that of Lesser Whitethroat. **Status** Once widespread in the S, now restricted to S Devon and Cornwall (*see* box). Favours low-intensity farmland with hedgerows.

Walk S Devon coastal paths in spring.

Reed Bunting

Emberiza schoeniclus Emberizidae
Length 14–15cm

Well-marked wetland bird. Forms flocks outside the breeding season. Sexes are dissimilar. **Summer adult male** Has a black head, throat and bib, and a white collar and 'moustache'. Underparts are whitish with faint streaking, back is dark and wings have reddish-brown feather margins. **All other plumages** Head has dark brown and buffish-brown stripes and a pale 'moustache'. Back has brown and buff stripes, wing feathers have brown margins, and pale underparts are streaked on flanks and breast. Males show a hint of summer head pattern. **Voice** A thin *seeu* call. Song is simple, chinking and repetitive. **Status** Locally common. In winter, flocks visit arable fields.

Easiest to find in spring.

TOP: **female**; ABOVE: **male**

TOP: **male**; ABOVE: **female**

TOP: **male summer**; ABOVE: **female**

Ortolan Bunting

Emberiza hortulana Emberizidae
Length 15–16cm

Subtly colourful bunting, usually seen here in first-winter plumage. All birds have a diagnostic combination of pale yellow eye-ring, pink bill, and yellow throat and sub-moustachial stripe. **Summer adult male** (Seldom seen here.) Also has a mainly greenish-grey head, neck and breast, orange-brown underparts and streaked brown upperparts. **All other plumages** Also have streaked pale orange-brown underparts, a streaked brown back and a streaked greyish head. **Voice** Calls include a thin *tsee* and tongue-clicking *tche*. **Status** Regular passage migrant, mainly in autumn, to coastal short grassland and stubble fields. Unobtrusive and tricky to find. Portland in Dorset in September can be a good location.

LEFT:
male summer
BELOW: **female**

Snow Bunting

Plectrophenax nivalis Emberizidae
Length 16–17cm

Plump-bodied bunting with extensive white in plumage. Sexes are dissimilar. **Summer adult male** Is mainly white with a blackish back, black on wings, and a black bill and legs. **Summer adult female** Similar but back is brownish and has brown and buff streaking on head, neck and sides of breast. **Winter and first-winter birds** Have mainly white underparts and buffish-orange upperparts; males are whiter than females. Bill is yellowish and legs are black. **Voice** Has a tinkling flight call. Song is twittering. **Status** Scarce breeder in Scottish mountains. Locally common winter visitor to saltmarshes, mainly on E coast. Easiest to find in winter on the N Norfolk coast.

RIGHT: **1st winter**
BELOW: **male summer**

Lapland Bunting

Calcarius lapponicus Emberizidae
Length 14–16cm

Nervous and wary bunting. Sexes are dissimilar. **Summer adult male** (seldom seen here) Has a black face and throat, defined by a white line; crown is black and nape is chestnut. Underparts are mainly white and back is streaked brown and black. Bill is yellow. **Summer adult female** (seldom seen here) Has a pale suggestion of male's head pattern. **Winter adult and juvenile** Have a reddish-brown face with a dark line defining ear coverts; crown is dark and back is streaked brown, and note the reddish-brown wing panel and whitish wingbars. Whitish underparts are streaked on flanks. **Voice** Has a rattling flight call. **Status** Scarce passage migrant and winter visitor to coastal fields and saltmarshes, mainly in E England. Visit N Norfolk in autumn and winter for a chance to see this species.

RIGHT:
male winter

ABOVE: **juvenile**; BELOW: **male summer**

Corn Bunting

Emberiza calandra Emberizidae
Length 16–18cm

Plump-bodied bunting with nondescript plumage but a distinctive song. Dangles legs when flying short distances. Forms flocks in winter. Sexes are similar. **Adult and juvenile** Have streaked brown upperparts and whitish underparts, these streaked on breast and flanks and flushed buff on breast. Bill is stout and pinkish buff. **Voice** A *tsit* call. Jingling song is sung from a fencepost or overhead wire. **Status** Local and declining bird of cereal fields, particularly Barley. Has suffered terribly from modern farming practices. Sadly, now hard to find in many areas. Parts of N Norfolk and the Outer Hebrides are still strongholds.

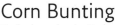

Migration, migrants and vagrants

Although some mammals and a few insects travel considerable distances during their lives, it is birds, with their ability to fly, that have mastered migration. Although some species may move only relatively short distances, many birds undertake lengthy migrations, travelling hundreds or even thousands of kilometres each year.

Migrating Curlews.

Britain and Ireland's seasonal climate has a profound influence on most birds found here. Some species vary their diet according to season, hardy mountain species descend to lower altitudes in bad weather, and others make longer distance movements outside the breeding season, wandering nomadically in search of food. A few switch habitats altogether within Britain between the breeding and non-breeding seasons, but many birds migrate and have different summer and winter grounds. So, some Arctic breeders may winter in Britain, while some British breeders winter in Africa.

Summer visitors

Summer breeding visitors are obvious migrants and many of our most familiar songbirds fall into this category; the majority of warbler species, for example, are with us for only a few brief months in spring and early summer. As a general rule, summer migrant visitors that breed in our region head south in autumn, many wintering in sub-Saharan Africa.

ABOVE: Having spent the winter in West Africa, this newly arrived Willow Warbler is keen to feed.
LEFT: Honey-buzzards spend the winter in sub-Saharan Africa and breed here in small numbers.
RIGHT: Most Snow Buntings seen in Britain are migrant winter visitors from Icelandic breeding grounds.

Winter visitors

Most of our wintering waders and wildfowl breed inside the Arctic Circle as far north as Svalbard or Greenland. Taking advantage of the continuous daylight and brief but plentiful supply of food during the short Arctic summer, they quickly raise their young before returning to our shores in autumn.

Brent Geese breed in the Arctic and winter on British and Irish coasts.

Passing through

Many migrating land birds stop off to feed in Britain for just a day or so. Such birds are called passage migrants. But many seabirds (shearwaters in particular) also visit and pass through our seas. In addition to the Manx Shearwater (*see* p.234), which breeds here, four other species are seen in good numbers. **Balearic Shearwater** *Puffinus mauretanicus* and **Cory's Shearwater** *Calonectris diomedea* visit to feed from the Mediterranean and Atlantic islands, which is quite a journey. More remarkable still are the travels of **Great Shearwater** *P. gravis* and **Sooty Shearwater** *P. griseus*, which breed in the southern Atlantic (on islands off New Zealand and South America); outside the breeding season they undertake a circum-Atlantic journey, passing through European waters mainly from July to September.

Balearic Shearwater
Great Shearwater

Cory's Shearwater
Sooty Shearwater

Migrating Brambling flock.

Swept off course

During periods of southeasterly winds in spring, a few individuals of migrant species that typically breed in southern Europe overshoot their usual breeding grounds and turn up in Britain. Unsurprisingly, south and east coasts receive the most records of these birds.

Easterly winds in autumn sometimes bring a scattering of unusual passerine birds from as far away as Siberia or central Asia. Typically, records relate to juveniles whose migratory instincts have failed them.

Occasionally, westerly gales that sweep across the Atlantic bring migrant North American birds to our shores. Waders – long-distance migrants – turn up regularly, which is perhaps not surprising, but the appearance of tiny songbirds is truly amazing.

Many Arctic birds stay put and endure the rigours of the northern winters, although the worse the weather, the more likely Glaucous and Iceland gulls (*see* p.277) are to turn up in Britain. Some species, such as the White-billed Diver *Gavia adamsii*, are more resilient and turn up only during or just after exceptionally poor weather.

RIGHT: Desert Wheatear *Oenanthe deserti*, usually seen in late autumn.
BELOW: Little Bunting *Emberiza pusilla*, an autumn visitor from Asia.

Citrine Wagtails *Motacilla citreola* nest in central Asia and winter in south Asia. Following strong easterly winds adults (shown here) are rare vagrants in spring while juveniles occur in autumn.

In the autumn of 2006, this Long-billed Murrelet *Brachyramphus perdix* appeared off Dawlish in Devon, the first time this northern Pacific seabird had been seen in Europe.

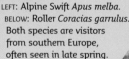

LEFT: Alpine Swift *Apus melba*.
BELOW: Roller *Coracias garrulus*. Both species are visitors from southern Europe, often seen in late spring.

This Spotted Sandpiper *Actitis macularia* spent a winter on the Hayle estuary in Cornwall instead of in the Caribbean.

White-billed Diver in Cornwall 2007.

Migration studies and the BTO

The British Trust for Ornithology (BTO) is the pivotal birding organisation when it comes to understanding the complexities of British birdlife, including many aspects of migration. BTO scientists and an army of passionate volunteers carry out survey work and ringing studies that reveal vital information about bird populations; it is the collaboration of professionals and amateurs that makes the BTO uniquely successful. In recent years, their research has assumed a new significance. Trends in bird populations, detected by survey work, serve as barometers of the state of the environment. Without this information, Government and other decision makers cannot respond in an informed way to environmental issues, notably climate change. So, even if they do not participate directly in survey work, every responsible birder should be a BTO member.

BTO ringing studies revealed the importance of the Sahel as a wintering ground for our Whitethroats. A decline in numbers of birds returning to Britain, demonstrated by surveys undertaken by the BTO, helped alert the world to the environmental significance of droughts being suffered by this region of Africa.

Unusual birds

Most birders enjoy seeing common species, but there is no denying the thrill of the unexpected and to whet your appetite for the unusual here are some birds on the wish lists of many British birders. Those shown here come from far-flung places (E=Europe; A=Asia; NA=North America).

White Stork

Ciconia ciconia Ciconiidae
Length 100-115cm

Large, unmistakable black and white bird. Winters in Africa, breeds in mainland Europe and turns up usually in spring and summer; favours wet grassland. (E)

Night Heron

Nycticorax nycticorax
Ardeidae Length 60-65cm

One of several southern European herons to appear occasionally. Adults are mainly black and white, and juveniles are brown and spotted. (E)

Glossy Ibis

Plegadis falcinellus Threskiornithidae
Length 55-65cm

Wetland bird with heron-like proportions and a Curlew-like bill. Its maroon and metallic plumage

is obvious only in good light. Breeds in S Europe and winters mainly in Africa. (E)

Red-breasted Goose

Branta ruficollis Anatidae
Length 55-60cm

Unmistakable winter visitor with red, black and white plumage. Associates with Brent or White-fronted geese. (E)

Ruddy Shelduck

Tadorna ferruginea Anatidae
Length 61-67cm

Related to Shelduck (*see* p.239) but with orange-brown body plumage and a paler buff head and upper neck. Occurs mainly in SE Europe but wanders. Also kept in captivity from which it escapes occasionally. (E)

American Wigeon

Anas americana Anatidae
Length 48-55cm

Similar to Wigeon (*see* p.240), but male has a green eye-stripe, creamy white forehead and crown, and speckled grey face. (NA)

Ring-necked Duck

Aythya collaris Anatidae
Length 38-45cm

Similar to Tufted Duck (*see* p.242) but with a peaked crown and tricoloured bill. Mainly black and white male has a vertical white line on its flanks and brownish female has a white 'spectacle' around eye. (NA)

Red-footed Falcon

Falco vespertinus Falconidae
Wingspan 65-75cm

Male is mainly sooty grey with thighs and feet red, while female (shown here) has orange-buff crown and underparts, grey back, white cheeks and a dark mask. Seen mostly in spring. (E)

Black-winged Stilt

Himantopus himantopus
Recurvirostridae Length 33-35cm

Unmistakable, with black and white plumage and ridiculously long, red legs. Usually breeds in S Europe and winters in Africa. (E)

Marsh Sandpiper

Tringa stagnatilis Scolopacidae
Length 22–25cm

Elegant Asiatic wader, resembling a miniature Greenshank (*see* p.267) but with thin, needle-like bill. (E)

White-winged Black Tern

Chlidonias leucopterus Sternidae
Length 20–24cm

European species with striking black, grey and white plumage in summer. At other times, mainly grey above, with white rump and underparts, and a dark ear spot. (E)

Subalpine Warbler

Sylvia cantillans Sylviidae
Length 12–13cm

Classic Mediterranean breeder. Well marked and colourful, and most birds have an obvious white 'moustache'. Turns up mainly in spring and autumn. (E)

Lesser Yellowlegs

Tringa flavipes Scolopacidae
Length 23–25cm

Elegant N American wader that recalls an outsized Wood Sandpiper (*see* p.268) but with long, bright yellow legs. (NA)

Ring-billed Gull

Larus delawarensis Laridae
Length 42–48cm

Regular N American gull visitor. Recalls Common Gull (*see* p.275), but adult has larger yellow bill with a bold black ring. (NA)

Great Reed Warbler

Acrocephalus arundinaceus Sylviidae
Length 17–20cm

Recalls Reed Warbler (p.300) but larger, with a stouter bill; favours large reedbeds and has a loud, frog-like song. (E)

Pectoral Sandpiper

Calidris melanotos Scolopacidae
Length 19–22cm

Autumn-visiting N American wader with yellow legs, white stripes on its back, and a clear demarcation between the clean white underparts and streaked breast. (NA)

Short-toed Lark

Calandrella brachydactyla Alaudidae
Length 15–16cm

Dumpy lark with a stubby bill, pale underparts and a dark patch on sides of breast. Feeds unobtrusively. Winters in Africa and breeds in S Europe. (E)

Richard's Pipit

Anthus richardi Motacillidae
Length 17–20cm

Larger than Meadow Pipit (*see* p.292) with much longer tail, bill, claws and legs. An Asian species and long distance migrant that turns up here in autumn; favours open grassland.(A)

Rose-coloured Starling

Sturnus roseus Sturnidae
Length 20–22cm

Mainly Asiatic species, sometime found with Starlings (*see* p.311). Adult has unmistakable pink and dark plumage. Juvenile is buffish grey with a yellow bill. (A)

Rose-coloured Starling

REPTILES and AMPHIBIANS

Although comparatively few reptile and amphibian species occur in Britain and Ireland, those that do live here lead intriguing lives. Being rather secretive and seasonal, most are tricky to observe, but for many enthusiasts the challenge only adds to the appeal of these fascinating creatures.

Common Toad.

What is an amphibian?

Amphibians have soft skins, making them vulnerable to damage and attacks by predators. Water loss can also be a problem, so most species are tied to damp environments and combat desiccation by secreting mucus. On the plus side, amphibians can absorb oxygen directly through their skin, whose colour they can also modify to match their surroundings – a useful ploy when it comes to avoiding detection by predators.

Amphibians cannot generate internal body heat like birds and mammals, so the environment around them determines their body temperature, although their behaviour ensures they avoid excessive heat and cold.

Reproduction

Amphibians reproduce in water through a process that is probably easiest to observe with the **Common Frog**. Layers of jelly surround the eggs, which, when laid, swell on contact with water to produce a protective, transparent bubble through which oxygen can enter and waste products exit. Inside, the developing embryo is nourished by a yolk sac and soon comes to resemble a dark 'comma'. After a couple of weeks or so, a tiny tadpole, with stubby external gills, hatches. Nourished by a diet of grazed algae and detritus, the tadpole grows, its external gills are absorbed and its tail lengthens, allowing it to become free-swimming. External legs then appear – the hind pair followed by the front pair – and the diet begins to incorporate more animal material. As larval development reaches its final stages, the tail shrinks and is absorbed, while the body shape itself metamorphoses into something resembling a miniature adult frog. By the time the transformed tadpole leaves the water (usually three to four months after spawning), the tail has all but disappeared and the froglets resemble miniature adults.

Smooth Newts, like others of their kind, wrap individual eggs in the leaf of a water plant for protection.

mating frogs

fertilised egg

tadpole developing inside egg

tadpole with both pairs of legs

newly metamorphosed froglet

newly hatched tadpole

tadpole with hind legs appearing

active, free-swimming tadpole

Reproduction in the Common Frog

What is a reptile?

A reptile's skin has a tough outer layer, which in snakes and lizards takes the form of scales. These overlap and are embedded in softer skin, which allows flexibility and speedy movement. Lizards with external legs can run at speed, while legless lizards and snakes employ muscular, S-shaped movements of the body to achieve forward propulsion.

The tough layer of skin provides reptiles with protection from predators and from physical damage. It also helps reduce water loss and, as a consequence, many species are able to survive in comparatively dry habitats. Snakes and lizards periodically shed or 'slough' the outer layer of their skin. While our terrestrial reptiles are unable to generate their own body heat, they are adept at gaining radiated heat from the sun and heat absorbed from the ground they are resting on.

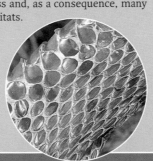

Periodically, reptiles shed their skin; the scales are clearly visible in this sloughed Grass Snake skin.

Smooth Snake.

Reptile reproduction

The developmental stages of reptiles occur inside an egg with an opaque shell and so, unlike amphibians, observing the processes that go on internally is impossible. In a few reptiles, the eggs develop inside the body of the mother, further compounding problems of observation. However, most species (such as the Grass Snake, shown right) lay eggs and typically these are deposited, and incubate, in the ground. Egg-laying reptiles are extremely selective and choosy about where they lay their eggs. Free-draining sites, where the dangers of predation are minimised, are important and the eggs can take a couple of months for development to proceed to the point where the young hatch.

A clutch of Grass Snake eggs incubating in a compost heap.

Reptile-watching

British reptiles are often hard to observe, partly because they are rather secretive but also because they hibernate for a third of the year or more. Understanding their lives and biology can greatly improve the chances of getting more than a fleeting glimpse of a disappearing tail. Most reptiles emerge from hibernation in March or April, and on sunny days they spend long periods basking. So, in spring it is generally easy to locate suitable sheltered, south-facing spots where Common Lizards or Adders like to bask. They are usually easier to approach at this time than later in the season, when they are far more active and inclined to avoid human encounters.

On dull days, reptiles will seek refuge under natural fallen objects such as tree trunks, as well as man-made artefacts such as sheets of iron or even pieces of old carpet. Herpetologists (reptile specialists), therefore, sometimes deliberately introduce foreign objects to natural environments for the benefit of reptiles. If you are in a position to do this, only do it if there is no chance that the unwelcome attentions of other people will be attracted, and be aware of the fine line between benefiting reptiles and fly-tipping. You must always obtain the permission of the landowner.

ABOVE: Reptile conservationists often place sheets of corrugated plastic or iron, or roofing felt, in natural habitats in order to create refuges for reptiles.

BELOW: All British reptiles slough their skin regularly and their discovery provides information about numbers and animal size. The skin shown here belonged to a 60cm-long Grass Snake.

Like other British reptiles, Adders hibernate in winter but emerge from the beginning of March onwards to bask in the sunshine. It is not uncommon to see several animals entwined.

Reptiles

Once much more widespread in Britain, reptiles are now threatened and vulnerable to habitat loss. Our reptile fauna is divided into three orders: Sauria (lizards), Serpentia (snakes) and Chelonia (turtles and terrapins). Lizards can close their eyelids, while snakes cannot. All British species have carnivorous diets.

Common Lizard

Lacerta vivipara Lacertidae
Length 10–15cm

Our most widespread legged lizard. Fond of sunbathing. Hibernates Oct–Apr. Gives birth to live young. **Adult** Has a rather slender body with an angular, pointed snout. Ground colour is variable but brown is usual. From above, note the vertebral row of dark spots or patches, and parallel rows of dark markings on flanks, bordered above by pale spots. Some have a green or reddish flush to head. Mature male has bright yellow or orange underparts studded with dark spots. **Juvenile** Resembles a miniature adult but with a relatively much shorter tail; uniformly black for first few weeks of life. **Voice** Silent. **Status** Has declined markedly owing to habitat loss but still very locally common in open habitats, notably heathland.

Sand Lizard

Lacerta agilis Lacertidae
Length 16–19cm

Bulky lizard. Threatened owing to precise habitat requirements; protected by law. Hibernates Sep–Mar. Females lay eggs. Sexes are dissimilar. **Adult male** Has a relatively large head. Ground colour of back is typically buffish brown but note the 3 rows of white-centred dark spots along length of body. Head and flanks are flushed bright green in spring and summer. **Adult female** Has a smaller head and more bulky body than male. Ground colour is pale to rich brown with 3 longitudinal rows of eye-spots. **Juvenile** Recalls a small, slender adult female with eye-spots on back and flanks. **Voice** Silent. **Status** Restricted to heathland sites in Dorset and Surrey, and coastal sand dunes in Merseyside.

LEFT: Common Lizard

BELOW: Sand Lizard male (on top) and female courtship

Unusual lizards

A couple of mainland European lizards have a toehold in our region. The **Wall Lizard** *Podarcis muralis* (length 14–17cm) has a longer tail than a Common Lizard. It is brown with a variably complete dark stripe down the back, and variable stripes and marbling on its flanks. Wall Lizards are native to Jersey and have been introduced to the Isle of Portland, the Isle of Wight and south Hampshire coast. The **Green Lizard** *Lacerta viridis* (length 30–40cm), a large-headed, long-tailed green species is native to Jersey and has been introduced to Guernsey, the Isle of Wight and south Hampshire.

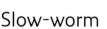

Wall Lizard

Slow-worm rear end is blunter after tail has been shed.

Slow-worm

Anguis fragilis Anguidae
Length 30–40cm

Rather snake-like legless lizard. Hibernates Oct–Mar. Length comprises 50–60% tail but this can be shed in dire distress. Female gives birth to live young. Sexes are similar. **Adult male** Has a slender, shiny body and no discernible 'neck'. Usually coppery brown or greyish brown, sometimes with darker markings on head. **Adult female** Similar but with a thin, dark vertebral stripe along back (may end in a faint 'V' at anterior end) and a broken black line on flanks. Belly is marbled bluish and some animals have blue spots on sides. **Juvenile** Has a golden or silvery back with a thin, dark vertebral stripe and dark flanks. **Voice** Silent. **Status** Favours sunny, open habitats with areas of grass and Bramble. Undisturbed grassland, heathland and railway embankments are ideal. Quickly exterminated by domestic cats.

Turtle visitor

Although turtles are usually thought of as tropical animals, **Leatherback Turtles** *Dermochelys coriacea* (length 1.5–2.5m) turn up regularly enough in our seas to be considered part of our native fauna. Unlike our terrestrial reptiles, they can generate body heat and hence remain active in cold waters. Leatherbacks are gigantic and unmistakable, with a ridged shell and huge front flippers; they feed mainly on jellyfish.

Grass Snake

Natrix natrix Colubridae
Length 60–90cm

Large, non-venomous snake. Hibernates Oct–Apr. Hunts on land but also active in water, feeding on frogs and fish. Female lays eggs, often in composting piles of vegetation. Sexes are similar. **Adult** Has a slender body, thickest towards middle and evenly tapering towards tail. Ground colour on upperparts is olive-green; has occasional dark vertical stripes on flanks and a double row of indistinct dark spots down back. Neck has black and yellow crescent-shaped markings on sides, forming an incomplete collar. Scales on undersurface are whitish with dark chequering. Has backward-curved teeth that retain struggling prey. Eye has a round pupil. **Juvenile** Has relatively larger head than adult. **Voice** Hisses if distressed. **Status** Locally common in grassland and heathland, usually near water. Sunbathes or rests under metal sheets.

LEFT: If attacked by a predator, a Grass Snake will sometimes pretend to be dead.

young animal

Did you know?

Reptiles and amphibians might slip off the conservation radar if not for dedicated enthusiasts. The Herpetological Conservation Trust (HCT) is their most significant advocate and manages reserves for endangered species. For more information, visit the HCT website or become a volunteer for your local branch of the Amphibian and Reptile Group (ARG-UK; *see* p. 466).

Smooth Snake

Coronella austriaca Colubridae
Length 50–70cm

Rare snake with vaguely Adder-like markings. Pupil is rounded (vertical slit in Adder). Sunbathes but typically partially hidden. Hibernates Oct–Apr. Diet includes other reptiles. Sexes are similar. **Adult** Has a slender body but relatively large head; scales are satiny smooth. Overall ground colour ranges from bluish grey to reddish brown. Has darkish spots along its length, sometimes combining to form Adder-like zigzag. Head has a dark patch; posterior margin is sometimes rather V-shaped. From side, note the dark eye-stripe. **Juvenile** Similar but has spots along flanks. **Voice** Mostly silent. **Status** Restricted to mature heathland with bushy Heather (*Calluna vulgaris*). Rare because of habitat loss, inappropriate heathland management and fires.

Adder

Vipera berus Viperidae
Length 45–60cm

Well-marked snake with a vertical pupil and red iris. Britain's only venomous reptile. Hibernates Oct–Mar and sunbathes in spring. Males perform wrestling court-ship 'dances'. Sexes are similar but females are larger than males. **Adult** Ground colour ranges from reddish brown, greenish yellow or grey to creamy buff. Most have a blackish zigzag line along back, the anterior end of which looks arrow-headed and framed by an inverted 'V' on head. Melanic 'Black Adders' also occur; commonest in the N. **Juvenile** Similar but slender and usually reddish brown. **Voice** Mostly silent. **Status** Widespread but local on heaths, moors, open woodlands and coastal dunes.

RIGHT: Black Adder

Identifying snakes

To the untrained eye, snakes can appear similar to one another. A closer look at the shape and markings on the head will provide you with vital clues. The Slow-worm (a legless lizard) is shown here too.

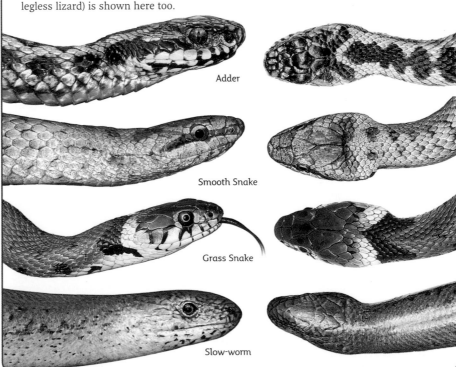

Adder

Smooth Snake

Grass Snake

Slow-worm

Amphibians

Amphibians need aquatic habitats in the breeding season for egg-laying and the development of their young. British amphibians comprise newts (belonging to the division Urodela, family Salamandridae) and the division Anura, containing frogs (family Ranidae) and toads (family Bufonidae).

Smooth Newt

Triturus vulgaris Salamandridae
Length 9–10cm

Widespread newt. Usually hibernates underground Nov-Feb. Returns to breeding ponds in early spring. Performs elaborate courtship displays prior to egg-laying. **Adult** Buffish brown with a whitish throat that is variably dark-spotted. Females have small black spots over rest of body that sometimes coalesce to form lines along body and onto tail. Males are marked with large, blotchy spots, most intense in the breeding season, when they acquire a wavy crest along back and around tail; belly is variably flushed orange and adorned with dark spots. **Juvenile** Similar to adult female. **Voice** Silent. **Status** Favours fish-free neutral to alkaline ponds for breeding. At other times, found in open woodland, commons and mature gardens.

Palmate Newt

Triturus helveticus Salamandridae
Length 8–9cm

Generally our smallest newt. Male is distinctive in the breeding season. Female is similar to female Smooth Newt at all times. **Adult** Has yellowish belly and pinkish, unspotted throat at all times. Note the hint of a pale vertical stripe above hind legs. Breeding male develops diagnostic palmations between toes on hind feet and thin filament projecting from tip of blunt tail. Body is olive-brown with dark marbling; an orange-buff band extends along flanks and side of tail. Colours are duller at other times but retains a dark eye-stripe. Female is yellowish brown. Unspotted throat allows separation from female Smooth Newt (throat is spotted). **Juvenile** Resembles adult female. **Voice** Mostly silent. **Status** Locally common in neutral to acid ponds in the breeding season, often on heaths and moors.

Great Crested Newt

Triturus cristatus Salamandridae
Length 14–16cm

Our largest newt. Present in breeding ponds Feb-Aug. **Adult male breeding** Mainly blackish brown with variable dark spots and patches, and white-tipped warts. Underparts, from neck to vent, are orange-yellow with black spots; blackish throat has smallish orange-yellow spots. Has a large, jagged dorsal crest and undulating crest on tail; a pale stripe runs along centre of tail. **Male non-breeding** Similar but lacks crest. **Female** Similar to non-breeding male but body is darker and note the yellowish stripe along lower edge of tail. **Juvenile** Similar to female but with a paler body. **Voice** Silent. **Status** Scarce and local. For breeding, favours neutral to slightly alkaline, fish-free ponds that seldom dry up. Woods and scrub are used at other times. Protected by law.

TOP: male; BOTTOM: female

TOP: male; MIDDLE: female; BOTTOM: hibernating adult

TOP: larva; MIDDLE: male; BOTTOM: female

Great Crested Newt

Triturus cristatus Salamandridae
Length 14–16cm

Our largest newt. Present in breeding ponds Feb–Aug. **Adult male breeding** Mainly blackish brown with variable dark spots and patches, and white-tipped warts. Underparts, from neck to vent, are orange-yellow with black spots; blackish throat has smallish orange-yellow spots. Has a large, jagged dorsal crest and undulating crest on tail; a pale stripe runs along centre of tail. **Male non-breeding** Similar but lacks crest. **Female** Similar to non-breeding male but body is darker and note the yellowish stripe along lower edge of tail. **Juvenile** Similar to female but with a paler body. **Voice** Silent. **Status** Scarce and local. For breeding, favours neutral to slightly alkaline, fish-free ponds that seldom dry up. Woods and scrub are used at other times. Protected by law.

INSET: froglet

A sad case

The **Pool Frog** *Rana lessonae* (length 5–6.5cm) has a confused past and an uncertain future. Only recently recognised as a genuine species this native frog is now extinct. The conservation world only awoke to its plight after it was too late, but introductions of Swedish Pool Frogs may allow its return. It is well-marked with dark spots and a pale vertebral stripe. Males utter a duck-like quacking *ou-Whack* when courting.

Common Frog

Rana temporaria Ranidae
Length 6–10cm

Our most familiar amphibian. **Adult male** Has smooth, moist Skin, olive-yellow or greyish brown with variable dark markings. Darker red animals occur in uplands. Eye has a yellow iris with a dark, oval pupil. Dark mask runs from eye to eardrum. Underparts are greyish white with faint darker marbling. Hind feet have 5 webbed toes. In breeding season, has a bluish throat and swollen nuptial pads on innermost digit of front feet, used to grip female when mating. **Adult female** Similar but larger with white granulations on flanks. **Juvenile** When newly metamorphosed resembles tiny, large-headed adult. **Voice** Courting male utters low-pitched croaking calls. **Status** Found in a wide range of watery habitats. **Similar species** The **Agile Frog** *R. dalmatina* is restricted to Jersey.

ABOVE: toadlet; BELOW: spawn

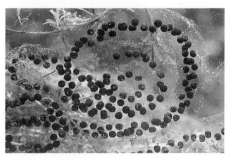

Common Toad

Bufo bufo Bufonidae
Length 5–9cm

Our commonest toad. Skin is covered in toxin-containing warts. Spawn is laid in double-rowed spawn strings. Gait consists of short hops. **Adult** Olive-brown to greenish buff (hue is influenced by ambient light). Has a red iris and webbed hind feet. Female is larger than male. **Juvenile** Recalls a tiny, large-headed adult. **Voice** Courting male utters croaking calls in spring. **Status** More terrestrial than most other amphibians, with adults spending much of their lives on land: woodland, scrub, grassland and moors are favoured. Must return to water to breed (Feb–Mar) and most are found within 2km or so of suitable ponds.

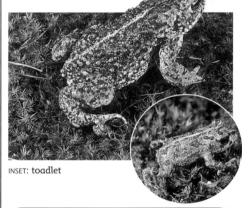

INSET: toadlet

Established aliens

Two introduced frog species are found in SE England. The **Marsh Frog** *Rana ridibunda* (**1**) (length 11–13cm) is large and green with warts and dark spots; when singing (a croaking *Whoa-aa-aa-aa*) a male shows grey vocal sacs. Marsh Frogs are tied to water and are found in coastal marshes in Sussex and Kent. The **Edible Frog** *Rana × esculenta* (**2**) (length 7–9cm) is long-legged, bright green or yellowish brown with dark spots and a pale greenish vertebral stripe. It is a hybrid between Pool and Marsh Frogs and is now restricted to a few garden ponds.

1

2

TERRESTRIAL INVERTEBRATES

Pardosa wolf spider female carrying egg sac.

In the animal kingdom, the most profound division is between vertebrates (animals with backbones) and invertebrates (animals without backbones). Invertebrates that live on land are represented by three main phyla: Annelida (segmented worms); Mollusca (snails and slugs); and Arthropoda (a large and complex group that includes most other terrestrial invertebrates). Arthropods include insects (class Insecta); centipedes (class Chilopoda); millipedes (class Diplopoda); crustaceans – including creatures like woodlice (class Crustacea); and spiders and allies, including false scorpions, ticks, mites and harvestmen (class Arachnida).

Peacock butterfly.

Annelid worms

Annelid worms have soft, segmented bodies and often have bristles to aid their movement. Most terrestrial annelid worms live in burrows in the soil, or among leaf litter. Many of their marine cousins (*see* p.464) have elaborate tentacles and external gills, while leeches (*see* p.440) have suckers.

Common Earthworm.

Molluscs

Molluscs are animals with soft bodies that are divided into three main regions: an obvious head; the main body mass, containing the organs; and a foot for movement. The group contains many freshwater and marine representatives (*see* pp.440–1 and 454–7, respectively), but on land the group is represented by slugs and snails, the latter with a hard shell into which they can retreat to protect themselves from predators and to help avoid losing water in dry conditions.

White-lipped Snail.

Leopard Slug.

Arthropods

Arthropods are invertebrate animals with an external skeleton and paired, jointed limbs. Muscles allowing movement of the limbs and body are internal, attached to the walls of the hardened exoskeleton. Terrestrial representatives include the following:

Crustaceans

Most crustaceans are aquatic, but woodlice are terrestrial examples. They prefer damp environments and have segmented bodies with a hard dorsal carapace and seven pairs of walking legs.

Millipedes

Superficially similar to centipedes but the elongate, segmented body has two pairs of legs per segment. Millipedes feed on plant tissue and detritus, and are relatively slow-moving.

Centipedes

These are elongate, segmented arthropods. Each body segment has a pair of legs. Centipedes are predators bearing a pair of sharp poison claws at the mouth.

Arachnids

Members of this varied group have four pairs of walking legs – this distinguishes them from insects, the adults of which have three pairs of legs. Their bodies are divided into three obvious sections.

Flat-backed Millipede.

A centipede, *Lithobius variegatus*.

House Spider.

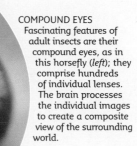

COMPOUND EYES
Fascinating features of adult insects are their compound eyes, as in this horsefly (*left*); they comprise hundreds of individual lenses. The brain processes the individual images to create a composite view of the surrounding world.

Insects

Visually extremely diverse group, many insects are highly colourful. An adult insect's body is divided into three main regions: the head, which supports many sensory organs as well as the mouthparts; the abdomen, to which three pairs of legs and paired wings are attached; and the segmented abdomen, within which many of the main body organs are contained. Insects' dietary habits range from the strictly vegetarian to voraciously predatory. *See* pages (pp.334–5) for more about terrestrial insects.

Large Marsh Grasshopper.

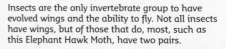

Insects are the only invertebrate group to have evolved wings and the ability to fly. Not all insects have wings, but of those that do, most, such as this Elephant Hawk Moth, have two pairs.

Introducing terrestrial insects

Insects are among the most numerous organisms on the planet and also the most diverse, the million-plus species known worldwide being found in almost all terrestrial and freshwater habitats; Britain and Ireland alone can probably boast more than 100,000 species. As a group, insects are divided into numerous subdivisions or orders, 18 of which are covered in this book; members of the remaining orders are either small, obscure, or both, and are seldom noticed by the average amateur naturalist. The more regularly found orders are as follows:

Insect measurements

Some insects typically rest with their wings spread out, while in others the wings are either held lengthways along the body or are not visible at rest. Measurements given for the species listed are wingspan in the case of the former and length in the case of the latter.

Lepidoptera
(butterflies and moths)

Attractive insects, the adults of which have wings covered in scales, allowing extra-ordinary colours and patterns to appear; the mouthparts are modified to form a long, sucking proboscis (*see* pp.336–73).

Marsh Fritillary

Thysanura (bristletails)

Flattened, elongate and wingless insects with bristly tail appendages (*see* p.374).

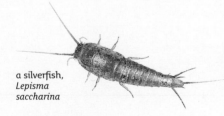

a silverfish,
Lepisma saccharina

Collembola (springtails)

Tiny, squat and wingless insects with a springing device at the rear end (*see* p. 374).

a springtail,
Tomocerus longicornis

Ephemeroptera
(mayflies)

Clear-winged, delicate and short-lived as adults, with two or three long tail appendages; the nymphs are aquatic (*see* pp.374–5).

a mayfly,
Ephemera vulgata

Plecoptera (stoneflies)

Flattened-bodied insects as adults, with weak flight or essentially flightless; the nymphs are aquatic (*see* p.375).

a stonefly, *Nemoura cinerea*

Phasmida
(stick insects)

Slender-bodied insects, all stages of which are slow-moving and closely resemble the twigs and foliage among which they live (*see* p.375).

Prickly Stick-insect

Orthoptera (grasshoppers, crickets and bush-crickets)

Robust insects with rather limited powers of flight as adults but hind legs that allow them to hop well (*see* pp.376–9).

Meadow Grasshopper

Odonata (dragonflies and damselflies)

Mainly clear-winged as adults, with a slender body and predatory habits; the nymphs are aquatic (*see* pp.380–5).

Ruddy Darter

Hemiptera (true bugs)

An extremely varied group of insects with sucking mouthparts; the adults of many species have wings, these either folded flat over the body or held tent-wise (*see* pp.386–9).

Hawthorn Shield Bug

Megaloptera (alderflies)

Superficially similar to lacewings, adults are waterside insects that have smoky wings and feed on pollen. The larvae are carnivorous and aquatic.

Alderfly

Dermaptera (earwigs)

Flattened-bodied insects with pincer-like appendages at the tail end; their forewings are modified to form protective plates, while the hind wings are either reduced or absent (see p.390).

Common Earwig

Dictyoptera (cockroaches)

Flattened insects with a scuttling gait, long antennae and spiny legs; the head is covered dorsally by a protective plate (see p.390).

Dusky Cockroach

Neuroptera
(lacewings and allies)

Slender-bodied insects with proportionately large wings as adults, these held in a tent-wise manner at rest (see pp.390–1).

Common Lacewing

Mecoptera (scorpion flies)

Peculiar, slender-bodied insects, the adults of which have a snout-like projection on the head; in males of some species the abdomen is curved upwards in a scorpion-like fashion (see p.391).

Scorpion Fly

Trichoptera (caddis flies)

Delicate insects whose wings are coated with fine hairs and are held in a tent-like manner at rest; the larvae are aquatic and construct elaborate shelters (see p.391).

a caddis fly, *Potamophylax cingulatus*

Diptera (true flies)

A varied and diverse group, the adults of which have one pair of wings (the hind wings are reduced to tiny, club-shaped stabilising organs); adults have mainly sucking mouthparts (see pp.392–5).

a hoverfly, *Episyrphus balteatus*

Hymenoptera (bees, wasps, ants and allies)

An extremely variable group, the adults of which have two pairs of membranous wings; some species live communal lives and a few have powerful stings (see pp.396–9).

Giant Wood-wasp

Coleoptera (beetles)

A diverse group, the adults of which have biting mouthparts and whose front pair of wings is modified to form protective plates called elytra; hind wings, when present, allow flight and are often folded beneath the elytra at rest (see pp.400–5).

Lesser Stag Beetle

Minor orders

There are several groups of tiny insects, sometimes collectively referred to as minor orders, that are not covered in this book. More noticeable examples include Psocoptera (psocids and booklice), Anoplura (sucking lice) and Siphonaptera (fleas).

ABOVE LEFT: Psocid
ABOVE: Human Head Louse
LEFT: a flea, *Ctenopthalmus nobilis*

Insect life cycles

Insect life starts with an egg laid by a female. The growing stage hatches and its rigid skin is moulted several times as it grows; each successive step is called an instar. Insect life cycles progress either by:

1. Incomplete metamorphosis, where the growing stage (nymph) becomes gradually more adult-like with each instar.
2. Complete metamorphosis, where a larva hatches from the egg and stays larva-like during the growing period, simply getting bigger each instar. The full grown larva metamorphoses into a pupa, an inactive stage inside which the adult forms and eventually hatches out.

1. Incomplete metamorphosis in the Great Green Bush-cricket. The nymph (left) increases in size and looks more like an adult (right) with each instar.

2. Complete metamorphosis in the Brown Hairstreak. The four stages in the life cycle (from left to right: egg, larva, pupa and adult) look very different from one another.

Whites & Swallowtail

Butterflies with white or extremely pale wings, belonging to the family Pieridae, are referred to simply as 'whites', but this family also includes more colourful species: Brimstone, Clouded Yellow and Orange-tip. The Swallowtail, our most spectacular butterfly, belongs to the family Papilionidae.

Swallowtail

Papilio machaon ssp. *britannicus*
Papilionidae Wingspan 70mm

Large and unmistakable. **Adult** Has mainly yellow and black wings; note the tail streamers on hind wings. Flies May-Jun and again in Aug. **Larva** Colourful; feeds on Milk-parsley. **Status** Rare. Now confined to a few fens and marshes in East Anglia, mainly in the Norfolk Broads; easiest to see at Hickling Broad.

upperwings

underwings

Large White

Pieris brassicae Pieridae
Wingspan 60mm

Our largest 'white' butterfly. **Adult** Has yellowish underwings. Upperwings are creamy white with a black tip to forewing; female also has 2 spots on forewing. Flies May–Sep. **Larva** Black and yellow; feeds on cabbages and other garden brassicas. **Status** Common and widespread.

Small White

Pieris rapae Pieridae
Wingspan 45mm

Smaller than similar Large White. One of the most common and familiar garden butterflies. **Adult** Has yellowish under-wings; upperwings are creamy white with a dark tip to fore-wing; female has 2 dark spots on forewing. Flies Apr-May and Jul-Aug. **Larva** Feeds on cabbages and other brassica species, and if unchecked they can devastate whole cabbage crops in the garden. **Status** Common and widespread throughout Britain and Ireland, although the species can be relatively scarce in some seasons, mainly if the summer weather is unsuitable.

Green-veined White

Pieris napi Pieridae
Wingspan 45-50mm

Wayside butterfly. **Adult** Similar to Small White but veins on upperwings are dark and those on underwings are greyish green, particularly on hind wing. Double-brooded; flies in spring and midsummer. **Larva** Feeds on Garlic Mustard and other wild crucifers. **Status** Locally common.

Orange-tip

Anthocharis cardamines Pieridae
Wingspan 40mm

Attractive spring butterfly. **Adult** Has rounded wings. Forewing is dark-tipped but only male has an orange patch; hind underwing of both sexes is marbled green and white. Flies Apr–Jun. **Larva** Feeds mainly on Cuckooflower. **Status** Widespread in S Britain and Ireland.

RIGHT: male upperwings
FAR RIGHT: underwings

Wood White

Leptidea sinapis Pieridae
Wingspan 40mm

Delicate-looking butterfly with rather feeble flight. **Adult** Has rounded whitish wings; forewings are dark-tipped, most noticeably on upper surface. Flies May–Jul in 2 broods. **Larva** Feeds on plants of the pea family. **Status** Very local in S and SW England; replaced in S Ireland by the very similar Réal's Wood White *L. reali*.

Clouded Yellow

Colias croceus Pieridae
Wingspan 50mm

Fast-flying, active butterfly. **Adult** Has dark-bordered upperwings that are yellow in female and orange-yellow in male. Both sexes have yellow underwings with few dark markings. **Larva** Feeds on Lucerne and other members of the pea family. **Status** Summer migrant in variable numbers. Sometimes breeds but does not survive winter.

Brimstone

Gonepteryx rhamni Pieridae
Wingspan 60mm

Herald of spring. **Adult** Has uniquely shaped wings. Male's brimstone-yellow colour is unmistakable; paler female can be mistaken for Large White in flight. Single-brooded – summer adults hibernate and emerge on sunny spring days. **Larva** Feeds on Buckthorn and Alder Buckthorn. **Status** Locally common.

TOP: male
ABOVE: newly emerged female

Watching for migrants

Keep an eye on the weather, because warm southerly and southeasterly winds in summer and early autumn often bring with them migrant butterflies from mainland Europe. Species such as Painted Lady, Red Admiral and Clouded Yellow are regular visitors, best looked for around the south coasts of England and Ireland. In good years, more unusual species sometimes turn up, such as the prized **Pale Clouded Yellow** *Colias hyale* shown here.

Nymphalid butterflies

Members of the family Nymphalidae – popularly referred to as nymphalids – are medium to large butterflies, many of which have brilliantly colourful upperwings; their underwings are normally drab in comparison. Some are common wayside and meadow species, while others are associated with undisturbed woodland.

Red Admiral

Vanessa atalanta Nymphalidae
Wingspan 60mm

 Stunning and familiar, sun-loving wayside butterfly. **Adult** Has marbled smoky-grey underwings and black upperwings with red bands and white spots. Seen in many months but commonest Jul–Aug. **Larva** Feeds on Common Nettle. **Status** Adults hibernate in small numbers, but it is seen mostly as a summer migrant, often in good numbers. Influxes arrive from mainland Europe and subsequently breed here.

Small Tortoiseshell

Nymphalis urticae Nymphalidae
Wingspan 42mm

 Familiar wayside species. Sun-loving. **Adult** Upperwings are marbled orange, yellow and black; underwings are smoky brown. Flies Mar–Oct with 2 or 3 broods. **Larva** Gregarious; feeds on Common Nettle. **Status** Fairly common and widespread, but much declined in recent years.

Painted Lady

Vanessa cardui Nymphalidae
Wingspan 60mm

 Subtly attractive butterfly. **Adult** Upperwings are marbled pinkish buff, white and black; underwings are buffish with a similar pattern to upperwing. **Larva** Feeds on thistles. **Status** Summer migrant in variable numbers; most numerous near coasts. Sometimes breeds but does not survive winter.

Hibernation

While many butterflies are short-lived and ephemeral, several members of the nymphalid family live for many months and indeed overwinter in a state of hibernation: Peacocks, Small Tortoiseshells and Commas all adopt this strategy. Typically, dense Ivy foliage or a tree hollow is chosen for hibernation, although Small Tortoiseshells sometimes use garden sheds or roof spaces.

Small Tortoiseshells clustering together in hibernation.

underwings

Woodland icon

The **Purple Emperor** *Apatura iris* (wingspan 65mm) is a magnificent and iconic butterfly of well-managed woodlands with tall oaks and mature Goat Willow (the larval food plant). The adult has brown upperwings that are marked with a white band; only the male has the purple sheen, seen at certain angles. The underwings of both sexes are chestnut. The Purple Emperor flies in July and August and is restricted to central southern England. Nowadays, it is rare and extremely local.

LEFT: male
BELOW: female
underwings

Peacock

Nymphalis io Nymphalidae
Wingspan 60mm

Distinctive and familiar visitor to garden flowers. **Adult** Has smoky-brown underwings and maroon upperwings with bold eye markings. Flies Jul–Sep and again in spring after hibernation. **Larva** Feeds on Common Nettle. **Status** Common and widespread except in the far N.

White Admiral

Limenitis camilla Nymphalidae
Wingspan 50mm

Superb flyer. Visits Bramble flowers along woodland rides. **Adult** Has sooty-black upperwings with white bands and chestnut underwings with a similar pattern of white to upperwings. Flies Jun–Jul. **Larva** Feeds on Honeysuckle. **Status** Locally common woodland species in S England.

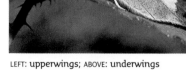

LEFT: **upperwings**; ABOVE: **underwings**

Comma

Polygonia c-album Nymphalidae
Wingspan 45mm

Attractive butterfly with distinctive, ragged-edged wings. **Adult** Has smoky-brown underwings with a white 'comma' mark; upperwings are orange-brown with dark markings. Double-brooded and hibernates; flies Mar–Sep. **Larva** Feeds on Common Nettle, elm and Hop. **Status** Locally fairly common.

RIGHT: **upperwings**
FAR RIGHT: **underwings**

Fritillaries

Fritillary butterflies are attractive species whose mainly orange wings are adorned with dark spots and chequerboard markings, the precise patterns of which are a help in identification. All but one are members of the family Nymphalidae; the Duke of Burgundy is the exception and belongs to the unrelated family Riodinidae.

ABOVE: underwings
BELOW: upperwings

Pearl-bordered Fritillary

Boloria euphrosyne Nymphalidae
Wingspan 42mm

 Sun-loving butterfly of woodland glades. **Adult** Has orange-brown upperwings; underside of hind wing has 7 silver spots on margin and 2 in centre. Flies May-Jun. **Larva** Feeds on violets. **Status** Widespread but local in the British Isles, mostly in the S; very local in W Ireland.

Dark Green Fritillary

Argynnis aglaja Nymphalidae
Wingspan 60mm

 Fast and powerful flyer, visiting thistles and knapweeds. **Adult** Has orange-brown upperwings; underside of hind wing has greenish scaling. Flies Jul-Aug. **Larva** Feeds on violets. **Status** Widespread but local on sand dunes and downs in Britain and Ireland.

Small Pearl-bordered Fritillary

Boloria selene Nymphalidae
Wingspan 40mm

Similar to Pearl-bordered. **Adult** Best separated by looking at underwing: shares Pearl-bordered's 7 silver marginal spots but has several silver central spots (2 in Pearl-bordered). Flies in Jun. **Larva** Black and hairy; feeds on violets. **Status** Local in woods and grassland where the larval food plants are common.

ABOVE: underwings; BELOW: upperwings

TOP: upperwings; ABOVE: underwings

High Brown Fritillary

Argynnis adippe Nymphalidae
Wingspan 60mm

Fast-flying open country species. **Adult** Has orange-brown upperwings with dark spots; underside of hind wing has brownish scaling. Flies Jul-Aug. **Larva** Feeds on violets. **Status** Scarce and rather endangered, found mainly in W and NW England; favours meadows and open, grassy woodlands.

BELOW: **upperwings**
INSET: **underwings**

Marsh Fritillary

Euphydryas aurinia Nymphalidae
Wingspan 40-50mm

Slow-flying butterfly, active only in sunshine. **Adult** Wings are beautifully marked with a mosaic of reddish orange, buff and yellow; upperwings are more colourful than underwings. Flies May-Jun. **Larva** Feeds on Devil's-bit Scabious and plantains. **Status** Local on heaths and chalk grassland.

BELOW: **upperwings**
INSET: **underwings**

Heath Fritillary

Melitaea athalia Nymphalidae
Wingspan 45mm

Small, slow-flying species. **Adult** Has well-marked dark orange-brown upperwings; underwings are creamy white and chestnut. Flies Jun-Jul. **Larva** Feeds on Common Cow-wheat, Wood Sage or plantains. **Status** Rare and local, mainly in SE and SW England; usually seen in sunny woodland rides.

BELOW: **upperwings**
INSET: **underwings**

Silver-washed Fritillary

Argynnis paphia
Nymphalidae
Wingspan 60mm

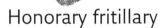

Sun-loving species; adults are fond of Bramble flowers. **Adult** Has well-marked orange-brown upperwings; underside of hind wing has a silvery sheen. Flies Jun-Aug. **Larva** Feeds on violets. **Status** Locally common woodland species in S and SW England and Ireland.

Honorary fritillary

The Duke of Burgundy *Hamearis lucina* (wingspan 25mm) is a tiny but charming butterfly that flies in May-June. With its beautifully marked orange and brown wings it looks for all the world like a miniature fritillary, but in fact it belongs to an entirely different family, Riodinidae. Its larva feeds mainly on Cowslip but also Primrose, and it is local and declining, confined to chalk downs in S England where the larval food plants grow.

Glanville Fritillary

Melitaea cinxia Nymphalidae
Wingspan 40mm

Beautifully marked butterfly. **Adult** Has orange-brown upperwings; underwings are creamy white and orange-buff. Flies May-Jun active only in sunshine. **Larva** Feeds gregariously on Sea Plantain. **Status** Very locally common on undercliffs on the Isle of Wight and in S Hampshire.

BELOW: **upperwings**
INSET: **underwings**

Studying butterflies

Butterflies are highly rewarding to watch and study. Many are colourful and most are associated with lovely habitats, which makes looking for them all the more pleasing. Although most species have declined in recent decades, some are still common and widespread. But, regardless of their status, all butterflies remain vulnerable to the way we, as a nation, use and abuse our countryside.

Location, location, location

Although some butterfly species are widespread and are found in a range of habitats, many have extremely specific requirements, usually linked to the presence of food plants for their larvae. The more unusual species tend to be found only in sites where the habitat, and the plants that grow there, are managed specifically with wildlife in mind. The best sites to look for butterflies are usually those owned or managed by conservation bodies such as the county wildlife trusts, which are protected to a degree by nature reserve status.

Flowery meadows little affected by agriculture are ideal for the Common Blue, Small Copper, skippers and brown butterflies. Appropriately timed cutting of grass for hay is a vital part of their management.

Timing

If you want to improve your chances of seeing a particular species, a knowledge of the flight periods for butterflies is essential. Although some species are long-lived, many are on the wing for only a brief few weeks and it is easy to miss them. Remember that no single year is the same as the next, and that the flight period for adult butterflies is strongly influenced by the prevailing weather in any given year. Experienced butterfly-watchers can gauge whether the spring is a late one by the timing of the appearance of such species as the Orange-tip and Speckled Wood. The time of day can also have a bearing on your chances of seeing a species. Most are active only in sunshine and will remain hidden in dense vegetation on dull days.

At the start and end of the day, some butterflies disappear quickly into the cover of vegetation. In contrast, many Common Blues and their relatives will often sit about and bask in the mornings and evenings, making observation of their delicately patterned underwings an easy prospect.

Using behaviour to identify butterflies

At first glance, butterflies may appear to fly around at random, but if you take the time to study them you will discern refined and complex behaviour patterns. These patterns are mostly linked to the location of members of the opposite sex, food plants for themselves and their larvae, and the avoidance of predators. A knowledge of some of these behaviours can help butterfly observers to get extremely good views.

Mating butterflies remain in tandem, linked at the abdomen tips, while fertilisation takes place. During this time they cannot fly well and rest on vegetation. These mating Duke of Burgundy butterflies were joined for nearly 30 minutes. Never disturb mating butterflies; if alarmed they will separate and fly away.

Sunbathing is a favourite occupation of the Marsh Fritillary and a number of other species of butterflies. Basking individuals can sometimes be approached to within a couple of metres while they are taking in the warmth of the late-afternoon sunshine.

Not all adult butterflies feed, but those that do often visit flowers to feed on nectar. Other liquid sources are sometimes used and this Purple Emperor is drinking the honeydew produced by aphids. It also has more unsavoury habits – it will drink the fluid from freshly deposited dung!

Studying life cycles

The life cycle of all butterfly species involves four different stages: the egg, laid by a female of the previous generation; the larva (sometimes called a caterpillar), which hatches from the egg and moults several times before reaching full size; the pupa (sometimes called a chrysalis), an inactive stage inside which adult organs and tissues develop; and the adult, which emerges from the pupa. Learn more about the butterfly life cycle on p.335 and about the moth life cycle (similar to that of a butterfly) on p.352. Once you have discovered adults of the common butterfly species, have a go at finding other stages in their life cycle; this is seldom easy because most are well camouflaged.

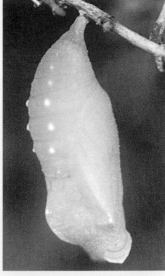

ABOVE: Being inactive and hence vulnerable to predators, most butterfly pupae are well concealed and tricky to find: this Small Heath pupa is tucked away beneath a tussock of grass.

The Purple Emperor larva emerges from its egg in summer and overwinters on the twigs of its food plant, Goat Willow. The next spring, when catkins appear on the plant, the tiny larva becomes active and can be found by careful searching.

BELOW: In spring, after they emerge from hibernation and mate, female Brimstones are preoccupied by the urge to lay their eggs. They lay only on Buckthorn or Alder-buckthorn, so if you sit beside a bush for long enough it is likely to be visited. Four eggs were laid on the underside of this Buckthorn twig.

ABOVE: It can be rewarding to rear butterflies in captivity and watch their life cycle at close quarters. The most ethical approach is to use common species that live in your garden – nobody is likely to object to you rearing Peacock larvae fed on Common Nettles.

RIGHT: Finding the pupa of a Purple Emperor is a real challenge: not only is the species rare and local, but the pupa is amazingly well camouflaged and brilliantly resembles the folded leaf of Goat Willow.

Butterfly conservation

Habitat loss and degradation, largely due to changes in farming practices, have had a devastating effect on butterfly numbers in Britain and Ireland. None has suffered more than the **Large Blue** *Glaucopsyche arion*, which became extinct in the UK in 1979. The larvae of this species can survive only if 'adopted' by colonies of a red ant called *Myrmica sabuleti*; inappropriate management can quickly kill off ant colonies and, with them, the butterflies. Fortunately, a better understanding of Large Blue ecology is allowing a reintroduction programme to go from strength to strength.

In the past, another human factor also had a significant impact on these delightful insects: butterfly collecting. During the Victorian and Edwardian eras in particular, hundreds of thousands of butterflies were killed to satisfy the urges of collectors and the rarer the species the greater the drive to collect one; this destructive practice is credited as being a contributory nail in the coffin of the Large Blue. Fortunately, mass collecting is now a thing of the past, thanks to the enlightened approach of modern naturalists and the rise in popularity of photography.

Brown butterflies

The term 'browns' is apt (and one used by butterfly enthusiasts) for the species on this page, since many have largely brown wings. They belong to the family Satyridae and are rather slow-flying species, often basking in the sun with their wings open. Their larvae feed on various grasses, so many are found in meadows and grassy woodland rides.

Speckled Wood

Pararge aegeria Satyridae
Wingspan 45mm

Favours woodland clearings and is fond of sunbathing. **Adult** Has dark brown upperwings with pale markings; underwings are rufous brown. Double-brooded; flies Apr–Jun and Jul–Sep. **Larva** Feeds on grasses. **Status** Widespread but common only in S England; local or absent further N; patchy but widespread in Ireland.

upperwings

underwings

Wall Brown

Pararge megera Satyridae
Wingspan 45mm

Colourful sun-loving butterfly. **Adult** Has orange-brown upperwings and a fritillary-like appearance, but shows small eye-spots on wings. Double-brooded; flies Apr–May and Jul–Sep. **Larva** Feeds on grasses. **Status** Widespread but declining and now rather scarce, least so on grassy heaths and coasts.

underwings

upperwings

Scotch Argus

Erebia aethiops Satyridae
Wingspan 40mm

Hardy upland butterfly. **Adult** Ringlet-like but has rich brown upperwings with an orange band and eye-spots. Flies Jul–Sep but only in sunny weather. **Larva** Feeds on Purple Moor-grass. **Status** Local in N England and S and central Scotland. Favours woodland margins and moors where the larval food plant is common.

Mountain Ringlet

Erebia epiphron Satyridae
Wingspan 32mm

Small and surprisingly delicate butterfly given the inhospitable nature of its upland habitat. **Adult** Has brown upperwings with an orange band in which small eye-spots are sited. Flies Jun–Jul but only when sunny. **Larva** Feeds on grasses. **Status** Very local on moors and mountains in the Lake District and central Scottish Highlands.

Marbled White

Melanargia galathea Satyridae
Wingspan 50mm

Distinctive meadow butterfly that visits knapweed and thistle flowers. **Adult** Has distinctive black and white patterns on wings. Flies Jul–Aug. **Larva** Feeds on grasses. **Status** Locally common in SE and central S England. Favours flower-rich, grassy meadows, often on chalk downs but also on neutral soils.

RIGHT: **underwings**
BELOW: **upperwings**

Ringlet

Aphantopus hyperantus Satyridae
Wingspan 48mm

Familiar, dark-looking meadow butterfly. **Adult** Has smoky-brown wings, darker on males than females, with variable numbers of small eye-spots. Flies Jun–Jul. **Larva** Feeds on grasses. **Status** Widespread and fairly common in Britain and Ireland as far N as S Scotland; found in grassy places.

TOP: **upperwings**; ABOVE: **underwings**

Gatekeeper

Pyronia tithonus Satyridae
Wingspan 40mm

Wayside and hedgerow butterfly that often feeds on Bramble flowers. **Adult** Has smoky-brown upperwings with orange markings and a paired eye-spot on forewing. Flies Jul–Aug. **Larva** Feeds on grasses. **Status** Found in central and S England and Wales, and S Ireland; locally common in most parts.

upperwings; INSET: underwings

Grayling

Hipparchia semele Satyridae
Wingspan 50mm

Sun-loving butterfly that sits with its wings folded and angled to the sun to cast the least shadow. **Adult** Has beautifully patterned underwings with an orange forewing showing 2 eye-spots. Flies Jun–Aug. **Larva** Feeds on grasses. **Status** Widespread, favouring warm, dry places, including sea cliffs, heaths and dunes.

Meadow Brown

Maniola jurtina Satyridae
Wingspan 50mm

Classic grassland butterfly. **Adult** Has brown upperwings; male has a small orange patch on forewing containing an eye-spot; orange patch is larger on female. Flies Jun–Aug. **Larva** Feeds on grasses. **Status** Common and widespread, least so in N Scotland and Ireland. Favours all kinds of grassy places.

LEFT: **upperwings**
TOP: **underwings**
ABOVE: **pupa**

Browns, hairstreaks & coppers

'Heaths' are small members of the family Satyridae, found in unimproved grassland. Hairstreaks have colourful or delicately patterned underwings, the hind wings with tail streamers in some species, while coppers have bright orange markings on their upperwings. Both hairstreaks and coppers are atypical members of the blue family, Lycaenidae.

Large Heath

Coenonympha tullia Satyridae
Wingspan 38mm

Hardy, upland butterfly. **Adult** Seldom reveals upperwings; underside of hind wing is grey-brown while orange-brown forewing has a small eye-spot. Flies Jun-Jul, only in sunny weather. **Larva** Feeds on White Beak-sedge. **Status** Local on acid moors from central Wales northwards, and occurs locally in Ireland.

Small Heath

Coenonympha pamphilus Satyridae
Wingspan 30mm

Small grassland butterfly. **Adult** Seldom reveals upperwings. Underside of forewing is orange with an eye-spot; hind wing is marbled grey, brown and buff. Double-brooded; flies May-Jun and Aug-Sep. **Larva** Feeds on grasses. **Status** Widespread, but restricted to unimproved grassland and locally common only in the S.

A life tied to an oak tree

Few butterflies are so closely tied to their larval food plant as the **Purple Hairstreak** *Favonius quercus* (wingspan 38mm). As an adult it seldom strays more than a couple of metres from the top of the mature tree on which the egg, larva and pupa from which it grew were located. Indeed, sizeable Purple Hairstreak colonies, perhaps of hundreds of individuals, live on the same tree and their courtship and territorial behaviour can be observed with the aid of binoculars from the ground. The species is locally common in only south England and Wales. Adults have a purple sheen on the upperwings; the underwings are grey with a hairstreak line.

RIGHT: underwings; BELOW: upperwings

underwings, illustrating regional variation in wing patterns

White-letter Hairstreak

Satyriuim w-album Lycaenidae
Wingspan 35mm

Flies around treetops but also feeds on Bramble flowers. **Adult** Seldom reveals upperwings. Underwings are brown with a jagged orange band and white 'W' on hind wing. Flies Jul–Aug. **Larva** Feeds on elms. **Status** Widespread loss of larval food plants from Dutch elm disease has caused its decline. Now very local.

TOP: underwings; ABOVE: pupa

Black Hairstreak

Satyriuim pruni Lycaenidae
Wingspan 35mm

Lethargic flyer that visits Privet flowers or feeds on honeydew on its leaves. **Adult** Seldom reveals upperwings. Underwings are rich brown, with an orange band and white line on both wings. Flies Jul. **Larva** Feeds on Blackthorn. **Status** Rare, in woods bordered with Blackthorn thickets; occurs mainly in the East Midlands.

TOP: underwings; ABOVE LEFT: larva; ABOVE RIGHT: pupa, camouflaged to resemble bird dropping

Green Hairstreak

Callophrys rubi Lycaenidae
Wingspan 25mm

Small, highly active species that is hard to follow in flight. **Adult** Seldom reveals brown upperwings; underwings are bright green and blend in with foliage when resting. Flies May–Jun. **Larva** Feeds on gorses, Heather and trefoils. **Status** Locally common on heaths, cliffs and downland scrub.

underwings

Small Copper

Lycaena phlaeas Lycaenidae
Wingspan 25mm

Attractive, open country species. **Adult** Has a variable orange and dark brown pattern on upperwings; underwings have a similar pattern to upperwings but dark brown is replaced by grey-buff. Flies May–Sep in 2 or 3 broods. **Larva** Feeds on Sheep's Sorrel. **Status** Locally common in undisturbed grassland, heaths and cliffs.

LEFT: underwings
BELOW: upperwings

Brown Hairstreak

Thecla betulae Lycaenidae
Wingspan 40–50mm

Sluggish flyer that often walks among foliage. **Adult** Has dark brown upperwings, the male with an orange patch on forewing; underwings are orange-brown with a white line. Flies Aug. **Larva** Feeds on Blackthorn. **Status** Very local in S and central England and Wales; favours Blackthorn thickets.

Blue butterflies

What blue butterflies (family Lycaenidae) lack in size, they make up for with their dazzling colours and intricate wing patterns. These are open country butterflies that favour meadows, chalk downs and heaths.

Common Blue

Polyommatus icarus Lycaenidae
Wingspan 32mm

Our most widespread blue butterfly. **Adult** Male has blue upperwings; female's are usually brown, but sometimes tinged blue in middle. Underside of both sexes is grey-brown with dark spots. Flies Apr–Sep in successive broods. **Larva** Feeds on trefoils. **Status** Common in grassy places.

male upperwings

ABOVE: female upperwings; BELOW: male underwings

Chalkhill Blue

Polyommatus coridon Lycaenidae
Wingspan 40mm

Beautiful downland butterfly. **Adult** Male has pale sky-blue upperwings; female's are dark brown with orange sub-marginal spots. Underwings of both sexes are grey-brown with spots. Flies Jul–Aug. **Larva** Feeds on Horseshoe Vetch. **Status** Restricted to chalk and limestone grassland in S England.

male upperwings

female upperwings

male and female underwings

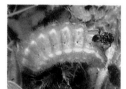
Adonis Blue

Polyommatus bellargus Lycaenidae
Wingspan 32mm

Stunning downland butterfly. **Adult** Male has iridescent blue upperwings with black and white margins; female's upperwings are brown with orange sub-marginal spots. Underwings of both sexes are grey-brown with spots. Flies May–Jun and Jul–Aug in 2 broods. **Larva** Feeds on Horseshoe Vetch. **Status** Local in S England on chalk downs.

TOP: male upperwings
ABOVE: female upperwings
LEFT: underwings

Holly Blue
Celastrina argiolus Lycaenidae
Wingspan 30mm

The most likely blue to be seen in gardens. Looks silvery in flight. **Adult** Has violet-blue upperwings (seldom revealed when resting) and black-dotted white underwings. Flies Apr–May and Aug–Sep in 2 broods. **Larva** Feeds on Holly in spring, Ivy in autumn. **Status** Fairly common in S England, S Wales and S Ireland.

TOP: **upperwings**; ABOVE: **underwings**

Small Blue
Cupido minimus Lycaenidae
Wingspan 25mm

Small, highly active blue. **Adult** Has smoky-brown upperwings, the male's with a purplish iridescence; underwings of both sexes are grey. Flies Jun–Jul. **Larva** Feeds on Kidney Vetch. **Status** Very locally common in England, Wales and S Ireland, often on chalk grassland.

TOP: **upperwings**; ABOVE: **underwings**

Brown Argus
Aricia agestis Lycaenidae
Wingspan 25mm

Attractive grassland butterfly. **Adult** Recalls a female Common Blue but wingspan is smaller and orange sub-marginal spots on brown upperwings are much more prominent. Flies May–Aug in 2 broods. **Larva** Feeds on Common Rock-rose and stork's-bills. **Status** Local in S and central England, often on chalk downs.

TOP: **upperwings**; ABOVE: **underwings**

Heathland jewels

Heathlands are not rich in butterfly species, but what they lack in quantity they make up for in quality. In June and July, **Silver-studded Blues** *Plebejus argus* (wingspan 25–30mm) emerge and sunbathing males adorn sprays of Heather like tiny jewels. The species' name derives from the shiny spots at the centre of the orange and black spots that adorn the underwings. The larvae feed on Heather and gorse.

male upperwings

male underwings

Northern Brown Argus
Aricia artaxerxes Lycaenidae
Wingspan 25mm

Northern counterpart of the Brown Argus. **Adult** Very similar to Brown Argus, with brown upperwings and orange sub-marginal spots, but note the white spot on upper forewing. Flies Jun–Jul. **Larva** Feeds on Common Rock-rose. **Status** Very local in calcareous grassland in N England and Scotland.

Skippers

Along with the 'browns', most skippers (family Hesperiidae) are archetypal grassland butterflies, restricted to sites that are undisturbed and traditionally managed. All are small, dainty species with a buzzing flight and superficially moth-like appearance when resting. They are particularly attracted to Composite flowers such as thistles.

Lulworth Skipper

Thymelicus acteon Hesperiidae
Wingspan 28mm

 Subtly attractive skipper with an active, buzzing flight. **Adult** Has khaki-brown upperwings. Forewing has a crescent of pale spots like a paw-print, the markings brighter on female than on male. Flies Jun–Jul. **Larva** Feeds on grasses. **Status** Very local in coastal grassland from Purbeck in Dorset to E Devon.

Small Skipper

Thymelicus sylvestris Hesperiidae
Wingspan 25mm

 Active little butterfly with a buzzing flight. Fond of sunbathing on grass stems and foliage, and often visits thistle flowers. Rests with wings held at an angle. **Adult** Has orange-brown upperwings and orange-buff underwings. Antennal tip colour allows separation from similar Essex Skipper (*see* box opposite). Flies Jul–Aug. **Larva** Feeds on various grasses. **Status** Common and widespread grassland species in England and Wales.

Essex Skipper

Thymelicus lineola Hesperiidae
Wingspan 25mm

 Similar to Small Skipper but underside of antennal tips is black. Has a buzzing flight and visits knapweed and thistle flowers. **Adult** Has orange-brown upperwings and orange-buff underwings. Flies Jun–Jul. **Larva** Feeds on various grasses. **Status** Locally common (often overlooked) in meadows in SE England.

Large Skipper

Ochlodes venatus Hesperiidae
Wingspan 34mm

 Often holds its wings at an angle at rest and can look rather moth-like. **Adult** Has dark brown upperwings with pale markings. Underwings are buffish orange with paler spots. Flies Jun–Jul. **Larva** Feeds on various grasses. **Status** Common and widespread in England and Wales in grassy places of all kinds.

TOP: **underwings**; ABOVE: **upperwings**

Distinguishing between Small and Essex skippers

It takes an expert eye to tell apart Small and Essex skippers in the field when they are active and on the wing in the middle of the day. Both are of a similar size and have almost identical wing colours and markings. However, there is one clear difference between the two, best observed in the late afternoon when the butterflies are going to roost on flower heads and grass stems. At this time of day, you can often creep up close to resting skippers and take a good look at the underside of the tip of their antennae - in Small Skippers this is rufous brown, while in Essex Skippers it is black.

Small Skipper

Essex Skipper

Grizzled Skipper

Pyrgus malvae Hesperiidae
Wingspan 20mm

 Attractively marked skipper. **Adult** Has dark grey-brown upperwings with conspicuous white spots; underwings are reddish brown with numerous pale spots. Flies May-Jun. **Larva** Feeds on Wild Strawberry and cinquefoils. **Status** Locally common in S England and S Wales. Favours rough grassland and woodland rides.

Silver-spotted Skipper

Hesperia comma Hesperiidae
Wingspan 34mm

 Late summer skipper. **Adult** Similar to Large Skipper but dark brown upperwings have distinctive pale spots; greenish-brown underwings have silvery-white spots, which give the species its name. Flies Aug-Sep. **Larva** Feeds on grasses. **Status** Very local on chalk downland in S England.

Dingy Skipper

Erynnis tages Hesperiidae
Wingspan 25mm

 Moth-like skipper with a buzzing flight. **Adult** Has dark grey-brown upperwings; under-wings are reddish brown. Flies May-Jun. **Larva** Feeds mainly on Common Bird's-foot Trefoil but also various vetch species. **Status** Locally common in England and Wales in meadows and rough grassy woodland rides.

Chequered Skipper

Carterocephalus palaemon Hesperiidae
Wingspan 25mm

 Its flight is active and fast. Fond of sunbathing; hard to find in dull weather. **Adult** Has rich brown upperwings with orange-yellow spots; underwings are paler and marked with pale spots. Flies May-Jun. **Larva** Feeds on various grasses. **Status** Very locally common in open birch woods in NW Scotland; formerly found in England but now extinct there.

RIGHT: **upperwings**
BELOW: **underwings**

TOP: **upperwings**; ABOVE: **underwings**

Studying moths

Moths are among the most diverse and fascinating of all British and Irish insects. Although most fly at night and are difficult to observe in action, it is relatively easy to find a significant number of species resting in the daytime. Using a moth trap opens up a new dimension to the study of moths: hundreds of different species that would otherwise probably remain undetected can be discovered by even a novice naturalist.

Once abundant in suburban areas, the Garden Tiger moth (*see* p.360) is now scarce. To help with the species' conservation, leave patches of native wild flowers such as docks and dandelions in your garden.

What is a moth?

Moths are closely related to butterflies and both belong to the order Lepidoptera. So how do you distinguish between the two? While all butterflies fly in daytime, not all moths fly at night – some species are also active in sunshine. The best distinction is that British butterflies all have clubbed antennae, while in almost all moths the antennae are either feathery (in the males of some species as in the Muslin Moth above) or slender and tapering; exceptions are the burnet moths (*see* p.356), which also have clubbed antennae, but whose wings are slender and are held pressed to the body at rest in an entirely different manner to butterflies.

Encouraging moths in the garden

These days, moths need all the friends they can get as many species are in serious decline. Many formerly common garden species have been hard hit too, often because allotment-type gardens are in decline and many people favour hard landscaping over gardening with plants. To encourage moths, leave wild corners in the garden where 'weeds' can flourish (many of these are food plants for moth larvae), grow as many native species of plants as you can (native plants support far more wildlife than alien ones) and do not use chemical sprays.

Moth life cycles

Like butterflies, moths have four stages in their life cycle: the egg, laid by a female of the previous generation; the larva, a growing stage that emerges from the egg; the pupa, an inactive stage inside which adult organs and tissues form; and the adult, which emerges from the pupa. In some species the life cycle is repeated two or three times each year, between spring and autumn. But in most, the life cycle is annual and adults emerge at the same time each year. If you find some moth eggs or larvae, you can try rearing them so you can follow the life cycle. Before you do so, identify the species so you know what to feed it. Daily supplies of fresh leaves, and the removal of old ones, is vital. At maturity, most larvae like to pupate in dry soil.

1 Single Poplar Hawkmoth egg.

2 Newly emerged from its egg, this young Poplar Hawkmoth larva's first task is to eat the eggshell.

3 This Poplar Hawkmoth larva has recently moulted its skin and is on the way to becoming fully grown.

4 Prior to metamorphosing into a pupa (*shown*), a Poplar Hawkmoth larva digs into the soil to create a chamber in which it then pupates.

5 This freshly emerged Poplar Hawkmoth adult is drying its wings prior to taking its first flight as evening approaches.

Overwintering pupae

It is fascinating to keep moth pupae over the winter until they emerge as adults the following spring. You may have dug one up while gardening or have reared larvae to the point where they have pupated. They are best placed in a box with air holes and containing dry soil, and kept in a cool, frost-free place over the winter. Spray them with water once in a while to stop them drying out and dying. Come spring, you need to check the box on a daily basis to see if anything has emerged.

Moths' daytime hide outs

A few moth species are active in the day and are relatively easy to find; these include species of burnet moths, the Hummingbird Hawkmoth and the Silver-Y. Most other species take to the wing only after dark and in the daytime remain hidden to avoid predation. Many moth species have remarkably camouflaged wings when sited on substrates their wing markings and colours have evolved to mimic. Some resemble tree bark or lichens that coat branches, while others are similar to leaves, either fallen or living. Moths can be hard to spot at first, but once you have your eye in, finding them becomes easier.

For much of their lives, Hornet Moths are very hard to find. But for a few hours after they appear as adults, they can sometimes be found resting near the base of the Black Poplar trees from which they emerged.

Buff-tip moths bear an uncanny resemblance to a snapped birch twig.

Spot the moth – a Lappet Moth looks just like the Beech leaves on which it is resting.

A Peppered Moth's wings are a good match for lichen growing on tree bark.

Attracting moths

Leave an outdoor light on after dark in summer and you will soon have plenty of moths fluttering around. To take this natural attraction a stage further, suspend a light over a white sheet. Or you can buy or make a light trap (*below left*) whereby moths are collected in a box beneath the light. The best light source is a bulb that emits ultraviolet light (to which moths are particularly receptive). Captured moths are unharmed by the experience and can be released after they have been studied and identified.

Another tried and tested technique for attracting certain species of moth is to lure them with food. The classic method involves dissolving a bag of sugar in half a litre or so of warmed red wine and soaking ropes in the mixture. The soggy ropes can then be hung up on branches to tempt passing moths. The technique is more of an art than a science, and catches are extremely variable; warm, muggy nights produce the best results.

A third method uses the moths' natural scent. Many animals locate members of the same species using chemicals called pheromones. In moths, it is typically females that emit these airborne scents and the males that are attracted to them. Captive females may attract the attentions of males if kept in a muslin container. With day-flying clearwing moths, there is a more direct approach: artificial pheromones can be bought (*below right*), and although their efficacy varies considerably, with some species the results are impressive.

Although a full-grown Privet Hawkmoth larva is large and colourful, in dappled light it is surprisingly difficult to spot among Privet foliage.

Finding larvae

Moth larvae are fascinating in their own right. They bear no resemblance to their adult counterparts and many have a bizarre appearance. Being juicy and succulent, moth larvae appeal to a large number of predators – they are the main food source for the young of many songbirds, for example. So many larvae go to great lengths to conceal or camouflage themselves. One technique used to collect larvae involves spreading a white sheet on the ground and giving the vegetation above a sharp tap to dislodge feeding insects. A less intrusive method is to search plant leaves and stems for the larvae, which often leave signs of their presence such as nibbled leaves.

Some larvae, such as this Cinnabar Moth larva, are easy to find – being poisonous, they advertise their presence with bright warning colours.

Although light traps can be left overnight, many true moth enthusiasts prefer to sit and watch the operation for a few hours, just in case something special turns up.

Clearwing pheromones are supplied in impregnated rubber bungs – these male Six-belted Clearwings are being driven mad with desire by the scent being given off.

Hawkmoths

Hawkmoths are spectacular insects of the family Sphingidae. Adults are fast-flying and, in many cases, large, and they have beautifully patterned wings. Even the larvae are intriguing: some are colourful and well marked, and most have the trademark hawkmoth 'horn' at the rear end.

Lime Hawkmoth

Mimas tiliae Sphingidae
Wingspan 65mm

 Well camouflaged when resting among dappled leaves. **Adult** Wing colour is variable but usually olive-green with darker markings. Flies May–Jun. **Larva** Pale green with diagonal stripes and numerous white dots; feeds mainly on lime. **Status** Common only in S England, becoming scarce further N.

Poplar Hawkmoth

Laothoe populi Sphingidae
Wingspan 70mm

 Rests by day among leaves and hence easy to overlook. **Adult** Has grey-brown forewings that, at rest, obscure reddish mark on hind wing; this is exposed if moth becomes alarmed. Flies May–Aug in 2 broods. **Larva** Bright green with diagonal stripes and a 'horn' at tail end; feeds on poplar and willows. **Status** Common and widespread.

Pine Hawkmoth

Hyloicus pinastri Sphingidae
Wingspan 80mm

 Superbly camouflaged when resting on pine bark. **Adult** Has grey-brown forewings with darker streaks and dots. Buffish-orange hind wings are seldom seen at rest. Flies Jun–Jul. **Larva** Green with diagonal stripes and a dark 'horn' at tail end; feeds on conifer needles. **Status** Locally common in S England.

Eyed Hawkmoth

Smerinthus ocellata Sphingidae
Wingspan 80mm

 Intriguing moth. **Adult** Has marbled grey-brown forewings that obscure hind wings at rest; when disturbed, it arches its body and wings to expose striking eye-spots on hind wings. Flies May–Aug in 2 broods. **Larva** Bright green with diagonal stripes and a 'horn' at tail end; feeds on willows and apple. **Status** Widespread and common in England, Wales and Ireland.

LEFT: pupa; ABOVE: adult; INSET: larva

Privet Hawkmoth

Sphinx ligustri Sphingidae
Wingspan 100mm

Large and impressive moth. **Adult** Well camouflaged at rest but exposes pink-striped abdomen and pale pink stripes on hind wing when alarmed. Flies Jun-Jul. **Larva** Bright green with purple and white diagonal stripes and a dark-tipped 'horn'; feeds on Privet and lilac. **Status** Common only in S England.

Elephant Hawkmoth

Dielephila elpenor Sphingidae
Wingspan 70mm

Beautiful moth that visits garden flowers such as Honeysuckle. **Adult** Has pink and olive-green wings and body. Flies May-Jun. **Larva** Brown or green; head end fancifully resembles an elephant's trunk; eye-spots deter would-be predators. Feeds on willowherbs. **Status** Widespread, commonest in England and Wales.

INSET: **larva**

Broad-bordered Bee Hawkmoth

Hemaris fuciformis Sphingidae
Wingspan 43mm

Day-flying moth that hovers at flowers such as Bugle to feed. **Adult** Wings have a covering of scales when newly emerged from pupa; these are soon lost and moth becomes strikingly similar to a bumblebee. Flies May-Jun. **Larva** Feeds mainly on Honeysuckle. **Status** Common only in S England along sunny woodland rides.

TOP LEFT: **adult hovering**; TOP RIGHT: **larva**; ABOVE: **adult**

Unusual migrants

Hawkmoths are strong flyers and many cover long distances in their comparatively brief lives. Warm southeasterly winds in summer usually bring with them good numbers of Hummingbird Hawkmoths from mainland Europe, but moth enthusiasts also look out for more unusual visitors. Prize finds include the **Death's-head Hawkmoth** *Acherontia atropos* shown below, **Convolvulus Hawkmoth** *Agrius convolvuli*, **Spurge Hawkmoth** *Hyles euphorbiae*, **Bedstraw Hawkmoth** *Hyles gallii* and **Striped Hawkmoth** *Hyles livornica*.

Small Elephant Hawkmoth

Dielephila porcellus Sphingidae
Wingspan 50mm

Attractive moth that is sometimes seen on the wing at dusk. **Adult** Has pink and yellow-buff wings. Flies Jun-Jul. **Larva** Grey-brown with eye-spots near head end; lacks a 'horn' at tail end; feeds on bedstraws. **Status** Local on heaths and downs where the larval food plants are common.

Hummingbird Hawkmoth

Macroglossum stellatarum Sphingidae
Wingspan 45mm

Day-flying species that hovers with an audible hum and collects nectar using its long tongue. **Adult** Has brown forewings and an orange patch on hind wings; note the white on sides of abdomen. Flies May-Oct. **Larva** Feeds on bedstraws but seldom seen in Britain. **Status** Migrant visitor from mainland Europe during summer, in variable numbers.

Burnet moths & clearwings

Not all moths are showy: many have muted wing colours and some are decidedly dowdy. Clearwings can be hard to recognise as moths at all, having only a limited coating of scales on their wings; most look and behave like tiny wasps.

Brown House Moth

Hofmannophila pseudospretella Oecophoridae
Length 10mm

Seldom flies but typically scuttles along floor into a crevice or cupboard. **Adult** Has shiny brown wings and is usually found in vicinity of stored natural fabrics and foods, on which the eggs are laid. **Larva** Feeds inside fabrics and foodstuffs. **Status** Frequent and unwelcome visitor to houses; can be found year-round.

White Plume Moth

Pterophorus pentadactyla Pterophoridae
Wingspan 28mm

Attractive and distinctive moth, seen resting on low vegetation in daytime. Flies from dusk onwards and is often attracted to light; sometimes comes indoors.
Adult Has white, dissected, feather-like wings. Flies May–Aug. **Larva** Feeds on Hedge Bindweed and lives inside a rolled-up leaf. **Status** Common and widespread.

Burnet moths

Burnet moths are day-flyers, belonging to the family Zygaenidae, and are found in meadows and other grassy places. Their bodies are toxic to birds and mammalian predators, and so rather than hide themselves away they actually advertise their presence with black and red warning colours.

Five-spot Burnet

Zygaena trifolii Zygaenidae
Wingspan 35mm

Distinctive moth. **Adult** Has red hind wings and 5 red spots on otherwise metallic greenish-blue forewings. Flies Jul–Aug. **Status** Favours damp meadows. Its larva feeds on Greater Bird's-foot Trefoil. **Similar species Six-spot Burnet** *Z. filipendulae* (wingspan 35mm) has 6 spots on its forewing; it favours drier ground and its larvae feed on Common Bird's-foot Trefoil. **Forester** *Adscita statices* (length 15mm) has shiny green wings and favours damp meadows.

Mother of Pearl

Pleuroptya ruralis Pyralidae
Wingspan 35mm.

Has a weak flight and frequently comes to light. **Adult** Has a mother-of-pearl sheen to wings at certain angles. Flies Jun–Aug. **Larva** Feeds on Common Nettle and lives inside a rolled-up leaf. **Status** Common and widespread. Found in meadows, waste ground and overgrown hedgerows.

Ghost Moth

Hepialus humuli Hepialidae
Length 25mm

Groups of males engage in dancing display flights at dusk. **Adult** Male is pure white, female has buffish-yellow wings with orange streaks. Flies Jun–Aug. **Larva** Lives underground and feeds on roots of various plants. **Status** Widespread and fairly common in meadows and grassy hedgerows.

BELOW LEFT:
Six-spot Burnet
BELOW RIGHT: Forester

Clearwings

Clearwings are remarkable day-flying moths that are related to burnet moths and belong to the family Sesiidae. They are so secretive and wary that many naturalists would go their entire lives without seeing one were it not for the advent of pheromone lures. These artificial scents attract males of the species, sometimes in considerable numbers (see p.353).

Six-belted Clearwing

Bembecia ichneumoniformis Sesiidae
Length 14mm

 Sun-loving meadow species. **Adult** Has 6 yellowish bands on abdomen and orange scaling on otherwise black-veined clear wings. Flies Jun–Aug. **Larva** Feeds on roots of Common Bird's-foot Trefoil and Kidney Vetch. **Status** Widespread and very locally common in short grassland.

Yellow-legged Clearwing

Synanthedon vespiformis Sesiidae
Length 14mm

 Well-marked woodland moth. **Adult** Has yellow bands on abdomen and bright yellow legs. Wings are black-veined and clear, but note the small red bar. Flies Jun–Aug. **Larva** Feeds beneath bark of Pedunculate Oak. **Status** Very locally common in S England.

Currant Clearwing

Synanthedon tipuliformis Sesiidae
Length 11mm

 Tiny, fly-like moth. **Adult** Has a bluish body with 3 yellow bands on abdomen, a yellow 'collar' and yellow lines on side of thorax. Flies Jun–Jul. **Larva** Feeds inside stems of Black and Red currants. **Status** Locally common but easily overlooked; often occurs in gardens.

Red-belted Clearwing

Synanthedon myopaeformis Sesiidae
Length 13mm

 Easily overlooked fly-like moth. **Adult** Has a mainly bluish-black body with a striking red band on abdomen. Wings are black-veined and clear. Flies Jun–Aug. **Larva** Feeds under bark of apple trees. **Status** Locally common in mature gardens and woodlands with neglected apple trees.

Goat Moth

Cossus cossus Cossidae
Length 5cm

larva

 Impressive and distinctive moth. **Adult** Has white, silvery-grey and buff forewings, beautifully patterned to resemble cracked tree bark. Flies Jun–Jul. **Larva** Feeds under bark of deciduous trees, including willows and poplars; sometimes found wandering in search of a pupation site. **Status** Local.

Large Red-belted Clearwing

Synanthedon culiciformis
Sesiidae Length 16mm

 Distinctive woodland clearwing. **Adult** Has a mainly blackish body but with a striking reddish band on abdomen and orange markings on head, side of thorax and otherwise clear wings. Flies May–Jun. **Larva** Feeds below bark of birches. **Status** Widespread and scarce, but probably overlooked.

Hornet mimicry

Wasps deliver painful stings and their yellow and black colours warn potential predators to keep clear. Many harmless insects mimic these markings and take advantage of the protection they afford. The best example is the **Hornet Moth** *Sesia apiformis* (**1**) (length 2.5cm) (see p.353), although the **Lunar Hornet Moth** *S. bembeciformis* (**2**) (length 2.5cm) seen here, whose larvae feed in Sallow timber, comes a close second.

Emperor, Puss & prominent moths

Most British moths have amazingly patterned wings. In some of the species shown here, the colours and markings camouflage them from predators. But a few have markings designed to scare potential enemies.

Lappet Moth

Gastropacha quercifolia Lasiocampidae
Length 40mm

Superbly camouflaged when resting among fallen leaves. **Adult** Has reddish wings with scalloped margins; head has a 'snout'. Flies Jun-Aug and comes to light. **Larva** Feeds on Hawthorn; camouflaged when resting lengthwise along a twig. **Status** Commonest in the S.

Lackey

Malacosoma neustria Lasiocampidae
Length 17mm

Well-marked wayside moth. **Adult** Has buffish or brown wings with 2 parallel transverse lines. Flies Jun-Aug. **Larva** Colourful; feeds on Hawthorn and Blackthorn, living in communal silken tents spun on branches. **Status** Common and widespread in the S.

Puss Moth

Cerura vinula Notodontidae
Length 35mm

Attractive, furry-looking moth. **Adult** Has pale grey and white wings. Flies May-Jul. **Larva** Squat and green with 2 whip-like tail appendages; feeds on willows and poplars. **Status** Common and widespread in much of Britain and Ireland.

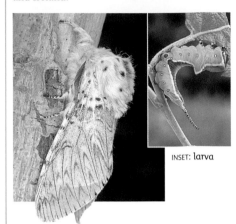

INSET: larva

Sallow Kitten

Furcula furcula
Notodontidae
Length 20mm

larva

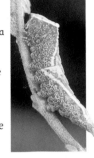

Charming little moth. **Adult** Has a grey band across otherwise white forewing. Flies May-Aug in 2 broods and comes to light. **Larva** Resembles a miniature Puss Moth larva with 2 tail appendages; its main foodplants are sallows. **Status** Common and widespread throughout.

Drinker

Euthrix potatoria Lasiocampidae
Length 30mm

Striking moth with a pronounced 'snout' when at rest. **Adult** Has broad wings with a bold diagonal stripe; ground colour is orange-yellow in males, yellow-buff in females. Flies Jul-Aug. **Larva** Feeds on grasses. **Status** Common in grassy habitats.

Emperor Moth

Saturnia pavonia
Saturniidae
Wingspan 50-60mm

larva

Large and impressive day-flying moth. **Adult** Has striking eye-markings on wings; male is grey and orange, female is grey and pink. Flies Apr-May. **Larva** Green with black markings and hair tufts; feeds on Heather and Bramble. **Status** Locally common on heaths and moors.

Emperor Moth female

Emperor Moth male

ABOVE: Sallow Kitten

Lobster Moth

Stauropus fagi Notodontidae
Length 32mm

 Named after its bizarre-looking larva, which is fancifully lobster-like (*see* p.372). **Adult** Has reddish-grey wings, usually with a pinkish flush; well camouflaged on tree bark. Flies May–Jul. **Larva** Feeds on Pedunculate Oak and Beech. **Status** Locally common in the S.

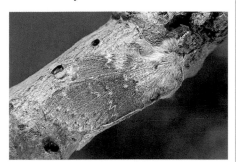

Pebble Prominent

Eligmodonta ziczac Notodontidae
Length 25mm

 Distinctive moth. **Adult** Has buffish-brown forewings with a grey-brown fingernail-like mark near tip and a white patch on leading edge. Flies May–Jun, and sometimes in Aug as 2nd brood. **Larva** Feeds on willows and Aspen. **Status** Widespread and locally common in woodlands.

Pale Prominent

Pterostoma palpina Notodontidae
Length 30mm

 Well-camouflaged moth. **Adult** Has greyish-buff wings; palps project at head end; tufted tip to abdomen protrudes beyond wings and in profile has prominences on back. Flies May–Aug. **Larva** Feeds on sallows and Aspen. **Status** Fairly common in woodlands and hedgerows.

Iron Prominent

Notodonta dromedarius Notodontidae
Length 26mm

 Richly coloured moth. **Adult** Has smoky-brown forewings marked with rusty brown and yellow; markings provide good camouflage when resting on tree bark and twigs. Flies May–Aug in 2 broods. **Larva** Feeds on birches, oaks, Alder and Hazel. **Status** Widespread and locally common.

Maple Prominent

Ptilodonta cuculina Notodontidae
Length 22mm

 Attractive species with distinct prominences when seen in profile. **Adult** Has marbled brown wings with a whitish patch on outer margin of forewing. Flies May–Jul. **Larva** Feeds mainly on Field Maple. **Status** Local and common only in S England, in old hedgerows and woods.

Great Prominent

Peridea anceps Notodontidae
Length 30mm

 Superbly camouflaged moth when resting on oak bark. **Adult** Has marbled brown and grey-brown wings, and hairy legs. Flies May–Jul. **Larva** Green with diagonal stripes; feeds on oaks. **Status** Locally common in mature oak woodland.

Lesser Swallow Prominent

Pheosia gnoma Notodontidae
Length 27mm

 Well-marked moth. **Adult** Has pale grey-buff wings with dark stripes; a distinct white wedge on outer margin of forewing allows separation from similar Swallow Prominent. Flies May–Jun and Aug as 2 broods. **Larva** Feeds on birches. **Status** Widespread and common.

Swallow Prominent

Pheosia tremula Notodontidae
Length 30mm

 Attractive moth. **Adult** Has grey-buff and brown wings; distinguished from Lesser Swallow Prominent by 2 or 3 narrow white stripes running from outer margin. Flies Apr–Aug in 2 broods. **Larva** Feeds on willows and poplars. **Status** Widespread and common in woods and gardens throughout the region.

Ermine & tiger moths

Hedgerows, flower-rich meadows and even roadside verges are rich hunting grounds for the moth enthusiast. The key to the importance of these habitats is the rich variety of herbaceous plants that grow there – food for the larvae of tiger moths, ermines and many more.

Buff-tip

Phalera bucephala Notodontidae
Length 30mm

 Intriguingly marked moth. **Adult** Has a buff head and silvery-grey wings with buff tips, giving resting moths a broken twig appearance. Flies May–Jul. **Larva** Yellow and black; feeds on deciduous trees, including oak and lime. **Status** Widespread and common in gardens and woodland.

Chocolate-tip

Clostera curtula Notodontidae
Length 17mm

 Beautifully marked moth. **Adult** Has grey-brown forewings with a chocolate-maroon tip; when alarmed, exposes a brown-tipped abdomen. Flies May–Sep in 2 or more broods and comes to light. **Larva** Feeds on Aspen, poplars and willows. **Status** Local in woodland, mainly in S.

Pale Tussock

Calliteara pudibunda Lymantriidae
Length 30mm

 Attractive, hairy moth. **Adult** Has grey or greyish-buff wings. Flies May–Jun. **Larva** Yellow and black and hairy; feeds on birches, oaks, limes and other deciduous trees (*see* p.373). **Status** Locally common in woodland in England, Wales and S Ireland.

Vapourer

Orgyia antiqua
Lymantriidae
Length 16mm (male)

 Sexes are dissimilar. **Adult** Male has a chestnut forewing with a white spot on trailing edge; flies Jul–Sep. Wingless female is sometimes seen near clusters of eggs laid on tree bark. **Larva** Has tufts of yellow and black hairs; feeds on deciduous trees. **Status** Common.

Yellow-tail

Euproctis similis Lymantriidae Length 24mm

 Treat with caution, since both adult and larval hairs cause irritation if handled or inhaled. **Adult** Has white wings but exposes yellow-tipped abdomen when alarmed. Flies Jun–Aug. **Larva** Hairy with black and red markings; feeds on deciduous shrubs. **Status** Locally common.

INSET: **larva**

A plea for Ragwort

 Few people who keep grazing animals have a good word to say for Common Ragwort *Senecio jacobaea*, whose foliage is toxic (*see* p.136). However, naturalists view it in a different light because it is a fantastic nectar source for insects and is also the food plant for the orange-and-black-striped larvae (*see* p.353) of the **Cinnabar Moth** *Tyria jacobaeae* (length 22mm). Adults fly in May–July and are recognised by their red and charcoal-grey wings.

Garden Tiger

Arctia caja Arctiidae Wingspan 65mm

 Familiar moth. **Adult** Has brown and white forewings that conceal dark-spotted orange hind wings. Flies Jul–Aug. **Larva** Hairy; feeds on a wide range of plants. **Status** Widespread but declining; found in most habitats and comes to outdoor lights around houses.

Cream-spot Tiger

Arctia villica Arctiidae
Wingspan 55mm

 Beautifully marked moth. **Adult** Has black and white forewings and yellow and black hind wings; flies Jun-Jul. **Larva** Feeds on a range of herbaceous plants. **Status** Typically a coastal or heathland species, but local and commonest in SW.

White Ermine

Spilosoma lubricipeda Arctiidae
Length 28mm

 Well-marked and distinctive moth. **Adult** Has white wings with numerous small black spots; the yellow and black abdomen is hidden by the wings at rest. Flies May-Jul. **Larva** Feeds on a range of herbaceous plants. **Status** Widespread and common in most habitats; comes to light.

Buff Ermine

Spilosoma luteum Arctiidae
Length 28mm

 Subtly attractive moth, sometimes seen resting on foliage in daytime. **Adult** Pale yellow-buff with variable black markings; flies May-Jul. **Larva** Feeds on a range of herbaceous plants. **Status** Widespread and common in most habitats; comes to light.

Scarlet Tiger

Callimorpha dominula Arctiidae
Wingspan 45mm

 Strikingly marked moth, sometimes active in daytime. **Adult** Has black, white and yellow forewings, and scarlet and black hind wings; flies Jun-Jul. **Larva** Feeds on wetland plants such as Common Comfrey. **Status** Locally common in marshy habitats.

Muslin Moth

Diaphora mendica Arctiidae
Length 23mm

 Superficially similar to the ermines. **Adult** Male is reddish buff, female is white and both have black spots on wings. Flies May-Jun. **Larva** Food plants include dandelions and plantains. **Status** Common and widespread in meadows. Night-flying male comes to light but females are often diurnal.

Ruby Tiger

Phragmatobia fuliginosa Arctiidae
Length 22mm

 Atypical tiger moth. **Adult** Has reddish forewings that conceal pink or grey hind wings at rest. Flies mainly May-Jun. **Larva** Food plants including Common Dandelion and docks. **Status** Widespread and often common; favours meadows and grassy hedgerows.

Scarce Footman

Eilema complana Arctiidae
Length 25mm

 Rests with wings rolled length-ways. **Adult** Has grey forewings with yellow leading edge; hind wings are yellow. Flies Jul-Aug. **Larva** Eats lichens. **Status** Locally common in woods.
Similar species Common Footman *E. lurideola* rests with less-rolled wings; common.

Rosy Footman

Miltochrista miniata
Arctiidae Length 14mm

 Attractive little moth. **Adult** Has rosy-orange forewings bearing black spots. Rests with wings pressed close to the surface on which it is clinging. Flies Jun-Aug. **Larva** Feeds on lichens on tree bark. **Status** Favours wooded areas; common only in S England and Wales.

Noctuid moths

Noctuid moths (family Noctuidae) form a diverse group with several very common representatives. At rest the wings are either held flat or tent-like over the body. The noctuids shown here regularly come to light and are frequently captured, providing wonderful opportunities for close-up study.

Heart and Dart

Agrotis exclamationis Noctuidae
Length 20mm

Its name derives from the forewing markings. **Adult** Has brown wings with variable dark markings but always a dark heart-shaped patch and line. Flies May-Jul. **Larva** Eats a wide range of herbaceous plants. **Status** Widespread and extremely common in gardens and waysides.

True Lover's Knot

Lycophotia porphyrea Noctuidae
Length 16mm

Small, well-marked moth. **Adult** Forewing has variably brown ground colour and intricate black and white markings. Flies Jun-Aug. **Larva** Feeds on Heather and Bell Heather. **Status** Widespread and locally common on heaths and in open woods.

Large Yellow Underwing

Noctua pronuba Noctuidae
Length 25mm

Familiar wayside moth. **Adult** Similar to Broad-bordered but forewings are more marbled and yellow hind wings have only a narrow black border. Flies Jun-Sep. **Larva** Eats almost any herbaceous plant. **Status** Common and widespread in gardens, meadows and woods.

Dark Sword-grass

Agrotis ipsilon Noctuidae
Length 25mm

Has elongated wings and rests with one overlapping the other. **Adult** Has variably brown forewings with a pale band and a kidney-shaped mark and dart-shaped line near outer edge. Flies Jul-Oct. **Larva** Feeds on low herbaceous plants. **Status** Common migrant.

Broad-bordered Yellow Underwing

Noctua fimbriata Noctuidae
Length 25mm

Easily disturbed, needing little encouragement to fly even in daytime. **Adult** At rest, has variably marked forewings that obscure yellow and black hind wings. Flies Jul-Sep. **Larva** Eats a wide range of herbaceous plants. **Status** Widespread and common.

Beautiful Yellow Underwing

Anarta myrtilli Noctuidae
Length 15mm

Beautifully marked moth. **Adult** Has rufous wings with an intricate pattern of darker brown and white markings; black-bordered yellow hind wings are hidden at rest. Flies Apr-Aug in 2 broods. **Larva** Feeds on Heather and Bell Heather. **Status** Widespread on heaths and moors.

Broom Moth

Melanchra pisi Noctuidae
Length 22mm

 Variable but distinctive moth. **Adult** Has a variable forewing ground colour but with a characteristic pale spot on hind edge; spots on both wings meet when moth is resting. Flies May-Jun. **Larva** Feeds on Broom and many other plants. **Status** Widespread and fairly common throughout the region.

larva

Hebrew Character

Orthosia gothica Noctuidae
Length 20mm

 Familiar moth. **Adult** Recognised by a dark mark on wings that has a semicircular section removed as if by a hole-punch. Flies Mar-Apr and comes to light. **Larva** Feeds on a variety of herbaceous plants. **Status** Common and widespread in gardens, hedgerows and meadows.

Brown-line Bright-eye

Mythimna conigera Noctuidae
Length 21mm

 Distinctive and well-named moth. **Adult** Has markings on reddish-brown forewings that are accurately described by its name. Flies Jun-Aug. **Larva** Feeds on various species of grasses. **Status** Widespread and often common in meadows and grassy hedgerows throughout the region.

Common Wainscot

Mythimna pallens Noctuidae
Length 20mm

 Rather undistinguished moth. **Adult** Has straw-coloured forewings, the veins of which are white; white hind wings are usually hidden at rest. Flies Jun-Oct in 2 broods. **Larva** Feeds on various grass species. **Status** Widespread and locally common. Favours meadows and hedges.

Shark

Cuculia umbratica Noctuidae
Length 32mm

 Unusually shaped moth, sometimes found resting on frayed wooden surfaces in daytime. **Adult** Has narrow, pointed wings, grey-brown with darker veins. Flies May-Jul. **Larva** Eats sow-thistles. **Status** Locally common in the S on rough meadows and waste ground.

Satellite

Eupsilia transversa Noctuidae
Length 23mm

 Distinctive, richly coloured moth. **Adult** Has orange-brown wings with a prominent white half-moon mark, around which 2 'satellite' white spots orbit. Flies Oct-Apr. **Larva** Feeds on deciduous shrubs and trees. **Status** Common in woods, hedges and gardens.

Noctuid moths

Poplar Grey
Acronicta megacephala Noctuidae
Length 20mm

 Subtly beautiful moth. **Adult** Has grey forewings that are well marked with darker lines and a well-defined pale circle; pale hind wings are usually hidden at rest. Flies May–Aug. **Larva** Eats poplars and willows. **Status** Common and widespread in parks, gardens and woods.

Miller
Acronicta leporina Noctuidae
Length 20mm

 Variable but distinctive moth. **Adult** Forewing ground colour varies from pale to dark grey and typically shows a jagged black transverse line from leading edge. Flies Apr–Jun. **Larva** Hairy; feeds on birches and other deciduous trees. **Status** Fairly common in woodland throughout.

Grey Dagger
Acronicta psi Noctuidae
Length 23mm

 Distinctively patterned moth. **Adult** Recognised by the distinctive black, dagger-like markings on pale grey forewings. Flies Jun–Aug. **Larva** Colourful; feeds on deciduous shrubs and trees. **Status** Common and widespread, mainly in woodland.

Nut-tree Tussock
Colocasia coryli Noctuidae
Length 18mm

 Well-marked moth. **Adult** Forewing has a variable ground colour but always shows a rich brown transverse band containing a small eye-like marking outlined in black. Flies May–Jul. **Larva** Feeds on Hazel. **Status** Locally common throughout, favouring deciduous woodland.

Lunar-spotted Pinion
Cosmia pyralina Noctuidae
Length 15mm

 Attractive moth that holds its wings in a tent-like manner when at rest. **Adult** Has reddish-brown wings with pale scalloping on leading edge. Flies Jul–Aug. **Larva** Feeds on Hawthorn, Blackthorn and other shrubs. **Status** Locally common only in the S, in woods and hedges.

Straw Underwing
Thalpophila matura Noctuidae
Length 22mm

 Well-marked moth. **Adult** Has marbled brown forewings with an eye-spot; at rest, these conceal dark-bordered pale yellow hind wings. Flies Jul–Aug. **Larva** Feeds on various grasses. **Status** Widespread and fairly common, favouring meadows and rough ground.

Green Silver Lines

Pseudoips fagana Noctuidae
Length 17mm

Stunning moth. **Adult** Has bright green forewings; hind wings are white in male and yellowish in female. Flies Jun–Jul. **Larva** Feeds on oaks, Hazel and other trees and shrubs. **Status** Widespread but local in England and Wales; found in deciduous woodland.

Scarce Silver Lines

Bena prasinana Noctuidae
Length 22mm

Forewings are held in a tent-like manner over body at rest. **Adult** Has green forewings with 2 white transverse lines. Flies Jul–Aug. **Larva** Feeds on oak. **Status** Despite its name, it is not uncommon in many parts of England and Wales; found in mature woodland.

Burnished Brass

Diachrisia chrysitis Noctuidae
Length 21mm

Stunning moth. **Adult** Has golden metallic areas on forewings. Flies Jun–Jul; a 2nd brood flies in Aug in the S. **Larva** Feeds on Common Nettle and other herbaceous plants. **Status** Common and widespread in gardens, hedgerows and rough ground.

House guests

Uninvited or otherwise, a surprising variety of wildlife enters our homes from the garden. Several moth species belonging to the family Noctuidae are more than casual visitors drawn in by the light; rather, they are determined insects that seek sanctuary in our homes. During the summer months, **Svensson's Copper Underwing** *Amphipyra berbera* (**1**) (length 26mm) and the closely related Copper Underwing *A. pyramidea* often roost in the daytime in sheds and sometimes behind curtains in open windows; the adults' coppery underwings are hidden at rest by the well-marked forewings and it flies in August–September. The **Herald** *Scoliopteryx libatrix* (**2**) (length 20mm) is unusual because it hibernates over the winter, often using sheds or cool doorways for shelter; its has brick-red forewings and flies in August–November, then again in March–June after waking from hibernation. The **Old Lady** *Mormo maura* (**3**) (length 32mm), which is superbly camouflaged when resting on wooden fences, also ventures indoors occasionally to roost in the daytime; it flies in July–August. The **Angle Shades** *Phlogophora meticulosa* (**4**) (length 27mm), recognised by the ragged margins and olive-green and pinkish markings of its forewings, flies in May–October; its larvae sometimes feed on potted geraniums and so are brought indoors in autumn, the adults emerging in midwinter.

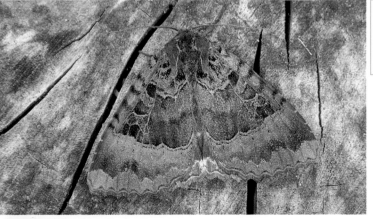

Noctuids to emeralds

Even if you are simply interested in moths for their aesthetics, you will never run out of inspiration. Many noctuids (family Noctuidae) and hook-tips (Drepanidae) have distinctive markings or striking wing shapes while emeralds (family Geometridae) are just plain colourful.

The Spectacle

Abrostola triplasia Noctuidae
Length 18mm

Well-named moth: viewed head on, it shows distinctive 'spectacle' markings. **Adult** Has grey forewings with a dark central band. Flies May-Aug. **Larva** Feeds on Common Nettle. **Status** Widespread and often common on rough ground, verges and woodland rides.

Migrant moths

Several species of moths are strong flyers and migrate each year in spring and summer from mainland Europe. One of the most regularly encountered and numerous of these is the **Silver Y** *Autographa gamma* (length 21mm), which occurs between May and October and is often seen on the wing in the daytime, as well as after dark. Its forewing bears a distinctive white 'Y' marking, hence its name.

Red Underwing

Catocala nupta Noctuidae
Wingspan 65mm

Impressive moth, camouflaged when resting on bark. **Adult** Has black-barred red underwings that are concealed at rest by marbled grey and brown forewings, and revealed when the moth is disturbed. Flies Aug-Sep. **Larva** Feeds on willows and poplars. **Status** Common only in S England, in open woodland.

Beautiful Golden Y

Autographa pulchrina Noctuidae
Length 21mm

Well-marked moth. **Adult** Superficially similar to Silver Y but forewings are marbled brown not grey and 'Y' marking is broken. Flies Jun-Jul. **Larva** Feeds on dead-nettles and other plants. **Status** Widespread and common in woodland and hedgerows.

Witch's namesake

Mother Shipton, England's best-known prophetess (or witch to some people) was born in a cave near Knaresborough, North Yorkshire, in 1488. Famed for her ability to foresee the future rather than for her looks, her hooked-nose profile is easy to discern on the wings of her namesake moth. The **Mother Shipton** *Callistege mi* (wingspan 23mm) flies in May-July and is locally common in meadows, woodland rides and on rough ground. Its larva feeds on various species of clover.

Straw Dot

Rivula sericealis Noctuidae
Length 15mm

Unmistakable moth. **Adult** Has buffish-yellow wings with a dark, central, kidney-shaped dot. Flies Jun–Jul. **Larva** Feeds on grasses, notably Purple Moorgrass. **Status** Common in rough grassland; resident numbers are boosted by influxes of migrants.

Large Emerald

Geometra papilionaria Geometridae
Wingspan 42mm

Colourful moth, brightest when newly emerged. **Adult** Has green wings with subtle white spots and lines. Flies Jul–Aug and comes to light. **Larva** Feeds on birches and Hazel. **Status** Locally common in England, Wales and Ireland, on heaths and in woods.

Beautiful Hooktip

Laspeyria flexula Noctuidae
Length 14mm

Aptly named moth. **Adult** Has hooked-tipped, purplish-grey forewings, stippled with black dots and with 2 pale transverse lines. Flies Jul–Aug. **Larva** Feeds on various trees and shrubs. **Status** Locally common in the S in woods and hedges.

Peach Blossom

Thyatira batis Drepanidae
Length 17mm

Beautifully marked moth. **Adult** Has brown forewings with conspicuous pinkish spots and blotches. Flies Jun–Jul; comes readily to light. **Larva** Feeds on Bramble. **Status** Widespread and fairly common along woodland rides and in hedgerows.

Blotched Emerald

Comibaena bajularia Geometridae
Wingspan 30mm

Beautifully marked moth. **Adult** Has bright green wings with brown and white blotches on margins. Flies Jun–Jul and comes to light. **Larva** Feeds on oaks and Hazel. **Status** Locally common only in S and central England and E Wales, in mature oak woodland.

Pebble Hooktip

Drepana falcataria Drepanidae
Wingspan 28mm

Similar to, but not closely related to, Beautiful Hooktip. **Adult** Hooked-tipped forewings have a variable ground colour that shows a dark transverse line. Flies May–Jun. **Larva** Feeds mainly on birches. **Status** Locally common in woodland and on heaths.

Figure of Eighty

Tethea ocularis Drepanidae
Length 24mm

Aptly named moth. **Adult** Has distinct white markings that resemble a figure 80 on otherwise grey-brown forewings. Flies May–Jul and often comes to light. **Larva** Feeds on Aspen and poplars. **Status** Locally common only in the S, favouring wooded habitats.

Geometer moths

Moth camouflage manifests itself in a range of ways, and the insects' behaviour often makes best use of the deception. Many of the geometer moths (family Geometricae) shown here have wings that resemble tree bark covered with lichens, and they have an uncanny ability to place themselves in the right spot to maximise the effect. Others look like dead leaves and rest during the day among dried-up foliage. Geometer moth larvae are called 'loopers' because of the way they walk (*see* p.372).

Carpets and related moths

Carpet moths are a varied group, belonging to the family Geometridae, and are characterised by rather rounded wings that are well marked and provide the moth with extremely good camouflage when resting on appropriate backgrounds. Many species are confusingly similar to one another and their identification is best left to experts, but a few are more striking and obvious.

Blood-vein
Timandra comae Geometridae
Wingspan 32mm

 Unmistakable moth. **Adult** Has straw-coloured wings with a dark red transverse line across forewings and red outer margin to both wings. Flies May-Nov in 2 broods. **Larva** Feeds on Common Sorrel, Common Orache and related plants. **Status** Locally common in damp waysides.

Green Carpet
Colostygia pectinataria Geometridae
Wingspan 20mm

 Colourful, well-marked moth. **Adult** Well camouflaged when resting on lichen-covered tree bark; flies May-Aug. **Larva** Feeds on bedstraws. **Status** Common in meadows and on heaths throughout the region.

Red Twin-spot Carpet
Xanthorhoe spadicearia Geometridae
Wingspan 20mm

 Subtly colourful moth with a descriptive name. **Adult** Has a reddish-brown band on forewings, with 2 reddish spots near tip; flies mainly May-Jul. **Larva** Feeds on low herbaceous plants. **Status** Common in wooded areas.

Silver Ground Carpet
Xanthorhoe montanata Geometridae
Wingspan 22mm

 Sometimes disturbed from wayside vegetation in daytime. **Adult** Has whitish forewings marked with a darker central transverse band; flies May-Aug. **Larva** Feeds on bedstraws. **Status** Widespread and common in grassy areas throughout the region.

July Highflyer
Hyrdriomena furcata Geometridae
Wingspan 18mm

 Variably marked moth. **Adult** Has marbled green and brown wings; flies Jul-Aug. **Larva** Feeds on Hazel. **Status** Widespread and common; associated mainly with Hazel, which limits its precise occurrence.

Red Twin-spot Carpet

Silver Ground Carpet

Barred Straw

Eulithis pyraliata Geometridae
Wingspan 35mm

Intriguing moth whose forewings obscure its hind wings at rest. **Adult** Has straw-yellow wings marked with dark brown bars. Flies Jul–Sep and comes to light. **Larva** Feeds on bedstraws and Common Cleavers. **Status** Common and widespread in wayside habitats.

Foxglove Pug

Eupithecia pulchellata Geometridae
Wingspan 18mm

Attractive and brightly marked moth. **Adult** Has marbled reddish-brown and dark grey forewings. Flies May–Jul. **Larva** Feeds within Foxglove flowers. **Status** Widespread and common, particularly in the S and W, on heaths, along woodland rides and on coasts.

Scalloped Oak

Crocallis elinguaria Geometridae
Length 17mm

Subtly attractive moth. **Adult** Has variably coloured forewings, although they are often buffish yellow with a broad, transverse brown band containing a single dark spot. Flies Jun–Aug. **Larva** Feeds on most deciduous trees and shrubs. **Status** Fairly common in woodland.

Leaf look-alikes

Tree and shrub leaves are such features of the countryside from spring to autumn that it is little wonder that many moths, including members of the family Geometridae, use the foliage as a place of refuge during the daytime. A few are particular mimics of the leaves themselves. The **Canary-shouldered Thorn** *Ennomos alniaria* (**1**) (wingspan 35mm) has a bright canary-yellow thorax and variably buff wings; it flies in August–September and favours woodland, and its larvae feed on birches, Hazel and other trees and shrubs. The **Purple Thorn** *Selenia tetralunaria* (**2**) (length 24mm) rests with its marbled purplish-brown and buff wings raised above the body. It flies in April–May and July–August, and favours woods and heaths. The resemblance of the **Lilac Beauty** *Apeira syringaria* (**3**) (wingspan 40mm) to a dead leaf is enhanced by the creased leading edge to its forewing; it flies in June–September in woodland, and its larvae eat Honeysuckle and Privet.

Magpie Moth

Abraxas grossulariata Geometridae
Wingspan 38mm

Unmistakable and easily recognised moth. **Adult** Has patterns of black spots and yellow on otherwise white wings. Flies Jul–Aug. **Larva** Feeds on various shrubs. **Status** Widespread and fairly common in England, Wales and Ireland in woods, hedgerows and gardens.

Brimstone Moth

Opisthographis luteolata
Geometridae Wingspan 28mm

Colourful, unmistakable moth. **Adult** Has bright yellow wings marked with chestnut blotches. Flies Apr–Oct in several broods. **Larva** Feeds on Blackthorn, Hawthorn and other shrubs. **Status** Common in the S in hedgerows, woodlands and mature gardens.

Scorched Wing

Plagodis dolabraria Geometridae
Wingspan 23mm

Intriguingly marked moth. **Adult** At rest, base of wings and tip of abdomen look scorched. Flies May–Jun. **Larva** Feeds on oaks, birches and other deciduous trees. **Status** Fairly common in England, Wales and Ireland in wooded habitats.

Geometer moths

Orange Moth
Angerona prunaria Geometridae
Wingspan 40mm

Attractive and colourful moth. **Adult** Male has orange-brown wings while those of female are yellowish. Forewings of both sexes have numerous short, transverse lines. Flies Jun-Jul. **Larva** Feeds on birches, Hawthorn and Heather. **Status** Locally common in the S in woodland and heaths.

Speckled Yellow
Pseudopanthera macularia Geometridae
Wingspan 30mm

Day-flying moth that is often disturbed from wayside vegetation. **Adult** Has deep yellow wings with grey-brown blotches. Flies May-Jun. **Larva** Feeds on Wood Sage, Yellow Archangel and related plants. **Status** Locally common in the S in woodland rides, rough ground and hedgerows.

Swallowtailed Moth
Ourapteryx sambucaria Geometridae
Wingspan 52mm

Distinctive moth. **Adult** Has pale yellow wings, an angular tip to forewing and a short tail streamer on hind wing. Flies Jun-Jul. **Larva** Feeds on Ivy, Hawthorn and other shrubs. **Status** Locally common throughout Britain and Ireland, except in the N, in woodland, gardens and hedgerows.

Moths in winter

If you have an outside light at home that remains on overnight, it can often be worth checking nearby walls in the morning for roosting moths. This can be surprisingly rewarding in winter, and you may encounter some of these whose names reflect their seasonal appearances.

The **November Moth** *Epirrita dilutata* (**1**) (length 20mm) has rather rounded wings that are variably marbled grey and brown; it flies mainly in November and is found in woodland, hedges and gardens, its larvae feeding on a variety of shrubs and trees. The **Winter Moth** *Operophtera brumata* (**2**) (wingspan of male 28mm) is similar but smaller, and only males have wings; it flies mainly in November-February and is often seen in car headlights. It is a woodland and garden species, the larva feeding on most deciduous trees and shrubs. Often coinciding with the appearance of the Winter Moth is the **December Moth** *Poecilocampa populi* (**3**) (length 16mm), a dark, hairy-bodied species whose larvae feed on deciduous trees. Next to appear is the **Early Moth** *Theria primaria* (**4**) (length of male 17mm), only males of which are winged; they fly in January-February in woods and hedgerows. The **March Moth** *Alsophila aescularia* (**5**) (length of male 20mm) appears in its namesake month and typically rests with one wing overlapping the other; it is a woodland and hedgerow species.

Mottled Umber

Erannis defoliaria Geometridae
Wingspan 40mm (male)

Autumnal moth with wingless females. **Adult** Male's forewings vary from pale to dark brown; a pale central band, bordered by a black line and containing a black dot, is usually visible. Flies Oct-Dec. **Larva** Feeds on deciduous trees and shrubs. **Status** Common in woods and gardens.

Waved Umber

Menophra abruptaria Geometridae
Wingspan 40mm

Attractive moth with wing markings that recall cut, decaying timber. **Adult** Has buffish wings marked with horizontal dark lines. Flies Apr-Jun. **Larva** Feeds on lilac and Garden Privet. **Status** Favours mature wooded gardens and parks, and is fairly common.

Oak Beauty

Biston strataria Geometridae
Length 23mm

Attractive moth with good camouflage on tree bark. **Adult** Has marbled reddish-brown, grey and black wings. Flies Mar-Apr. **Larva** Feeds on various deciduous trees. **Status** Widespread and common only in England and Wales in wooded areas, hedgerows and gardens.

Pollution, melanism and the Peppered Moth

In the 1980s and 1990s, biology textbooks often used the **Peppered Moth** *Biston betularia* (wingspan 48mm) to illustrate natural selection in action. The wing colour varies naturally but is usually one of two extremes: an all-black melanic form; and a form with dark-speckled white wings that are camouflaged on tree-bark lichens. As 20th-century atmospheric pollution caused lichens to die out in urban and industrial areas, so the speckled form dwindled (it was easily picked off by birds) and the black form predominated. In rural areas, there was always a more even proportion of the two forms. Today, as some forms of atmospheric pollution have been reduced, lichens are faring better in urban sites and hence black Peppered Moth forms are less frequent. However, we now pollute our world in more subtle ways and, in common with most other moth species, all forms of the Peppered Moth are in decline.

Brindled Beauty

Lycia hirtaria Geometridae
Wingspan 40mm

Variable colours and markings. **Adult** Usually grey-brown, the forewings with black lines and stippling, and a yellow-buff suffusion. Flies Mar-Apr and males come to light. **Larva** Feeds on deciduous trees. **Status** Common in the S in wooded areas.

Mottled Beauty

Boarmia repandata Geometridae
Wingspan 38mm

Well camouflaged on tree bark. **Adult** Has pale grey-brown or sometimes darker wings with fine black lines and stippling. Flies Jun-Jul. **Larva** Feeds on birches, oaks and Bramble. **Status** Widespread and often common in woods and gardens.

Moth & butterfly larvae & pupae

The larval and pupal stages of butterflies and moths can be just as fascinating as the adults. Some are colourful, many are superbly camouflaged and a few have an extraordinary appearance. Some species can be a challenge to find, but once discovered they do have the advantage over their parents that they cannot fly away!

Larvae

Moth and butterfly larvae are sometimes referred to as caterpillars, and their role in life is to eat, grow and finally pupate. Because they have relatively inelastic skins but grow considerably, they moult several times, the body size increasing by increments with each moult. The larvae of some moth and butterfly species will eat only one particular food plant, which makes them especially vulnerable to changes in the environment. But many species – moths in particular – eat the leaves of a wide range of plants.

Looper caterpillars The larvae of many moth species in the family Geometridae are camouflaged to resemble twigs; when resting on their food plants they can be hard to spot, such is the accuracy of the deception. They move in a shuffling manner, with arched bodies, and are often called 'looper' caterpillars because of their style of movement.

Swallowtailed Moth larva.

The **Puss Moth's** plump-bodied larva is an extraordinary creature. It has two whiplash tails that help ward off parasitic wasps and a head end that can be swollen when alarmed, exaggerating the false eye-spots; these are presumed to serve as a deterrent to small birds.

Hawkmoth larvae are always impressive, being strikingly marked and often rather large. The larva of the Death's-head Hawkmoth when full grown, can reach a length of 14cm. Although seldom seen in this country (it is a rare migrant), a sighting is a prize worth seeking.

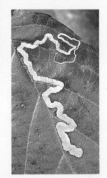

Leaf-miners In addition to the larger British and Irish moths, many tiny species are found, collectively known as 'micro-moths'. Their larvae are tiny and often inconspicuous. That of the Bramble Leaf-miner *Stigmella aurella* leaves a telltale sign of its presence: it lives and feeds inside Bramble leaves and its home shows up on the leaf as a white streak.

The **Lobster Moth larva** bears a resemblance to its crustacean namesake when fully grown. Smaller larvae are rather ant-like, and when resting on Beech leaves they can look rather like the fallen remains of the bractsthat encased the tree's flowers.

Warning colours

Taking the principle of 'you are what you eat' to the extreme, many larvae are distasteful or poisonous as a result of accumulating chemicals from the leaves they consume. Generally, such larvae go out of their way to be noticeable, advertising their distasteful natures with warning markings. In addition to such markings, the Alder Moth has lobed bristles to discourage unwelcome interest from parasites and predators.

INSET: Mullein Moth larva
BELOW: Alder Moth larva

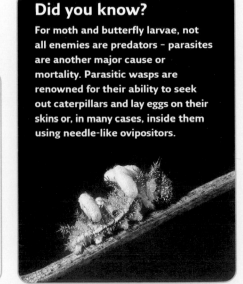

Did you know?

For moth and butterfly larvae, not all enemies are predators – parasites are another major cause or mortality. Parasitic wasps are renowned for their ability to seek out caterpillars and lay eggs on their skins or, in many cases, inside them using needle-like ovipositors.

Hairy caterpillars

As larvae have soft, juicy bodies and make succulent meals for insect-eating birds and other predators, many species use camouflage to avoid detection. Some tackle the problem of predation by producing larvae that are armed with irritating hairs that cause considerable discomfort if ingested. A bird or small mammal may try this once, but soon learns to avoid such hairy caterpillars.

Drinker larva

Pale Tussock larva

Garden Tiger larvae, sometimes called 'woolly bears'

Hanging by a strand

Many moth larvae produce silk, either to help attach themselves to the leaf on which they are feeding, or in some cases to spin a

cocoon in which to pupate. But some tiny larvae use it in a more active manner: in order to find new leaves on which to feed, or to reach the ground safely for pupation, they suspend themselves in silken strands and 'abseil' in the breeze.

Pupae

The pupa of a butterfly or moth is sometimes referred to as a chrysalis. Although it may appear to be a resting stage in the insect's life cycle, miraculous changes are occurring internally, with larval tissues and organs metamorphosing into adult structures. Being externally inactive, pupae are vulnerable to attack by predators and parasites, and to the weather, which is why many species pupate in sheltered places. With butterflies, the location may be the underside of a leaf, or among basal plant stems, while the larvae of many moths burrow underground before pupating in a silk-lined chamber.

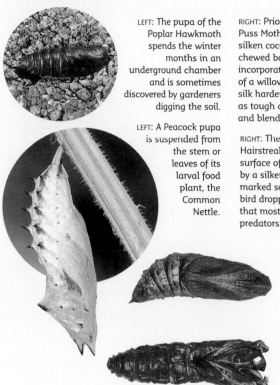

LEFT: The pupa of the Poplar Hawkmoth spends the winter months in an underground chamber and is sometimes discovered by gardeners digging the soil.

LEFT: A Peacock pupa is suspended from the stem or leaves of its larval food plant, the Common Nettle.

RIGHT: Prior to pupation, the Puss Moth larva spins a silken cocoon, in which chewed bark is incorporated, on the trunk of a willow tree. Once the silk hardens, the cocoon is as tough as the bark itself and blends in superbly well.

RIGHT: The pupa of the Black Hairstreak is attached to the surface of a Blackthorn leaf by a silken girdle, and is marked so as to resemble a bird dropping, something that most potential predators would avoid.

LEFT: This Elephant Hawkmoth pupa shows the outlines of adult organs and structures including the wings, proboscis and segmented abdomen

LEFT: Hornet Moth larvae live and feed inside the trunks and roots of Black Poplar trees and pupate just below the bark, making them difficult to discover. Empty pupal cases, left on the ground where moths have emerged, persist for some time and can give a good clue to the species' presence in an area.

The miracle of metamorphosis

If you follow the life cycle of butterflies or moths by breeding them in captivity, you will sometimes have the privilege of watching an adult emerge from a pupa. Typically, this happens in the early morning. The process takes about 10–15 minutes: first, the pupa case splits near the head end and the adult head and thorax appear; the adult then frees its legs and pulls its soft wings and abdomen free; finally, resting on the empty pupal case, the adult inflates its wings by pumping fluid into the veins before allowing them to dry and harden.

INSET: Comma emerging from a split pupal case.
MAIN PIC: adult resting on an empty pupal case.

A Marbled White pupa seen a few hours before the adult emerges.

Did you know?

For a few hours or so prior to the emergence of the adult butterfly or moth, the outer case of the pupa becomes partly transparent and the wing colours of the adult can be discerned.

Bristletails, springtails, mayflies & stoneflies

Even the smallest and seemingly most insignificant of insects, such as bristletails and springtails, have features and habits to engage our interest. And, if nothing else, short-lived groups such as mayflies are fascinating for their ephemeral adult lives. The nymphs of mayflies and stoneflies are aquatic and are covered on p.438.

Bristletails

Bristletails belong to the order Thysanura and are wingless, scuttling insects with elongate, rather flattened and strongly segmented bodies; a close look reveals the body to be covered in scales, with three long bristly 'tails' at the rear end. They feed on detritus and two species are regularly encountered. The Silverfish *Lepisma saccharina* (length 10mm) is a familiar, if not altogether welcome, household resident that favours damp sites such as behind sinks, baths and kitchen units; it feeds on spilt starchy food. Petrobius maritimus (length 10mm) is found on rocky shores above the high-tide line and in sea caves.

Silverfish

Petrobius maritimus

 Silverfish

 Petrobius maritimus

Springtails

These tiny wingless insects (order Collembola) are so called because of the forked appendage at their rear end, which allows them to spring into the air and hence escape danger; if you turn over leaf litter on a woodland floor you are highly likely to see some examples. Tiny they may be, but springtails are among the most abundant animals on the planet. Their abdomens have fewer segments than those of other insects. Specific identification can be hard unless you are an expert, but species of *Tomocerus* (length 2–4mm) are common in gardens and woodlands.

ABOVE: *Tomocerus* sp., typical springtails
BELOW: mass of springtails

Mayflies

Best known for their brief adult lives, mayflies belong to the order Ephemeroptera. Adults are delicate-looking with long tail streamers and two pairs of wings in most species; the forewings are much larger than the hind wings and are held above the body at rest. Their flight is typically weak, although males of some species do perform elegant 'dances' on the wing. Mayfly nymphs are aquatic.

Green Drake

Ephemera danica Ephemeridae
Length 20mm

Familiar mayfly. **Adult** Has brown bands and spots on wings, a creamy abdomen with darker dorsal markings towards rear, and 3 'tails'; flies May–Aug. **Nymph** Aquatic. **Status** Widespread and common; favours rivers and gravel pits. **Similar species** *E. vulgata* (length 20mm) has usually triangular (not oblong) dark markings on upper surface of yellowish abdomen that are also more extensive; it favours slow-flowing rivers, is common only in S and central England, and flies May–Aug.

Cloeon dipterum
Baetidae Length 5mm

Commonly occurring mayfly, seen as an adult in spring and summer. **Adult** Has 2 'tails', 2 pairs of wings (clear forewings appear sheened at some angles) and a yellowish-brown abdomen. **Nymph** Aquatic. **Status** Common and widespread in ponds, streams and canals.

Stoneflies

Stoneflies belong to the order Plecoptera. The adults are rather flat-bodied, weak-flying insects that are invariably found in the vicinity of fresh water and are often seen creeping or scuttling over waterside stones. They have two pairs of wings that are almost rolled around the body length-ways at rest, and there are two projecting 'tails' at the end of the abdomen. Specific identification is best left to the experts, but common examples include the following.

Nemoura cinerea
Nemouridae Length 10mm

Classic waterside stonefly. **Adult** Wings are smoky-brown, and tail appendages do not project beyond them at rest; flies May–Jul. **Nymph** Aquatic. **Status** Widespread and fairly common. Found in vegetation beside clean, fast-flowing streams and rivers.

Dinocras cephalotes
Perlidae Length 22mm

Large, robust stonefly. **Adult** Has a dark brown body and long tail appendages that project beyond smoky-grey wings at rest. Flies May–Jul. **Nymph** Lives in fast-flowing stony upland rivers and streams. **Status** Common in parts of N and W Britain.

Mayfly nymphs

Mayfly nymphs are always aquatic and, depending on their favoured water body and body form, can be free-living (clinging to stones) or bury themselves in mud or silt. Three tail appendages can be seen at the rear end. Representative species can be found in all types of fresh water, from fast-flowing streams to the still waters of lakes and streams. Most are sensitive to pollution. Find out more about mayfly nymphs, and other aquatic insects, on pp.436–8.

Ecdyonorus venosus nymph

Stonefly nymphs

Stonefly nymphs are aquatic, with flattened bodies and two projecting 'tails'. **Perla bipunctata** nymphs (length 16mm; shown here) are found in stony rivers, mainly in N and W Britain. They are sometimes seen clinging to stones in the water, and shed nymphal skins can be found among waterside vegetation from May–Jul after the adult (which has a yellowish body and smoky wings) has emerged. Find out more about stonefly nymphs and other aquatic insects on pp.436–8.

Did you know?

Although stick insects are not native to Britain and Ireland, five species (one from S Europe, four from New Zealand) have been recorded here, mainly in SW England and as a result of accidental introductions on imported plants. Colonies of some species are now established – the **Prickly Stick-insect** *Acanthoxyla geisovii* (length 10cm) is probably the most regularly encountered, especially on the Isles of Scilly. It is found on Brambles and garden shrubs.

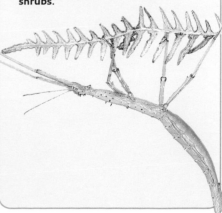

Crickets & grasshoppers

Crickets and grasshoppers belong to the order Orthoptera, a group that also includes bush-crickets and groundhoppers (*see* pp.378–9). They are squat insects with powerful hind legs that allow them to hop or, in some cases, leap. Wings, when present, comprise two pairs and the thorax is protected by a hardened shield called the pronotum.

Recognising orthopterans

It is usually relatively easy to know if you are looking at an orthopteran, but deciding which of the four common family groups it belongs to can be more tricky.

True crickets (family Gryllidae) (1) have flattened bodies, long antennae, wings that lie flat over the abdomen and two projecting appendages at the tail end (called 'cerci'). Males 'sing' by rubbing their wings together, and females have a needle-like ovipositor at the tail end for laying eggs.

Grasshoppers (family Acrididae) (2) have rather cylindrical bodies, short antennae and powerful hind legs that are adapted for jumping; the wings are held in a tent-like manner over the abdomen, and males 'sing' by rubbing the hind legs against the forewing. Females are larger than males.

Groundhoppers (family Tetrigidae) (3) resemble tiny grasshoppers, but the shield-like pronotum extends back over most of the abdomen (it just covers the thorax in grasshoppers). The antennae are short and only the hind wings are used for flight.

Bush-crickets (family Tettigoniidae) (4) have extremely long antennae and long, slender legs, the hind pair of which enables them to hop well. Males 'sing' by rubbing their wings together and females have a sword-like ovipositor for laying eggs.

Field Cricket

Gryllus campestris Gryllidae
Length 24mm

Impressive large-headed cricket. Male sings chirping song from burrow entrance in warm weather. **Adult** Has a blackish body with yellow at base of wings. **Status** Favours short grassland and lives in burrows. Rare; found in a few sites in S England. Adults seen May–Jun, nymphs Jul–Apr.

Mottled Grasshopper

Myrmeleotettix maculatus Acrididae
Length 16mm

Small, stout grasshopper. **Adult** Ground colour varies from green to brown but the body always looks marbled. Has inflected, angular lines on pronotum. Tips of antennae are clubbed in male, swollen in female. **Status** Fairly common throughout on dry dunes, heaths and chalk downs.

Wood Cricket

Nemobius sylvestris Gryllidae
Length 7mm

Unobtrusive cricket. Male produces a soft, warbling song. **Adult** Marbled yellowish brown and black. **Status** Local, in a few sites in S England; easiest to see in the New Forest. Lives in colonies among leaf litter. Has a 2-year life cycle, so either nymphs or adults are seen at most times of year.

Meadow Grasshopper

Chorthippus parallelus Acrididae
Length 17–23mm

Fairly distinctive grasshopper. **Adult** Lacks hind wings and has rather short forewings that do not reach tip of abdomen; proportionately much shorter in female. Pronotum is gently incurved. **Status** Widespread and common in grassy meadows.

BELOW: male; INSET: female

Lesser Marsh Grasshopper

Chorthippus albomarginatus
Acrididae Length 21mm

Distinctive coastal species. **Adult** Recalls Meadow Grasshopper with its gently incurved pronotum, but forewing is much longer (still does not reach tip of abdomen) and shows bulge near base on anterior margin. **Status** Restricted to coastal grassland and dunes, mainly in the S.

Common Field Grasshopper

Chorthippus brunneus Acrididae
Length 18–24mm

 Familiar species. **Adult** Ground colour is variable, but marbled grey, brown or grey are common. Note the bulge at base of forewing; pronotum is inflected and angular, and black wedge markings do not reach hind edge. **Status** Common throughout in dry, grassy places.

Woodland Grasshopper

Omocestus rufipes Acrididae
Length 15–18mm

 Distinctive species. **Adult** Has a proportionately large head, conspicuous white tips to palps, and (in maturity) a bright red tip and underside of abdomen; head and thorax are often almost black. **Status** Rather local in open woods and grassy heaths in S England and S Wales.

Large Marsh Grasshopper

Stethophyma grossum Acrididae
Length 28–32mm

 Our largest grasshopper. **Adult** Body and wings are well marked with lime green, yellow and black. Hind legs are marked with black and yellow bands. **Status** Local and restricted to floating acid bogs with *Sphagnum* moss in S and E England and W Ireland.

Heathland specialist

The **Heath Grasshopper** *Chorthippus vagans* (length 15–18mm) is confined to heathland and is known only from Dorset and the New Forest. It is superficially similar to the Common Field Grasshopper but is mottled grey, and black wedge markings on the pronotum reach the hind edge. It is usually found in Heather clumps, into which it dives if alarmed.

Common Green Grasshopper

Omocestus viridulus Acrididae
Length 17–20mm

 Widespread meadow species. **Adult** Usually has a pure green body. Note the keel on top of head; pronotum is incurved and forewings do not have a bulge near base on anterior margin. **Status** Locally common in Britain and Ireland in undisturbed grassy places.

Stripe-winged Grasshopper

Stenobothrus lineatus Acrididae
Length 17–19mm

 Beautifully marked species. **Adult** Has a greenish ground colour to body and wings. Forewing has a white comma-like mark and white stripe along anterior margin; the latter continues along pronotum and around head. **Status** Local on dry chalk and limestone grassland.

Rufous Grasshopper

Gomphocerippus rufus Acrididae
Length 16–18mm

 Distinctive species. **Adult** Recognised by rufous-brown colour and white-tipped, clubbed antennae. Adults do not appear until Aug–Sep. **Status** Rather local and restricted to chalk and limestone grassland in S England and S Wales; favours S-facing slopes and North Downs is its stronghold.

Bush-crickets & groundhoppers

Bush-crickets (family Tettigoniidae) are insects of wayside habitats. Males 'sing' high-pitched songs to attract mates and females use their fear-some-looking (but harmless) ovipositors to insert eggs into plant stems. Groundhoppers resemble miniature grasshoppers; they are hard to find.

Speckled Bush-cricket

Leptophyes punctatissima Tettigoniidae
Length 14mm

 Dainty plump-bodied insect. **Adult** Green, speckled with black dots, and with long, spindly legs. Female has a scimitar-shaped ovipositor. **Status** Common in S England and S Wales but scarce elsewhere. Favours hedges and scrub. Often on Bramble leaves.

Bog Bush-cricket

Metrioptera brachyptera Tettigoniidae
Length 17mm

 Well-marked species. **Adult** Has a brown body except for bright green underside to abdomen. Top of pronotum and wings are either brown or bright green; note the pale line on hind margin of pronotum side-flap. **Status** Local and restricted to bogs and wet heaths.

Grey Bush-cricket

Platycleis albopunctata Tettigoniidae
Length 24mm

 Distinctive insect. **Adult** Has a mainly marbled grey-brown body with a yellow underside to abdomen; female has an upcurved ovipositor. **Status** Favours warm coastal grassland, often on S-facing slopes. Restricted mainly to S England and S Wales. Often remains hidden during daytime.

Dark Bush-cricket

Pholidoptera griseoaptera Tettigoniidae
Length 15–17mm

 Familiar wayside insect. Male sings a chirping 'song'. **Adult** Has a marbled dark brown body and thick 'thighed' hind legs. Female has an upcurved ovipositor. Fore-wings vestigial in female and just flaps in male. **Status** Common except in Scotland, in hedgerows and rough grassland.

Oak Bush-cricket

Meconema thalassinum Tettigoniidae
Length 15mm

 Most active after dark and attracted indoors by lighted windows. **Adult** Has a slender green body and long, spindly legs. Female has a narrow, slightly upcurved ovipositor. **Status** Fairly common in S and central England and Wales. Favours woodland and gardens.

BELOW: Oak Bush-cricket

Spreading bush-crickets

As recently as 15 years ago, if you wanted to see a **Roesel's Bush-cricket** *Metrioptera roeselii* (**1**) (length 15–18mm) or a **Long-winged Conehead** *Conocephalus discolor* (**2**) (length 17–19mm) you would have had to visit coastal marshes in SE England. In recent years, both species have spread inland, and their progress northwards seems unstoppable. Today, they are found in all manner of grassy habitats and their ascendancy is usually ascribed to global warming rather than being a positive conservation story. Roesel's Bush-cricket has a marbled brown body with a pale margin to the entire pronotum side-flap; the forewings reach halfway along the abdomen. The Long-winged Conehead has a slender body that is bright green except for a brown stripe on the dorsal surface of the head and pronotum; its brown wings are as long as the abdomen and the female's ovipositor is almost straight.

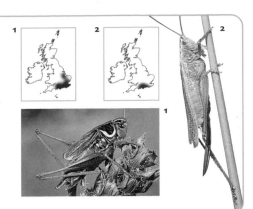

Save the Wartbiter

Another member of the Tettigoniidae family is the intriguingly named **Wartbiter** *Decticus verrucivorus* (length 35mm), so called because in the past it was supposedly used to remove warts. It is a bulky, impressive insect whose body and wings are marbled green and brown; the female has a long ovipositor. Habitat destruction and degradation caused by farming have led to the species' decline, and it is now restricted to a few sites on chalk grassland in S England. Despite its size, it is easily overlooked, although males can sometimes be detected by their loud mechanical songs.

Short-winged Conehead
Conocephalus dorsalis Tettigoniidae
Length 16–18mm

Slender, elongate insect. When alarmed, it aligns its body, antennae and legs along stems. **Adult** Has a green body with a brown dorsal stripe; forewings are reduced. Female has a curved ovipositor. **Status** Locally common only in S England and S Wales; found mainly in coastal grassland.

Groundhoppers

These intriguing little insects (family Tetrigidae) are easily overlooked because of their camouflaged appearance and small size. Two species are common, particularly in the south.

Common Groundhopper
Tetrix undulata Tetrigidae
Length 10mm

Unobtrusive insect. **Adult** Variably coloured but usually marbled brown; wings are shorter than pronotum, which extends length of abdomen. **Status** Common within range in areas of short vegetation. **Similar species Slender Groundhopper** *T. subulata* (length 10mm) has wings that are longer than pronotum; favours damper sites and swims and flies well.

Great Green Bush-cricket
Tettigonia viridissima Tettigoniidae
Length 46mm

Our largest bush-cricket. The male's song is loud but hard to pinpoint. **Adult** Bright green except for a brown dorsal stripe. Female has a long, straight ovipositor. **Status** Local in scrub in S England and S Wales; commonest on the S coast from Dorset to Cornwall.

Great Green Bush-cricket

Large dragonflies

RIGHT: The compound eyes of this Southern Hawker ensure that very little around it goes unnoticed.

Dragonflies belong to the order Odonata, a group that also contains the closely related damselflies. As adults, dragonflies are fast-flying predatory insects with keen eyesight. Their nymphs are aquatic (*see* p.437).

Dragonflies

In flight, dragonflies are remarkable for their speed and manoeuvrability, skills they need in order to avoid predators and to capture flying prey. They are also greatly aided by their keen eyesight – their wrap-around compound eyes give them virtually all-round vision (*see* photograph above right) – and they are extremely sensitive to movement. Hawker dragonflies hunt while on the wing, while darters (*see* p.383) perch for long periods until they spot their prey and launch an attack. Dragonfly mating is an unusual affair, in which the male's accessory reproductive organs at the front of the abdomen play an important role. Females lay their eggs in water, and the predatory nymphs that hatch from them live an aquatic life until emerging as adults the following spring.

Like other dragonflies, this Migrant Hawker can fly for hours on end, hover with consummate ease and outmanoeuvre almost all potential prey insects in flight.

Dragon or damsel?

The order Odonata contains dragonflies (**1**) and damselflies (**2**). Both have two pairs of wings, but in damselflies the forewing and hind wing are usually similar in outline and at rest are typically held folded above the body, whereas in dragonflies the hind wing is usually broader than the forewing and the wings are held out flat at rest. Damselflies have slender bodies while dragonflies have thicker ones.

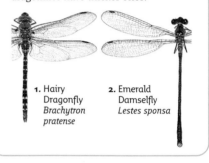

1. Hairy Dragonfly *Brachytron pratense*

2. Emerald Damselfly *Lestes sponsa*

Common Hawker

Aeshna juncea Aeshnidae
Length 70mm

Active, fast-flying species. **Adult** Has narrow yellow lines on thorax, and blue eyes and blue markings on otherwise dark brown abdomen; dorsal spots at rear of abdomen are dissected by a dark line. Flies Jun–Oct. **Status** Common and widespread. Associated with ponds, lakes and canals, although adults often feed kilometres away from water.

Migrant Hawker

Aeshna mixta Aeshnidae
Length 60mm

Similar to Common Hawker but smaller and with subtly different markings. **Adult** Has blue markings on abdomen but dark upper surface to thorax (stripes are seen in Common); note also the yellow triangle at front end of abdomen. Flies Jun–Oct. **Status** Local resident and common migrant visitor, found mainly in coastal areas in the S.

Southern Hawker

Aeshna cyanea Aeshnidae
Length 70mm

Large, active species. **Adult** Has broad green stripes on thorax, with markings of similar coloration on abdomen except for last 3 segments on male, where markings are blue. Flies Jun–Oct. **Status** Commonest in S and central England. Favours ponds, lakes and canals, but also hunts away from water.

Brown Hawker

Aeshna grandis Aeshnidae
Length 74mm

Easily recognised, even in flight, by its brown body and bronze wings. **Adult** Male also has blue spots on 2nd and 3rd segments of abdomen. Flies Jul–Sep. **Status** Commonest in SE England. Found on well-vegetated ponds, lakes and canals, patrolling regular hunting territory around margins.

Dragonfly emergence

Emergence involves the nymph leaving the water by climbing up and gripping a stem. The skin of the head and thorax splits, and the adult gradually pulls itself free before pumping up its wings with internal fluids.

Emperor Dragonfly

Anax imperator Aeshnidae
Length 78mm

Large, active and wary dragonfly. **Adult** Recognised by dark dorsal line running length of abdomen; abdomen colour is sky blue in male, greenish blue in female. Flies Jun–Aug. **Status** Locally common only in S England. Favours lakes, ponds and canals, and usually hunts over open water.

Golden-ringed Dragonfly

Cordulegaster boltonii Cordulegasteridae
Length 78–80mm

Well-marked dragonfly. Fast-flying in sunny weather, lethargic on dull days. **Adult** Recognised by its dark body marked with bright yellow rings on abdomen. Flies Jun–Aug. **Status** Commonest in the W and N, in clean, fast-flowing rivers and streams; locally common further E and S on heathland streams.

Smaller dragonflies

Many common names for dragonfly species reflect their aerial habits, as in the case of the fast-flying hawkers discussed on pp.380-1. Here, smaller, less energetic dragonflies are shown, the most active of which are the chasers. But, like darters and skimmers, even chasers spend long periods resting on prominent perches, ever alert for danger or prey.

Broad-bodied Chaser

Libellula depressa Libellulidae
Length 43mm

Actively hawks for insects but also perches for long periods. **Adult** Has a broad, flattened abdomen; sky blue with small yellow spots on sides in mature male, brown with yellow spots on sides in female and immature male. Wings have dark brown bases. Flies May-Aug. **Status** Common only in S England in ponds and canals.

TOP: female; ABOVE: male

Scarce Chaser

Libellula fulva Libellulidae
Length 40mm

Active streamside species. **Adult** Similar to Broad-bodied Chaser but has a narrower abdomen and no yellow spots on sides. Abdomen tip is dark (as in Four-spotted), but dark spots on leading edge are absent and only hind wings have a dark basal patch. Flies May-Jul. **Status** Local in well-vegetated streams.

BELOW: male
INSET: female

Four-spotted Chaser

Libellula quadrimaculata Libellulidae
Length 40mm

Distinctively marked dragonfly. **Adult** Has a mainly orange-brown body with yellow spots on sides of broad, dark-tipped abdomen. Each wing has a brown patch at base and a dark spot at midpoint on leading edge. Flies May-Aug. **Status** Locally common in S and central England and Wales; favours bogs and marshes.

Keeled Skimmer

Orthetrum coerulescens Libellulidae
Length 40mm

Similar to Black-tailed Skimmer but separable with care. **Adult** Has a more slender body. Mature male has a sky-blue abdomen without black tip. Female and immature male have yellowish abdomen without obvious black lines. Flies Jun-Sep. **Status** Locally common in the W and SW, in bogs and marshes.

Dragonfly nymphs

Dragonfly nymphs are aquatic and live in freshwater habitats. Some live buried in mud and silt and are squat and often hairy; free-living ones often have elongated bodies. Like adult dragonflies, nymphs are predators and prey on small invertebrates; these are captured using fang-like 'jaws' attached to a jaw extension called the 'mask'. When the adult has emerged, the empty nymphal skin can be examined and the nymphal characteristics can be clearly seen. Find out more about dragonfly nymphs on p.437.

Emperor Dragonfly nymphal skin

Black-tailed Skimmer

Orthetrum cancellatum Libellulidae
Length 50mm

Skims low over water and frequently uses a regular perch. **Adult** Male has blue eyes and a black-tipped blue abdomen with orange-yellow spots on sides in maturity. Female and immature male are yellow-brown with black lines on abdomen. Flies Jun-Aug. **Status** Locally common in S on lakes and flooded gravel pits.

ABOVE: female; BELOW: mating pair

Common Darter

Sympetrum striolatum Libellulidae
Length 36mm

 Frequently rests on ground but also uses perches. **Adult** Has a narrow body. Abdomen is deep red in mature male, and orange-brown in immature male and female. Flies Jun–early Nov. **Status** Commonest dragonfly in many parts. Breeds in still and slow-flowing fresh water; often feeds well away from water.

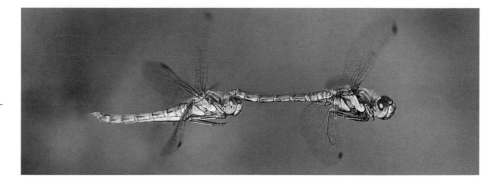

Ruddy Darter

Sympetrum sanguineum Libellulidae
Length 35mm

 Often perches with wings slightly depressed. **Adult** Similar to Common Darter but mature male has a markedly constricted bright red abdomen; female and immature male have black legs (variably brown in Common) and yellow patch at base of hind wing. Flies Jul–Sep. **Status** Common near ponds and lakes.

Downy Emerald

Cordulia aenea Corduliidae
Length 48mm

 Alert dragonfly. Flies fast and low over water, and perches high in bushes. **Adult** Has a green head and thorax, and a metallic bronze-green abdomen. Thorax is hairy and bases of wings are yellowish. Front of abdomen is constricted in male. Flies in Jun–Jul. **Status** Very locally common in the S near ponds, lakes and canals.

Club-tailed Dragonfly

Gomphus vulgatissimus Gomphidae
Length 50mm

 Distinctive species with widely spaced eyes. **Adult** Male has a swollen-tipped abdomen. Both sexes have a mainly blackish body with conspicuous pale markings: yellow in immatures, lime green when mature. Flies May–Jun. **Status** Very local, along stretches of the Severn and Thames rivers, and in a few sites in Sussex.

Black Darter

Sympetrum danae Libellulidae
Length 30mm

 Small, active darter. **Adult** Has a slender body. Abdomen has a blackish dorsal surface in mature male and is swollen towards tip; dull yellow in immature male and female, with dark triangle at front. Flies Jul–Oct. **Status** Locally common in heathland bogs and acid waters.

Regional specialities

Several dragonfly species have restricted regional distributions: the **Brilliant Emerald** *Somatochlora metallica* (**1**) (length 50mm), which is brighter and less hairy than the Downy Emerald; the **Northern Emerald** *S. arctica* (**2**) (length 50mm), which is darker than the Brilliant Emerald; the **White-faced Darter** *Leucorrhinia dubia* (**3**) (length 40mm), with a white face, and red markings in the male and yellow in the female; and the **Norfolk Hawker** *Aeshna isosceles* (**4**) (length 65mm), which has a bright orange-buff body and clear wings. Distribution should be used as a factor when identifying them (*see* maps).

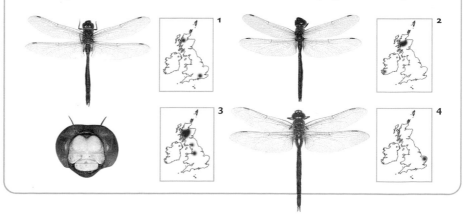

Damselflies

Damselflies may lack the stature and powers of flight of dragonflies but their delicate structure belies their predatory powers: they are great consumers of midges and other small flying insects. Their nymphs are aquatic (*see* p.437).

Azure Damselfly

Coenagrion puella Coenagriidae
Length 33mm

 Striking species. **Adult** Male is blue with black bands on abdomen and black 'U' marking on segment 2; female is blackish with blue tip to abdomen. Flies May–Aug. **Status** Common except in N, in canals and ponds.

abdominal segment 2

Banded Demoiselle

Calopteryx splendens Calopterygidae
Length 45mm

 Rests among waterside vegetation. Males fly in fluttering groups over water. **Adult** Male has a blue body with a metallic sheen; smoky wings have a blue 'thumbprint' mark. Female has a metallic green body and greenish-brown wings. Flies May–Aug. **Status** Locally common near clean streams; the aquatic nymph lives buried in sediment.

Red-eyed Damselfly

Erythromma najas Coenagriidae
Length 35mm

 Distinctive species. **Adult** Has striking red eyes and blackish, blue-tipped abdomen; thorax of male is black above and blue on sides, that of female is black and yellow. Flies May–Sep. **Status** Locally common only in central and S England, favouring ponds and canals.

tip of abdomen

Variable Damselfly

Coenagrion pulchellum Coenagriidae
Length 35mm

 Similar to Azure Damselfly. **Adult** Male is sky blue with black bands on abdomen; the black 'U' marking on segment 2 is linked by a stalk to black band, the marking goblet-like; female abdomen is more black than blue. Flies May–Aug. **Status** Common in S Britain and Ireland, in ponds and lakes.

ABOVE: male
BELOW: female

abdominal segment 2

Beautiful Demoiselle

Calopteryx virgo Calopterygidae
Length 45mm

 Similar to Banded Demoiselle but separable on wing pattern. **Adult** Male has almost wholly dark bluish-brown wings; those of female are duller brown than in female Banded. Body of male is metallic bluish, female is metallic green. Flies May–Aug. **Status** Locally common in the S in fast-flowing clear streams; the nymph lives buried in sand or gravel.

male

abdomen

Blue-tailed Damselfly

Ischnura elegans Coenagriidae
Length 32mm

tip of
abdomen

Distinctive damselfly. **Adult** Both sexes easily identified by mainly black body, with segment 8 of abdomen sky blue. Flies May–Aug. **Status** Common and widespread except in the far N. Favours ponds, lakes, canals and ditches. Fairly tolerant of moderate pollution.

mating pair

Common Blue Damselfly

Enallagma cyathigerum
Coenagriidae
Length 32mm

abdomen

abdominal
segment 2

Similar to *Coenagrion* sp. **Adult** Has a single black line on side of thorax (2 in *Coenagrion*). Male is blue with black bands on abdomen and mushroom-cloud-shaped dot on segment 2. Green and black female has diagnostic spine on the underside, near tip of abdomen. Flies May–Sep. **Status** Common throughout near vegetated still water.

Pair depositing eggs

Large Red Damselfly

Pyrrhosoma nymphula Coenagriidae
Length 35mm

Distinctive damselfly with rather weak flight. **Adult** Mainly bright red but abdomen is marked with black, this more extensive on female than male. Note the black legs. Flies May–Aug. **Status** Common and widespread in a wide range of freshwater habitats, including ponds, canals and bogs.

Damselfly nymphs

The immature stage of a damselfly is a slender aquatic nymph whose delicate appearance belies the fact that it is a voracious predator of small freshwater invertebrates. At the rear end are three flattened 'tails' that are, in fact, gills; in some species these are broad and leaf-like. Prior to their emergence as adults, damselfly nymphs leave the water and climb emergent vegetation before metamorphosing. Find out more about damselfly nymphs on p.437.

nymph of
Banded
Demoiselle

Small Red Damselfly

Ceriagrion tenellum Coenagriidae
Length 30mm

Small, extremely slender and delicate-looking species. **Adult** Male has a mainly red body while female abdomen is mainly black. Note the red legs (black in the Large Red). Flies Jun–Sep. **Status** Local, restricted to heathland bogs in the S.

White-legged Damselfly

Platycnemis pennipes Platycnemididae
Length 35mm

Dainty, pale-looking species. **Adult** Has a pale blue body with dark markings down centre of abdomen and dark upper surface to thorax. Middle and hind legs are white with dark central line. Flies May–Aug. **Status** Local in the S, favouring well-vegetated still waters.

Regional specialities

Five of our damselflies are very local. The **Southern Damselfly** *Coenagrion mercuriale* (**1**) (length 30mm), **Irish Damselfly** *C. lunulatum* (**2**) (length 30mm) and **Northern Damselfly** *C. hastulatum* (**3**) (length 30mm) are best separated from their common cousins using geographical range (*see* maps) and by examining the markings on segment two of the abdomen. The **Scarce Emerald Damselfly** *Lestes dryas* (**4**) (length 30mm), an Irish speciality, is similar to the commoner Emerald Damselfly (*see* box on p.380); males have less extensive bloom on abdomen. The **Scarce Blue-tailed Damselfly** *Ischnura pumilio* (**5**) (length 30mm) is similar to Blue-tailed but segment 9, not 8, creates the blue tail-light.

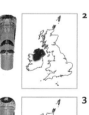

Bugs

Bugs belong to the order Hemiptera and range in size from tiny aphids to impressive shield bugs. All bugs have similar mouthparts comprising a piercing needle-like 'beak', with which they suck fluids. Some are plant-feeders and suck sap, while others are predatory and suck the life out of other invertebrates; a few are even blood-suckers of vertebrates. Terrestrial bugs are covered here; freshwater bugs are dealt with on p.436.

Did you know?

The word 'bug' is sometimes misused as a description for all insects, typically by people who don't understand the differences between the various insect groups. 'Bugs' in the true sense refers solely to members of the family Hemiptera, and to avoid confusion and misunderstanding the word should be used only in this context, except perhaps with small children.

Shield bugs

These bugs have shield-like bodies and feed mainly on plants, although some will suck the liquid contents from a juicy caterpillar. Most appear as adults in summer and hibernate in winter, re-emerging the following spring.

Hawthorn Shield Bug

Acanthosoma haemorrhoidale
Acanthosomatidae Length 13mm

Attractive and familiar shield bug that feeds mainly on Hawthorn berries and it is most frequently associated with this shrub. But it also consumes the leaves of deciduous trees and shrubs generally. **Adult** Shiny green with black and deep red markings; wings are pale and membranous at tip. Autumn adults hibernate, reappearing Apr–Jul; larvae seen Jun–Aug. **Status** Common and widespread except in N Scotland, and found in hedgerows, woods and mature rural gardens.

Hawthorn Shield Bug

Sloe Bug

Dolycoris baccarum Pentatomidae
Length 12mm

Associated with Blackthorn and other shrubs, feeding on sloes and other berries. **Adult** Dark reddish brown, stippled with black dots; banded reddish-yellow and black abdomen segments are visible beyond outer margin of wings. Pronotum lacks the lateral projections seen in some other shield bugs. Seen Jun–Aug. **Status** Common in hedgerows, except in the N.

Green Shield Bug

Palomena prasina Pentatomidae
Length 13mm

Oval-shaped bug. Feeds on Hazel and other shrubs. **Adult** Mainly green but stippled with tiny black dots; tips of wings are dark. Dull when hibernating but bright green when newly emerged in May. Nymphs are seen in summer and new adults appear in Sep. **Status** Common in hedgerows, except in the N.

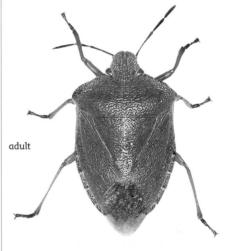

adult

nymph

Forest Bug

Pentatoma rufipes
Pentatomidae Length 14mm

Shield-shaped with lateral processes on pronotum. **Adult** Reddish brown and shiny, with orange-red legs and body markings; abdomen is banded orange-red and black. Overwintering larvae appear in Apr; adults seen Aug–Oct. **Status** Widespread in woodland; favours oak, on which it feeds.

Forest Bug

Bronze Shield Bug

Eysarcoris fabricii Pentatomidae
Length 6mm

Tiny wayside bug. **Adult** Grey-brown, with areas of bronze sheen on the head and pronotum. Most obvious May–Jul. **Status** Locally common in damp hedgerows and waysides, often on Hedge Woundwort.

Pied Shield Bug

Sehirus bicolor Cydnidae
Length 7mm

Hedgerow species. **Adult** Oval with black and white markings on wings and abdomen (nymphs are similarly pied but lack fully formed wings). Adults emerge in autumn, hibernate, then reappear May–Jul. **Status** Widespread in hedgerows and waysides; often feeds on White Dead-nettle and Black Horehound.

Dock Bug

Coreus marginatus Coreidae
Length 15mm

Distinctive bug whose body outline vaguely recalls that of a violin. **Adult** Has a mainly buffish-brown body with red antennal segments and markings on abdomen margins. Hibernates; seen autumn and spring. **Status** Found mainly in the S, in meadows and hedgerows, where it feeds predominantly on dock fruits.

Assassin bugs

These predatory insects (family Reduviidae) feed on the body fluids of other invertebrates – usually caterpillars – using a long, piercing beak. ***Rhinocoris iracundus*** (**1**) (length 13mm) is often found on meadow flowers in May–Sep and has a red and black body. The **Fly Bug** *Reduvius personatus* (**2**) (length 18mm) is associated with wooded meadows but also comes indoors; its strange flattened nymph covers itself with detritus for camouflage; it is local in S England.

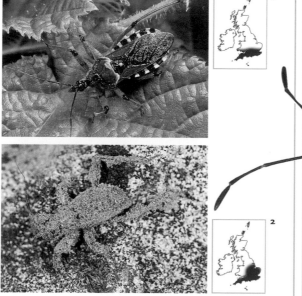

nymph

Bugs

Froghoppers

Known for their hopping skills – they leap a metre or more when threatened by danger – and vaguely frog-like appearance, these tiny bugs (family Cicadidae) have hardened forewings. They are plant-suckers and their nymphs live in protective masses of froth, popularly known as 'cuckoo-spit'.

Common Froghopper

Philaenus spumarius Cicadidae
Length 6mm

Familiar froghopper. **Adult** Oval in outline and usually marbled brown. **Nymph** Green; lives in 'cuckoo-spit' froth. **Status** Common and widespread in Jun-Aug.

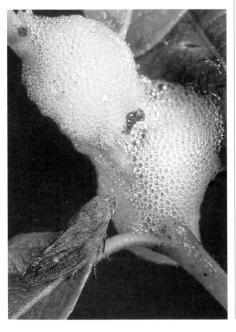

Capsid bugs

Capsids (family Miridae) are delicate-looking bugs with long antennae. Some feed on plants while others are predators, and they are found in woodland, scrub and meadows.

Deraeocoris ruber

Miridae Length 7mm

Feeds on developing fruits and seeds, as well as on aphids and other small insects. **Adult** Usually dark reddish brown, with an orange-red patch near tip of forewing. Seen Jul-Aug. **Status** Common in S Britain.

Campyloneura virgula

Miridae Length 5mm

Tiny capsid that feeds on aphids and Red Spider Mites. **Adult** Mainly buff and dark brown, but has orange-yellow on pronotum and near tip of the forewing. Seen Jul-Oct. **Status** Common, usually in deciduous trees and shrubs.

Orthops campestris

Miridae Length 4mm

Colourful capsid that is usually found on umbellifer flower heads. **Adult** Usually green overall but sometimes light brown, with bold dark markings on pronotum and forewings. Tips of hind wings are pale. Overwinters as an adult and seen Jul-Apr. Larva feeds on buds and flowers of umbellifers. **Status** Widespread and common in grassy places and verges, and particularly associated with the flowers of Wild Parsnip and Hogweed.

Cercopis vulnerata

Cicadidae Length 9mm

Distinctive froghopper, adorned with striking warning colours. **Adult** Has unmistakable red and black markings; rests on low vegetation, jumping to escape danger. Feeds on plant sap. **Nymph** Feeds on plant roots. **Status** Common and widespread in hedgerows, woodland rides and meadows; seen May-Aug. Makes no attempt to hide itself and hence easy to spot.

Leafhoppers

Leafhoppers (family Cicadellidae) look mainly rather like froghoppers or miniature cicadas. Their forewings are usually soft, they can hop well, and they feed on the foliage of trees and shrubs.

Rhododendron Leafhopper

Graphocephala coccinea Cicadellidae
Length 9mm

 Colourful leafhopper. **Adult** Has red forewing stripes on an otherwise green body. **Nymph** Feeds on rhododendron. **Status** Introduced from North America, but now firmly established on rhododendrons in S England.

Warning colours

Many insects are distasteful to predators, typically because they accumulate toxins from the plants they eat, and most of them advertise this fact with bright colours – usually red and black. Over time, potential predators associate these warnings with unpalatable prey and avoid the insects concerned. A good example of a bug with warning coloration is **Corizus hyoscyami** (length 10mm), a plant-feeding species whose range in S England is expanding northwards.

Ledra aurita

Cicadellidae
Length 14mm

 Largest and most bizarre-looking leafhopper. **Adult** Brown, winged and with thoracic projections. **Nymph** Oval, pale, squat and strangely flattened. **Status** Fairly common, living on oak, though its excellent camouflage makes it hard to spot.

Aphids

Although recognisable as a group, aphids are hard to identify to species level except by experts. They suck sap and are reviled by gardeners, but they are important food for predators and so are significant in ecological terms.

Rose Aphid

Macrosiphum rosae Aphididae
Length 2mm

 Familiar garden 'greenfly'. **Adult** Can either be green or pink with 2 black horn-like projections near abdomen tip. **Status** Widespread and often abundant, forming colonies on roses in spring but found on other plants by the summer months.

Horned Treehopper

Centrotus cornutus Membracidae
Length 6mm

 As in other treehoppers, the pronotum extends back the length of the abdomen. **Adult** Brown with lateral 'horns' on pronotum; wings are yellowish and membranous. Seen Jun-Jul. **Status** Common in meadows and woods.

Black Bean Aphid

Aphis fabae Aphididae
Length 2mm

 Familiar garden aphid. **Adult** Eaten by ladybirds but guarded and 'milked' for honeydew by ants. Eggs overwinter on Spindle, wingless females appear on beans in spring, and huge colonies form by summer, when winged adults appear. **Status** Widespread and seasonally abundant.

Colony being tended by ant

Cockroaches to caddis flies

The species covered here represent several different insect orders. What they have in common are the unobtrusive lives that they lead, either being largely nocturnal or preferring to remain hidden for much of the time, or both. Cockroaches scuttle among the ground layer of plants and debris; earwigs are sometimes found in flowers but more often hide under logs and stones; and alderflies, lacewings, scorpion flies, snakeflies and caddis flies usually skulk in vegetation. The larvae of caddis flies and alderflies are aquatic and are dealt with separately on p.438.

Neuroptera

Members of the order Neuroptera include lacewings, scorpion flies and snakeflies. All have soft, slender bodies and delicate, membranous wings. Adult and larval neuropterans are carnivorous. Typical members of the group include the following.

Lacewing

Chrysoperla carnea
Chrysopidae
Length 15mm

larva

Familiar insect of house and garden. **Adult** Wings transparent and well veined. **Larva** Feeds mainly on aphids and creates a camouflaged home from their empty skins. **Status** Common and widespread. Found among vegetation in summer and attracted to lighted windows at night. In autumn, often hibernates indoors, turning from green to pink.

Alderflies

Although they are closely related to lacewings, alderflies are placed in a separate order, the Megaloptera. They have smoky wings and adults feed on pollen; their larvae are carnivorous and live in water (find out more about alderfly larvae on p.438).

Alderfly

Sialis lutaria Sialidae
Length 14mm

Typical member of the group. **Adult** Wings have clearly defined black veins and are held in a tent-like manner at rest. It is a poor flyer, often found resting on waterside vegetation in May–Jun. **Larva** Aquatic and predatory; its brown, tapering abdomen is fringed with gills. **Status** Widespread and locally common.

Earwigs

These members of the order Dermaptera are characterised by pincer-like appendages (called cerci) at their rear end. Their bodies are slender and slightly flattened.

Common Earwig

Forficula auricularia Forficulidae
Length 13mm

Typical earwig. **Adult** Flightless species with a shiny chestnut-brown body; pincer-like cerci are curved in males but rather straight in females. **Status** Common and widespread. Hides in leaf litter and under stones or logs during the daytime; most active after dark, when it emerges to feed mainly on dead organic matter.

Cockroaches

Cockroaches (order Dictyoptera) are flattened insects with tough forewings that cover the abdomen, long antennae and a scuttling gait. The **Dusky Cockroach** *Ectobius lapponicus* (family Blattidae) (length 9mm) is one of three native species; its grey-brown body has a pale translucent margin to the pronotum and forewing. It flies in warm summer months and is locally common only in S England in low vegetation. Several introduced species are pests of kitchens and warehouses.

Scorpion Fly

Panorpa communis
Panorpidae Length 14mm

Strange-looking insect. **Adult** Males have an upturned scorpion-like abdomen. Head has a beak-like downward projection that is used in feeding: it scavenges on dead animals, including contents of spiders' webs, and also feeds on ripe fruit. **Status** Common and widespread. Seen May-Jul in hedgerows and among Bramble patches.

female

male

Snakefly

Phaeostigma notata Raphodiidae
Length 14mm

Bizarre and distinctive insect of the order Neuroptera. **Adult** Has transparent wings, and an elongated thorax that enables its head to be raised in a fancifully snake-like manner. Feeds mainly on aphids. **Status** Locally common in mature oak woodlands in May-Jul.

Caddis flies

Caddis flies (order Trichoptera) are weak-flying insects with hairy wings and slender antennae that are held in front of the insect at rest. Adults are found near water and their larvae are aquatic (*see also* p.438); they live inside tubes constructed from materials around them. There are nearly 200 species in Britain and the identification of most is best left to experts. The following are common and reasonably distinctive examples.

Potamophylax cingulatus

Limnephilidae Length 15mm

Widespread and familiar species. **Adult** Has orange-brown hairy wings and comes to light. Flies Apr-Jun. **Larva** Makes a case from tiny stone particles and wood debris. **Status** Locally common in streams, particularly ones that are not subject to maintenance and including those prone to partial seasonal drying.

Agapetus fuscipes

Glossosomatidae Length 9mm

Relatively small caddis fly. Masses often congregate on waterside vegetation. **Adult** Has yellow-brown hairy wings. Flies Apr-Jun. **Larva** Builds a case from small stones, including relatively sizeable pieces of flint. **Status** Associated with fast-flowing streams with stony bottoms, and often extremely common in chalk streams.

Glyphotaelius pellucidus

Limnephilidae Length 16mm

Widespread standing-water species. **Adult** Appears Apr-Jun and has marbled brown and white wings, the forewing with a notched outer margin. **Larva** Constructs a case from dead leaves and is difficult to detect in pond samples until it decides to move. **Status** Common throughout in ponds and lakes.

Phryganea striata

Phryganeidae Length 25mm

One of our largest caddis flies. **Adult** Has marbled brown and buff wings; female has broken black line on forewing. Flies Jun-Aug and sometimes comes to light. **Larva** Builds a case made from leaves, stems and other plant material, arranged spirally to form a cylinder. **Status** Widespread, but commonest in the N and W.

Limnephilus marmoratus

Limnephilidae Length 13mm

An upland and N species. **Adult** Has well-marked forewings and is seen resting among emergent vegetation in May-Jul. **Larva** Constructs a case from fragments of plant material; sometimes observed moving around in clear shallow water. **Status** Locally common in well-vegetated upland and N tarns and lakes.

Limnephilus elegans

Limnephilidae Length 12mm

A well-marked caddis fly. **Adult** Has grey-brown hairy wings, the forewings bearing black lines. Flies May-Jul. **Larva** Constructs a rather cumbersome case from plant material. **Status** Widespread, but associated mainly with upland pools and slow-flowing rivers.

Flies

Flies belong to the family Diptera, a name that means 'two wings', and a character they all share. The hind wings are modified into tiny club-shaped structures called 'halteres'; these act as stabilisers in flight. Flies feed on liquids; some have piercing mouthparts while others suck up surface fluids. Mosquito and midge larvae are aquatic (*see* p.438).

Crane-flies

Also known as daddy-long-legs, crane-flies (family Tipulidae) are rather gangly-looking insects with incredibly long, slender legs that break easily. Their larvae are mainly soil-dwellers and most feed on decaying organic matter.

Tipula maxima
Tipulidae Length 30mm

One of the largest crane-flies. **Adult** Wings have striking brown patches; flies mainly May-Jul. **Larva** Lives in watery margins. **Status** Widespread and fairly common in damp woodland and beside wooded streams.

Tipula paludosa
Tipulidae Length 16mm

Familiar insect. **Adult** Has weak flight and dangling legs; commonest Aug-Oct. **Larva** Known as a leatherjacket; lives in soil and eats the roots and stems of plants. **Status** Widespread and common, on garden lawns and grassland.

leather-jacket

Mosquitoes and midges

Although many people try to ignore them, mosquitoes and midges are ever-present in the countryside, with a range of species representing the group at different times of year. Mosquitoes belong to the family Culicidae, but 'midge' is an imprecise term and includes several groups of flies. Several species of *Culex*, *Anopheles* and *Aedes* mosquitoes (length 8-15mm) are common in the region. Only females suck blood while males feed on nectar; larval and pupal stages are found in standing water and swim with a wriggling motion.

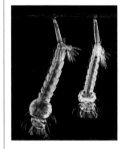

ABOVE: Only female mosquitoes bite – their abdomen swells to accommodate the blood meal.

Culex larvae

Phantom Midge
Chaoborus crystallinus Chaoboridae
Length 16mm

Familiar non-biting midge. **Adult** Male has plumed antennae and wings that do not reach abdomen tip. **Larva** Transparent and hence known as 'ghost worms'. Aquatic, feeding on small animals. **Status** Common; associated with standing fresh water.

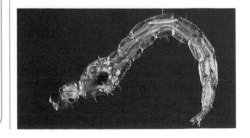

Life in rotting wood

Ctenophora atrata is an atypical, unusual-looking crane-fly whose larvae live in decaying wood; it is locally common in S Britain. The adult, which flies mainly in May-Jul, is found mostly in oak woodland and is strikingly black and red. The female has a fearsome-looking (but harmless) sickle-shaped ovipositor at the tail end, which she uses to lay her eggs in timber.

Chironomus plumosus

Chironomidae Length 10mm

Typical non-biting midge. **Adult** Males have plumed antennae and wings that are shorter than abdomen, and they form large swarms; females have relatively longer wings and simple antennae. Seen May–Sep. **Larva** Aquatic; often known as a 'bloodworm'. **Status** Common and widespread.

LEFT: Male *Chironomus plumosus* have plumed antennae.
BELOW: *Chironomus* sp. larva

St Mark's Fly

Bibio marci Bibionidae Length 11mm

Often appears around St Mark's Day (25 Apr). Sluggish when resting on vegetation. Male flies with dangling legs. **Adult** Has short antennae and a hairy black body. **Larva** Lives in soil. **Status** Locally abundant in areas of short grass.

Bee-flies

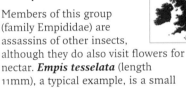

These flies (family Bombyliidae) look like small bees and lay their eggs in the nests of solitary bees and wasps; their larvae feed on the larvae of their hosts. Adults have a long proboscis that is used to suck nectar. *Bombylius major* (length 10mm) is common and produces a high-pitched hum in flight; it flies in Apr–May.

Horse-flies

The bane of picnickers' lives, horse-flies (family Tabanidae) are robust insects with prominent (for a fly) antennae and, in the case of females, a blood-sucking diet; males feed on nectar. Their larvae live mainly in damp soil and are almost impossible to identify.

Chrysops relictus

Tabanidae Length 10mm

Attractive horse-fly. **Adult** Has patterned wings, yellow markings on abdomen and iridescent green eyes. Flies Jun–Aug and can inflict a painful bite. **Status** Common and widespread; associated with damp woodland and heaths.

Cleg-fly

Haematopota pluvialis Tabanidae Length 10mm

Horse-fly whose biting attentions are difficult to repel. **Adult** Has a grey-brown body, mottled brown wings held in a roof-like pose at rest, and iridescent eyes. Flies May–Sep. **Status** Common.

Empid flies

Members of this group (family Empididae) are assassins of other insects, although they do also visit flowers for nectar. *Empis tesselata* (length 11mm), a typical example, is a small fly with a strongly downcurved abdomen and is seen in Apr–Jul. It is sometimes observed carrying victim insects in flight, which it then sucks dry using its long, downward-pointing proboscis.

Haematopota crassicornis

Tabanidae Length 10mm

Also sometimes called a 'cleg'. **Adult** Has amazingly iridescent eyes; inflicts a painful bite and approaches its victim silently and stealthily; flies Jul–Aug. **Status** Common.

Tabanus bromius

Tabanidae Length 14mm

Robust horse-fly. **Adult** Has yellowish-brown and black markings on abdomen; flies Jul–Aug. **Status** Widespread; commonly found around cattle and ponies.

Robber-flies

Belonging to the family Asilidae, robber-flies are predators of other insects, which they catch in flight. *Asilus crabroniformis* (length 27mm) is the largest of its kind in our region. It often sits on vegetation on the look-out for passing prey, and it flies in Jul–Oct.

Flies

Adult flies are liquid-feeders. Hover-flies consume fairly innocuous fluids rich in sugars while house-fly relatives have unsavoury tastes and visit dung and decaying carrion. The larvae of many are important agents of decay, feeding inside dead animals or other decomposing organic matter.

Tachina fera

Tachinidae Length 10mm

 Well-marked fly. **Adult** Has a bristly body with orange and black on abdomen and yellow hairs on head; wings are flushed yellow towards base and legs are orange with yellow 'toes'. Flies May–Sep. **Larva** A parasite of caterpillars. **Status** Common and widespread in marshes and damp meadows.

Flesh-fly

Sarcophaga carnaria Sarcophagidae Length 15mm

 Attracted to carrion and carcasses. Seldom ventures indoors. **Adult** Has a greyish body but with chequered markings on abdomen; eyes are red and feet are proportionately large. **Larva** Live-born (eggs are not laid) on decaying flesh. **Status** Common and widespread.

Common House-fly

Musca domestica Muscidae Length 8mm

 Familiar visitor to houses. Attracted to rubbish, where eggs are laid. **Adult** Has red eyes and a mostly dark body except for orange patches on abdomen. Note the sharp bend in 4th long vein of wing. **Larva** Maggot-like. **Status** Extremely common and widespread throughout Britain and Ireland.

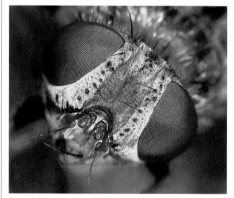

Greenbottle

Lucilia caesar Calliphoridae Length 9mm

 Colourful fly that is attracted to flowers, dung and carrion. **Adult** Has a shiny green or bronze-green body and red eyes. **Larva** Lives and feeds in rotting carcasses. **Status** Common and widespread in Britain and Ireland. Found in all sorts of habitats, and often occurs in gardens.

Bluebottle

Calliphora erythrocephala Calliphoridae Length 11mm

 Extremely familiar visitor to homes. Found in most months but commonest in summer. Makes a loud buzzing sound in flight. **Adult** Has a shiny blue body and reddish eyes and jowls. Female is attracted to meat, on which eggs are laid and larvae feed. **Larva** Maggot-like. **Status** Common and widespread.

Lesser House-fly

Fannia canicularis Fanniidae Length 5mm

 Familiar fly that is often found indoors; males fly repetitive circuits beneath ceiling lights and fixtures. **Adult** Similar to Common House-fly but smaller, and 4th long vein of wing is straight, without sharp bend. **Larva** Lives in putrefying corpses and dung. **Status** Widespread and common throughout.

Yellow Dung-fly

Scatophaga stercoraria Scatophagidae
Length 9mm

Adult preys on other flies. **Adult** Furry and golden. Swarms of males collect on cowpats in Mar–Oct, greeting arrival of a female with a flurry of activity; mating pairs linger in tandem. **Larva** Develops inside cowpat. **Status** Widespread and common.

Snipe-fly

Rhagio scolopaceus Rhagionidae
Length 12mm

Often sunbathes on vegetation, but also sits head-down on tree trunks and will sometimes suddenly fly straight at observer. **Adult** Has a rather slender, tapering brown body; flies May–Jul. **Larva** Carnivorous and lives in leaf litter. **Status** Common and widespread.

Hover-flies

Aptly named for their ability to hover with consummate ease, hover-flies belong to the family Syrphidae. Adults of many species are fabulous mimics of bees or wasps, not just in terms of their appearance but also in their behaviour and the buzzing sounds they make. The larvae of many species are important predators of aphids and other small insects. There are numerous species in our region, of which the following are typical.

Helophilus pendulus

Syrphidae Length 10mm

Common hover-fly. **Adult** Male often hovers over water. Frequently sunbathes; flies May–Sep, often seen visiting garden flowers and wayside species such as Hogweed and Ragwort. **Larva** Lives in stagnant water.

Status Common and widespread, favouring damp wooded sites.

Syrphus ribesii

Syrphidae Length 12mm

Familiar insect. **Adult** Visits flowers to feed on nectar. Multiple-brooded; flies Apr–Oct. **Larva** Lives on leaves and feeds on aphids. **Status** Extremely common and widespread in gardens and hedgerows.

Volucella bombylans

Syrphidae Length 14mm

Superb bumblebee mimic. **Adult** Abdomen is white-tipped in some individuals and red-tipped in others. Frequent visitor to flowers; seen May–Sep. **Larva** Scavenges in nests of bumblebees and wasps. **Status** Common and widespread; favours gardens, hedgerows and woods.

Picture-winged flies

This intriguing group of flies belong to the family Tephritidae. Adults have marbled or mottled wings and fly mainly in May–Aug, while the larvae of many species form plant galls. One of the most striking species is **Urophora cardui** (length 5mm), the adults of which, although small, are beautifully marked. Their larvae form large pear-shaped galls in thistle stems.

Sericomyia silentis

Syrphidae Length 16mm

Striking wasp-like hover-fly. **Adult** Often visits flowers to feed on nectar; flies May–Aug. **Larva** Lives in damp, peaty soil. **Status** Common and widespread in wet habitats.

Drone-fly

Eristalis tenax Syrphidae
Length 12mm

Superficially very similar to a Honey Bee drone. **Adult** Has relatively large eyes; these, and a body that is not 'waisted' between thorax and abdomen, confirm its dipteran status. Visits flowers to feed on nectar. **Larva** One of the so-called 'rat-tailed maggots' (it has a long breathing tube); lives in stagnant water (*see* p.438). **Status** Common.

Bees & wasps

Bees and wasps belong to the order Hymenoptera, which also includes ants, sawflies and gall-forming wasps (*see* pp.398–9). Bees and wasps have a 'waisted' abdomen and two pairs of wings linked by minute hooks. Some species are solitary but many are extremely social, living in complex colonies whose success relies on collective cooperation.

Honey Bee

Apis mellifera Apidae
Length 12mm

Widely kept in hives for honey. Inside the colony, a network of wax cells forms a comb in which honey is stored and the young are raised. **Adult** Female workers comprise the bulk of the colony, which is ruled by a single queen. Typical colony comprises 1 queen, 200 or so drones (fertile males) and 20,000+ workers (sterile females). **Larva** Grub-like. **Status** In the wild, colonies nest in holes in trees in wooded areas.

INSET: worker with pollen sacs on legs
BELOW: inside a hive, with the queen central

Bumblebees

These members of the family Bombidae are easily recognised by their plump furry bodies and typically rather striking yellow, orange, red or white bands on an otherwise blackish body. They are social insects, the queens hibernating over winter and forming new colonies in spring, often sited in a burrow. There are 20 or so species in our region, of which a handful are common and widespread. They are found in gardens, open woodland, hedgerows and grassland. ***Bombus lapidarius*** (**1**) (length 20mm) has a red tip to its otherwise black abdomen. ***B. lucorum*** (**2**) (length 24mm) has buffish-yellow bands on the thorax and abdomen, which is white-tipped. ***B. terrestris*** (**3**) (length 24mm) is similar but has buffish-orange bands, and the queen has a buff-tipped abdomen. ***B. pratorum*** (**4**) (length 16mm) has a red-tipped abdomen that is also marked with buffish yellow.

1

2

3

4

Solitary bees

Many hymenopterans are social insects, but some bees are solitary. They feed on nectar, collect pollen and lay their eggs in nests, either underground or in crevices.

Leaf-cutter Bee

Megachile centuncularis Megachilidae
Length 13mm

Recognised by neat semicircular holes cut in leaf margins of garden roses and other plants. **Adult** Tawny-buff, female has orange pollen baskets; flies Jun–Jul. **Larva** Lives inside rolled, cut leaf sections. **Status** Common and widespread.

Red Mason Bee

Osmia rufa Megachilidae
Length 10mm

Familiar garden insect. **Adult** Has orange hair on abdomen. Excavates nest in mortar; flies Apr–Jul. **Larva** Hidden within chamber inside mortar. **Status** Common in the S.

Tawny Mining Bee

Andrena fulva Andrenidae
Length 13mm

Distinctive small bee. **Adult** Female has orange-buff fur covering body; appears Apr–Jun and nests in excavated burrows in soil. **Larva** Lives in an underground chamber. **Status** Widespread and common.

Wasps

Members of the family Vespidae, wasps are social insects that can inflict a painful sting when alarmed. They are recognised by their black and yellow markings, wings that are folded lengthways when at rest, buzzing sound made in flight and rather jerky movements when walking. Adult wasps are active May–September and feed primarily on nectar and ripe fruit, although they feed their young on other insects. Several species are common in our region; important identification features include markings on the face.

Wasp nests

Nests built by wasps are papery and are constructed from chewed wood pulp mixed with saliva. As with Honey Bee nests, layers of hexagonal cells are found inside, each cell containing a growing larva that eventually pupates there.

layer of Common Wasp nest

German Wasp

Vespula germanica Vespidae
Length 18mm

Familiar garden wasp. **Adult** Has typical wasp colours and body markings, and a face with 3 black dots when seen head on. Sometimes makes its nest in a loft space. **Larva** Lives in a chamber inside nest. **Status** Common and widespread.

Hornet

Vespa crabro Vespidae
Length 30mm

Formidable-looking insect. **Adult** Larger than other wasps, with tawny-brown and dull yellow colours. Colony is most active Jun-Sep. **Larva** Grub-like. **Status** Range has expanded in recent years and now locally common in S England in a gardens, and usually nests in tree holes.

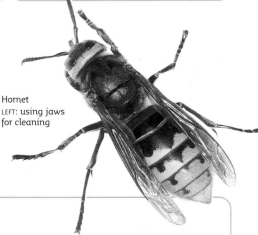

Hornet
LEFT: using jaws for cleaning

Common Wasp

Vespula vulgaris Vespidae
Length 17mm

Active social insect. **Adult** Similar to German Wasp, but seen head on has a black anchor mark on face. Constructs its nest underground or in buildings. **Larva** Lives in a chamber inside nest. **Status** Common and widespread.

Construction workers

Tireless activity is a characteristic of many hymenopteran species and, for their size, among the most industrious are solitary species that build a nest and supply food for their young on their own. Potter and mason wasps create nests of cemented sand, while digger wasps, as their name suggests, excavate a deep burrow in the ground. Many species catch and immobilise large insect prey, with which they stock their nests to feed their developing young. The **Field Digger Wasp** *Mellinus arvensis* (**1**) (length 12mm) has yellow and black colours similar to those of true wasps, but its 'waisted' abdomen is more pronounced; it stocks its nest with flies, and adults appear May-Aug. The **Sand Digger Wasp** *Ammophila sabulosa* (**2**) (length 20mm) favours sandy areas, particularly coasts, and is often seen dragging immobilised caterpillars larger than itself back to its nest burrow, the entrance to which is plugged with mud; adults appear May-Aug. The **Bee-killer** *Philanthus triangulum* (**3**) (length 15mm) catches Honey Bees, which it carries back to its nest slung underneath its body; adults fly Jun-Sep and are common only in E England.

Median Wasp

Dolichovespula media Vespidae Length 20mm

Boldly marked, robust wasp. **Adult** Has a dark line down centre of head. **Larva** Lives in a chamber inside nest. **Status** A recent arrival from Europe but now common in the S.

Ants, gall-forming wasps, parasitic wasps & sawflies

Sawflies, wasps and gall-forming wasps belong to the order Hymenoptera, along with the species on pp.396–7. Most have biting mouthparts, although in some these are modified for supping liquids. Some species are social while others have completely extraordinary lifestyles, either parasitising insects or creating often bizarre-looking plant galls in which their larvae live and feed.

Ants

These members of the family Formicidae are tiny but phenomenally industrious social insects. As with other social hymenopterans, the ant colony contains different forms, or 'castes', of individuals whose roles are specific. Worker ants – whose job it is to collect food and build and repair the nest – are wingless, whereas sexual forms are winged, allowing for dispersal and the colonisation of new habitats. There are several different ant species in our region, and the precise identification of many of these can be difficult; but a few are reasonably distinctive.

Wood Ant
Formica rufa Formicidae
Length 10mm

Our largest ant. **Adult** Reddish-brown workers can be seen collecting caterpillars and other insects, and will spray formic acid from rear end if alarmed. **Larva** Protected deep inside colony. **Status** Locally common in the S. Forms large colonies in woodland clearings, recognised as mounds of dry plant material.

Red Ant
Myrmica rubra Formicidae
Length 4mm

Often abundant garden ant. **Adult** Yellowish-red workers can deliver a painful bite. Active throughout year, swarming in late summer. Its varied diet includes many garden pests. **Larva** Lives protected inside colony. **Status** Common and widespread. Favours garden soil and lawns.

Black Garden Ant
Lasius niger Formicidae
Length 3mm

Familiar garden ant. **Adult** Uniformly blackish brown. Diet is varied but 'milks' aphids for honeydew. **Larva** Lives protected inside colony. **Status** Common in a range of habitats but perhaps most familiar in gardens, where nests are formed under paving stones and brickwork. Flying forms swarm in summer.

Black Garden Ant 'milking' aphids

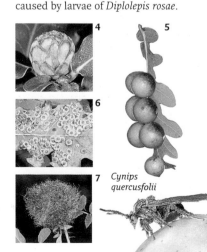

Parasitic wasps

Parasitic wasps are the stuff of nightmares: they typically lay their eggs inside another insect (often a caterpillar) and the wasps' own larvae then eat the internal organs and tissues of the living host before pupating and subsequently emerging as adults.

Rhyssa persuasoria

Ichneumonidae
Length 35mm

Our largest parasitic wasp. **Adult** Female uses her long ovipositor to parasitise larvae of Horntail, located deep in living timber; flies Jul–Aug. **Status** Locally common; associated with mature pine woods.

Yellow Ophion

Ophion luteus Ichneumonidae
Length 20mm

Intimidating-looking insect. **Adult** Has an orange-yellow body and constantly twitching antennae. Flies Jul–Sep and parasitises caterpillars. **Status** Common and widespread.

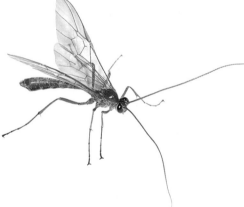

Pteromalus puparum

Pteromalidae Length 2mm

Tiny wasp. **Adult** Parasitises pupae of Large and Small white butterflies (shown); emerge en masse from host. **Status** Common and widespread throughout the region.

Ruby-tailed Wasp

Chrysis ignita Chrysididae
Length 11mm

Small but attractive insect. Female searches diligently on walls and banks for mason wasp nests; if owner is absent she enters and parasitises the larvae. **Adult** Has a shiny green head and thorax, and a ruby-red abdomen. Flies Jun–Aug. **Status** Locally common throughout.

Sawflies

Sawflies (suborder Symphyta) do not have a 'waisted' abdomen, but adults do have the group's characteristic wing arrangement. A female lays eggs inside plant tissue, cutting slits with her ovipositor, which is shaped like a miniature sharp-tipped breadknife. The typical adult diet is nectar and insect prey.

Sawfly larvae

Sawfly larvae eat plant material and are often found feeding on leaves. Superficially they are very similar to the larvae of moths and butterflies, but a close inspection reveals that, in addition to the three pairs of true legs near the head that both groups possess, sawfly larvae have six or more fleshy, suckered 'prolegs' at the rear end; butterfly and moth larvae have five or fewer of these prolegs. They are often gregarious and, when alarmed, typically curl up their tail ends. Specific identification is difficult.

Birch Sawfly

Cimbex femoratus Cimbicidae
Length 21mm

One of our most impressive sawflies. **Adult** Has yellow-tipped antennae, yellow on base of abdomen and smoky, dark-bordered wings. Appears May–Jun and makes a buzzing sound in flight. **Status** Rather local; favours birch woods, its larvae feeding on the leaves of these trees.

Horntail

Urocerus gigas Siricidae
Length 30mm

Formidable-looking, albeit harmless, sawfly. **Adult** Has black and yellow markings, and female has a long ovipositor that will penetrate timber for egg-laying; flies May–Aug. **Status** Fairly common in pine forests.

Rhogogaster viridis

Tenthredinidae Length 12mm

Strikingly colourful insect. **Adult** Vivid green and black; seen in May–Aug, often on flowers. **Status** Common. Found in meadows, hedgerows and woods.

Beetles

Beetles belong to the order Coleoptera and comprise the largest group of insects: more than 4,000 species are found in Britain and Ireland alone. Although they vary tremendously in terms of size and appearance, most beetles are easily recognisable, mainly because the front wings (called 'elytra') are hardened and form a protective casing over the abdomen and folded hind wings.

Beetle life cycles

Beetles undergo complete metamorphosis and have four stages to their life cycle. Adult females lay eggs, which hatch into larvae; when full grown these pupate and eventually emerge as the next generation of adults.

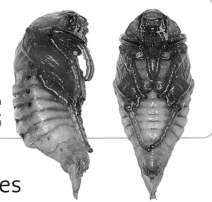

beetle pupa
(Summer Chafer,
Amphimallon solstitialis)

Ground and tiger beetles

Belonging to the family Carabidae, ground and tiger beetles are highly active predators with powerful, often easily visible jaws that make short work of their insect prey. They have long legs and can run fast, and they are easiest to find in spring and summer. Tiger beetles are typically diurnal and have large eyes and keen vision. Ground beetles are mainly nocturnal and rely to a greater extent on their senses of touch and smell to find their prey.

Recognising beetles

Pterostichus cupreus, a common ground beetle, is a characteristic species of the group. In addition to the elytra (common to most beetles) that meet in the middle, and the three pairs of jointed legs typical of almost all adult insects, it has long antennae, well-developed compound eyes and powerful biting mouthparts. The thorax is covered by a protective plate called the pronotum.

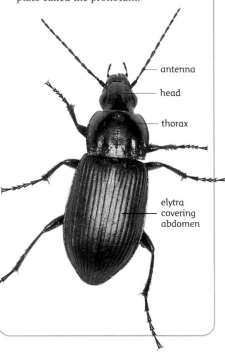

antenna

head

thorax

elytra
covering
abdomen

Pterostichus madidus

Carabidae Length 14mm

Common ground beetle. **Adult** Shiny black with finely grooved elytra and reddish legs; diet includes fruit as well as insects. **Status** Very common and widespread throughout; usually found under stones or logs in the daytime, often in gardens.

Violet Ground Beetle

Carabus violaceus Carabidae
Length 28mm

Distinctive species. **Adult** Has a beautiful violet sheen on thorax and elytra margins. **Status** Common and widespread throughout; lives in gardens and hedgerows, and hides under stones in the daytime.

Carabus nitens

Carabidae Length 35mm

Stunning ground beetle. **Adult** Metallic red and green, with lengthways grooves on elytra. **Status** Scarce; restricted to areas of *Sphagnum* bog on heathlands and moors.

Green Tiger Beetle

Cicindela campestris Carabidae
Length 14mm

Active ground-dwelling beetle. **Adult** Has green upperparts with pale spots on elytra, and shiny bronze legs and thorax margins. **Status** Locally common, found in dry heaths and many sandy coastal districts.

RIGHT: male Stag Beetles fighting

Did you know?

There are many species of beetle that live in fresh water. Their immature stages – egg, larva and pupa – are tied to water and most absorb oxygen directly from it. Adult water beetles are able to leave the water but do so only occasionally, and then simply to disperse. Find out more about water beetles on p.437.

Stag Beetle

Lucanus cervus Lucanidae
Length 40mm

female

Large and impressive beetle. **Adult** Has reddish-brown elytra and a black head and thorax. Male has enlarged antler-like jaws, used for battling with rivals for right to mate with antler-less female. Found May–Jul. **Larva** Lives in buried rotting wood. **Status** Local in S and central England, often in suburbia.

Lesser Stag Beetle

Dorcus parallelipipedus Lucanidae
Length 28mm

Flat-bodied beetle. **Adult** Recalls a female Stag Beetle; has an all-black body but proportionally large and broad head and thorax. Found May–Sep but mainly in spring. **Larva** Lives in rotting wood. **Status** Widespread but local in S and central England, Wales and S Ireland, favouring deciduous woodland.

Sinodendron cylindricum

Lucanidae Length 15mm

Small but distinctive beetle, sometimes found under rotting wood. **Adult** Male resembles a miniature tropical rhinoceros beetle, having a horn-like projection on head. Found May–Aug. **Larva** Lives in rotting wood. **Status** Local but widespread in mature Beech and oak woodland.

Burying beetles

Burying beetles (family Silphidae) perform an invaluable service and are vital in the cycle of life and death. They are sometimes discovered when an animal carcass is overturned, and can even be seen in the act of interring a dead mouse or small bird. Having buried the corpse, the female lays her eggs beside it and the developing larvae use it as a source of food. *Nicrophorus humator* (**1**) (length 22mm) is an all-black

species with orange tips to the antennae; it is often covered in mites. ***N. vespilloides*** (**2**) (length 16mm) is the commonest of several species with orange-red markings on otherwise black elytra.

Rove beetles

These members of the family Staphylinidae have short elytra that leave the slender, elongated abdomen exposed. Many rove beetles have ferocious-looking jaws that reflect their often predatory habits. The **Devil's Coach-horse** *Staphylinus olens* (**1**) (length 24mm) is a spectacular, large, all-black species; if threatened, it curls up its abdomen and opens its jaws in a menacing fashion. It hides under stones in the daytime, emerging after dark to feed on invertebrates, and is

common in hedgerows and gardens. ***Paederus littoralis*** (**2**) (length 9mm) is one of several small but striking red and black species; it is common in damp habitats.

Beetles

Chafers

Members of the family Scarabaeidae, chafers are beetles with squat bodies and antennae that are clubbed or fanned at the tip. Typically they can fly well.

Rose Chafer

Cetonia aurata Scarabaeidae
Length 17mm

 Attractive beetle. **Adult** Shiny bronzy green, the parallel-sided elytra flecked with white lines and marks. Active May–Sep and moves in a cumbersome manner. **Larva** Lives in rotting wood. **Status** Widespread and locally common; often found in flowers, including roses.

Bee Beetle

Trichius fasciatus
Scarabaeidae Length 14mm

 Extremely hairy chafer. **Adult** Has wasp- or bee-like black and orange-yellow markings on elytra; seen Jun–Sep, often visiting flowers. **Larva** Lives in rotting birch stumps. **Status** Seldom common; associated mainly with N and W regions.

Cockchafer

Melolontha melolontha Scarabaeidae
Length 35mm

 Familiar beetle. **Adult** Has hairy, rufous-brown elytra and a pointed-tipped abdomen; seen May–Jun. **Larva** Lives in soil for several years, feeding on roots. **Status** Common enough in some areas to form swarms around treetops at dusk.

Beetle larvae

The larvae of beetles are usually simple and grub-like. Most have three small pairs of legs at the front end, a slightly armoured head with powerful biting mouthparts and an undifferentiated, segmented body. The larva of the Cockchafer (*see* below) lives in soil, spending several years there feeding on the roots of grasses, herbaceous plants and trees.

Soldier beetles

These brightly coloured insects (family Cantharidae) are active diurnal predators that are often found on flowers. Several species occur in our region.

Rhagonycha fulva

Cantharidae Length 11mm

 Familiar soldier beetle. **Adult** Mainly orange-red with dark-tipped elytra; flies well in sunny weather, mainly May–Aug. **Larva** Seldom seen. **Status** Extremely common throughout; mating pairs are a regular sight.

Rhagonycha fulva   *Cantharis rustica*

Cantharis rustica

Cantharidae Length 14mm

 Well-marked beetle. **Adult** Has a reddish abdomen, black elytra and a reddish pronotum that is marked with a dark spot. **Larva** Seldom seen. **Status** Common in S, in hedges and meadows.

Bioluminescence

A select band of creatures can actually produce their own light by a process known as bioluminescence. Although the phenomenon is more common in the tropics, we do have one classic example in Britain. The grub-like females of the **Glow-worm** *Lampyris noctiluca* (length 14mm) can be located after dark by the greenish light they emit from the underside of the abdomen tip; this serves to attract winged males. Adult Glow-worms do not feed but their larvae, which can also emit light, eat snails. It is local in meadows and along forest rides.

Click beetles

Click beetles (family Elateridae) have elongated bodies; they are able to leap into the air – accompanied by an audible click – and can right themselves if they land upside-down.

Athous haemorrhoidalis

Elateridae Length 14mm

 Our most familiar click beetle. **Adult** Has hairy reddish-brown elytra and a darker head and thorax; seen May-Jun. **Larva** Lives in rotting wood. **Status** Common and widespread, found in woodland, scrub and hedgerows, often on Hazel.

Cardinal beetles

Dressed in ecclesiastically rich colours, cardinal beetles (family Pyrochroidae) are showy and conspicuous insects.

Pyrochroa serraticornis

Pyrochroidae Length 14mm

 Our most familiar cardinal beetle. **Adult** Recognisable by its scarlet body. Active May-Jul, visiting flowers to hunt for small insects on sunny days. **Larva** Carnivorous; lives in rotting wood. **Status** Locally common species in England and Ireland, often found under flaking bark or in rotting timber. **Similar species** *P. coccinea* has a black head.

Oedemera nobilis

Oedemeridae Length 10mm

 Distinctive little beetle. **Adult** Shiny green and surprisingly slender. Elytra taper and splay towards tip of abdomen and do not completely cover wings. Male has a distinctive swollen hind femora. Seen May-Aug. **Larva** Lives in plant stems. **Status** Common in S England and Wales in flower-rich grassland.

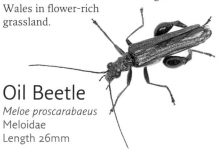

Oil Beetle

Meloe proscarabaeus
Meloidae
Length 26mm

 Produces pungent oil when alarmed. **Adult** Has a shiny bluish-black body with small elytra that do not cover the swollen abdomen. Seen Apr-Jun. **Larva** After hatching, climbs a flower and attaches itself to a passing solitary bee; back at the bee's nest, it develops inside. **Status** Local and widespread in grassy places.

Lagria hirta

Tenebrionidae Length 10mm

 Soft-bodied and rather sluggish beetle, sometimes found crawling over flowers and leaves. **Adult** Has an elongated blackish body, although the elytra are covered in yellowish hairs. Seen May-Aug. **Larva** Lives in leaf litter. **Status** Local, occurring mainly in the S and favouring woods and hedgerows.

Rhagium mordax

Cerambycidae Length 21mm

 Robust longhorn beetle. **Adult** Well-marked downy beetle. Active May-Jul; sometimes forages among flowers, or in rotting wood and stumps. **Larva** Lives in rotting wood. **Status** Widespread and fairly common in oak woods.

Longhorn beetles

These members of the family Cerambycidae usually have long antennae and elongated, flattened bodies. The larvae mainly live inside wood, the larger species taking several years to reach maturity.

Wasp Beetle

Clytus arietus Cerambycidae
Length 16mm

 Strikingly colourful beetle. **Adult** Extremely wasp-like, both in terms of its black and yellow markings and its behaviour; flies well in sunny weather in May-Jul, often visiting flowers. **Larva** Lives in rotting wood. **Status** Widespread and often common, in gardens and hedgerows.

Strangalia maculata

Cerambycidae Length 16mm

 Familiar wayside longhorn beetle. **Adult** Has variable yellow markings on elytra and legs but is otherwise black; visits flowers to feed on pollen and is active Jun-Aug. **Larva** Lives in rotting wood. **Status** Common and widespread. Found among leaves of trees and shrubs.

BELOW: *Rhagium mordax*

Beetles

Many of our smaller beetles are boldly patterned and colourful; the markings are often a warning to predators that they are distasteful. Ladybirds are recognised by their domed bodies; they scuttle along on tiny legs and feed on aphids and other small insects. Leaf beetles have shiny, plump bodies and feed mainly on plant material, and weevils are characterised by having a long 'snout'.

Eyed Ladybird

Anatis ocellata Coccinellidae
Length 8mm

 Distinctive ladybird. **Adult** Has orange-red elytra marked with white-ringed black spots; seen Jun-Jul. **Status** Widespread and locally common, usually associated with conifers.

Two-spot Ladybird

Adalia bipunctata Coccinellidae
Length 6mm

 Confusingly variable ladybird. **Adult** Usually has orange-red elytra with 2 black spots, although many are variably black with red spots; active Apr-Oct. **Status** Common.

dark form

Seven-spot Ladybird

Coccinella 7-punctata Coccinellidae Length 6mm

 Our commonest ladybird. **Adult** Has reddish-orange elytra and, seen at rest, 7 black spots; hibernates and is active Mar-Oct. **Status** Widespread and often abundant.

Fourteen-spot Ladybird

Propylea 14-punctata Coccinellidae
Length 5mm

 Boldly marked ladybird. **Adult** Has variable black and yellow markings, although suture between elytra is always black; active Apr-Sep. **Status** Common in S Britain.

Harlequin Ladybird

Harmonia axyridis Coccinellidae Length 9mm

 Striking, variable ladybird. **Adult** Usually deep red with variable number of black spots; much darker individuals are also found; active Apr-Sep. **Status** Native of Asia, recently arrived and spreading fast N and W across England. Predator of other ladybirds.

Poplar Leaf Beetle

Chrysomela populi Chrysomelidae
Length 10mm

 Striking rounded beetle. **Adult** Superficially ladybird-like but with unmarked bright red elytra and a black head, thorax and legs; seen Apr-Aug. **Status** Common on poplars and willows.

Mint Leaf Beetle

Chrysolina menthastri Chrysomelidae
Length 9mm

 Attractive rounded beetle **Adult** Extremely shiny bronzy green colour; usually found on mint leaves and sometimes also on hemp-nettles in May-Aug. **Status** Widespread and locally common in damp meadows.

Gastrophysa viridula

Chrysomelidae Length 6mm

 Rounded green beetle. **Adult** Elytra are pitted with a golden sheen; female has a distended abdomen. Seen May-Aug. **Status** Common and widespread; favours damp grassland and is usually associated with docks.

mating pair

Lily Beetle

Lilioceris lilii
Chrysomelidae Length 8mm

 Notorious garden species, both adults and larvae chewing lily buds and leaves. **Adult** Has a mainly red body with black legs and head; seen May–Aug. **Status** Common and widespread.

Dor Beetle

Geotrupes stercorarius Geotrupidae
Length 15mm

 Rotund beetle, often seen trundling across paths. **Adult** Has a shiny blue-black lozenge-shaped body and spiky-margined legs. **Larva** Feeds in horse or cow dung. Seen May–Jul. **Status** Locally common throughout, in grassland and woodland where grazing animals are kept.

Figwort Weevil

Cionus scrophulariae
Curculionidae Length 4mm

 Tiny, dumpy species. **Adult** Has a colourful, rather ornamented body; seen mainly May–Jul. **Larva** Lives inside plant stems. **Status** Locally common, often found feeding on figwort flowers.

Figwort Weevil

Bloody-nosed Beetle

Timarcha tenebricosa Chrysomelidae
Length 20mm

 Lumbering, flightless leaf beetle, often seen plodding across paths. **Adult** When disturbed, exudes drops of bright red blood-like fluid from mouth; seen Apr–Jun. **Status** Fairly common throughout in grassland.

Phyllobius pomaceus

Curculionidae Length 9mm

 Small weevil. **Adult** Black body is covered with greenish scales that are easily rubbed off; seen Apr–Aug. **Larva** Lives inside plant stems. **Status** Widespread and fairly common, often found on Common Nettle.

Vine Weevil

Otiorhynchus sulcatus Curculionidae
Length 11mm

 Destructive garden weevil. **Adult** Has a grey-brown body variably marked with patches of yellowish scales; seen mainly in May–Aug. **Larva** Eats the roots of potted plants. **Status** Often common in gardens. Widespread.

Vine Weevil

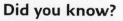

Tortoise Beetle

Cassida rubiginosa Chrysomelidae
Length 7mm

 Atypical leaf beetle. **Adult** Has broad, flattened green elytra and pronotum that overlap body and provide camouflage and protection when insect is clamped against a leaf; seen Jun–Aug. **Status** Fairly common throughout in hedgerows and meadows.

Hazel Weevil

Curculio nucum Curculionidae
Length 6mm

 Long-'snouted' weevil. **Adult** Seen Apr–Jun; female lays an egg inside embryo Hazel nut. **Larva** Feeds on developing nut inside the case until it falls in autumn. **Status** Widespread and locally common wherever Hazel occurs.

Platystomos albinos

Anthibidae Length 9mm

 Well-marked weevil. **Adult** Brown and white markings give it a passing resemblance to a bird dropping; seen May–Jun. **Larva** Lives in dead wood. **Status** Local woodland species.

Did you know?

Weevils are a varied group of beetles belonging to several families; there are numerous species in Britian. Most usually have a characteristic 'snout' that bears the mouthparts at the tip; this gives them an almost comical appearance.

Introducing spiders & their allies

Spiders and their close relatives inspire mixed emotions in many people but, love or hate them, they are an ever-present part of the British natural history scene. Aside from the marine environment, all British habitats have at least some representatives of the group and in many places they are positively abundant. A number of species have even ventured indoors and are now familiar house guests.

The spider's closest relatives

A number of other invertebrates are closely related to spiders and are also included in the class Arachnida. In Britain and Ireland, these are represented by false scorpions (order Pseudoscorpiones), harvestmen (order Opiliones) and ticks and mites (order Acarina). For more information *see* p.411.

What is a spider?

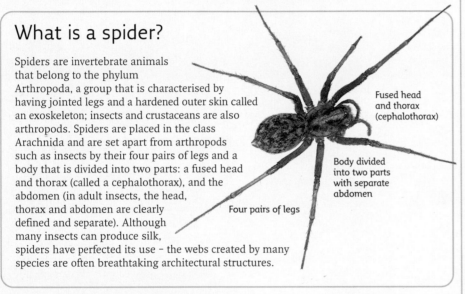

Spiders are invertebrate animals that belong to the phylum Arthropoda, a group that is characterised by having jointed legs and a hardened outer skin called an exoskeleton; insects and crustaceans are also arthropods. Spiders are placed in the class Arachnida and are set apart from arthropods such as insects by their four pairs of legs and a body that is divided into two parts: a fused head and thorax (called a cephalothorax), and the abdomen (in adult insects, the head, thorax and abdomen are clearly defined and separate). Although many insects can produce silk, spiders have perfected its use - the webs created by many species are often breathtaking architectural structures.

Fused head and thorax (cephalothorax)

Body divided into two parts with separate abdomen

Four pairs of legs

Hunting strategies

The ways that spiders catch their prey are as varied as their appearances. At the simplest level, opportunistic hunters such as wolf spiders roam in leaf litter, tackling almost anything of a suitable size; their excellent eyesight and sensitivity to vibrations make them acutely aware of what is going on around them. Other spiders favour the ambush approach: some employ camouflage and lie in wait for passing prey, while many spin silken webs to entrap their victims.

As predators, spiders' most valuable assets are their paired fangs, sited on either side of the mouth. These are plunged into the victim in the manner of hypodermic needles, injecting a paralysing venom; in many species the venom has the added property of partially digesting the internal organs of the prey. Some spiders then simply suck their victim dry, while others macerate the body and suck up the resulting nutritious soup.

Risky reproduction

Given that spiders are such unhesitating predators, complex behaviour patterns have evolved to ensure that members of the opposite sex can mate without one of them becoming a meal for the other, at least before mating has taken place. It is likely that pheromones - chemical attractants - have a role to play in this, but subtle tapping rhythms are important in the first instance. Having mated, female spiders of most species protect their eggs in silken cocoons; some guard the eggs while they develop and a few show a degree of parental care to the young when hatched.

left: A female Common Cross Spider guards her substantial egg sac.
below: Newly hatched Garden Spider hatchlings remain in their mother's brood web for several days before they venture out into the world.

A spider's fangs are concealed for much of the time and are only exposed when brought out to kill prey. This House Spider *Tegenaria domestica* is poised for action!

Silk and webs

Silk is a protein that spiders produce in glands in their abdomen; the glands open via pores on the spinnerets at the tip of the abdomen. Silk is produced in liquid form but hardens as it is stretched and has incredible tensile strength. At the simplest level, spiders use silk as a lifeline, produced constantly as they move through vegetation. They also use it to create a protective sac for their eggs, but most significantly many species spin webs in which their prey is trapped.

The intricate patterns in spider silk are most apparent when outlined by dewdrops on a misty day.

As a testament to the strength of silk, a single strand easily supports the weight of this plump female Common Cross Spider.

Silk traps

One of the simplest uses of silk is seen in the **Daddy-long-legs Spider** *Pholcus phalangioides*, which creates a tangled mass of threads – it hardly qualifies as a proper web – in which potential prey simply becomes trapped; despite the web's untidy appearance, it is nevertheless efficient at its job. Other species construct hammock-like webs into which prey falls, while some create a web funnel, surveying the world around them from its centre. However, the apogee of web construction is seen in the orb-web spiders.

At the centre of a large and untidy sheet web, this spider has created a funnel into which it retreats when danger threatens. If it senses vibrations from a passing insect it will rush out and try to grab its victim.

Orb webs

Orb webs are marvels of construction, especially considering the relatively small size of the manufacturer. They are created by members of the family Araneidae and typically are hung vertically in order to catch flying insects. Construction begins with the spider releasing a silk thread from a secure elevated piece of vegetation. Pulled by the wind, the end eventually touches another leaf or twig and the spider has its first point of contact. It then crawls along this thread, laying a more substantial partner to form the fundamental uppermost horizontal thread in the web. Thereafter, a 'Y' shape is formed to connect the two upper attachments to one below, and this forms the basis for the web's outer framework. An initial silk spiral is then created, after which a replacement with sticky silk completes the web. All that now remains is for the spider to sit and wait for its first meal.

An orb-web spider at the centre of its web.

Spiders

You hardly have to travel any distance to find a range of interesting spiders, since several species find our gardens and dwellings to their liking. If you venture further afield to almost any terrestrial habitat you will find still more, the common factor being an abundance of invertebrate prey for them to feed on.

House and garden spiders

Most houses have nooks and crannies that several species of spider find irresistible as refuges, and where they retreat to while human householders go about their daily business. Most venture out after dark and feast on other invertebrates – including other spiders – that are more occasional house guests. Outside in the garden, different spider species are often abundant especially in autumn, when their large webs appear.

House Spider

Tegenaria domestica Agelenidae
Length 10mm

 Large, long-legged spider that appears intimidating when trapped in a bath or scurrying across the floor. **Adult** Body is rather hairy and varies from pale to dark brown. Females can survive for several years. Spins an untidy web with a tubular retreat in the corner of a room. **Status** Widespread, often found in and around houses.

Daddy-long-legs Spider

Pholcus phalangioides Pholcidae
Length 8mm

 Unmistakable spider. **Adult** Narrow-bodied and long-legged. Can be seen at most times of year and is typically found hanging upside-down from ceilings, where it spins an untidy web. **Status** Relatively recent arrival to the British household, encouraged no doubt by the spread of central heating (it cannot survive in locations where the temperature dips below 12°C). Widespread.

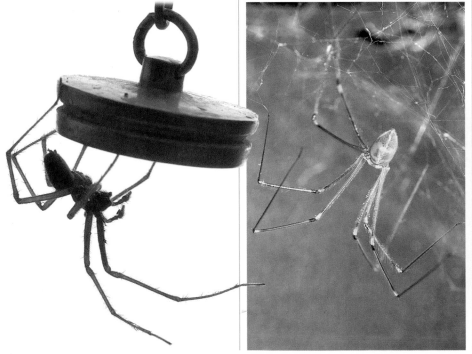

Garden Spider

Araneus diadematus Araneidae
Length 12mm

 Arguably our most familiar spider. **Adult** Female is much larger than male, but both can vary in colour from grey-brown to reddish brown and abdomen has a row of white dots down centre and transverse white streaks that form a distinct cross. Seen Jul–Oct. Has a sophisticated web. **Status** Often abundant in gardens, as well as in meadows, hedgerows and woods.

Common Cross Spider

Araneus quadratus Araneidae
Length 20mm

 Well-marked spider. **Adult** Similar to the Garden Spider but abdomen has 4 large white spots arranged in a square and a white anterior stripe. Abdomen colour varies from nut brown to bright red. Female is much larger than male. Seen Jul–Oct. **Status** Common throughout in gardens, scrub, hedgerows and meadows.

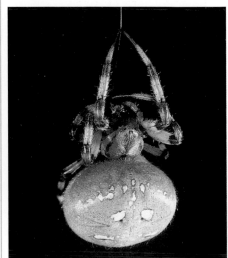

Araniella cucurbitina

Araneidae Length 6mm

Small but attractive spider that spins an untidy orb web among low wayside vegetation. **Adult** Has a lime-green abdomen with yellow bands; cephalothorax and legs are reddish brown. Seen mainly May–Sep. **Status** Common in rough meadows, hedgerows and gardens, and often found among thistles and Brambles.

Amaurobius similis

Amaurobiidae Length 12mm

Familiar spider, often found on walls, behind bark and on fences, where it constructs a tangled bluish-white web leading back to a retreat in a crevice. **Adult** Has a buffish abdomen with paired dark markings on the dorsal surface and chevrons towards the rear end. Cephalothorax and legs are usually orange-brown. Seen mainly May–Sep. **Status** Common throughout in gardens, woods and hedgerows.

Spiders and water

Fresh water is a prime breeding habitat for many insects, and the nymphs and larvae of these species often emerge from their aquatic homes to live terrestrial lives as adults. With all this potential food available, spiders are often abundant in the vicinity of ponds and streams, their webs ensnaring the likes of damselflies and mayflies. A few species have gone a step further, conquering the aquatic environment.

Swamp Spider

Dolomedes fimbriatus Pisauridae
Length 25mm

Without doubt our most impressive spider. **Adult** Has a chestnut-brown body and legs, and a yellow line around margins of cephalothorax and abdomen; most evident May–Aug. Typically, sits with front legs touching water surface, alert to any distress movements of insects trapped in surface film; in response, it skates over surface to catch its prey. Can submerge if alarmed. **Status** Restricted to heaths in S England, mainly in the New Forest and Surrey, and lives beside boggy pools.

Water Spider

Argyroneta aquatica Argyronetidae
Length 14mm

Our only truly aquatic spider. **Adult** Looks silvery underwater owing to a film of air trapped around the abdomen. Constructs a domed, air-filled web among water plants, in which it spends much of daytime; silk threads radiating from web alert occupant to passing prey. **Status** Very locally common year-round in weedy lakes and ponds.

The strange case of the Purse-web Spider

The **Purse-web Spider** *Atypus affinis* (length 12mm) is an extraordinary spider that lives inside a mainly subterranean silken tube, a small part of which lies on the soil surface like the finger of a glove. When an unwary insect walks over the tube, the spider rushes to the source of the disturbance and, walking upside-down, thrusts its long fangs through the silk to grab the prey; damage to the tube is repaired after the meal has been consumed. It is scarce and local, occurring in dry, well-draining soil, and the silken tubes are easily overlooked.

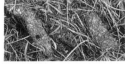

ABOVE: **silken tubes**

Dysdera crocata

Dysderidae Length 12mm

Attractive and distinctive spider that hides under stones during daytime. **Adult** Has reddish legs and cephalothorax and buffish-brown abdomen. Fangs are huge relative to body size, opposable and used to good effect when capturing woodlice. **Status** Widespread and fairly common in most habitats that support good numbers of woodlice.

Spiders & allies

Life for small invertebrates is full of perils. Apart from birds, probably the biggest threat they face comes from spiders, all of which are voracious predators. Spiders dispatch their prey with a bite from their fangs followed by the injection of toxic venom.

Wolf and hunting spiders

Although many spiders make silken webs to catch their prey, there are plenty of species that employ a more active approach to hunting than their ambushing cousins. Among them are the appropriately named wolf and hunting spiders, which roam at ground level or in low vegetation, locate prey with their keen eyesight, and leap on the victim in an assault attack.

Pardosa spp.
Pisauridae
Length 6mm

RIGHT: *Pardosa* female with young

Pardosa female carrying egg sac

A group of common wolf spiders. **Adult** Most active Apr-Jun, when females can sometimes be seen carrying around an egg sac attached to tip of abdomen; after young hatch, they remain clinging to mother's abdomen for several weeks. **Status** Common and widespread, found among woodland leaf litter.

Pisaura mirabilis
Pisauridae Length 14mm

Common hunting spider. **Adult** Has a buffish-brown body with a dark-bordered yellow stripe on carapace. Active mainly May-Jul; females sometimes seen carrying an egg sac underneath body, secured by mouthparts, and build a nursery tent just before eggs are due to hatch. **Status** Common and widespread; found in hedgerows, woodland rides and grassland.

Cave spiders

While many spiders are active mainly after dark, a few species shun the light completely and spend their entire lives in the dark recesses of caves and cellars.

Metellina merianae
Metidae Length 9mm

Typical cave spider. **Adult** Has a marbled brown and black abdomen, and shiny legs marked with irregular bands of reddish brown and black. Seen mainly May-Jul. **Status** Widespread and fairly common in suitable habitats; often particularly common near coasts.

Crab spiders

These amazing little creatures (family Thomisidae) bear a superficial resemblance to their crustacean namesakes, and when placed on flat ground they tend to scuttle in a crab-like manner. A range of different species is found in the countryside but all are masters of disguise, typically lurking unseen on flowers and other vegetation, where they capture unsuspecting passing insects, sometimes as large as butterflies.

Xysticus cristatus

Misumena vatia
Thomisidae Length 10mm

Well-camouflaged spider. **Adult** Males are tiny and easily overlooked; females are larger and often found sitting on flowers in May-Aug, their changeable colour matching the petals. **Status** Widespread and often common in gardens, hedgerows and meadows.

Xysticus cristatus
Thomisidae Length 6mm

Boldly patterned spider. **Adult** Has patterns of pale and dark brown on abdomen that form a series of overlapping triangles. Seen May-Jul; waits motionless for a passing insect with legs outstretched on a bare stalk or flower. **Status** Common and widespread in meadows and hedgerows.

Jumping spiders

Jumping spiders have amazing leaping powers, used to capture prey and evade danger. They can leap several centimetres, further if wind assisted – in relative terms like a person jumping over a house.

Zebra Spider

Salticus scenicus Salticidae
Length 7mm

Our most familiar jumping spider, named after its black and white stripes. **Adult** Seen May-Sep, often moving restlessly up sunny fences and walls; spots potential prey using large eyes and then stalks it to within leaping range. **Status** Common and widespread in most areas.

Tetragnatha extensa

Tetragnathidae Length 10mm

Long-legged spider with an elongated, sausage-like abdomen. Often aligns itself along a plant stem with legs outstretched. **Adult** Has reddish-brown legs and cephalothorax, and a marbled yellow, brown and white abdomen. Seen mainly Jun-Aug. **Status** Common throughout in damp meadows and hedgerows.

Wasp Spider

Argiope bruennichi Argiopidae
Length 24mm

Distinctive spider, the females of which are striking. Makes an orb web with a diagnostic zigzag band. **Adult** Female has black, yellow and white markings; male is much smaller and browner. **Status** Relative newcomer to Britain; common only in the SE, in grassland, mainly coastal.

Spider relatives

Spiders are the most familiar, and often the most numerous, members of the arachnids, but several other groups of invertebrates share the same classification. Different representatives of all groups are found throughout Britain and Ireland, although precise identification is best left to experts.

Harvestmen belong to the order Opiliones and are superficially spider-like, with four pairs of long, gangly legs. A close look at the body, however, reveals it to be undivided (spiders have two distinctly separate parts – the cephalothorax and the abdomen). Harvestmen do not produce silk and their diet includes a mixture of scavenged dead animals and live prey.

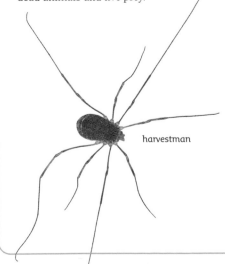

harvestman

Ticks belong to the order Acarina and have rather small legs and an abdomen that can become grossly distended after feasting on a blood meal.

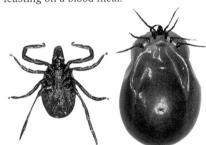

larval tick bloated adult tick

False scorpions are tiny Arachnids that belong in the order Pseudoscorpiones. They are predators with poison claws, and most species are found in leaf litter and compost heaps.

false scorpion

Mites also belong to the order Acarina. They have similarly short legs and a globular body but are much more active than ticks. There are several species in our region, their diets ranging from even smaller invertebrates to plant material.

Colony of Gorse Mites *Tetranychus lintearius*

Red Spider Mite

Life in soil & leaf litter

Many naturalists do not realise that beneath their feet is a world in miniature whose inhabitants are every bit as exciting as butterflies in a meadow or birds on the seashore. These tiny creatures live out their lives in the soil and leaf litter. If you look closely, you will be amazed at the diversity of life you find, from tiny snails and woodlice to many-legged millipedes and centipedes.

Woodlice

Related to aquatic species such as crabs and water fleas, woodlice are unusual among crustaceans for living terrestrial lives. They belong to the order Isopoda, other representatives of which live in water, and have a flattened, segmented body protected by armoured dorsal plates. As woodlice easily become desiccated they usually hide under stones or in leaf litter during the daytime. Of the British species, the following are regularly encountered.

Common Woodlouse

Oniscus asellus Oniscidae
Length 14mm

Familiar woodlouse. **Adult** Body is marbled yellow and brown, head has lateral lobes, and jointed end of antennae comprises 3 sections. **Status** Common and widespread across most of Britain and Ireland; often numerous in gardens but also found in woodland and hedgerows.

Porcellio scaber

Porcellionidae Length 10mm

Another common woodlouse. **Adult** Body is narrower than that of Common Woodlouse and jointed end of antennae comprises 2 sections. **Status** Common and widespread in lowland Britain and Ireland; often numerous in gardens.

Pill Woodlouse

Armadillidium vulgare
Armadillididae Length 11mm

Intriguing species. **Adult** Slate grey; easily recognised by its ability to roll into a ball when disturbed (*above*). **Status** Widespread and fairly common in S England, usually in woods, hedgerows and rural gardens.

Trichoniscus pusillus

Trichoniscidae Length 5mm

Small woodlouse. **Adult** Body is elongated and rather rounded in cross-section; usually pinkish purple. **Status** Widespread and fairly common among damp leaves and debris in ditches or around pond margins.

Philoscia muscorum

Oniscidae Length 11mm

Active woodlouse. **Adult** Has a dark head and dorsal stripe. **Status** Common and widespread in woodland and grassland.

Blind Woodlouse

Platyarthrus hoffmannseggi Squamiferidae
Length 3mm

Tiny species. **Adult** Has an essentially pure white body. **Status** Widespread and fairly common but easily overlooked. Lives a subterranean life in ants' nests.

Millipedes

Often discovered under logs or among leaf litter, millipedes belong to the class Diplopoda and can be recognised by the presence of two pairs of legs on each of the segments of their elongated bodies. They feed mainly on plant material.

Tachypodoiulus niger
Julidae Length 6cm

Typical woodland millipede. **Adult** Has a blackish cylindrical body. **Status** Common and widespread.

Cylindroiulus punctatus
Julidae Length 27mm

Typical woodland millipede. **Adult** Has a pale reddish-brown, cylindrical body. **Status** Locally common in wooded areas; often found under the bark of rotting wood or among leaf litter.

Pill Millipede

Pill Millipede
Glomeris marginata Glomeridae
Length 20mm

Distinctive millipede. **Adult** Has an armoured, shiny chestnut-brown body that gives it a rather woodlouse-like appearance; when disturbed, rolls into a protective ball. **Status** Locally common in mature woodland.

Flat-backed Millipede
Polydesmus angustus Polydesmidae
Length 24mm

Unusual millipede whose flat profile often allows it to creep under bark. **Adult** Has a flattened, golden-brown body with laterally projecting legs. **Status** Widespread and common in soils rich in organic matter, in compost heaps and under the bark of rotting wood.

The humble earthworm

Earthworms are members of the phylum Annelida, which comprises invertebrates with elongated segmented bodies. Charles Darwin rightly pointed out that earthworms are among the most important animals in Britain: the soil's health depends on the aerating effects of their tunnelling, and its fertility is improved by the burying and recycling of organic matter. Several species are found in Britain and Ireland, of which the **Common Earthworm** *Lumbricus terrestris* (length 8cm) is a typical representative. It is abundant in suitable soils, and its presence can be detected by conspicuous deposits of digested soil casts on the surface at burrow entrances.

Centipedes

Although superficially similar to some millipedes, centipedes can be recognised by having just one pair of legs on each of segment of their elongated, flattened bodies. They belong to the class Chilopoda and are voracious predators of other invertebrates.

Lithobius variegatus
Lithobiidae Length 30mm

Probably our most typical centipede. **Adult** Has a shiny orange-brown body. **Status** Widespread and common; familiar garden resident, also found in hedgerows and woods, hiding under stones or logs in the daytime.

ABOVE: *Lithobius variegatus*
BELOW: *Haplophilus subterraneus*

Haplophilus subterraneus
Geophilidae Length 35mm

Burrowing species. **Adult** Has an extremely slender orange body. **Status** Widespread and often common in garden soils and compost heaps.

worm cast

Slugs & snails

Like their cousins the snails, slugs are molluscs (members of the phylum Mollusca) and are readily identified by their large muscular foot, which they use to move by means of waves of muscular contractions, lubricated by copious quantities of slime. Unlike snails, slugs do not have a large obvious shell into which they can retreat, although a few species do have a tiny external shell, the size and shape of a miniature fingernail, and others have a minute internal shell.

Large Red Slug

Arion ater Arionidae
Length 12cm

Our most familiar large slug. **Adult** Often orange-red (commonest in the S and in gardens), but black forms prevail in the N and in upland areas. When alarmed, contracts into an almost spherical ball and often rocks from side to side. Mucus is colourless. Lays clusters of pale eggs under logs. **Status** Common throughout in almost all terrestrial habitats.

orange-red form

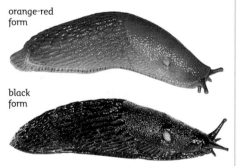

black form

Common Garden Slug

Arion distinctus Arionidae
Length 3cm

Rather small slug. **Adult** Has a yellowish-orange sole and orange body mucus. Body is usually striped and covered in tiny gold dots. **Status** Commonest in N England, the lowlands of Scotland and parts of Ireland; found in gardens and agricultural land, where it damages crops.

Southern Garden Slug

Arion hortensis Arionidae
Length 3cm

Southern counterpart of Common Garden Slug. **Adult** Has a grey body with dark longitudinal lines, and is typically palest on flanks; sole is orange and mucus is yellow. **Status** Common in gardens in S England; also found in hedgerows and on farmland.

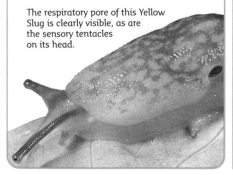

Slug respiration

Terrestrial slugs are air-breathers, and if you look closely at the head end you will see a conspicuous opening called the respiratory pore (be warned, however, that slugs can close this opening). The pore leads to an internal chamber called the mantle cavity, which is lined with blood vessels that take up oxygen and so acts as the slug's lung.

The respiratory pore of this Yellow Slug is clearly visible, as are the sensory tentacles on its head.

Dusky Slug

Arion subfuscus Arionidae
Length 7cm

Well-marked slug. **Adult** Has a mainly pale brown body, but with a dark dorsal surface and a single longitudinal dark stripe on each side. Often looks golden owing to orange body mucus; sole is yellow but sole mucus is colourless. **Status** Common in woodland and hedgerows across most of Britain and Ireland except E England.

Yellow Slug

Limax flavus Limacidae
Length 10cm

Large, colourful slug. **Adult** Has a yellowish body that is marbled and mottled with olive-brown; tentacles are blue and mantle has a fingerprint-like pattern of concentric rings typical of all *Limax* species. **Status** Widespread and usually associated with gardens and houses; ventures indoors after dark and into cellars. Feeds voraciously on seedlings and vegetables.

Slugs of ancient woodlands

Although many slug species are associated with disturbed sites, such as gardens and agricultural land, a few are restricted to pristine ancient woodland and are sometimes used as indicators of particularly good locations. One of these is the **Lemon Slug** *Limax tenellus* (**1**) (length 4cm), a small, bright yellow slug with dark tentacles; it is extremely local and is hard to find except in late summer and early autumn, when it feeds on fungi. Another is the **Ashy-grey Slug** *Limax cinereoniger* (**2**) (length 25cm), our largest slug when fully grown. Its ash-grey body is marked with a pale yellowish keel from mantle to tail; it is local and is found under logs in the daytime.

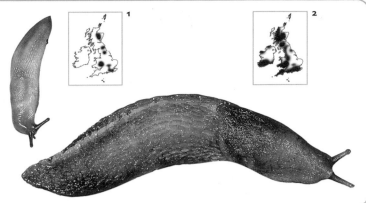

Leopard Slug

Limax maximus Limacidae
Length 16cm

Large, well-marked slug. **Adult** Usually mainly pinkish grey, but body is covered with numerous dark blotches and spots. Has a pronounced keel running along rear part of body to tail. Mucus is sticky and colourless, and sole is whitish. **Status** Widespread and common in woodland and gardens.

Tree Slug

Limax marginatus Limacidae
Length 7cm

Pale, translucent-looking slug that climbs trees, usually in wet weather. **Adult** Body is pale greyish buff, marked with 2 dark lines on both sides from mantle to tail. Produces copious quantities of watery mucus when disturbed. **Status** Widespread in woodland in W Britain and Ireland.

Carnivorous slugs

Most slugs feed on plant material – either living or decaying – but a few have a carnivorous diet. Shelled slugs, of which ***Testacella scutulum*** (length 10cm) is the most common species, are subterranean slugs. They eat earthworms and, sometimes, other slugs, and are recognised by the fingernail-like shell covering the mantle at the rear end. *T. scutulum* is easiest to find in garden compost heaps, occurs locally and is found mainly in the S.

Problem slugs

Although most slugs feed on decaying plant material, a few species give the group a bad name, feeding on living plant tissue and assuming pest proportions. The worst offenders are species that are spread by man's activities, and were probably introduced to Britain and Ireland. The following three species cause more problems than any others on farmland and in gardens.

The **Netted Slug** *Deroceras reticulatum* (**1**) (length 5cm) has a variable body colour but is usually buffish brown with a network of darker brown veins and blotches; the body often looks lumpy, the keel is truncated at the tail end and the slug produces large quantities of clear mucus when irritated. **Sowerby's Slug** *Milax sowerbyi* (**2**) (length 7cm) is usually brown or tan with a truncated yellowish keel, mucus that is thick, sticky and yellowish, and a pale rim to the respiratory pore. The **Budapest Slug** *Milax budapestensis* (**3**) (length 6cm) is greyish with a yellow keel and colourless mucus; it often contracts into a 'C' shape when disturbed.

Land snails

Land snails have shells, the shape and markings of which usually lead to a definite identification. Hence they are diagnostic. The shells are composed mainly of calcium and so persist long after their owners have died; when empty, they make a fascinating collection.

Garden Snail

Helix aspersa Helicidae
Diameter 4cm

 Our most familiar large snail. **Adult** Has a marbled brown and black shell, often rather worn in older specimens. **Status** Common and widespread in lowlands of England, Wales and S Ireland; absent from much of Scotland. Found in gardens, woods and hedgerows.

Chalk specialists

Calcium is an important component in snail shells, so it is hardly surprising that snail diversity and numbers are greatest on chalky soils. The following species are more or less restricted to calcareous habitats.

Round-mouthed Snail

Pomatias elegans Pomatiidae
Length 1cm

 Unmistakable snail. **Adult** Has a ribbed, rather conical shell and, uniquely among British land snails, a calcareous 'trapdoor' that can seal shell mouth. Whole shell is often encrusted with mud. **Status** Found locally in chalk scrub in S England.

Heath Snail

Helicella itala Helicidae
Diameter 18mm

 Attractive and distinctive snail. **Adult** Has a flattened shell with spiral brown bands. **Status** Despite its name, it is found on chalk grassland and dunes; it is locally common in S England.

Banded Snail

Cernuella virgata Helicidae
Diameter 18mm

 Well-marked snail with a rather conical shell. **Adult** Shell is marked with dark brown spiral bands and is more rounded than that of Heath Snail. **Status** Locally common in S England and found in similar locations to the Heath Snail; favours chalk grassland and dunes. Can be very locally abundant.

Copse Snail

Arianta arbustorum Helicidae
Diameter 25mm

 A snail of damp habitats, with a rather spherical shell. **Adult** Orange-brown with a dark spiral band. **Status** Widespread but local in England, Wales and Scotland; scarce in N Ireland. Found in damp lowland areas, including hedgerows, meadows and woods.

Pointed Snail

Cochlicella acuta Helicidae
Length 10mm

 Narrowly conical snail. **Adult** Typical shell is whitish with dark brown band on whorl nearest mouth. **Status** Very locally common on coasts of W Britain on dunes and chalk grassland.

Lapidary Snail

Helicigona lapicida Helicidae
Width 15mm

 Attractive, well-marked snail. **Adult** Has a flattened shell with a sharply keeled margin and radial bands of reddish orange. **Status** Local and declining; found among stones and in walls in calcareous areas.

Kentish Snail

Monacha cantiana Helicidae
Diameter 15mm

 Familiar wayside snail. **Adult** Has a mainly pale buff shell, typically flushed darker brown towards mouth (both inside and out). **Status** Common in a variety of habitats, including grassland, scrub and hedgerows; also on waste-ground.

Strawberry Snail

Trichia striolata Helicidae
Diameter 12mm

Familiar garden species. **Adult** Has a buffish-brown shell that is rather flattened and ridged. **Status** Widespread and fairly common in England, Wales and Ireland, but scarce and local in Scotland. Favours lowland habitats, including gardens, woodland and hedgerows.

Cellar Snail

Oxychilus cellarius Zonitidae
Diameter 12mm

Distinctive snail. **Adult** Has a rather translucent amber-brown shell that is a flattened spiral; body of snail is bluish grey. **Status** Widespread and common, although absent from most upland areas of Scotland. Favours gardens, woods and hedgerows.

Garlic Snail

Oxychilus alliarius Zonitidae
Diameter 6mm

Similar to, but smaller than, Cellar Snail and easily identified by strong smell of garlic that is emitted when handled. **Adult** Has a translucent orange-brown shell and blackish body. **Status** Widespread and often common throughout Britain and Ireland. Found in most terrestrial habitats.

Lip-reading

Two similar species of *Cepaea* are locally common throughout the region in gardens, woods and hedgerows, sometimes living side by side. Their shell markings vary considerably, but typical forms of both species can either be uniform yellow or yellow with dark spiral bands. The way to tell them apart is by the colour of the shell lip. The **White-lipped Snail** *Cepaea hortensis* (**1**) (diameter 18mm) has a white lip, while the **Brown-lipped Snail** *C. nemoralis* (**2**) (diameter 21mm) has a dark brown lip and is often found in drier, warmer sites than its cousin.

Rounded Snail

Discus rotundatus Endodontidae
Diameter 7mm

Distinctive snail. **Adult** Has a rather flattened shell with tightly packed whorls; usually shows conspicuous ridges and is marked with bands. **Status** Widespread throughout lowland Britain and Ireland. Common in gardens; also occurs in woods and stony places.

Amber Snail

Succinea putris Succineidae
Length 15mm

Delicate little snail, often observed climbing among leaves of waterside vegetation such as Yellow Iris. **Adult** Has a translucent orange-brown shell and a darker body. **Status** Widespread in lowland wetlands throughout much of England, Wales and S Ireland.

Woodland snails

A few snail species are found only in woodland. As they are sensitive to disturbance, ecologists use them as indicators of ancient and important sites. One such species is the **Lesser Bulin** *Ena obscura* (**1**) (length 8mm), which is easily overlooked as its shell is often coated with mud. It is very local, commonest in S and E England, and is found in leaf litter, although it climbs trees in wet weather. Another indicator species is the **Plaited Door Snail** *Cochlodina laminata* (**2**) (length 16mm), with a long, narrow, orange-brown shell; it also climbs trees in wet weather and at night, and is locally common only in England.

Freshwater habitats

For wildlife and naturalist alike, freshwater habitats have a magnetic appeal and there can be few people with an interest in the countryside who have not spent time beside one of our many ponds, lakes, streams or rivers. In Britain and Ireland there are a wealth of such habitats, and few people have to travel far to visit one of these sites.

Water-lilies and pondweeds grow in profusion at the surface of many lowland ponds and lakes, Canadian Pondweed flourishes at the expense of native species in the submerged world, and emergent plants such as Reed Sweet-grass, Bulrush and Common Reed thrive around the margins. In acidic upland waters, Bogbean and bladderworts can often be found.

Lowland streams support marginal Yellow Iris, Purple Loosestrife and Great Willowherb, while water-crowfoots grow in the water itself. Chalk streams

You know spring has arrived when **Common Frogs** (*see also* p.331) return to their breeding ponds. Like our other amphibians, they are tied to water for reproduction – their eggs and tadpoles cannot survive on land. Once spawning is complete, most adults leave the ponds and live in nearby terrestrial habitats.

Caddis flies are known for their case-making aquatic larvae. Numerous species are found in rivers and streams, and a few in ponds and lakes; the larvae of some make cases from sand and small stones, while others use plant debris. Adult caddis flies such as ***Potamophylax cingulatus*** (*see also* p.391) are weak-flying insects.

Pondweeds are members of the genus *Potamogeton*. Many species are found in still and flowing waters, some solely with submerged leaves and others with floating leaves as well. The presence or absence of any given species is influenced by water chemistry, although **Broad-leaved Pondweed** (*see also* p.426) is widespread.

support Watercress, and stream and river margins of all types are home to water-starworts, Water Mint and Lesser Spearwort.

Invertebrate life abounds in most unpolluted bodies of water, although the species vary according to the chemistry of the water and whether it is flowing or standing. Mayflies, stoneflies, dragonflies and damselflies are most familiar as airborne adults, but they also occur in large numbers as aquatic larvae or nymphs.

Frogs, toads and newts are seasonally abundant in ponds during their spring breeding season, but in larger bodies of water fish are the dominant force. Depending on the depth and quality of the water and its chemistry, species such as Pike, Roach, Gudgeon and Minnow are all widespread; the selective stocking or removal of certain species means that it is almost impossible to determine their natural distributions. Shallow, fast-flowing streams are the haunt of Three-spined Stickleback, Bullhead, Stone

Loach and Brown Trout, while slow-flowing deeper waters harbour Perch, Chub and many others.

Birdlife abounds on fresh water, too, and Mute Swans, Mallards and Grey Herons are ubiquitous. Small lowland streams are favoured by Kingfishers and Little Grebes, while in the north and west Dippers and Grey Wagtails are characteristic species. Sedge and Reed warblers and Reed Buntings often nest in emergent vegetation, and winter lakes support Tufted Ducks and Pochards.

Once widespread, the **Water Vole** (*see also* p.202) is now rare because of habitat destruction and predation by Mink. Nowadays, it favours unpolluted rivers, streams and ponds, and still fares well in parts of central England and Scotland. Water Voles lead an amphibious life: they burrow on land but swim well.

If a river or stream is clear it is usually possible to see fish beneath the surface. Brown Trout in clear chalk or moorland streams are usually easy to recognise. In lowland backwaters and ponds, **Three-spined Sticklebacks** (*see also* p.430) can also be seen and identified; the males are colourful in spring.

More than any other bird, the **Dipper** (*see also* p.294) is found near fast-flowing boulder-strewn rivers and streams, and its distribution reflects this – it is found mainly in the north and west of Britain. Stonefly nymphs are important in its diet and are caught during underwater forays. Resting birds perch on boulders.

Freshwater habitats & hotspots

Freshwater enthusiasts will tell you that each river, stream, pond or lake has its own unique qualities. While this is true, there are nevertheless broad habitat categories into which most can be placed according to their rate of flow and geographical location. Some of the best examples of these are included here.

Flooded gravel pits & reservoirs

The extraction of gravel and sand for building works and the subsequent flooding of the resultant pits has vastly increased the number of large bodies of standing fresh water in lowland England. Mature examples of these pits, along with some reservoirs, are often havens for wildlife, being full of fish and invertebrates, and with birdlife to match throughout the year.

Hotspots *see map:*
1 **Rutland Water** - www.rutlandwater.org.uk
2 **Thames Valley region generally**, particularly between Theale and Brimpton in Berkshire - www.thames-explorer.org.uk/about_the_river.html

Flooded gravel pit, near Theale, Berkshire.

Lowland ponds & lakes

Often nearly covered by water-lilies by late summer, lowland ponds abound with invertebrate life and many are notable for their dragonflies and damselflies. In spring, undisturbed sites are often full of breeding amphibians, the calls of Common Frogs and Common Toads attracting the attention of visitors. Sympathetically managed canals harbour similar wildlife.

Hotspots *see map:*
6 **Lough Neagh** - www.loughneagh.com
7 **Basingstoke Canal** - www.basingstoke-canal.org.uk/wildlife.html

Pond in the New Forest, Hampshire. INSET: Ruddy Darter

Upland lakes and tarns

With their clear peaty waters, upland lakes, lochs and tarns harbour specialist plants and animals, often including an abundance of biting midges and mosquitoes. In larger stream-fed water bodies, aquatic freshwater invertebrates support healthy Brown Trout populations, which in turn feed specialist birds. In the breeding season, bird highlights from Scottish lochs and tarns include Slavonian Grebe, and Red- and Black-throated Divers.

Hotspots *see map:*
3 **Snowdonia National Park** - www.eryri-npa.co.uk
4 **Lake District National Park** - www.lake-district.gov.uk
5 **Cairngorms National Park** - www.cairngorms.co.uk

Loch Garten, Scotland. INSET: Slavonian Grebe

Lowland streams & rivers

Lowland streams and rivers are the gentler cousins of their upland counterparts, and are usually relatively slow-flowing or even almost sluggish. Their water chemistry has a profound influence on the wildlife they support; arguably the most distinctive are those flowing over chalk in southern England.

Hotspots *see map:*
8 **Rivers Test and Itchen, Hampshire** - www.waterscape.com/features/wildlife
9 **River Erne** - www.habitas.org.uk/flora/habitats/rivers.htm
10 **Norfolk Broads** - www.broads-authority.gov.uk

Hampshire chalk stream. INSET: Grey Wagtail

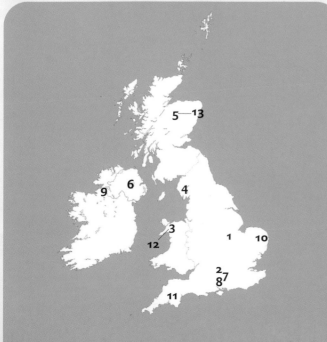

Upland streams & rivers

Mainly found in the mountains in the west and north of Britain, upland streams and rivers tend to be fast-flowing and boulder-strewn, and although there are exceptions, most have water that is rather acidic. As a consequence, the plants and animals that are found in these habitats are tough and resilient.

Hotspots *see map:*
11 **Dartmoor National Park** - www.dartmoor-npa.gov.uk
12 **Snowdonia National Park** - www.eryri-npa.co.uk
13 **Cairngorm National Park** - www.cairngorms.co.uk

Fast-flowing stream in Snowdonia. INSET: a stonefly, *Nemoura cinerea*

Find out more about...

- Reedbeds on **p.164**
- Mires, bogs and fens on **pp.170–1**
- Amphibians on **pp.330–31**
- Dragonflies and damselflies on **pp.380–5**
- Fish on **pp.430–33**
- Freshwater invertebrates on **pp.434–41**

Water plants

Water plants are represented by a wide range of families. While some of these plants have terrestrial relatives, many are exclusively adapted to the aquatic environment, and in the absence of permanent inundation and water of the right quality and chemistry, they will not survive.

Water-lilies

Recognised by their floating rounded leaves, water-lilies are members of the families Nymphaeaceae and Menyanthaceae.

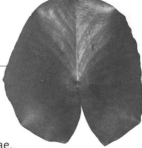

White Water-lily leaf

White Water-lily
Nymphaea alba Nymphaeaceae
Aquatic

Conspicuous when its large floating leaves are visible (they die back in winter). **Flowers** Fragrant, pinkish white and 15–20cm across, opening fully only in bright sunshine; borne on stalks just above water surface (Jun–Aug). **Fruits** Spongy. **Leaves** 10–30cm across, rounded and floating. **Status** Widespread and locally common; grows in still or slow-flowing fresh water to a depth of 3m.

Fringed Water-lily
Nymphoides peltata Menyanthaceae
Aquatic

Dainty water-lily. **Flowers** Fringed, yellow and 30–35mm across; borne on stalks rising just above water's surface (Jun–Sep). **Fruits** Spongy. **Leaves** 3–8cm across and rounded or kidney-shaped. **Status** Locally common in S England and naturalised elsewhere; grows in still or slow-flowing water.

Yellow Water-lily
Nuphar lutea Nymphaeaceae
Aquatic

Distinctive in bloom. **Flowers** Yellow, alcohol-scented and 6cm across; borne on stalks that rise well above water's surface (Jun–Sep). **Fruits** Spongy. **Leaves** Up to 40cm across and, unlike those of White Water-lily, basal lobes usually touch or overlap. **Status** Widespread and locally common except in N Scotland; grows in nutrient-rich still or slow-flowing fresh water to a depth of 5m.

Common Water-starwort
Callitriche stagnalis Callitrichaceae
Aquatic

Variable straggly water plant with slender stems. **Flowers** Minute, green, petal-less and borne at base of leaves (May–Aug). **Fruits** Tiny with 4 segments. **Leaves** Narrowly oval; those at water's surface form a floating rosette. **Status** Common in still and slow-flowing fresh water, and on the drying muddy margins of pools in summer.

Frogbit
Hydrocharis morsus-ranae Hydrocharitaceae
Aquatic

Floating perennial. **Flowers** 2cm across, the 3 white petals each with a yellow basal spot; borne on emergent stalks, the males and females separate (Jun–Aug). **Fruits** Capsules. **Leaves** 2–3cm across, floating and rounded or kidney-shaped. **Status** Local, mainly in England, in still waters of canals, ponds and ditches.

Water-soldier
Stratiotes aloides Hydrocharitaceae
Aquatic

Submerged perennial for most of year but floats during the summer. **Flowers** 3–4cm across, white and 3-petalled; females are solitary, males are borne in clusters of 2–3 (Jun–Aug). **Fruits** Capsules. **Leaves** 30–40cm long, lanceolate and toothed; borne in a tufted rosette. **Status** Local native in East Anglia; scarce introduction elsewhere.

Bladderworts

Bladderworts are intriguing water plants that belong to the family Lentibulariaceae. Small bladders, found along the submerged stems, trap tiny invertebrates and supplement the plants' nutrient supply.

Greater Bladderwort

Utricularia vulgaris Lentibulariaceae
Aquatic

Attractive aquatic plant. **Flowers** Deep yellow, 12–18mm long and carried in clusters of 4–10 on emergent stems (Jun–Jul). **Fruits** Capsules. **Leaves** Finely divided with bristled teeth. **Status** Locally common only in E England and N Ireland; found in still, mainly calcareous waters.

Lesser Bladderwort

Utricularia minor Lentibulariaceae
Aquatic

Dainty aquatic plant. **Flowers** Pale yellow, 6–8mm long and borne in clusters of 2–5 on slender stems (Jun–Jul). **Fruits** Capsules. **Leaves** Finely divided, untoothed and lack bristles. **Status** Local, mainly in N and W Britain; favours acid waters.

microscopic view of bladder and trigger hairs

From the horse's mouth

Mare's-tail *Hippuris vulgaris* is a distinctive aquatic plant that is locally common in streams, ponds and lakes, although it avoids acid conditions. Its upright emergent stems are produced from the submerged part of the plant more readily in still waters, and they carry narrow leaves that are borne in whorls of 6–12, giving it an appearance similar to a horse's tail. It produces minute pink petal-less flowers at the base of the leaves in Jun–Jul, and the fruits are tiny greenish nuts.

Water-plantains

Water-plantains are emergent aquatic plants that belong to the family Alismataceae. As their name suggests, their leaves are superficially plantain-like (*see* pp.126–7) and their flowers comprise three petals.

Common Water-plantain

Alisma plantago-aquatica
Alismataceae Height to 1m

Robust emergent plant. **Flowers** Whitish lilac, 1cm across and borne in branched whorls (Jun–Sep). **Fruits** Greenish and nut-like. **Leaves** Oval and long-stalked with parallel veins. **Status** Locally common in shallow ponds and lakes.

Lesser Water-plantain

Baldellia ranunculoides Alismataceae
Height to 25cm

Spreading perennial. **Flowers** Pale pink, 12–16mm across and usually solitary (Jun–Aug). **Fruits** Greenish and nut-like. **Leaves** Narrow and long-stalked. **Status** Locally common only in S and W Britain, where it thrives best in still, calcareous waters.

Arrowhead

Sagittaria sagittifolia Alismataceae
Height to 80cm

Distinctive aquatic perennial. **Flowers** 2cm across, the 3 petals white with a purple basal patch; borne in whorled spikes (Jul–Aug). **Fruits** Borne in globular heads. **Leaves** Comprise arrow-shaped emergent leaves, oval floating leaves and narrow submerged leaves. **Status** Locally common in the S in still or slow-flowing fresh water; scarce or absent elsewhere.

Emergent, floating & submerged water plants

The appearance of water plants is as varied as their family origins. At one extreme, some of the duckweeds are among the smallest of all our plants; at the other, species of trailing water plants such as Canadian Waterweed form huge tangled masses. Some species are perennial and evergreen, while others disappear or die back completely in winter.

Duckweeds

Members of the family Lemnaceae, duckweeds are relatively tiny floating water plants. What they lack in size, they make up for in numbers, and where conditions suit them millions often form a carpet over the surface of the water by late summer, growing by vegetative division. Their flowers are minute and seldom present, and their fruits are rarely seen.

Common Duckweed

Lemna minor Lemnaceae
Aquatic

Our most familiar duckweed. **Leaves** Round, flat, 5mm across and with a single dangling root. **Status** Widespread and locally common throughout.

Fat Duckweed

Lemna gibba Lemnaceae
Aquatic

Distinctive, fleshy duckweed. **Leaves** Swollen, spongy, 5mm across and 5mm deep. **Status** Local in the S.

Ivy-leaved Duckweed

Lemna trisulca Lemnaceae
Aquatic

Floats just below the water's surface. **Leaves** 10-15mm long, translucent and narrowly ovate to ivy-shaped, linked in a chain-like fashion. **Status** Widespread but local.

Greater Duckweed

Spirodela polyrhiza Lemnaceae
Aquatic

Our largest duckweed. **Leaves** Flat, oval to rounded, 10mm across and with several dangling roots. **Status** Locally common, mainly in the S.

Greater Duckweed leaf underside

Flowering-rush

Butomus umbellatus Butomaceae
Height to 1m

Stately, distant relative of grasses, with showy flowers. **Flowers** 25-30mm across and pink; borne in umbels (Jul-Aug). **Fruits** Purple. **Leaves** Rush-like, 3-angled and very long, arising from base of plant. **Status** Locally common only in England and Wales, growing in the vegetated margins of still or slow-flowing fresh water; often found in drainage ditches.

A coastal speciality

Beaked Tasselweed *Ruppia maritima* is a submerged aquatic perennial with slender stems that grows in brackish coastal pools and ditches. Because it is sensitive to disturbance and to pollution from agricultural run-off, its welfare is of considerable interest to conservationists. The tiny flowers comprise two greenish stamens and no petals, and are borne in pairs, arranged in umbels on stalks that rise to the surface (Jul-Sep). Its fruits are swollen, asymmetrical and long-stalked, and the leaves are hair-like and less than 1mm wide. The plant is local and declining.

Trailing water plants

By late summer, ponds and slow-flowing streams sometimes appear almost choked with trailing aquatic plants. Several species belonging to three families are involved, each suited to subtly different water conditions.

Alternate Water-milfoil

Myriophyllum alterniflorum Haloragaceae
Aquatic

Bushy water plant. **Flowers** Inconspicuous and yellow; borne in leafy spikes with tiny bracts (May–Aug). **Fruits** Warty and ovoid. **Leaves** Pinnate and feathery, the segments up to 25mm long; borne in whorls of 3–4 along the stems. **Status** Widespread but local, favouring acid conditions.

Spiked Water-milfoil

Myriophyllum spicatum Haloragaceae
Aquatic

Bushy, long-stemmed water plant. **Flowers** Inconspicuous, greenish; borne in leafy spikes (Jun–Sep). **Fruits** Rounded and warty. **Leaves** Pinnate and feathery, the segments up to 3cm long; usually borne in whorls of 4 along stems. **Status** Widespread and locally common in slow-flowing or still fresh water.

Canadian Waterweed

Elodea canadensis Hydrocharitaceae
Aquatic

Submerged perennial with trailing brittle stems. **Flowers** Tiny and floating; on slender stalks (Jul–Sep); plants seldom flower. **Fruits** Capsules. **Leaves** Narrow, back-curved and borne in whorls of 3. **Status** Introduced from North America but now widely naturalised. Grows in ponds, lakes and canals.

Rigid Hornwort

Ceratophyllum demersum Ceratophyllaceae
Aquatic

Submerged perennial with rather brittle stems; by late summer these acquire a coating of chalk and silt (as do the leaves to a lesser extent). **Flowers** Minute and borne at leaf nodes (Jul–Sep). **Fruits** Warty and beaked, with 2 basal spines. **Leaves** Rigid, toothed and forked 2–3 times; borne in whorls. **Status** Locally common only in England in still and slow-flowing fresh water.

Eelgrasses

Although seaweeds, which are algae, thrive in the marine environment, comparatively few flowering plants are able to tolerate the saline conditions of seawater. In Britain and Ireland, eelgrasses (members of the family Zosteraceae) are our only truly marine flowering plants.

INSET: **flowers**

Dwarf Eelgrass

Zostera noltei Zosteraceae
Aquatic

Similar to Eelgrass but smaller in all respects. An important food for Brent Geese (*see* p.237). **Flowers** Greenish clusters, enclosed by sheaths (Jun–Sep). **Fruits** Spongy. **Leaves** 4mm wide and 5cm long. **Status** Grows in estuaries and is often exposed at low tide.

Eelgrass

Zostera marina Zosteraceae
Aquatic

Grass-like perennial. Beds of the plant are important nursery sites for small fish. **Flowers** Small, greenish and borne in branched clusters, these enclosed by sheaths (Jun–Sep). **Fruits** Spongy. **Leaves** 1cm wide, up to 5cm long and bristle-tipped. **Status** Local; grows in sand and silt substrates, typically below the low-water mark and so is seldom exposed to air.

Submerged & floating water plants

With the exception of Horned Pondweed (family Zannichelliaceae), the true pondweed species here belong to the family Potamogetonaceae, all of which are aquatic plants. They are true flowering plants, unlike the aquatic fern and liverwort species also depicted here.

Pondweeds with broad oval leaves

Different species of pondweeds have leaves whose size and shape have evolved to suit their favoured growing environment. The following species have broad oval leaves.

Broad-leaved Pondweed

Potamogeton natans
Potamogetonaceae Aquatic

Widespread and familiar pondweed. **Flowers** Small, 4-parted and greenish; borne in 8cm-long spikes on stalks rising above the water (May–Sep). **Fruits** Round and short-beaked. **Leaves** Floating leaves are oval, up to 12cm long; stalk has a flexible joint near blade. Submerged leaves are long and narrow. **Status** Widespread and common in still or slow-flowing water.

Bog Pondweed

Potamogeton polygonifolius
Potamogetonaceae
Aquatic

Typical bog species. **Flowers** Greenish; borne in 4cm-long spikes on long stalks (May–Oct). **Fruits** Round with a tiny beak. **Leaves** Floating leaves are narrowly oval, up to 10cm long, sometimes tinged red and lack a flexible joint; submerged leaves are narrow. **Status** Locally common in heath and moorland pools with acid waters; commonest in N and W.

Fen Pondweed

Potamogeton coloratus Potamogetonaceae
Aquatic

Typical fenland pondweed. **Flowers** Greenish and borne in spikes on short stalks (Jun–Sep). **Fruits** Greenish. **Leaves** Ovate and borne on stalks that are shorter than blade (those of Broad-leaved are roughly equal to, or longer than, blade). **Status** Local in still and slow-flowing calcareous waters.

Perfoliate Pondweed

Potamogeton perfoliatus
Potamogetonaceae
Aquatic

Distinctive broad-leaved pondweed. **Flowers** Borne in small, few-flowered spikes (Jun–Sep). **Fruits** Rounded. **Leaves** All are submerged, dark green and translucent; oval, tapering and unstalked, with heart-shaped bases that clasp the stem. **Status** Widespread and locally common in still or slow-flowing waters.

Pondweeds with strap-like leaves

Two species of submerged pondweed have strap leaves; they can be distinguished by looking at the leaves, particularly the way they are arranged on the stems.

Curled Pondweed

Potamogeton crispus Potamogetonaceae
Aquatic

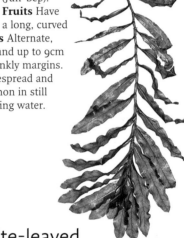

Has 4-angled stems. **Flowers** Borne in small spikes (Jun–Sep). **Fruits** Have a long, curved beak. **Leaves** Alternate, translucent and up to 9cm long with crinkly margins. **Status** Widespread and locally common in still or slow-flowing water.

Opposite-leaved Pondweed

Groenlandia densa Potamogetonaceae
Aquatic

Reasonably distinctive pondweed. **Flowers** Petal-less, greenish and borne in small submerged clusters (May–Sep). **Fruits** Greenish. **Leaves** Pointed-tipped and borne in opposite pairs. **Status** Widespread and locally common in ponds, ditches and streams.

Various-leaved Pondweed

Potamogeton gramineus Potamogetonaceae
Aquatic

Variable pondweed. **Flowers** Greenish and borne in spikes on short stalks (Jun–Sep). **Fruits** Greenish. **Leaves** Floating leaves are ovate and long-stalked; submerged leaves are narrow and unstalked. **Status** Locally common in still and slow-flowing acid waters; found mainly in the N.

INSET: flowers

Aquatic lower plants

Although the aquatic vegetation associated with fresh water is dominated by flowering plants, a few of their more primitive relatives also favour these habitats. The **Water Fern** *Azolla filiculoides* (**1**) (length 2cm) is a surface-floating species that was introduced from North America and is now naturalised in still waters, mainly in S England; its fronds are yellowish green but are often tinged red, and comprise overlapping leaves and dangling thread-like roots. The **Floating Liverwort** *Riccia fluitans* (**2**) (diameter 5cm) is a native surface-floating species found locally in still waters of ditches and canals, where it forms dense mats; individual fronds are narrow and regularly forked, with internal air chambers providing buoyancy.

Fine-leaved pondweeds

Four fairly widespread species of pondweed have extremely fine leaves. Individual strands of these are easily overlooked, both when viewed from above the water's surface and as samples in a net. However, by late summer the plants have often grown to sizeable clumps.

Small Pondweed

Potamogeton berchtoldii
Potamogetonaceae Aquatic

Freshwater pondweed with slightly flattened stems. **Flowers** Borne in small spikes on short stalks (Jun–Sep). **Fruits** Rounded. **Leaves** All submerged, narrow (50mm long × 1.5–2mm wide), 3-veined and bristle-tipped; air spaces are present on either side of midrib. **Status** Locally common in still and slow-flowing water.

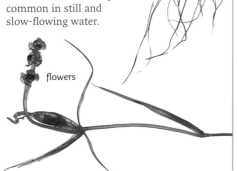

flowers

Slender-leaved Pondweed

Potamogeton filiformis Potamogetonaceae
Aquatic

Delicate-looking perennial. **Flowers** Borne in a spike comprising evenly spaced clusters. **Fruits** Reddish and borne in spikes. **Leaves** Extremely slender and all submerged. **Status** Local, found mainly in the N; grows in slow-flowing fresh and brackish water.

fruits

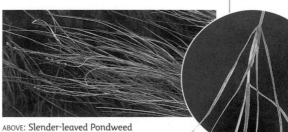

ABOVE: Slender-leaved Pondweed
INSET RIGHT: Horned Pondweed

Fennel Pondweed

Potamogeton pectinatus
Potamogetonaceae Aquatic

Extremely slender perennial. **Flowers** Borne as distinct whorls in short spikes (Jun–Sep). **Fruits** Rounded. **Leaves** All submerged, long and narrow (less than 1mm wide), and pointed at tip. **Status** Widespread and locally common; grows in still and slow-flowing waters, often in brackish conditions.

Horned Pondweed

Zannichellia palustris Zannichelliaceae
Aquatic

Slender, submerged perennial. **Flowers** Minute, greenish and borne in short-stalked clusters in leaf axils (May–Aug). **Fruits** Have a slender beak. **Leaves** 1.5mm wide and up to 5cm long, pointed and translucent. **Status** Widespread throughout, but local; found in still or slow-flowing fresh, or slightly brackish, water.

fruits

FISH

Fish are the dominant animals in most healthy freshwater habitats in Britain and Ireland and are superbly suited to the aquatic environment. Each species is subtly or clearly adapted to a particular niche, which means that in many locations several different species will coexist quite happily.

Common Eel.

Fish structure

Perfectly designed for swift movement in water, most fish are the absolute apogee of streamlining. Typically, their body is torpedo-shaped, deeper in profile than it is wide, and the fins are extremely thin in cross-section and can be raised or lowered to suit the animal's requirements. The caudal, or tail, fin at the rear end, is the main means by which a fish propels itself through the water. The remaining fins aid directional control and stability: along the dorsal surface of the body is the dorsal fin (or fins); on the ventral surface are the pelvic fins at the front and an anal fin at the back; and on the side of the body, just behind the gill opening, are the pectoral fins.

Like air-breathing animals, fish need oxygen in order to survive. They obtain it by extracting it, in dissolved form, from the water in which they live using gills; these are sited in chambers that connect with the back of the mouth and exit via openings on the side of the body. The comb-like structures that comprise the gills are covered in blood-filled tissues that absorb oxygen; and continual gulping by the fish ensures a constant flow of water over them.

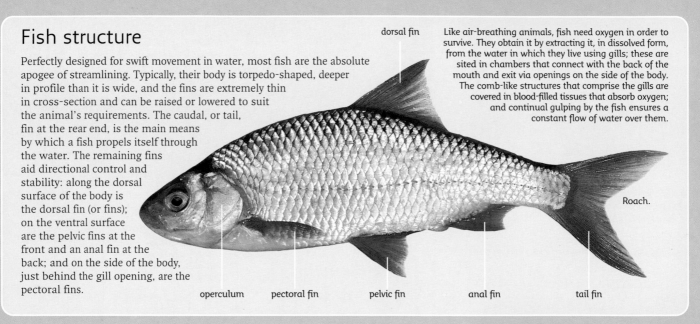

dorsal fin

Roach.

operculum   pectoral fin   pelvic fin   anal fin   tail fin

Skin and scales

The body of a fish is covered with a tough skin, and in most species small bony plates – called scales – are embedded in this to form an additional protective layer. Because the scales are not connected to one another and typically overlap, the body's flexibility is not compromised. Fish scales get bigger as the animal grows. Viewed close up, the layers on a scale are defined by a series of concentric rings, each of which corresponds to a growing season. These rings can be counted, and a fish aged, in the same way as growth rings in a cut tree trunk can be used to age the tree.

Roach scale.

Fish reproduction

Fish lay eggs, which are produced and fertilised externally in a process referred to as 'spawning'. In the run-up to breeding, most species engage in courtship displays, the aim being to ensure the optimum timing and location of spawning and the best chances of survival for the next generation. Some species simply scatter their eggs, but many create special nests and a few even engage in parental care. The Three-spined Stickleback is a well-known example of the latter – the brightly coloured male builds a nest in which the eggs are laid, and keeps it clean and oxygenated until the young fish hatch. Bullheads also care for their young, laying their eggs under stones in flowing water and the male then guarding his brood until the eggs hatch.

Most open-water marine fish form shoals when breeding, producing vast numbers of eggs to allow for heavy predation. Inshore species often attach their eggs to rocks or weeds, and some make a nest, while cartilaginous fish (dogfish and skates) protect their eggs and developing young within a tough, horny case attached to seaweed. Although the eggs of these species are laid in deep water, the empty 'mermaid's purses' are often found washed up on the shoreline.

Bullhead eggs.

ABOVE: Brook Lamprey courtship.
RIGHT: Newly hatched Lesser Spotted Dogfish and 'mermaid's purse'.

Diet and behaviour

Freshwater habitats are often extremely productive, and fish species have evolved to exploit a whole range of different foods. Most fish are at least partly predatory. At one extreme, species such as sticklebacks and the young of larger fish eat tiny freshwater invertebrates such as water fleas, and even species such as Carp and Tench, which engulf mouthfuls of sediment, will inadvertently ingest invertebrates. Other species have a more opportunistic and omnivorous diet, and will snatch anything suitable that comes within reach. A few species such as Pike and Perch are voracious predators, mainly of other fish. Fish are also important as food for other predatory animals, notably birds, some species of which feed on little else.

Although some fish species are solitary throughout their lives (predators such as Pike, for example), many generalist feeders live in single-species shoals, particularly when young; each shoal usually contains fish of a similar size. This behaviour presumably benefits the individuals on the principle that there are more eyes on the lookout for potential danger and if a predator attacks a shoal the chances of an individual becoming its meal are reduced. Shoaling is also noticeable during the breeding season, when sizeable congregations of certain species are seen in weedy shallows.

TOP RIGHT: Little Egret catching a Three-spined Stickleback.
RIGHT: There is no doubt that this Pike is a predatory fish with it's wide gape and sharp teeth.

Fish-watching

Of all native animals, fish can be the most tricky to observe well. We are obliged to view them from above the surface of the water, only ever glimpsing how they live their lives. And seeing a fish out of water is exactly that – you may be able to view it in close up and identify it with certainty, but you can learn nothing about the way the animal functions in its true element. But watching fish from a bridge or a tree overhanging a river bank can be surprisingly revealing, although your view will be distorted by refraction at the water's surface. Patience and stealth are key elements – fish are typically very wary of sudden movements.

LEFT: Rainbow Trout in a chalk stream.
BELOW: For the ultimate fish-watching experience, visit Cornwall during the summer months, which is when Basking Sharks *Cetorhinus maximus* (length to 15m) – the largest fish found in British waters – can be found. Despite the species' awesome size and capacious mouth, it feeds entirely on plankton, which it filters from the water using specialised gills. Little is known about the life history of the species and it is seen only during the summer. The Land's End and Lizard peninsulas in Cornwall, and the Isle of Man, are hotspots, and on calm days feeding animals can be seen from the shore. Boat trips operate out of Penzance and from the Isle of Man, offering close-up views of the fins, tails and snouts of these amazing animals.

Fish

Left alone in unpolluted productive habitats, some freshwater fish grow to a considerable size and may live for several decades. Some become top predators in the lakes or rivers in which they live, while others can have a significant impact on the topography of a lake or river bed.

Pike

Esox lucius Esocidae
Length 30–120cm

Superb predator. Takes invertebrates when small, but large Pike eat other fish. **Adult** Has marbled green and brown markings, hence camouflaged among water plants. Streamlined shape and broad tail allow lightning attacks; prey is engulfed in huge mouth. **Status** Common in weedy lowland lakes, flooded gravel pits and rivers.

Perch

Perca fluviatilis Percidae
Length 25–40cm

Well-marked predator. Forms shoals when young but usually solitary when large. **Adult** Has a greenish body with broad, vertical dark stripes; well camouflaged among water plants. Has 2 separate dorsal fins, the 1st very spiny; other fins are tinged red. **Status** Common in rivers and lakes except N Scotland.

Brown Trout

Salmo trutta Salmonidae
Length 30–50cm

Familiar sportfish, known in 2 forms: the Brown Trout, which spends its entire life in fresh water; and the so-called Sea Trout, which ventures up rivers only to breed. Both forms spawn in gravel beds in shallow water. **Adult** Brown Trout has an orange-brown body adorned with red and black spots; Sea Trout is pale and silvery with a few dark spots. **Status** Widespread and often common in fast-flowing unpolluted rivers and streams.

Brown Trout

Pike

Perch

Sticklebacks

Sticklebacks (family Gasterosteidae) are true tiddlers, smaller than many aquatic insects. The **Three-spined Stickleback** *Gasterosteus aculeatus* (**1**) (length 4–7cm) is found in streams and brackish water, and has three dorsal spines; it is silvery for most of the year but, in the breeding season, the male has a red belly. The **Ten-spined Stickleback** *Pungitius pungitius* (**2**) (length 2–4cm) usually has ten dorsal spines (sometimes nine) and an elongated tailstock; common in fresh and brackish waters.

A glacial relict

The **Arctic Charr** *Salvelinus alpinus* (length 50–70cm) is a member of the salmon family that, outside its limited British range, is restricted to N and Arctic regions. It is a relict species from the last post-glacial era, and is now mostly confined to land-locked and isolated populations in deep lakes in upland districts in N Wales (where it is called Torgoch), the Lake District, Scotland and Ireland. Adult male Arctic Charr have greenish-grey upperparts and a bright red belly; female fish have subdued colours.

Rainbow Trout

Oncorhynchus mykiss Salmonidae
Length 30–60cm

North American species that has been widely introduced and is farmed commercially. **Adult** Has greenish upperparts separated from bluish underparts by a pink band along flanks; whole body is covered in small dark spots. **Status** Alien but now widespread and locally common in rivers, lakes and flooded gravel pits owing to continual introductions.

Rainbow Trout

Grayling

Thymallus thymallus Thymallidae
Length 30–50cm

Attractive and distinctive fish of unpolluted flowing water. **Adult** Has a streamlined outline and silvery scales, and a large and diagnostic dorsal fin; also has a small adipose fin, a character shared by salmon and trout. **Status** Widespread and locally common in England and Wales (and introduced elsewhere), favouring fast-flowing streams and shallow rivers.

Grayling

Fish migration

Although many fish of freshwater habitats lead relatively sedentary lives within a lake or short stretch of river, a few undertake phenomenal journeys. That undertaken by the **Common Eel** *Anguilla anguilla* (**1**) (length 0.5–1m) is perhaps the strangest. It spawns in the Sargasso Sea, on the other side of the Atlantic Ocean, and young larvae drift across in the Gulf Stream for three years or so. On reaching our shores, the so-called elvers swim up rivers, where after several years they become familiar yellow-bodied snake-like eels; on reaching maturity they turn silvery, migrate down to the sea and are presumed to attempt their oceanic journey in reverse. During their time in Britain, Common Eels live among silt and debris at the bottom of ponds, rivers and canals.

The migration of **Atlantic Salmon** *Salmo salar* (**2**) (length 1–1.2m) is spectacularly visible in certain locations. This impressive fish spends much of its adult life at sea but returns to breed in clean, fast-flowing rivers, mainly in W and N Britain; adult fish can be seen swimming upstream and leaping obstacles when watercourses are in full spate, mainly in Nov–Feb. They spawn in shallow gravel beds, after which most die; the young fish migrate to the sea after two years or so.

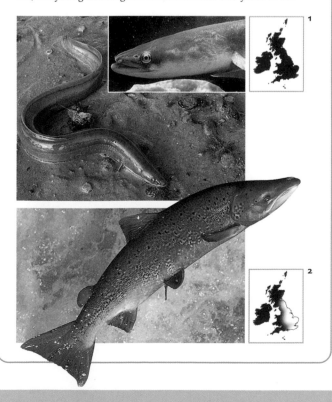

Lampreys

Members of the family Petromyzonidae, lampreys are intriguing fish that are easily recognised by their snake-like bodies and sucker-like mouthparts. The larger of our two freshwater species is the **River Lamprey** *Lampetra fluviatilis* (**1**) (length 30cm), which is migratory, the adults moving upstream from the sea in winter and spring to spawn on gravel beds in rivers. After spawning, the adults die and the larvae that hatch from the eggs live for several years buried in river silt; following metamorphosis to the adult stage, they move to the sea until maturity, when the cycle is repeated. Adult River Lampreys attach to other fish using their sucker and feed on blood; they also feed on carrion. The **Brook Lamprey** *L. planeri* (**2**) (length 15cm) is resident in unpolluted streams and rivers and spends most of its three- to five-year lifespan as a larva buried in silt; here, it filters organic matter. Following metamorphosis, adults are sometimes seen spawning in Apr–May in shallow gravel beds; the sucker is used to move stones to create an egg-laying site. Adults die after spawning.

Cyprinids

Some cyprinid fish (family Cyprinidae) reach a large size in maturity while many species remain as 'tiddlers' even as adults. All species grind their food using rough bony plates in their mouths (not teeth). Many cyprinids shoal when they are young and most small species remain in shoals throughout their lives; larger cyprinids become solitary with age.

Tench

Tinca tinca Cyprinidae
Length 20–40cm

Subtly colourful fish. **Adult** Thick-bodied fish with a proportionately large tail and rounded fins; body is mainly greenish but belly is flushed red. **Status** Widespread native of sluggish rivers and weedy lakes.

Carp

Cyprinus carpio Cyprinidae
Length 25–80cm

Can attain huge sizes as adults. **Adult** Ancestral form has a golden-olive colour and even-sized scales; so-called Leather and Mirror carp are different types of this form, with variable size or missing scales. All feed on bottom-living invertebrates and plants. **Status** Introduced but has long been established in lakes and flooded gravel pits across Britain and Ireland.

Tench

Carp

Roach

Rutilus rutilus Cyprinidae
Length 10–25cm

Familiar deep-bodied fish. **Adult** Has a mainly silvery body, palest below, red pelvic and anal fins, and reddish-brown dorsal, pectoral and tail fins. Dorsal fin is sited above pelvic fins (*cf.* Rudd). **Status** Common and widespread in England, but less so in Wales and Scotland; rare in Ireland, where, confusingly, the more common Rudd is sometimes called Roach.

Rudd

Scardinius erythrophthalmus Cyprinidae
Length 20–35cm

Superficially similar to Roach, but usually deeper-bodied and separable by noting fin location. Shoal-forming. **Adult** Has a rather silvery body but typically with a golden tinge to flanks. Fins are all reddish and note that the dorsal fin lies behind point of origin of pelvic fins. **Status** Locally common in lakes and rivers in England, Wales and Ireland.

Barbel

Barbus barbus Cyprinidae
Length 20–70cm

Slim-bodied, streamlined and attractive fish. Its mouth is sited for bottom-feeding and is bordered by sensory feeler-like barbels. **Adult** Has a silvery brown body and reddish-brown fins; dorsal fin is arched with an incurved outer margin. **Status** Locally common in moderate flows of larger rivers, mainly in central and S England.

Roach

Barbel

Rudd

Common Bream

Abramis brama Cyprinidae
Length 30–50cm

 Distinctive, extremely deep-bodied fish with a 'humpback' profile behind head; body is laterally compressed when viewed head-on. **Adult** Has a golden-brown body, palest below, and dark reddish-grey fins. **Status** Locally common in weedy lowland lakes and slow-flowing rivers, mainly in England.

Chub

Squalius cephalus Cyprinidae
Length 30–40cm

 Streamlined fish with relatively large scales. **Adult** Has a bronze sheen to body, with some silvery scaling on dorsal surface. Pectoral, dorsal and tail fins are dark, while pelvic and anal fins are red. Dorsal and anal fins are convex (concave in similar cyprinids). **Status** Locally common in rivers and sizeable streams in lowland England.

Dace

Leuciscus leuciscus Cyprinidae
Length 15–25cm

 Streamlined shoaling fish that is superficially similar to, but smaller than, Chub: size and body colour are useful in identification. **Adult** Has a silvery-green body, darkest above and palest below. Dorsal and tail fins are dark, while other fins are red. **Status** Locally common in lowland rivers and streams.

Common Bream

Chub

Dace

Bleak

Alburnus alburnus Cyprinidae
Length 12–15cm

 Small shoaling fish with a slim, streamlined body and an upward-opening mouth suited to feeding on insects at the water's surface. **Adult** Has a silvery body, darkest above and palest below. Pectoral fins are brownish but other fins are pale pinkish green. **Status** Locally common in lowland lakes and slow-flowing rivers.

Minnow

Phoxinus phoxinus Cyprinidae
Length 4–10cm

 Small but attractively marked fish with a slim, streamlined body and rounded fins. Shoals are seen in shallows in spring but move to deeper water in winter. **Adult** At most times of year has a silvery body, darkest above and with dark blotches along flanks; breeding male has a red belly. **Status** Widespread and common in rivers and lakes.

Gudgeon

Gobio gobio Cyprinidae
Length 7–15cm

 A distinctive fish with an extremely slim, streamlined body and sensory barbels around the mouth that help it detect prey in sediment and sand on the bed of the river; it forms shoals in summer. **Adult** Has a body that is bluish above and silvery below. **Status** Locally common in streams and rivers.

Bleak

Minnow

Gudgeon

Bottom-dwellers

Aquatic invertebrates thrive among stones and silt at the bottom of rivers and streams and several fish exploit this food resource. Gudgeon and Barbel feed here but two species also make the stream-bed their home, seldom willingly leaving this zone. The **Stone Loach** *Noemacheilus barbatulus* (1) (length 5–10cm) is a slim-bodied fish with barbels around its mouth; it is widespread in unpolluted gravel-bottomed streams and rivers. The **Bullhead** *Cottus gobio* (2) (length 8–15cm) has a large and broad head, and its fins have spine-tipped rays. It is camouflaged on gravel stream beds, often hiding under larger stones, and is common in England and Wales.

Introducing freshwater invertebrates

There can be few people who have not dipped a net into a freshwater pond or stream and been fascinated by the wealth of invertebrate life they discover. Even the smallest unpolluted pools are likely to produce plenty of interest, while large and long-established ponds, lakes, canals and rivers positively teem with life.

Among the freshwater invertebrates that are visible to the naked eye, most major groups are found. Some of these are exclusively aquatic and also have representatives in the marine environment. Others have terrestrial counterparts or have stages in their life cycles (in the case of many insects) that live on land.

Classification made simple

Biological classification involves the creation of a hierarchy of divisions and subdivisions in the living world that places different organisms in appropriate groups based on the similarities and differences between them and other living things (*see also* p.11). Animals and plants are placed in separate kingdoms and these are broken down into major divisions called phyla (singular phylum). Each phylum contains subdivisions called classes and within these are groups called orders. Each order contains a number of genera (singular genus), within which are contained different species; among members of the same genus, the first word in each species' binomial scientific name is the same.

Sponges

Sponges (phylum Porifera) are more commonly found in sea water. A few British species are found in ponds and streams, typically encrusting stones or rooted plants and producing finger-like projections to increase their surface area. Pores connect internal chambers to the water outside, and the beating of tiny hairs ensures a current that brings in food and removes waste products.

Pond Sponge *Spongilla lacustris*

Flatworms

Planarian worms are relatives of parasitic tapeworms and flukes, and are free-living freshwater members of the phylum Platyhelminthes. As their name suggests, they are extremely flat creatures with reasonably well-defined head and tail ends. They glide over surfaces, aided by the beating of tiny hairs, and most species are carnivorous. Find out more about planarian worms on p.440.

Polycelis nigra

Coelenterates

Marine jellyfish and corals are the best-known coelenterates and belong to the phylum Cnidaria. In fresh water, the group is represented by Hydra, which resembles a miniature sea anemone and has a jelly-like body and stinging tentacles. Find out more about Hydra on p.440.

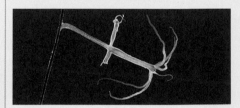

Hydra

Studying freshwater life

While many fascinating insights into the freshwater world can be gained by sitting quietly beside a pond or stream and watching, a whole new aspect will open up if you collect samples of pondweed or sediment in a net. Place the contents into a shallow tray (ideally one that is white), cover with fresh water and wait to see what wriggles or moves. Bear in mind that many aquatic animals are sensitive to disturbance and will remain motionless for several minutes before beginning to move. The water in your sample tray is likely to be murky, so you may want to move specimens to another containing clean water; this is often best done using a plastic spoon.

True worms

The phylum Annelida contains what are popularly referred to as 'true worms' – animals with segmented bodies, each segment of which is nearly identical. The worms have a distinct head end with a mouth and are hermaphrodite. Many freshwater annelids are strikingly similar to miniature earthworms and are closely related to them. Also included in the group are leeches, which have flattened bodies and a sucker at both ends. Find out more about annelid worms on p.440.

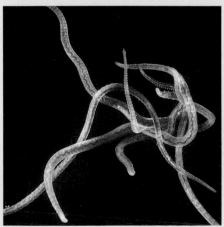

Lumbriculus spp

Arthropods

Members of the phylum Arthopoda (arthropods) are probably the most important group of freshwater invertebrates, and they are the most numerous in many places. They are characterised by having an exoskeleton and jointed legs. In fresh water there are:

Crustaceans

Animals belonging to the class Crustacea and whose exoskeleton is impregnated with calcium salts. British freshwater representatives spend their whole lives in water, and they include water fleas, ostracods, copepods, water lice, freshwater shrimps and crayfish. Find out more about these crustaceans on p.439.

Freshwater louse

Spiders and mites

Animals of the class Arachnida, most of which are terrestrial. They are characterised by having four pairs of legs. The Water Spider *Argyroneta aquatica* is our only truly aquatic spider (*see* p.409) and several species of tiny mites, some of which are bright red, can be seen scurrying over pond plants; immature stages of some mites are parasites of aquatic insects.

Water Spider

Insects

Animals belonging to the class Insecta, with segmented bodies that are divided (in adults at least) into three distinct regions: head, thorax and abdomen. With the exception of bugs and beetles, it is the immature stages (larvae or nymphs) of insects that are found in water, while adults live terrestrial lives. The group includes dragonflies, damselflies, stoneflies, mayflies, alderflies, caddis flies, bugs, true flies and beetles. Find out more about aquatic insects on pp.436–8.

Brown Hawker nymph

Molluscs

Members of the phylum Mollusca are diverse in terms of their size and shape, and they are represented in British freshwater systems by true snails (most with curled or spiral shells) and bivalves, whose body is protected by paired shells. Apart from their shell, molluscs are characterised by having a powerful, muscular foot. Find out more about freshwater molluscs on pp.440–1.

Ramshorn

Microscopic life

Many of the simplest freshwater organisms are also the smallest, and most of these are invisible to the naked eye. What they lack in size, they more than make up for in terms of numbers and diversity in appearance, and with the aid of a microscope a rich and varied world in miniature is revealed. At the microscopic level, it is often hard to distinguish tiny animals from miniature plants. Both are covered here, categorised under the most commonly encountered groups

Diatoms

Curious single-celled organisms that live inside silica cases; they photosynthesise like algae and can move in a gliding manner.

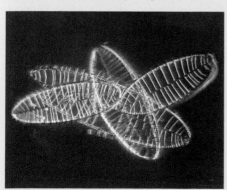

Diatom

Rotifers

Bizarre multi-cellular animals that have a wheel-like arrangement of beating hairs around the mouth, which creates a current for feeding and movement.

Rotifer

Algae

Tiny plants, some of which are single-celled while others form simple colonies or filamentous strands.

Spirogyra, a filamentous alga

Protozoans

Single-celled animals whose forms are extremely varied: *Amoeba* is free-living and moves by oozing over substrates; *Paramecium* is free-living but moves by means of beating hairs; *Vorticella* is sessile and is attached to substrates by means of a stalk.

Amoeba

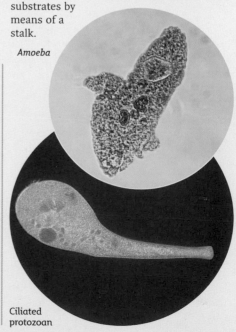

Ciliated protozoan

Freshwater insects

Aquatic insect life abounds in most freshwater habitats and the naturalist is seldom short of things to study. In the still or slow-flowing waters of ponds, lakes and canals, four insect groups are usually the most conspicuous representatives: water bugs, water beetles, and the nymphs of dragonflies and damselflies.

Water bugs

In common with their terrestrial cousins, water bugs are members of the order Hemiptera and have piercing mouthparts with which they suck fluids; predatory habits are common in the group. Wings, when present, comprise two pairs, and some aquatic bugs readily leave the water after dark and fly in search of new habitats. Some species are surface-dwellers while others live submerged lives. Body form among the group is extremely variable, ranging from slender and stick-like to rounded and squat.

Pond Skater
Gerris lacustris Gerridae
Length 10mm

Surface-dweller that skates over the water surface on tips of legs. Responds to distress movements of insects trapped in surface film; feeds on these using proboscis. **Adult** Has a narrow brown body and slender legs; hibernates on land. **Status** Common on ponds and lakes.

Water Boatman
Notonecta glauca Notonectidae
Length 14mm

Typically hangs at the water surface and swims upside-down using its fringed hind legs as paddles. Predator of other aquatic creatures and insects trapped in the surface film. **Adult** Has a tapering greyish body but looks silvery owing to an air bubble trapped on ventral surface of body. **Status** Common in ponds and lakes.

Lesser Water Boatman
Corixa punctata Corixidae
Length 10mm

Active bug that swims the right way up using fringed hind legs as paddles. Feeds on algae and detritus on bottom of pond. **Adult** Has a brownish body and a well-developed head and eyes. **Status** Common in weedy ponds.

Saucer Bug
Ilyocoris cimicoides
Naucoridae Length 12mm

Fierce predator of aquatic creatures. **Adult** Oval in outline, yellowish brown and stippled with numerous tiny dark dots. **Status** Widespread and fairly common in most clean weedy ponds and lakes. It can cause a painful bite.

Water Scorpion
Nepa cinerea Nepidae
Length 30mm

Predator of tadpoles, small fish and other aquatic insects. Its movements are slow and creeping. **Adult** Easily recognised by flattened body and leaf-like outline, long breathing siphon at tail end and pincer-like front legs. **Status** Common in weedy ponds and lakes year-round.

Water Stick Insect
Ranatra linearis Nepidae
Length 50mm

Distinctive species, although easy to overlook in pond samples as it remains motionless and stick-like when out of water. Adopts a mantid-like pose when submerged and captures passing aquatic creatures. **Adult** Has a slender body, grappling front legs and a breathing siphon at rear end. **Status** Fairly common in the S in weedy ponds and lakes.

with *Daphnia* prey

Water beetles

In many respects, water beetles are no different from their terrestrial counterparts: they share the same body form, having oval, domed bodies whose shape is largely defined by the hardened elytra (modified forewings) that protect the abdomen and folded hind wings. They are air-breathers but generally their bodies are more streamlined than those of their land-based cousins, and in many species the limbs are modified for swimming.

Acilius sulcatus
Dytiscidae Length 16mm

Active carnivore that swims well using fringed hind legs as paddles. **Adult** Has an oval body, the male with shiny golden elytra that are finely marked; those of female are grooved. **Status** Common and widespread in England but local elsewhere. Favours weedy ponds and canals.

Great Silver Beetle
Hydrophilus piceus Hydrophilidae
Length 40mm

Impressive beetle and a contender for Britain's largest insect. Adult is vegetarian but larva eats water snails. **Adult** Upper surface is shiny black while underside looks silvery in water owing to air film. **Status** Extremely local and now confined to a few areas in S England; favours weedy drainage ditches.

INSET: Great Silver Beetle underside, showing trapped air

Whirlygig Beetle
Gyrinus natator Gyrinidae
Length 5mm

Small but distinctive water beetle, seen whirling around on the water surface, often in groups. **Adult** Has a black bead-shaped body and well-developed front legs, used for swimming; middle and hind legs are small. **Status** Widespread and common in ponds, lakes and slow-flowing streams.

Great Diving Beetle
Dytiscus marginalis Dytiscidae
Length 30mm

Large and impressive water beetle, both the adults and larvae of which are fierce predators. **Adult** Has a mainly black, shiny body with orange-brown legs, margins of elytra and thorax. Male has smooth, shiny elytra; female's are grooved. **Larva** Has awesome piercing jaws and is voracious predator, tackling anything of a suitable size, including tadpoles and even small newts and fish. Periodically returns to surface of pond to replenish its air supply via tip of its abdomen. **Status** Widespread and fairly common in ponds and lakes.

jaws of larva

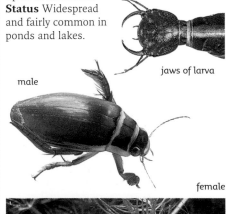

male

female

Dragonfly nymphs

The immature stage of a dragonfly is called a nymph (rather than a larva), because it does not pupate and after each moult it becomes closer in appearance to an adult. Dragonfly nymphs have good eyesight, which allows them to spot movements of potential prey; small creatures are caught using a fanged extensible flap called the 'mask'. The body form is variable, but free-living species (most hawkers) typically have slender bodies while nymphs that live among debris and silt at the bottom are usually squat and hairy. Empty nymphal cases (called exuviae) can be found on waterside vegetation after adults emerge.

RIGHT: Four-spotted Chaser, empty nymphal case (exuvia)
BELOW LEFT: Emperor Dragonfly, empty nymphal case
BELOW RIGHT: Hairy Dragonfly nymph

Damselfly nymphs

Like their cousins the dragonflies, immature damselflies are called nymphs. The nymphal form is fairly standard across the group: the body is rather slim and cylindrical, two antennae project forwards at the head end, and at the rear end are three narrowly flattened, projecting gills. Damselfly nymphs live rather unobtrusive lives, and in colour terms are often a good match for the pondweeds and debris among which they make their home. Though diminutive, they are active predators of small invertebrates.

Freshwater insects & crustaceans

The aquatic nymphs or larvae of many smaller insect species and certain crustaceans are often abundant where conditions suit them. They underpin the food webs of most freshwater habitats, serving as prey for all manner of other creatures, from larger insects to fish and amphibians. Their ecological importance cannot be overstated.

Freshwater insects

Many immature insects are soft-bodied and avoid predators by living unobtrusive lives. Caddis fly larvae protect themselves with a case while mayfly and stonefly nymphs, and alderfly larvae, often hide under stones or in silt. Many fly larvae are air-breathers and are able to live in stagnant waters that few predators can tolerate.

Caddis fly larvae

Caddis fly (order Trichoptera) adults are described on p.391. Their active immature stages are called larvae and, like certain moth larvae, they produce silk, which they use to create a tube to surround and protect the body. A few species – typically those that live in flowing water – spin silken webs to trap food. Others are free-living and adorn the tube with a protective case of small stones or plant material that varies with each species.

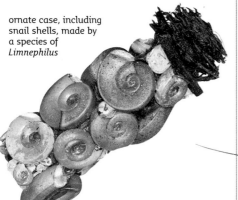

ornate case, including snail shells, made by a species of *Limnephilus*

Mayfly nymphs

Mayfly (order Ephemeroptera) adults (*see* pp.374–5) are a familiar sight in spring. Immature stages are aquatic and are called nymphs. These have gills along the side of the abdomen and three projecting 'tails' (called cerci) at the rear end. Some live for as long as three years in water and most lie buried, or part-buried, in silt at the bottom.

ABOVE: *Ephemerella* sp.

ABOVE: *Ecdyonurus venosus*, a flowing water mayfly species

Stonefly nymphs

Stonefly (order Plecoptera) adults are described on p.375. Their immature stages are called nymphs and are rather flattened creatures, typically seen clinging to stones and rocks in flowing water. They have long projecting antennae at the head end and two long 'tails' at the rear, a feature that allows separation from mayfly nymphs, which have three 'tails'.

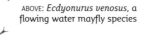

Stonefly nymph

Alderfly larvae

Alderfly (order Megaloptera) adults (*see* p.390) are weak-flying insects. Their active immature stages are called larvae, recognised by their slender, tapering body with gills along the flanks. Aside from its gills, an alderfly larva could perhaps be mistaken for an aquatic beetle larva were it not for the pointed 'tail' at the rear end; in beetle species that have tail projections, two is the usual number.

Fly larvae

The adult stages of flies (order Diptera) are described on pp.392–5. Many species have larvae that live in water and that are filter-feeders or detritivores. As a group, larval body forms are extremely variable, although many species are rather maggot-like. Mosquito and midge larvae are exceptional in being well developed. Perhaps the most bizarre larva is that of the Drone-fly, the so-called 'rat-tailed maggot', which has an extremely long, extensible breathing tube that allows it to survive in stagnant water.

ABOVE: Phantom midge larva/ghost worm
RIGHT: Mosquito larvae
BELOW: Rat-tailed maggot

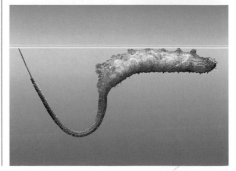

Freshwater crustaceans

Although crustaceans (members of the class Crustacea) are mainly a marine group, they have plenty of representatives in fresh water, ranging in size from almost microscopic water fleas to large and bulky crayfish. Crustaceans have an exoskeleton that is hardened by impregnated calcium salts, and they have a variable number of limbs that are used for feeding and movement according to the species.

Freshwater Louse

Asellus aquaticus Asellidae
Length 15mm

Rather like a woodlouse in appearance. Scuttles among water plants and scavenges organic matter from the sediment. **Adult** Has a flattened, segmented body and very long antennae. In spring, female carries whitish eggs in a brood pouch. **Status** Widespread and often extremely common in ponds, lakes and slow-flowing streams.

Freshwater Shrimp

Gammarus spp. Gammaridae
Length 11mm

Often found under stones and around the base of rooted water plants. Intolerant of polluted waters or those with a low oxygen content. **Adult** Has a laterally flattened body and swims on its side. **Status** Widespread and common in fast-flowing streams and rivers, and near inflows or outflows in stream-fed lakes; several species occur in the region.

White-clawed Crayfish

Austropotamobius pallipes Astacidae
Length 40mm

Once common, now scarce and endangered. **Adult** Resembles miniature lobster with large pincers and long antennae. **Status** Requires clean, fast-flowing water; during daytime, hides under stones. Threatened by pollution and disease from introduced species, notably **North American Signal Crayfish** *Pacifastacus leniusculus* (length 40mm).

North American
Signal Crayfish

White-clawed
Crayfish

Fairy Shrimp

Chirocephalus diaphanus
Chirocephalidae Length 16mm

Enigmatic creature that sometimes appears in sites as small and ephemeral as puddles and tyre-ruts. **Adult** Hatches after autumn rains. Shrimp-like in appearance; swims upside-down. **Status** Extremely local but often abundant where it does occur.

Tadpole Shrimp

Triops cancriformis Triopsidae
Length 30mm

Almost prehistoric-looking crustacean that resembles a miniature Horseshoe Crab. **Adult** Has a shield-shaped carapace, segmented abdomen and 2 tail-like projections. **Status** Restricted to a few seasonal pools in the New Forest.

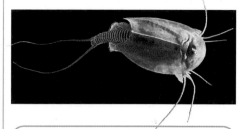

Minute crustaceans

If you hold a jar of pond water up to the light you will probably notice lots of small animals moving around, some of them swimming and others crawling over the surface of the glass. Many of these will be tiny crustaceans, and using a hand lens three different groups can usually be recognised.

Water-fleas *Daphnia* spp. (**1**) (length 1mm) have a single eye and a pinkish body that is so transparent that its internal organs can be seen through the carapace; in summer they can be abundant. They swim jerkily by beating their antennae. Copepods, such as **Cyclops** spp. (**2**) (length 0.5mm), have a pear-shaped body and a single dark eye-spot; they swim jerkily and often carry paired egg sacs. **Ostracods** (**3**) (length 0.5–1mm) are pea-shaped animals whose body is protected by paired shells; they scurry around with the aid of protruding antennae and legs.

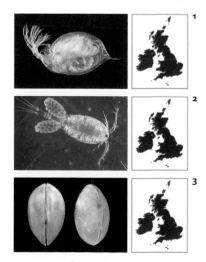

Freshwater worms & molluscs

Humble creatures such as worms and molluscs are worth more than a passing glance because many have bizarre lifestyles and all are extremely important to the ecology of the freshwater systems in which they live.

Freshwater worms

The term 'worm' popularly describes almost anything that is long, thin and wriggly. In the freshwater environment, two phyla qualify as worms: the Annelida comprising segmented true worms; and Platyhelminthes, represented in water by planarian flatworms.

Lumbriculus spp.

Lumbriculidae Length 30mm

Aquatic annelid worm. **Adults** Bear a striking resemblance to miniature terrestrial earthworms; their transparent body reveals many of their internal organs. **Status** Found in sediment and silt at the bottom of ponds and lakes, where they feed on detritus.

Erpobdella octoculata

Erpobdellidae Length 3cm

Free-living, active leech that preys on small invertebrates. **Adult** Has an extremely flexible body with a sucker at both ends. Will contract into a blob if disturbed. **Status** Widespread and locally common in ponds, lakes and canals. **Similar species** Several superficially similar species of leech are found in British and Irish freshwater; the identification of most of these is best left to experts.

Fish Leech

Piscicola geometra Piscicolidae
Length 10mm

External parasite of freshwater fish, notably sticklebacks, which usually attaches itself to their gills. **Adult** Body looks rather banded. **Status** Widespread and locally common.

Dugesia lugubris

Dugesiidae Length 20mm

Aquatic planarian worm. **Adult** Has a wafer-thin grey-brown body with 2 pale spots at head end next to eyes. Glides over surfaces by means of tiny beating hairs and feeds on small invertebrates. **Status** Common and widespread in slow-flowing streams and weedy ponds and lakes. **Similar species** Polycelis nigra (Planariidae; length 10mm) is much smaller and lives in still waters.

Polycelis nigra

Hydra

Hydra spp. Hydridae
Length 5mm

Freshwater relative of sea anemones and jellyfish (phylum Cnidaria), usually found attached to stems of water plants. **Adult** Has a stalk-like body with a terminal mouth surrounded by tentacles; if disturbed, it contracts into a blob and is difficult to see. Buds new animals in summer. **Status** Common throughout in ponds, canals and lakes.

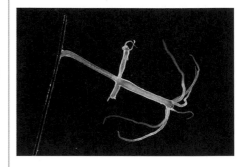

Freshwater molluscs

Freshwater molluscs (phylum Mollusca) are a diverse group. All have a muscular foot and a soft, unsegmented body that is protected by a hard shell composed of protein and calcium carbonate. Like their terrestrial cousins, some freshwater snails are air-breathers. But many take oxygen directly from the water, as do all our bivalve molluscs.

Common Bithynia

Bithynia tentaculata Bithyniidae
Length 15mm

Distinctive snail. **Adult** Has a conical brown shell and long, slender tentacles. When moving among water plants, a small plate (operculum) is seen on upper side of foot; disturbed animal retreats into shell and seals itself off with operculum. **Status** Fairly common locally in ponds, canals and lakes.

Great Pond Snail

Lymnaea stagnalis Lymnaeidae
Length 45mm

 Moves among water plants or is seen at the surface, gliding along the underside of the surface film. Visits the surface periodically to replenish its air supply. Feeds on encrusting algae using its rasping tongue. **Adult** Has a conical brown shell. Lays sausage-shaped gelatinous masses of eggs on water plants. **Status** Common except in the N, in ponds, lakes and canals.

Great Ramshorn

Planorbis corneus Planorbidae
Diameter 25mm

 Large spiral-shelled water snail. **Adult** Has a flattened dark brown shell comprising 5 or so whorls that become ridged with age. **Status** Common only in S and central England. Favours ponds, lakes and canals, sometimes living in stagnant water, where its oxygen uptake is aided by the presence of haemoglobin in its blood.

Great Ramshorn eggs

Wandering Snail

Lymnaea peregra Lymnaeidae
Length 10mm

 Familiar water snail. **Adult** Has an oval or rounded shell, pale brown with dark blotches; last whorl is relatively large and expanded. Tentacles are broad, flattened and ear-like. **Status** Common and widespread in lowland ponds, lakes and ditches; tolerates slightly brackish conditions.

Ramshorn

Planorbis planorbis Planorbidae
Diameter 12mm

 Water snail with a spirally coiled shell. **Adult** Appreciably smaller than Great Ramshorn, with narrower, more tightly packed whorls; shell is flattened on one side. **Status** Common in lowland ponds and ditches; can tolerate quite stagnant conditions.

River Limpet

Ancylus fluviatilis Ancylidae Length 8mm

 Freshwater limpet, typically found attached to stones and rocks. **Adult** Has a streamlined, flattened conical shell that allows it to cling on in fast currents. **Status** Widespread and locally common; occurs in clear, fast-flowing rivers and streams, where it grazes algae.

Freshwater bivalves

The soft bodies of bivalve molluscs are protected by paired shell valves that are hinged with a ligament at the base. The valves clamp shut if the animal is alarmed, but during feeding they gape, so that a flow of water can be siphoned through the body. A muscular foot can be extended, allowing the animal to adjust its position.

The smallest examples of freshwater bivalves in the region are the **orb mussels** *Sphaerium* spp. (**1**) (length 6mm) and **pea mussels** *Pisidium* spp. (length 10mm), both of which are almost spherical when the two valves are clamped together; when they are open, two short white siphons are seen in orb mussels, while pea mussels have just one. Several much larger bivalve molluscs are also found in our rivers and lakes, of which the **Swan Mussel** *Anodonta cygnea* (**2**) (length 12cm) is a typical example. It lives part-buried in silt, filter-feeds organic particles, and is widespread but local, mainly in the S. The **Zebra Mussel** *Dreissena polymorpha* (**3**) (length 3cm) is an introduced species that is spreading at an alarming rate.

Seashore habitats

Whether you want to witness the spectacle of teeming flocks of waders and wildfowl on our winter estuaries, fancy a spot of summer rock-pooling, where you will find a host of tiny fish, anemones and prawns or just to investigate the creatures that live in the sand beneath your feet, the British coast is the place for you.

Most marine bivalve molluscs live buried in the sand or mud. At low tide you would never know they were present, although they sometimes create a slight depression in the sand above them. The **Common Cockle** (*see also* p.456) is a widespread and typical member of the group.

Many rocky shore invertebrates either attach themselves firmly to rocks or hide in crevices or under stones. Opportunities for feeding birds are thus limited. Turnstones and Oystercatchers are regularly seen, but the **Purple Sandpiper** (*see also* p.266) can usually only be found on rocky shores during winter in Britain.

Sea anemones are colourful rocky-shore residents. They are soft-bodied and attach themselves to rocks using a basal sucker-like disc. Tentacles, armed with stinging cells, surround the mouth and catch small planktonic prey. The **Snakelocks Anemone** (*see also* p.463) can be found in rock pools and gullies on the mid-shore.

The large numbers of birds that feed on mudflats during winter and the huge quantity of shell debris washed up on sandy beaches testifies to the unseen productivity of these habitats. Incredible numbers of marine worms and tiny molluscs thrive in the oozing substrate, supported by organic matter deposited where river meets sea. The Lugworm is the typical marine worm of sand and mud, marine molluscs are represented by razorshells, cockles and clams, and specialist crabs and sea urchins spend their lives buried beneath the sand. Waders are drawn to estuaries, each species having bill lengths and feeding strategies adapted to a particular food source. Wildfowl include Shelduck that filter minute animals from the mud with their bills, while Brent Geese and Wigeon favour plant material.

For marine biologists, the rocky shore's intertidal zone offers a wealth of opportunities for exploration. At low water, rock pools and gullies positively teem with life, and scores of crabs, molluscs and small fish await discovery. Limpets and barnacles overcome the battering waves by protecting themselves with a hard shell; limpets clamp themselves down using a muscular foot, while barnacles 'glue' their shells to rock. In sheltered gullies and rock pools, sea anemones, blennies, prawns and Shore Crabs are found. A search under stones may reveal retiring hermit crabs living in the empty shells of periwinkles, and Dog Whelks. Many animals found on rocky shores are year-round residents, although during the summer months more unusual species visit from deeper waters.

When studying life on the rocky shore, you can make many amazing discoveries simply by turning over stones and boulders. It is vital that you replace these as you found them: by doing so you restore the shaded, sheltered niches that the inhabitants beneath the stones rely on and without which they would die.

Seaweeds are the dominant plants on the rocky shore. They are structurally simple, lacking proper roots and flowers. But many species are large and abundant, becoming important refuges for marine animals. **Serrated Wrack** (*see also* p.446) is common on rocky coasts, growing in a distinct zone on the lower middle shore.

Crustaceans are well represented on the rocky shore, with crabs probably being the most familiar and common examples of the group. Among their numbers, the **Green Shore Crab** (*see also* p.458) is one of the more distinctive species. Crabs are mainly opportunistic feeders on detritus.

Although numerous fish can be found around our rocky coasts, most larger species live below the low-tide level as they must remain in water at all times. However, some smaller species such as the **Common Blenny** (*see also* p.452) are easily found in rock pools when these are exposed at low tide.

Seashore habitats & hotspots

Nowhere in Britain and Ireland is far from the sea, so it is relatively easy for those interested in marine life to access the coast. The region has some outstanding coastal habitats, ranging from muddy estuaries to sandy beaches and dramatic cliffs. Some of the best examples of each are listed here.

Cliffs

Dominating the western and northern coasts of Britain and Ireland are towering sea cliffs, dramatic places where land meets ocean. Where the substrate is suitable, cliffs are often carpeted with colourful flowers in spring and are home to colonies of seabirds.

Hotspots *see map:*
1 **Lundy, Devon** –
   www.nationaltrust.org.uk
2 **Hermaness, Unst, Shetland Isles** – www.nature-shetland.co.uk
3 **Downpatrick Head, Mayo** –
   www.mayo-ireland.ie

Lundy, Devon. INSET: Thrift

Sandy shores

Beloved of holidaymakers, sandy shores are also havens for wildlife. Beneath the surface of the sand lives an abundance of marine worms and molluscs whose presence would go largely undetected were it not for the feeding activities of birds and the profusion of dead shells found along the strandline.

Hotspots *see map:*
4 **Holkham, Norfolk** –
   www.holkham.co.uk/naturereserve
5 **St Martins, Isles of Scilly** –
   www.isles-of-scilly.co.uk
6 **Dunnet Sands, Caithness** – www.caithness.org

Razorshells on a Norfolk beach. INSET: Peppery Furrow Shell

Find out more about...

- Estuary flowers on **pp.38–9**
- Sand and shingle flowers on **pp.40–1**
- Coastal colonisers on **p.169**
- Lichens on **pp.182–3**
- Auks on **pp.280–1**

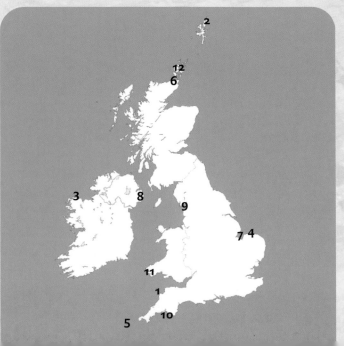

Tides & zonation

The tides are caused by the gravitational pull of the moon and have a profound influence on British shores – many seashore creatures will either be exposed to air or covered by seawater every eight hours or so. Different plants and animals have evolved to cope with different degrees of exposure or inundation, and on rocky shores the zones in which they grow are usually obvious. The occurrence of different seaweeds provides the most immediate evidence of zonation, as they appear as bands of different colours. This effect is most noticeable on western British coasts, where the tidal range is large and where rocky shores are often steep.

Rocky shore at high tide. INSET: *Lithophyllum incrustans*   Rocky shore at low tide.

Rocky shores

Hard, rocky substrates provide a permanent footing to which seaweeds and a variety of marine animals can attach themselves. Among the tangled weeds and in sheltered rock pools and gullies can be found a wealth of life, such as crabs, molluscs and sea anemones.

Hotspots *see map:*
10 **Wembury, Devon** - www.nationaltrust.org.uk
11 **Pembrokeshire Coast National Park** - www.pcnpa.org.uk
12 **Orkney Islands** - www.orkney.org

Hebridean rocky shore. INSET: Edible Crab

Mudflats & estuaries

Our estuaries and mudflats are among the finest in the world, home to internationally important concentrations of waders and wildfowl in winter. Vast numbers of invertebrates live buried beneath the surface of the sediment, which can be treacherous for walkers.

Hotspots *see map:*
7 **The Wash, Norfolk** - www.washestuary.org.uk
8 **Strangford Lough** - www.strangfordlough.org
9 **Morecambe Bay** - www.morecambebay.com

Lugworms at Budle Bay, Northumberland. INSET: Dunlin, winter adult

Seaweeds

Seaweeds are algae, with simple bodies that lack proper roots, stems, leaves and a vascular system. They photosynthesise and a few species are indeed bright green. However, other colours are also common: red and brown seaweeds all contain chlorophyll but its green colour is masked by other pigments.

Egg Wrack

Egg Wrack

Ascophyllum nodosum Phaeophyceae
Length to 150cm

Distinctive seaweed, growing in the zone between the upper and middle shore. **Fronds** Comprise long greenish stems that branch regularly and are tough and leathery, flattening towards tip. Air bladders are found at regular intervals; the yellowish-green reproductive bodies resemble sultanas. **Status** Widespread and locally abundant.

Channelled Wrack

Pelvetia canaliculata Phaeophyceae
Length to 15cm

Distinctive seaweed that forms a zone on the upper shore of rocky coasts. **Fronds** Olive-brown and much branched. Inrolled margins help conserve water when exposed to air for long periods at neap tides. Lacks air bladders and reproductive bodies are at frond tips. **Status** Widespread and common.

Bladder Wrack

Fucus vesiculosus Phaeophyceae
Length to 1m

Tough seaweed, growing on rocks on middle shore. **Fronds** Olive- or greenish brown, branching regularly. Air bladders are found in groups of 2 or 3 along entire length, and spongy reproductive bodies occur at frond tips. **Status** Widespread and common, although absent from exposed shores.

Spiral Wrack

Fucus spiralis Phaeophyceae
Length to 35cm

Familiar seaweed, growing attached to rocks on upper shore. **Fronds** Branch regularly and typically twist in a spiral fashion towards tip; margin is not serrated and air bladders are absent. Reproductive bodies are found at frond tips. **Status** Widespread and common, although absent from exposed locations.

Serrated Wrack

Fucus serratus Phaeophyceae
Length to 65cm

Distinctive seaweed, growing on rocks on lower middle shore. **Fronds** Greenish brown and flattened, with a distinct midrib; branch regularly along length and have serrated margins. Air bladders are absent and reproductive bodies are found in pitted, swollen frond tips. **Status** Widespread and common.

Thongweed

Himanthalia elongata
Phaeophyceae Length to 2m

Starts life as a button-like structure attached to rocks on lower shore. **Frond** Develops from this base; olive-green, long, slightly flattened and strap-like. **Status** Common only in S and W.

Sea-oak

Halidrys siliquosa
Phaeophyceae Length to 50cm

Much-divided seaweed whose branches are borne alternately on stem. **Fronds** Yellowish brown and rather flattened; note the pod-like, seemingly segmented air bladders. **Status** Widespread and locally common on middle and lower rocky shores.

Kelp

Laminaria digitata Laminariaceae
Length to 1m

Impressive brown seaweed that forms dense beds, with only floating frond exposed at most low tides. **Fronds** Olive-brown and comprise a tough, branched holdfast (home to small marine animals), a robust, flexible stipe, and a broad blade, divided into strap-like fronds. **Status** Common, but mainly in deep water.

Codiuim tomentosum

Chlorophyceae Length to 20cm

Distinctive seaweed that grows attached to rocks on the mid- to lower shore. **Fronds** Dark greenish, branching dichotomously and ending in rounded tips. Texture is rather felt-like. **Status** Widespread and locally common, especially in S and W Britain.

Sea-lettuce

Ulva lactuca Chlorophyceae
Length to 40cm

Membranous seaweed. **Fronds** Just 2 cells thick; often tattered, making their precise shape hard to describe. **Status** Widespread and common, growing on rocks on very sheltered shores; also thrives in rock pools on upper shore, even if detached from substrate.

Gutweed

Enteromorpha intestinalis Chlorophyceae
Length to 75cm

Aptly named seaweed. **Fronds** Membranous and comprising long tubes that inflate with oxygen when photosynthesising; constrictions along their length add to their already gut-like appearance. **Status** Common and widespread in sheltered estuaries, brackish lagoons and rock pools on the upper shore.

Laver

Porphyra umbilicalis
Rhodophyceae Length to 20cm

Distinctive seaweed that grows attached to rocks that lie among sand. **Frond** Reddish purple, sometimes greenish towards the centre, and forming a gelatinous sheet. **Status** Widespread and common in fairly sheltered locations.

Bryozoans

Walk along the strandline after an autumn gale and you may find specimens of what looks like pale crusty seaweed. In fact, this is likely to be **Hornwrack** *Flustra foliacea* (length 15cm), a colonial animal called a bryozoan, and not a plant at all. In life it appears darker and attaches itself to rocks below the low-tide mark. Individual members of the colony perform different functions for the good of the whole organism.

Pepper Dulse

Laurencia pinnatifida Rhodophyceae
Length to 15cm

Tufted seaweed that branches alternately and sometimes forms dense mats. **Fronds** Slender and purplish brown or olive; they are extremely tough and rather rubbery. **Status** Widespread and locally common, especially in the S, on sheltered rocky coasts from mid- to lower shore.

Corallina officinalis

Rhodophyceae Length to 6cm

Mat-forming coralline seaweed. **Fronds** Comprise branched pinkish stems that bear numerous opposite calcified segments. **Status** Common and widespread in rock pools on the middle shore.

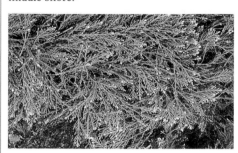

Lithophyllum incrustans

Rhodophyceae Width to 5cm

Encrusting coralline seaweed. **Fronds** Form flattish calcified plates, and you might be forgiven for failing to recognise it as a plant at all. **Status** Widespread but most common in the S.

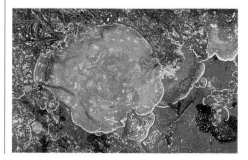

Introducing marine animals

There can be few people who have not succumbed to the fascination of the seaside rock pool at some time in their lives, whiling away many hours in search of crabs, shrimps, sea anemones and small fish. With so much life just waiting to be discovered, these habitats are entrancing to adults as well as to children.

Representatives of almost all forms of life can be found on the seashore, and at low tide a rock pool is full of natural wonders. Searching among the tangled seaweeds and under stones is like leafing through a book on animal life, and most of the main phyla – the major subdivision of classification in the animal kingdom – can be discovered within the space of a few hours. Among the most obvious and characteristic animal groups on the seashore are shown here.

Studying seashore life

At the simplest level, all you need to explore the seashore are your hands and eyes, and perhaps a good book to aid identification. A few basic items will increase the number of discoveries made and allow you to observe your animals better. A robust net will help you to capture active animals more easily. To avoid harming your finds, use a tray filled with clean seawater – many marine animals just look like blobs out of water and come to life only when immersed in their natural medium. If you have a white tray, try painting one half of it black before you visit the seashore – some marine animals are pale and translucent, and so show up much better against a dark background.

Sponges

On the evolutionary scale, sponges are almost the simplest of animals and belong to the phylum Porifera. The soft body of a typical sponge is given structure and strength by the presence of embedded calcareous fibres. The body is covered by pores through which water passes into an internal chamber; special cells lining these openings collect food and oxygen, and water passes out through a conspicuous exit passage.

Sea anemones and jellyfish

These soft-bodied animals belong the phylum Cnidaria and are sometimes referred to as coelenterates. They have simple, radially symmetrical bodies. Most jellyfish are free-living and their bodies have long, trailing tentacles armed with stinging cells that aid the capture of prey. Sea anemones typically attach themselves to rocks or other substrates, catching food and passing it to the mouth by means of sturdy tentacles. Corals are also members of the phylum Cnidaria; a few examples are found in Britain, although most are associated with tropical seas.

Snakelocks Anemone

True worms

The most common kinds of worms on the seashore are segmented worms belonging to the phylum Annelida; annelids are sometimes called true worms, distinguishing them from other creatures (not covered here) with worm-like bodies. An annelid's elongated body is divided into segments, each of which usually carries bristles. The head is often well developed, especially in predatory species. They tend to be free-living animals, but there are also plenty of filter-feeding and scavenging marine annelid worms, and these are often rather sedentary, living in burrows or under stones. True worms are extremely important in ecological terms because they are eaten by so many other marine creatures.

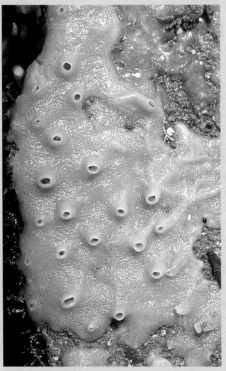
Breadcrumb Sponge

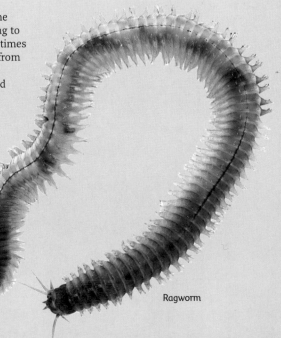
Ragworm

Molluscs

Molluscs are a varied group of animals with many marine forms. They are placed in the phylum Mollusca, and common examples include snails such as whelks and periwinkles, limpets, bivalves, and octopuses and squid. They have unsegmented soft bodies, a well-developed head (some with extraordinary eyes) and, in most species, a characteristic shell. Molluscs are found in almost all marine environments and exploit almost all sources of food.

Flat Periwinkle

Crustaceans

Crustaceans are members of the class Crustacea, a group that is a subdivision of the phylum Arthropoda. On land, arthropod representatives include insects and spiders, but in the marine world crustaceans predominate; examples include crabs, shrimps and lobsters. Most crustaceans are free-living, and they have well-developed heads (with striking eyes and mouthparts) and jointed, paired limbs on most of their body segments; some limbs are responsible for movement, while others are used in feeding. Most crustaceans have an exoskeleton that is hardened with calcium salts. Barnacles are unusual and exceptional crustaceans that live sedentary lives attached to rocks.

Starfish and sea urchins

Among the animals found on the seashore, members of the phylum Echinodermata (usually referred to simply as echinoderms) are almost unique in being restricted to the marine environment. Common examples include starfish and sea urchins. Echinoderms have bodies that are radially symmetrical, and in a typical animal there is a mouth located centrally on one side of the body (usually facing downwards) and a central anus on the other side. Tube-feet allow the animals to move. Starfish are predatory while sea urchins are filter-feeders.

Ascidians

Ascidians are peculiar marine animals whose commonest representatives are called sea squirts. As adults (the stage in which they are usually encountered) they appear to have primitive, undifferentiated bodies that look most like sponges. However, their tadpole-like larval stages reveal their true affiliations in evolutionary terms: the presence of a single hollow nerve cord means that they are more closely related to vertebrates than invertebrates. They are usually placed in the phylum Chordata (to which mammals, birds, fish, reptiles and amphibians also belong), separated from their more advanced cousins by placement in the subphyllum Ascidiacea.

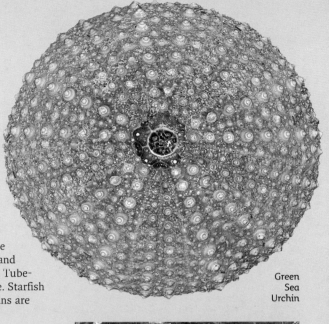

Green Sea Urchin

Star Ascidian

Fish

Fish are the most obvious and characteristic vertebrates found in the marine environment and are placed in the phylum Chordata. They have a well-developed head, with striking eyes and other sense organs, and a large mouth. Fish are good swimmers whose movement and directional control is usually achieved by means of fins. The main fish group encountered on the seashore includes species with a skeleton of hardened bones, an equal tail fin and a swim bladder to provide buoyancy. Dogfish, skates and rays are considered to be more primitive; they have cartilaginous (not hardened) bones and an unequal tail, and they lack a swim bladder.

Grey Gurnard

Velvet Swimming Crab

Fish

Although not always easy to see well, fish are present on the seashore in large numbers, and some small species are easily found by searching among seaweeds and under stones at low tide. A few surprisingly large species, usually associated with deeper offshore waters, venture close inshore during summer, providing a spectacle for keen observers.

Conger Eel

Conger conger Anguillidae
Length to 2m

Extremely muscular snake-like fish whose skin lacks scales. **Adult** Variable in colour, with an upper jaw that is just longer than lower one. **Status** Common, favouring rocky shores around Britain and Ireland; often in deep water but can also be found among rocks at low tide.

Pipefish

These curious little fish are related to seahorses (family Syngnathidae), and as their name suggests they have very slender, elongate bodies that make them difficult to spot among tangles of seaweeds. Several species occur in our waters.

Greater Pipefish

Worm Pipefish

Nerophis lumbriciformis Syngnathidae
Length 10cm

Very slender pipefish. **Adult** Even more worm-like than Greater Pipefish, and has an upturned snout. **Status** Found under stones and among seaweeds at low water on rocky shores.

Greater Pipefish

Syngnathus acus Syngnathidae
Length 50cm

Bizarre-looking fish. **Adult** Has a worm-like body, which appears segmented because of the body rings, and a snout-like mouth that is tapering, curved and longer than the rest of the head. Has prominent crest in neck region. **Status** Common among seaweeds in shallow coastal waters and estuaries, on sandy, muddy or stony bottoms. **Similar species** The **Lesser Pipefish** *S. rostellatus* (length 15cm) has a shorter snout and no crest. Common on sandy shores around British coast.

Cod

Many members of the cod family (Gadidae) are commercially important. Although Cod *Gadus morhua* and Whiting *Merlangius merlangus* are deep-water species and seldom encountered alive by land-based naturalists, two of their relatives are found close to the shore.

Pollack

Pollachius pollachius Gadidae
Length 1m

Common species, sometimes seen swimming in clear gullies or encountered when snorkelling. **Adult** Has a streamlined appearance; lateral line curves smoothly over pectoral fins. **Status** Widespread on rocky shores.

Five-bearded Rockling

Ciliata mustela Gadidae
Length 20cm

Found by turning over stones and weed at low tide. **Adult** Has an elongate body, long dorsal and anal fins, and 5 barbels around the mouth. **Status** Widespread in shallow water on muddy and sandy shores, often in the intertidal zone.

Open-water visitors

Mackerel and Bass are open-water fish for much of the year but come very close to rocky shores in the summer months.

Bass

Bass

Dicentrarchus labrax Moronidae Length 60cm

Powerful, fast-swimming fish. **Adult** Has a streamlined body and greyish scales; young fish sometimes have a pinkish tinge. **Status** Widespread and common. Small fish are found in shoals near mouths of estuaries; larger specimens are usually solitary and are found off shingle or rocky shores, sometimes in deep water.

Mackerel

Scomber scombrus Scombridae Length 30cm

Streamlined fish. **Adult** Has marbled blue-green upperparts and silvery underparts; has small 'finlets' on tailstock, as well as usual fins. **Status** Common and pelagic; comes inshore in summer.

Mackerel

Wrasse

Wrasse (family Labridae) are colourful fish with a long dorsal fin. Of several species found here, some are hermaphrodites, starting life female and becoming male as they age.

Ballan Wrasse

Labrus bergylta Labridae Length 30cm

Variably marked wrasse. **Adult** Often either marbled blue-green and black, or reddish brown with greenish spots; has a proportionately large head. **Status** Common and widespread.

Goldsinny

Centrolabrus rupestris Labridae Length 15cm

Colourful, well-marked fish. **Adult** Has a reddish body and a dark spot in front of the tail fin. **Status** Common, except in the N, favouring sheltered, rocky shores.

Corkwing Wrasse

Crenilabrus melops Labridae Length 15cm

Variably marked but beautiful fish. **Adult** Usually mainly blue with deep pink patterns and lines. **Status** Common except in the N; usually easy to see in deep rocky gullies.

Cartilaginous fish

Cartilaginous fish lack the hard bones of bony fish; instead, these structures are made of tough cartilage. Sharks are members of this group but are deep-water fish; dogfish, skates and rays are sometimes found in inshore waters.

Thornback Ray

Lesser Spotted Dogfish

Scyliorhinus canicula Scyliorhinidae Length 60cm

Common cartilaginous fish whose egg cases (so-called 'mermaid's purses') are often washed up. **Adult** Has well-marked rough skin and recalls a miniature slender shark. **Status** Widespread and fairly common; sometimes discovered in rock pools at very low tides.

Thornback Ray

Raja clavata Rajidae Length 50cm

Distinctive ray. **Adult** Has a diamond-shaped outline, rough skin, and conspicuous spines down its back and along its muscular tail. **Status** Widespread on muddy and sandy seabeds, and sometimes found trapped in pools at estuary mouths at very low tides.

Lesser Spotted Dogfish

Small fish & flatfish

Common Blenny
Lipophrys pholis Blennidae
Length 10cm

Delightful little fish, found in rocky gullies on the lower shore, or trapped in rock pools at low tide. **Adult** Has a proportionately large head and tapering body, marked with marbled bands along length. **Status** Common and widespread on rocky shores throughout Britain and Ireland; particularly abundant in the W.

Food for Black Guillemots

The Black Guillemot (*see* p.280) is an endearing seabird that frequents inshore waters off rocky coasts. Its northerly range corresponds to areas where one of its favourite prey species, the **Butterfish** *Pholis gunnellus* (length 15cm), is most abundant. The Butterfish is a bottom-dwelling species that lives among rocks on otherwise sandy seabeds. Its body is elongate, with long dorsal and anal fins that are continuous at the tail end; it is reddish with striking eye-spot markings along its length.

Black Guillemot with Butterfish

Rock Goby
Gobius paganellus Gobiidae
Length 15cm

Large-headed rock-pool fish. **Adult** Has characteristic dark bands along the length of its otherwise pale brown body. **Status** Common and widespread but is hard to spot until it moves.

Blennies and Gobies
Along with the Common Blenny, our various goby species (family Gobiidae) are probably the most familiar of all small inshore fish, and are often found in a variety of coastal habitats. Their diminutive size makes them easy to overlook, but what they lack in size they make up for in numbers.

Sand Goby
Pomatoschistus minutus Gobiidae
Length 8cm

Well-camouflaged fish when sitting on a sandy seabed. **Adult** Small, and rather pale and translucent. **Status** Common and widespread, favouring sandy shores.

Common Goby
Pomatoschistus microps Gobiidae
Length 8cm

Resembles the Sand Goby. **Adult** Has a well-marked but overall pale body. **Status** Common and widespread; often found in similar habitats to the Sand Goby, although it usually favours more sheltered locations.

Lesser Sand Eel
Ammodytes tobianus Ammodytidae
Length 20cm

Distinctive shoal-forming fish, much favoured as food by terns. **Adult** Has a long, slender body, silvery blue above and pale below, with very long dorsal and anal fins. **Status** Widespread and locally abundant in estuaries and shallow, sheltered coastal seas with sandy beds.

Grey Gurnard
Eutrigla gurnardus Triglidae
Length 30cm

Distinctive bottom-dwelling fish. **Adult** Has a proportionately large, angular head and tapering body; pectoral fins are partly divided into 3 feeler-like rays, used in sensory detection. **Status** Locally common on suitable shores, favouring rocky outcrops on sandy or muddy seabeds; sometimes seen near jetty supports.

Short-spined Bullhead

Myoxocephalus scorpius Cottidae
Length 15cm

Squat little bottom-dwelling fish. Previously known as Father Lasher. **Adult** Has a proportionately large, spiny head and tapering body. Body is usually mottled grey-brown but sometimes darker. Mouth is large and gill covers are spiny. **Status** Locally common among seaweeds and stones on otherwise muddy or sandy seabeds.

Flatfish

With extremely laterally flattened bodies, flatfish actually live on their sides. During their development their heads become distorted so that both eyes point upwards when they are resting on the seabed.

Sole

Solea solea Soleidae Length 25cm

Classic flatfish. **Adult** Has oval outline. Short tail fin is not separated from the dorsal and anal fins by any distinct tailstock. **Status** Common on sandy and muddy seabeds, often in shallow water in estuary and river mouths.

Flounder

Platichthys flesus Pleuronectidae Length 17cm

Oval to diamond-shaped flatfish. **Adult** Has long dorsal and anal fins that are separated from tail fin by a distinct tailstock; typically mottled brown above and pale below. **Status** Common and widespread on sandy and muddy seabeds.

Plaice

Pleuronectes platessa Pleuronectidae
Length 50cm

Well-marked flatfish. **Adult** Oval in outline and also has a distinct tailstock separating tail fin from dorsal and anal fins. Upper surface is buffish brown with distinct orange spots but colour and markings can be altered subtly to match surroundings. **Status** Common and widespread on sandy and muddy substrates.

Suckers

In the turbulent waters of exposed rocky shores, some species of small fish hide under stones as a way of protecting themselves from predators, but a select band of fish avoid being swept away by attaching themselves to rocks using a sucker (modified pelvic fins).

Sea Snail

Liparis liparis
Liparidae Length 10cm

Strange tadpole-like fish. **Adult** Has a smooth oval outline, a reddish body, and fused anal and tail fins. **Status** Found on rocky outcrops on otherwise sandy or muddy seabeds.

Cornish Sucker

Lepadogaster lepadogaster Liparidae
Length 6cm

Another tadpole-like fish. **Adult** Has a pointed snout, reddish body and 2 bluish eye-spot markings on back of head. **Status** Commonest in the turbulent waters of SW Britain.

Lumpsucker

Cyclopterus lumpus Cyclopteridae
Length 30cm

Round-bodied fish. **Adult** Older fish are grey above and red below. Young specimens are mottled or marbled yellow, buff and brown. **Status** Commonest on S coasts.

Seashore molluscs

Molluscs are among the most conspicuous and varied inhabitants of the seashore, and almost every habitat and niche has a mollusc species that it supports. In life, the behaviour and appearance of molluscs are fascinating to observe, while their empty shells can form the basis of interesting natural history collections.

Common Periwinkle

Littorina littorea Littorinidae
Length 25mm

Familiar mollusc of rocky shores. **Adult** Shell is rather rounded but with a pointed conical apex and a thick lip; usually dark brown with concentric ridges and dark lines. **Status** Common among seaweeds on which it feeds, between the low- and high-tide marks.

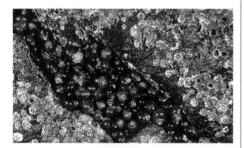

Flat Periwinkle

Littorina obtusata Littorinidae
Length 10mm

Familiar mollusc, found on Egg Wrack and Bladder Wrack (*see* p.446). **Adult** Has a smooth oval shell; colour varies from bright yellow to reddish brown. **Status** Abundant at the mid-tide level on rocky shores around Britain and Ireland. **Similar species** *L. mariae* shares the same common name and is also abundant in similar habitats; its shell is drop-shaped, largest at the opening.

colour forms

Rough Periwinkle

Littorina saxatilis Littorinidae
Length 8mm

Congregates in rock crevices. **Adult** Has a pointed oval shell marked with rough ridges and grooves. **Status** Common and widespread on the mid- to upper shore. **Similar species Small Periwinkle** *Littorina neritoides* (length 5mm) is tiny and clusters in rock crevices on the upper shore, feeding on lichens; it is common and widespread in the W.

Small Periwinkle

Common Necklace Shell

Polinices polianus Naticidae
Length 15mm

Burrowing predatory sea snail. **Adult** Has a large-mouthed, rather spherical shell that is flushed orange. **Status** Common on sandy shores but occurs most in the S.

Laver Spire Shell

Hydrobia ulvae
Hydrobiidae
Length 6mm

Sometimes climbs Glasswort at low tide. Important source of food for Shelduck and other birds. **Adult** Has a dark brown conical shell; apex is usually blunt and worn. **Status** Widespread and often abundant on estuary mud; found mostly between low- and high-tide levels.

Common Necklace Shell

Laver Spire Shell

Toothed Top Shell

Osilinus lineatus Trochidea
Length 24mm

Colourful seashell. **Adult** Shell has close-packed concentric ridges and beautiful purple zigzag patterns; mouth has a 'toothed' lip. **Status** Common on SW rocky coasts at the mid-tide level. **Similar species Painted Top Shell** *Calliostoma zizyphinum* (width 25mm) has a conical reddish shell with tight-packed ridges; it is found under stones on lower shores.

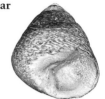

Grey Top Shell

Gibbula cineraria Trochidea
Length 12mm

Rather rounded seashell. **Adult** Shell narrow with dark grey or reddish bands. **Status** Common on lower shores among seaweeds and under stones. **Similar species Flat Top Shell** *G. umbilicalis* (length 12mm) is grey-buff with striking purple bands; common in the W, on the middle shore among rocks and seaweeds.

Cowries

It is always a delight to find cowries, beautifully ornate bean-shaped seashells of the family Triviidae. The **Common Cowrie** *Trivia monacha* (length 11mm) (shown below) is marked with pinkish-purple lines and three dark spots; in life, the mottled mantle edges envelop the shell margins; it is widespread and occurs on the lower shore. The **Arctic Cowrie** *T. arctica* (length 11mm) (not illustrated) is similar but unspotted; it is also widespread but is usually found in deeper water.

Common Whelk

Buccinum undatum Buccinidae
Length 8cm

 Common tideline species; empty shells are often occupied by hermit crabs. **Adult** Has a buffish conical shell with a large mouth; shells of larger animals are often coated with encrusting sponges and algae. **Status** Common and widespread on sandy and muddy seabeds.

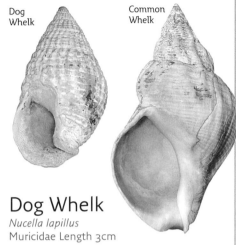

Dog Whelk

Common Whelk

Dog Whelk

Nucella lapillus
Muricidae Length 3cm

 Familiar sea snail of rocky shores. **Adult** Has a pointed oval shell whose colour (creamy white, grey-brown or banded) is influenced by its diet that includes barnacles and mussels. **Status** Widespread on rocky shores where its prey is common; sometimes seen beside egg masses on rocks. **Similar species Thick-lipped Dog Whelk** *Hinia incrassata* (Nassariidae, length 10mm) is smaller with ribbed whorls and thick 'lips'. Common on rocky shores.

Common Limpet

Patella vulgata Patellidae
Diameter 6cm

 Grazes algae when covered by water; feeding trails are visible at low tide. **Adult** Has a conical, ridged shell; older specimens are encrusted with barnacles and algae. **Status** Common on exposed rocky shores around mid- and high-tide levels. **Similar species Black-footed Limpet** *Patella depressa* (width 5cm) has a flatter shell and a black foot (orange in Common Limpet); it is restricted to SW Britain.

Blue-rayed Limpet

Helcion pellucidum Patellidae
Length 15mm

 Delicate little limpet. **Adult** Has a conical brown shell marked with radiating iridescent blue lines; young specimens are brightest. **Status** Locally common on lower rocky shores. In life, lives on fronds, stalks or holdfasts of Kelp, usually inside an excavated depression or cavity.

Blue-rayed Limpet

Keyhole Limpet

Keyhole Limpet

Diodora graeca
Fissurellidae Diameter 20mm

 Distinctive mollusc whose conical shell has a keyhole opening at the apex, through which a siphon protrudes in life. **Adult** Has a grey-buff shell with radial ridges. **Status** Widespread but local on lower rocky shores.

Slipper Limpet

Crepidula fornicata Calyptraeidae
Length 3cm

 Familiar tideline species. **Adult** Has a greyish ear-shaped shell. **Status** Introduced alien from North America, now widespread on sheltered coasts in the S, sometimes in large enough numbers that empty shells dominate beach debris. In life, attaches itself to other molluscs.

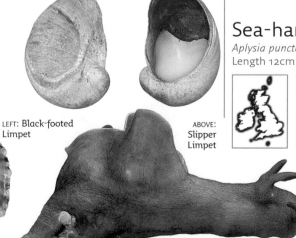

LEFT: Black-footed Limpet

ABOVE: Slipper Limpet

Sea-slugs

With soft bodies, sea-slugs lack the external shells of their snail cousins. Many are extremely brightly coloured and most are seen on the seashore only in summer, moving to deeper water in winter.

Sea-lemon

Archidoris pseudoargus Archidorididae
Length 6cm

 Attractive mollusc. **Adult** Has a warty yellowish body with olive-brown blotches. When undisturbed, looks sausage-like with 2 tentacles at head end and frilly gills at rear. Contracts body when alarmed. **Status** Seasonally locally common on rocky shores, mainly in the S.

ABOVE: Sea Lemon
RIGHT: *Greilada elegans*

Greilada elegans

Polyceridae Length 3cm

 Sea-slug whose bright colours indicate its unpleasant taste to potential predators. **Adult** Orange with purple spots. **Status** Intolerant of cold waters and restricted to SW Britain and Ireland. Occurs mostly in deep water but can be trapped in low-tide rock pools in summer.

Sea-hare

Aplysia punctata Aplysiidae
Length 12cm

 Bizarre-looking slow-moving mollusc whose body encloses its soft shell. Ejects a purple dye if alarmed. **Adult** Has a rather lumpy body, blotched grey and brown, with 4 tentacles at head end. **Status** Widespread on S and W coasts, found among seaweeds in rock pools on very low tides.

Sea-hare

Marine molluscs

Bivalves are among the most numerous of all molluscs, although this is not always apparent since many species live buried in sediment – they are right under our feet as we walk across a sandy beach. In comparison to their sedentary, buried cousins, octopuses, squid and cuttlefish are by far the most active of all molluscs.

Bivalve molluscs

Belonging to the class Bivalvia, bivalve molluscs have laterally compressed bodies that are enclosed and protected by a shell in two halves, or valves; these are connected at the base by a hinged ligament, and internal muscles allow the shell to be opened or clamped shut. Bivalves are filter- or sediment-feeders.

Peppery Furrow Shell
Scrobicularia plana Scrobiculariidae
Length 5cm

Common bivalve mollusc that filter-feeds using its long siphons. **Adult** Has a pale grey-brown shell, rounded in outline, thin in profile and marked with many close-packed concentric ridges. **Status** Widespread and common in sand and mud on shores and estuaries.

Thin Tellin
Tellina tenuis Tellinidae
Length 20mm

Locally abundant bivalve. **Adult** Has a delicate pinkish or orange shell with concentric darker bands. **Status** Common on clean sandy shores; dead shells are often washed up in quantity.
Similar species Baltic Tellin *Macoma balthica* (length 2.5cm) is a typical, widespread estuary species.

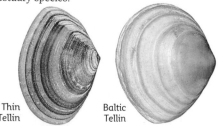

Thin Tellin

Baltic Tellin

Striped Venus
Chamelea gallina Veneridae
Length 25mm

Robust bivalve with a rather rounded shell. **Adult** Has an orange-buff shell with radiating dark bands and concentric ridges. **Status** Widespread and locally very common, burrowing in sand from the lower shore downwards.

Common Cockle
Cerastoderma edule Cardiidae
Length 5cm

Familiar seashore shell. **Adult** Has identical shell valves, grey-buff with radiating ridges outside and shiny and white inside. **Status** Common throughout, living buried in mud and sand in estuaries and beaches. **Similar species Spiny Cockle** *Acanthocardia aculeata* (length 10cm) is larger with spiny radiating ridges.

Common Cockle

Boring molluscs

Enquire of a wit about boring molluscs and you may get the facetious response that all molluscs are boring. In reality, only one group – the piddocks (family Pholadidae) – live up to the name, inhabiting tunnels they have bored into sandstone and other soft rocks. The **Common Piddock** *Pholas dactylus* (length 12cm), whose shell is pale buffish brown, lives at high densities in suitable rocks, and is widespread in SW Britain and Ireland.

Piddocks in holes in rock

Common Mussel
Mytilus edulis Mytilidae
Length 9cm

Easily recognised bivalve. **Adult** Has identical valves, rounded at one end and pointed at the other. These are dark grey-brown outside but with mother-of-pearl inside. **Status** Common throughout, found in large groups. **Similar species Bearded Horse Mussel** *Modiolus barbatus* (length 6cm) is covered in bristly hairs and is locally common throughout.

Bearded Horse Mussel

Great Scallop

Pecten maximus Pectinidae
Length 15cm

Familiar bivalve with iconic shell outline. **Adult** Has a flat pinkish-white upper valve and convex orange-brown lower one; both are ridged. **Status** Common on sand and gravel in deep water, but empty shells are often washed up. **Similar species Variegated Scallop** *Chlamys varia* (length 6cm) has a variegated shell with one lateral projection much longer than the other.

Common Oyster

Ostrea edulis Ostreidae
Length 10cm

Familiar bivalve. **Adult** Shell valves are rough and grey-brown outside, smooth with a mother-of-pearl coating inside. Lower valve is saucer-like and sits on substrate; upper valve is flattened. **Status** Formerly abundant and still common in shallow water with a gently shelving seabed.

Common Oyster

Variegated Scallop

Pod Razorshell

Ensis siliqua Pharidae
Length 18cm

Elongated bivalve that lives buried in sand. **Adult** Has olive-brown flaky outer shell valves. **Status** Widespread and common around the low-tide mark on beaches. Empty shells are often washed up. **Similar species** *E. ensis* (length 11cm) has narrow and distinctly curved shell valves. **Grooved Razorshell** *Solen marginatus* (length 12cm) has a deep shell with a groove one end.

Ensis ensis

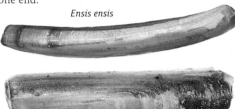

ABOVE: Grooved Razorshell; BELOW: Pod Razorshell

Cephalopod molluscs

Octopuses, squid and cuttlefish are the most advanced molluscs, being active predators that are renowned for their well-developed eyes. They belong to the class Cephalopoda.

Common Octopus

Octopus vulgaris Octopodidae
Length 50cm

Active predator of marine invertebrates. Can change colour quickly and dramatically. **Adult** Recognised by its bulbous, bag-like body, large eyes and 8 arms bearing powerful suckers. **Status** Locally common off rocky coasts of SW Britain and Ireland; sometimes found in rock pools at very low tide.

Little Cuttlefish

Sepiola atlantica Sepiolidae
Length 5cm

Dainty little cephalopod. **Adult** Has a short cylindrical body and proportionately large head, and 10 tentacles, 2 of which are long. Colour is variable and changeable, but typically mottled reddish brown. **Status** Favours sandy seabeds and is locally common in SW Britain and Ireland.

Chitons

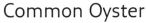

Sometimes known as coat-of-mail shells, chitons are oval molluscs whose shell comprises eight interlocking curved plates. They graze algae and live on rocks, holding on with their muscular foot. *Acanthochitona fascicularis* (length 5cm) is common in the S and is found on the lower shore.

Common Octopus

Little Cuttlefish

Common Cuttlefish

Sepia officinalis Sepiidae
Length 25cm

Best known from cuttlebones (shown here), washed up on beaches. **Adult** Has a flattened oval body with paired lateral fins. Has 10 tentacles, 2 of which are long. Changes colour to match surroundings. **Status** Common over sandy seabeds.

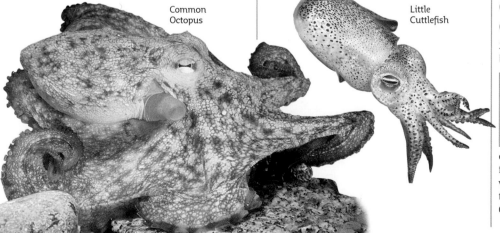

Seashore crustaceans

Crustaceans are extremely well represented in the marine environment and whether you visit a rocky shore or a sandy estuary you are sure to find examples. Among their numbers are such seashore favourites as crabs, lobsters, shrimps and prawns, most of which have a pugnacious will to live and are equipped with fearsome-looking claws to help them.

Edible Crab

Cancer pagurus Cancridae
Width 15cm

Familiar crab, recognised by 'piecrust' margin to carapace. **Adult** Pinkish orange with black pincer tips. **Status** Widespread and common on rocky coasts. Small specimens are sometimes found in rock pools; larger individuals live further offshore.

Green Shore Crab

Carcinus maenas Portunidae
Width 5cm

Our commonest crab. **Adult** Has 3 blunt teeth between eyes and is usually greenish or olive-brown. **Status** Widespread and common in all kinds of marine habitats. Found on the middle and lower shore; usually hides beneath seaweed or stones at low tide.

Velvet Swimming Crab

Necora puber Portunidae
Width 7cm

Aggressive crab that brandishes pincers when cornered. **Adult** Has beady red eyes, reddish leg joints and 8-10 teeth between eyes. Hairy upper surface of carapace collects silt. Tips of back legs are paddle-like. **Status** Widespread and common on rocky shores.

Masked Crab

Corystes cassivelaunus Corystidae
Length 4cm

Bizarre-looking burrowing crab; its long antennae create a passage along which seawater flows to gills. **Adult** Carapace is much longer than it is broad, and pincer-bearing front pair of legs is long. **Status** Widespread and locally common on suitable sandy beaches.

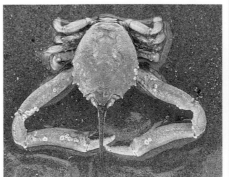

Spiny Spider Crab

Maja squinado Majidae
Length 15cm

Distinctive long-legged crab. **Adult** Has a pinkish-orange, rather triangular carapace covered in spines and bristles. **Status** Locally common on rocky shores and sandy seabeds; sometimes found in rock pools at extreme low tides.

Spider Crab

Macropodia rostrata Majidae
Length 10mm

Body is usually coated with, and camouflaged by, sponges and seaweeds. **Adult** Has long spider-like legs and a roughly triangular carapace. **Status** Widespread and fairly common in rock pools on the lower shore; easily overlooked owing to camouflage.

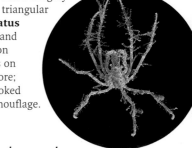

Broad-clawed Porcelain Crab

Porcellana platycheles Porcellanidae
Width 13mm

Has a flattened, rounded carapace and very broad pincers. **Adult** Has a sandy-brown upper surface that is hairy and traps silt, improving camouflage; underside is white and porcelain-like. **Status** Widespread and common under stones on sheltered rocky shores.

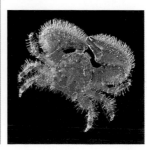

Hermit crabs

Hermit crabs (family Paguridae) have rather soft, lobster-like bodies that they protect by living inside empty mollusc shells; these are periodically exchanged for larger ones as the animal grows. The **Common Hermit Crab** *Eupagurus bernhardus* (length 9cm) is widespread. It is reddish brown; the larger right-hand pincer is used to block the shell entrance when the crab retreats inside. It is found on sheltered shores and in rock pools.

Common Lobster

Homarus gammarus Nephropidae
Length 40cm

Familiar crustacean that hides in crevices, defending itself using powerful pincers. **Adult** Blue in life. **Status** Mainly a deep-water species, but smaller individuals are sometimes found in rock pools at extreme low water.

Squat Lobster

Galathea squamifera Galatheidae
Length 4cm

Related to hermit crabs and with a dumpy body; abdomen is folded under thorax. **Adult** Has a greenish-brown body and long, slender pincers. **Status** Common, found under stones on the lower shore of rocky coasts.

Prawns and shrimps

These swimming crustaceans have semi-transparent bodies and long antennae. Of the many species, prawns are likely to be found in rock pools and shrimps on sandy seabeds.

Common Prawn

Palaemon serratus Palaemonidae
Length 6cm

Delicate crustacean. **Adult** Has a near-transparent body marked with purplish-brown dots and lines; very long antennae and a toothed rostrum extending forward between eyes. **Status** Common on rocky shores in SW Britain and Ireland.

Common Prawn

Common Shrimp

Crangon vulgaris Crangonidae
Length 5cm

Common Shrimp

Familiar estuary crustacean. **Adult** Has a translucent body stippled with dark dots, providing good camouflage when resting on sand; has long antennae, but rostrum is reduced to a tiny tooth. **Status** Common on sandy shores.

Sea-slater

Ligia oceanica Ligiidae
Length 25mm

Large woodlouse relative. Alert and fast-moving. Most active after dark, when it scavenges for organic matter. **Adult** Has a greyish, segmented and flattened body with 2 long antennae. **Status** Common and widespread among rocks in the splash zone and in sea walls.

Sea Slater

Amphipods

If you turn over stones or rotting seaweed washed up on the shore you are likely to see large numbers of small, laterally flattened creatures hopping away from you and scurrying around in search of shelter. These are amphipod crustaceans (order Amphipoda), a group that includes sandhoppers such as ***Talitrus saltator*** (length 15mm).

Talitrus saltator

Barnacles & sponges

Many marine creatures are sedentary, and most barnacles and sponges spend their entire adult lives attached to rocks. Because they have evolved their own specific tolerances to exposure to air at low tide, each species is found in a precise zone on the beach. Being fixed in one spot means that these marine creatures have to rely on currents and tides to bring them their food, which they filter from the water.

Barnacles

Specialised filter-feeding crustaceans, adult barnacles live sedentary lives, their bodies attached to a substrate. They are a familiar sight on rocky shores, and several species form encrusting greyish-white carpets on suitable surfaces. Each has evolved to favour different degrees of inundation, and exposure to air and wave action, and grows in subtly different zones on the shore. The body of rocky shore barnacles resembles a miniature volcano comprising (in native species) six plates, which together form a protective shell; feathery appendages appear through the central opening when it is immersed in seawater, but this is sealed at low tide by additional plates. Three species (*see* photographs below) dominate the upper and middle shore and collectively they are sometimes referred to as acorn barnacles. The shape of the central opening and its sealing plates (the ventral one is called the rostral plate) are the key to identification.

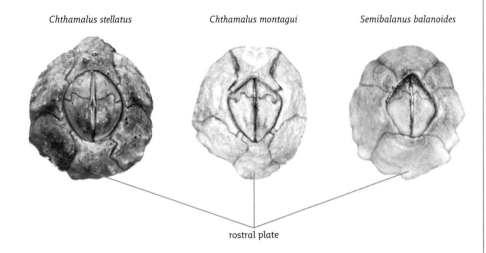

Chthamalus stellatus   *Chthamalus montagui*   *Semibalanus balanoides*

rostral plate

Chthamalus montagui
Chthamalidae Width 11mm

A barnacle that tolerates only moderate wave action. **Adult** Central opening is kite-shaped and rostral plate is narrow. **Status** Range extends to Scotland; lives on the middle to upper shore.

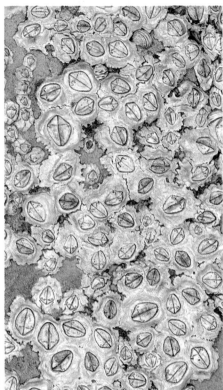

Semibalanus balanoides
Archaeobalanidae Width 12mm

A barnacle that requires longer periods of seawater immersion than other similar species. **Adult** Central opening is diamond-shaped and rostral plate is broad. **Status** Absent from SW England but common elsewhere; found on the lower shore, always below *Chthamalus* barnacles.

Chthamalus stellatus
Chthamalidae Width 10mm

A barnacle that tolerates considerable wave action. **Adult** Central opening is oval and rostral plate is narrow. **Status** Commonest in W and SW Britain, between the middle shore and splash zone.

Wartime arrival

The barnacle **Elminius modestus** (width 10mm) was introduced inadvertently from Australia to the Solent during the Second World War. It has since become established and has spread widely along coasts of S England and Wales. Its shell comprises four smooth greyish-white plates (native acorn barnacle species have six), and it favours the middle shore in sheltered waters, where it attaches itself to stones and tolerates a freshwater influence.

The strange case of the Goose Barnacle

If you walk along an exposed west coast beach in autumn you may find a strange creature called a **Goose Barnacle** *Lepas anatifera* (length 4cm). This pelagic species is sometimes washed up after gales, and is typically found in sizeable groups attached to driftwood by a 15cm-long retractable stalk. The five translucent bluish plates, and the feathery appendages that often project from them, led in the past to the fanciful suggestion that Goose Barnacles were the embryonic stages of the Barnacle Goose.

Breadcrumb Sponge

Halichondria panicea Halichondriidae
Width 10cm

Encrusting, patch-forming sponge. **Body** Usually bright orange, sometimes greenish brown. Surface is pitted with crater-like openings, through which seawater passes. **Status** Widespread on rocky coasts, on overhangs or in shady crevices from the middle shore downwards.

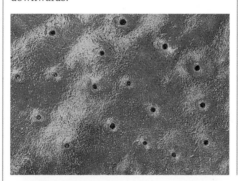

Star Ascidian

Botryllus schlosseri Botryllidae
Diameter 5mm

Encrusting colonial animal that forms tough mats. **Individuals** Arranged in a star-like fashion around a common opening. Colour is variable, but white 'stars' on a purplish-brown or bluish background is typical. **Status** Widespread, commonest off W coasts; occurs on shaded rocks on the lower shore.

Dead Man's Fingers

Alcyonium digitatum Alcyoniidae
Length 15cm

Coral-like colonial animals that inhabit a branched skeleton. **Colony** Colour is usually pale pink or yellowish white. When feeding, numerous tiny polyps emerge, giving surface a fuzzy appearance (fancifully like a decomposing hand). **Status** Widespread on rocky coasts from the lower shore downwards.

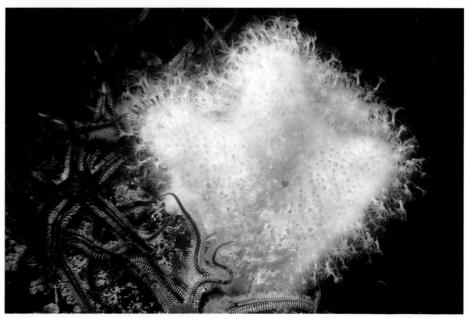

Jellyfish & sea anemones

Battering waves, powerful tides and a host of potential predators make the sea a challenging environment in which to live. Many marine animals rise to the challenge by arming themselves with protective shells. Not so members of the phyla Cnidaria (jellyfish and sea anemones) and Ctenophora (sea gooseberries) whose bodies are soft and, in terms of volume, almost entirely composed of seawater.

Jellyfish

Although jellyfish are competent swimmers, they often fall foul of wind and tide and get washed up on beaches, where they inevitably die. Sadly, this is how most people encounter these strange animals, although sometimes it is possible to get excellent views of them in life from jetties, piers and boats. Jellyfish belong to the phylum Cnidaria, and are sometimes referred to as coelenterates. Several species are found around British and Irish coasts, but only a few reach inshore waters.

Compass Jellyfish

Chrysaora hysoscella Pelgidae
Diameter 25cm

 Well-marked and familiar jellyfish. **Adult** Has an umbrella-shaped body with a dark central spot, dark red radiating lines and long tentacles suspended from lobed margin. **Status** Seen mainly off S and W Britain and Ireland; sometimes washed inshore.

Common Jellyfish

Aurelia aurita Ulmaridae
Diameter 20cm

 Pale, translucent jellyfish. **Adult** Body colour is pale yellowish brown, tinged bluish purple as it pulsates through water; underside of 'umbrella' has numerous short tentacles and 4 frilly mouth arms. **Status** Generally our commonest inshore species, seen throughout the region.

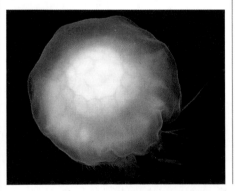

By-the-wind-sailor

Velella velella Velellidae
Diameter 3cm

 Colourful, surface-floating jellyfish relative. **Adult** has bluish, transparent and horny float below which tentacles dangle. **Status** Mainly oceanic but washed ashore by strong westerly gales, usually in autumn.

Sea Gooseberries

Resembling miniature jellyfish, Sea Gooseberries are unrelated and belong to the phylum Ctenophora. ***Pleurobrachia pileus*** (length 2cm), shown here, is the commonest species in inshore waters, and is sometimes stranded in pools. Its transparent body is the size and shape of a gooseberry and it feeds by catching plankton with its two long tentacles. After dark, it is bioluminescent.

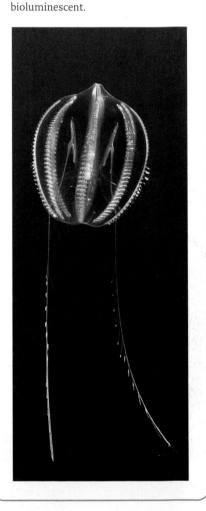

Sea anemones

Related to jellyfish, sea anemones are soft-bodied members of the phylum Cnidaria that, as adults, spend their lives attached to rocks and other substrates with a sucker-like base (a few, rare species burrow into sand). Many of our species favour deep water, and so are beyond the scope of this book, but several live in the intertidal zone.

Beadlet Anemone

A jewel of an anemone

Despite its appearance and common name, the **Jewel Anemone** *Corynactis viridis* (height 2cm) is more closely related to corals than to true sea anemones. It is found in small colonies attached to rocks on lower shores around south-west and west Britain and Ireland, often in gullies or crevices. Its body colour is usually pale buffish white, although sometimes pale grey-green, and the tentacles are tipped with iridescent pinkish purple.

Beadlet Anemone
Actinia equina Actiniidae
Height 5cm

Familiar, colourful sea anemone. **Adult** Red and green forms occur, and their tentacles are extended when immersed and retracted when exposed. **Status** Widespread and sometimes abundant on middle and lower shores of rocky coasts around Britain and Ireland.

Snakelocks Anemone
Anemonia sulcata Actiniidae
Height 10cm

Rather squat anemone. **Adult** Body colour is variable but is often grey-brown or purplish green; tentacles (which can't be retracted) are often purple-tipped. **Status** Found mainly on W and SW coasts, where it attaches itself to rocks on middle and lower shores.

BELOW: Snakelocks Anemone

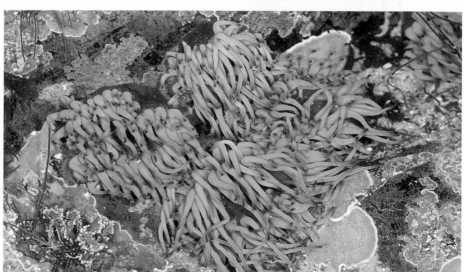

Plumose Anemone
Metridium senile Metridiidae
Height 7cm

Pale, delicate-looking anemone. **Adult** Occurs in both pink and white forms and has numerous feathery tentacles. **Status** Widespread; lives below the middle shore.

Dahlia Anemone
Urticina felina Actiniidae
Height 12cm

Well-marked and distinctive anemone. **Adult** Occurs in red and green forms, both with banded tentacles. **Status** Widespread and found from the lower shore into deep water.

Marine worms & echinoderms

Because they live buried in mud or conceal themselves, most marine worms are easy to overlook. However, signs of their presence can be found (worm casts, for example) and they show fascinating adaptations to seashore life. Echinoderms are easier to spot, and like many marine worms they possess an extraordinary functional beauty.

Peacock Worm

Sabella pavonia Sabellidae
Height 20cm

Intriguing worm. **Adult** Has radiating banded gills surrounding a mouth (seen only when immersed), these used for feeding and breathing. Lives in a tube constructed of sand and mud particles glued together by mucus; these stand proud of substrate. **Status** Locally common throughout on suitable sandy shores.

Amphitritides gracilis

Terebellidae Length 10cm

Bizarre-looking worm. Lives in a shallow burrow that protects its soft body, often sited under a stone. **Adult** Has numerous long tentacles and branched blood-red gills at head end. Curls into a spiral when removed from burrow. **Status** Locally common throughout on sheltered muddy and sandy shores.

Lugworm

Arenicola marina Arenicolidae
Length 18cm

Burrowing worm that lives in a U-shaped burrow; entrances are often indicated by holes and casts. **Adult** Greenish with red frilly gills on middle segments. **Status** Widespread and locally common from the middle shore downwards in muddy estuaries and on sandy beaches.

Spirorbis borealis

Serpulidae Diameter 3mm

Curious little worm. **Adult** Lives inside a white calcareous spiral tube. Clusters of tubes are found on fronds of Serrated Wrack and other seaweeds, and on rocks and mollusc shells. **Status** Common and widespread on lower shores.

Ragworm

Nereis diversicolor Nereidae
Length 10cm

Active predatory worm that burrows freely. **Adult** Has a toothed proboscis and jaws at head end, with tufts of hair-like bristles on each of the 100 or so segments. **Status** Widespread and often abundant in estuaries and on sandy or muddy beaches.

Ragworm

Common Starfish

Asterias rubens Asteriidae
Diameter 40cm

Familiar starfish. **Adult** Orange-red upper surface has pale warts. **Status** Widespread and common wherever its prey (bivalve molluscs) occur.

Spiny Starfish

Marthasterias glacialis Asteriidae
Diameter 30cm

Rough-textured starfish. **Adult** Has rather slender, often upturned arms that are usually olive-brown with large pink spines; underside is much paler. **Status** Widespread, but is commonest off W and SW coasts of Britain and Ireland on exposed and sheltered rocky shores.

Cushion-stars

Cushion-stars are blunt-armed starfish where only short tips project beyond the otherwise pentagonal outline. ***Asterina gibbosa*** (diameter 5cm) is commonest, often found off south and west British and Irish coasts under rocks on the lower shore. Its upper surface is rough and usually blotched pinkish yellow and grey-brown, while the underside is yellowish grey with numerous tube-feet.

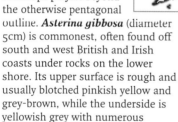

upperside

underside

Common Brittle-star

Ophiothrix fragilis Ophiotrichidae
Disc diameter 15mm

Echinoderm with brittle arms. **Adult** Colour is variable, although it is usually purplish brown with paler bands visible along the arms. **Status** Widespread; often found at low tide among seaweeds and under stones, especially where silt and sediment collect.

Common Sea Urchin

Echinus esculentus Echinidae
Diameter 10cm

Almost spherical echinoderm. **Adult** In life, is red and purplish brown, covered by spines and tube-feet; dead animal loses spines to reveal reddish test with radiating paler lines. **Status** Widespread, commonest in W and N. Found mainly in deep waters, but also in rock pools at low tide.

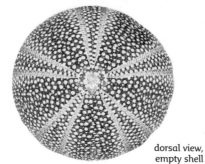

dorsal view, empty shell

Green Sea Urchin

Psammechinus miliaris Echinidae
Diameter 4cm

Rather flattened, circular echinoderm. **Adult** In life, is green overall, covered by tube-feet and purple-tipped spines; dead animal is green with radiating pale lines. **Status** Widespread and locally common on rocky shores.

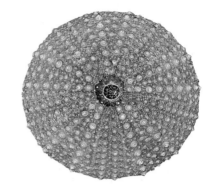

Sea Potato

Echinocardium cordatum
Loveniidae Length 8cm

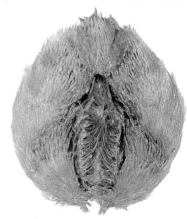

Distinctive burrowing sea urchin. **Adult** In life, is covered by a dense mat of fine spines, most pointing backwards. Most familiar as a smooth, pale brown potato-like shell of dead animal. **Status** Widespread and locally common on sandy seabeds around Britain and Ireland.

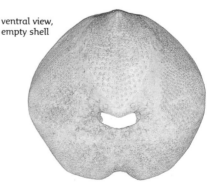

ventral view, empty shell

Further reading

British Wildlife. A magazine for the modern naturalist (6 issues per year). British Wildlife Publishing.

Brown, A. and Grice, P. (2004). *Birds in England*. Christopher Helm.

BSBI Handbooks. A series of identification guides to selected plant groups or families. Botanical Society of the British Isles.

Campbell, B. and Lack, E. (eds) (1985) *A Dictionary of Birds*. Poyser.

Chinery, M. (2005). *Complete British Insects*. HarperCollins.

Edlin, H.L. (1970). *Trees , Woods and Man*. Collins New Naturalist 32. Collins.

Fitter, A. (1978). *An Atlas of the Wild Flowers of Britain and Northern Europe*. Collins.

Fitter, R., Fitter, A. and Blamey, M. (1974). *The Wild Flowers of Britain and Northern Europe*. Collins.

Fitter, R., Fitter, A. and Farrer, A. (1989). *Collins Guide to the Grasses, Sedges, Rushes and Ferns of Britain and Northern Europe*. Collins.

Garrard, I. and Streeter, D. (1983). *The Wild Flowers of the British Isles*. MacMillan London.

Gibbons, R. and Brough, P. (1992). *The Hamlyn Photographic Guide to the Wild Flowers of Britain and Northern Europe*. Hamlyn.

Greenhaigh, M. and Ovenden, D. (2007). *Freshwater Life of Britain and Northern Europe*. HarperCollins.

Johnson, O. and More, D. (2004). *Collins Tree Guide*. Collins.

Lack, P. (1986). *The Atlas of Wintering Birds in Britain and Ireland*. Poyser.

Preston, C.D., Pearman, D.A. and Dines, T.D. (2002). *New Atlas of the British and Irish Flora*. Oxford University Press.

Rose, F. (1981). *The Wild Flower Key* (illustrated by R.B. Davis, L. Mason, N. Barber and J. Derrick). Warne.

Rose, F. (1989). *Colour Identification Guide to the Grasses, Sedges, Rushes and Ferns of the British Isles and North-western Europe* (illustrated by L. Mason, C. Dalby and R.B. Davis). Viking.

Spooner, B. and Roberts, P. (2005). *Fungi*. Collins New Naturalist 96. Collins.

Stace, C. (1997). *New Flora of the British Isles*. 2nd edn. Cambridge University Press.

Sterry, P.R. (2004). *Complete British Birds*. HarperCollins.

Sterry, P.R. (2005). *Complete British Animals*. HarperCollins.

Sterry, P.R. (2006). *Complete British Wild Flowers*. HarperCollins.

Sterry, P.R. (2007). *Complete British Trees*. HarperCollins.

Tubbs, C.R. (1986). *The New Forest*. Collins New Naturalist 73. Collins.

Wernham, C., Toms, M., Marchant, J., Clark, J., Siriwardnea, G. and Baillie, S. (eds) (2002). *The Migration Atlas*. Christopher Helm.

Wingfield Gibbons, D., Reide, J.B. and Chapman, R.A. (1993). *The New Atlas of Breeding Birds in Britain and Ireland: 1988–1991*. Poyser.

Useful websites

Botanical Society of the British Isles (BSBI) – www.bsbi.org.uk

British Trust for Ornithology (BTO) – www.bto.org

Campaign to Protect Rural England (CPRE) – www.cpre.org.uk

Countryside Council for Wales – www.ccw.gov.uk

The Forestry Commission – www.forestry.gov.uk

Herpetological Conservation Trust – www.herpconstruct.org.uk

Joint Nature Conservation Committee – www.jncc.gov.uk

The National Trust – www.nationaltrust.org.uk

The National Trust for Scotland – www.nts.org.uk

Natural England – www.naturalengland.org.uk

Plantlife International – www.plantlife.org.uk

Royal Society for the Protection of Birds (RSPB) – www.rspb.org.uk

Scottish Natural Heritage – www.snh.org.uk

The Wildlife Trusts – www.wildlifetrusts.org.uk

The Woodland Trust – www.woodland-trust.org.uk

Glossary

Abdomen - hind section of an insect's body; usually appears segmented.

Achene - one-seeded dry fruit that does not split.

Acute - sharply pointed.

Alien - introduced by Man from another part of the world.

Alternate - not opposite.

Amplexus - mating embrace adopted by paired frogs and toads, in which the male grasps the female with his front legs.

Annelid - segmented worm, phylum Annelida.

Annual - plant that germinates, grows and sets seed within a single growing season.

Antennae - slender, paired sensory organs on the head of an insect, or other arthropod.

Anther - pollen-bearing tip of the stamen.

Aquatic - living in water.

Arboreal - tree-dwelling.

Arthropod - invertebrate characterised by having a hard exoskeleton and jointed legs.

Awn - stiff, bristle-like projection, seen mainly in grass flowers.

Axil - angle between the upper surface of a leaf, or its stalk, and the stem on which it is carried.

Basal - appearing at the base of a plant, at ground level.

Basic - soil that is rich in alkaline (mainly calcium) salts.

Berry - fleshy, soft coated fruit containing several seeds.

Biennial - plant that takes two years to complete its life cycle.

Bivalve - mollusc whose shell comprises two halves.

Blow - spout of a whale as air is expelled from the lungs via the nostrils.

Bog - wetland habitat where the soil and water is acid.

Bract - modified, often scale-like leaf found at the base of flower stalks in some species.

Bulb - fleshy underground structure found in certain plants, comprising leaf bases and the next year's buds.

Bulbil - small bulb-like structure.

Calcareous - soil containing calcium, the source typically being chalk or limestone.

Calyx - outer part of a flower, comprising the sepals.

Cap - structure seen in fungi under which spore-bearing structures, usually gills or pores, are suspended.

Capsule - fruiting structure within which seeds are formed in flowering plants and which splits to liberate them; structure inside which spores develop in mosses and liverworts.

Carapace - hard upper surface of a crustacean's shell.

Carnivore - animal that feeds on meat and whose teeth, digestive system and behaviour are adapted to this lifestyle.

Carpal - area on a bird's wing corresponding to the 'wrist' joint.

Caterpillar - larval stage of a butterfly or moth.

Catkin - hanging spike of tiny flowers.

Caudal - relating to the tail.

Cephalothorax - fused head and thorax found in spiders.

Cerci - paired appendages at the hind end of an insect's body.

Cere - bare skin at the base of the bill and around the nostrils in birds.

Chlorophyll - green pigment found in plant tissue and essential for photosynthesis.

Clasping - referring to leaf bases that have backward-pointing lobes that wrap around the stem.

Compound - leaf that is divided into a number of leaflets.

Compound eye - eye structure typical of insects and some other invertebrates, comprising numerous cells and lenses, as opposed to a single lens.

Cone - structure bearing reproductive parts in conifers.

Conifer - tree that bears its reproductive elements in cones.

Cordate - heart-shaped at the base.

Corm - swollen underground stem.

Corolla - collective term for the petals.

Crepuscular - active at dusk.

Cultivar - plant variety created by cultivation.

Deciduous - plant whose leaves fall in autumn.

Decurrent - with the leaf base running down the stem.

Dentate - toothed.

Detritivore - animal that feeds on decomposing organic matter.

Dioecious - having male and female flowers on separate plants.

Disc florets - small flowers found at the centre of an inflorescence of members of the daisy family.

Diurnal - active during daylight.

Dorsal - upper surface.

Double-brooded - rearing two broods (birds) or completing two life-cycles (insects) in a year.

Downland - grassland habitat found on chalky soil.

Ear-coverts - the feathers cloaking the ears of a bird.

Echolocation - the method by which a bat perceives its surroundings, involving the interpretation of reflected high-frequency sound, the source of the sound being the bat itself.

Eclipse - female-like plumage acquired by many male ducks during their summer moult.

Elytra - hardened forewings of a beetle.

Emergent - plant growing with its base and roots in the water, the rest of it emerging above the surface.

Entire - in the context of a leaf, a margin that is untoothed.

Epiphyte - plant that grows on another plant, on which it is not a parasite.

Evergreen - plant that retains its leaves throughout the year.

Eye-stripe - stripe through the eye of a bird, from the base of the bill to the ear coverts.

Fen - wetland habitat where the inundated soil and water has a pH that is neutral to alkaline.

Feral - having returned to the wild.

Filament - stalk-like part of a flower's stamen that carries the anther.

First-autumn - bird in its first autumn, whose plumage may be juvenile or first-winter, depending on when moult occurs.

First-winter - plumage acquired after a bird's juvenile feathers have been moulted.

Flight feathers - long feathers (primaries, secondaries and tertials) on the trailing half of the wing.

Floret - small flower.

Frond - leaf-like structure found in some lower plants.

Fruits - seeds of a plant and their associated structures.

Gall - plant growth induced by another organism, often a gall wasp or fungus.

Genus (plural **genera**) - group of closely related species, which share the same genus name.

Glabrous - lacking hairs.

Globose - spherical or globular.

Glume - stiffened bract found in grass flowers.

Haemoglobin - red pigment in blood that absorbs oxygen.

Herbivore - animal that feeds on plant material.

Hibernation - state of winter dormancy in which the activity, body temperature and energy requirements of the animal are greatly reduced.

Holdfast - root-like structure that anchors seaweeds to rocks.

Hybrid - organism derived from the cross-fertilisation of two different species.

Immature - young bird whose plumage is not adult. Depending on the species, this stage may last months or years.

Inflorescence - flowering structure in its entirety, including bracts.

Insectivore - animal that feeds on insects.

Introduced - not native to the region.

Invertebrate - animal without a backbone.

Juvenile - newly fledged bird with its first set of feathers.

Lanceolate - narrow, lance-shaped leaf.

Larva - soft-bodied, pre-adult growing stage in the life cycle of certain insect species.

Leaflet - leaf-like segment or lobe of a leaf.

Lek - communal display area used by certain bird species.

Ligule - membranous leaf sheath found in grasses.

Linear - slender and parallel-sided.

Lobe - division of a leaf.

Lore - area of feathering between the eye and the base of the upper mandible of a bird.

Malar stripe - narrow stripe of feathers that borders the throat of a bird.

Mantle - area of feathers on the upper back of a bird.

Marine - living in the sea.

Melanic - showing dark pigmentation.

Metamorphosis - process whereby an animal is transformed morphologically from an immature form into an adult, e.g. pupa to adult insect or tadpole to frog.

Midrib - central vein of a leaf.

Migrant - bird that spends the summer and winter in different places.

Mire - wetland habitat where the soil is typically permanently inundated.

Moult - process seen in birds during which old feathers are lost and replaced by new ones; also applied to certain invertebrates that shed their skins.

Moustachial stripe - line of feathers running from the base of the lower mandible to the cheeks in birds.

Mucus - slimy, viscous fluid secretion.

Native - occurring naturally in the region and not known to have been introduced.

Needle - narrow leaves found in conifer trees.

Nocturnal - active after dark.

Node - part of the stem at which leaves arise.

Nut - dry and often hard fruit containing a single seed.

Nymph - immature growing stage found in certain insect groups, notably bugs, that has some characters in common with the adult.

Oblong - leaf whose sides are at least partly parallel.

Obtuse - blunt-tipped (usually refers to a leaf).

Operculum - plate found in some molluscs and used to seal off the entrance to their shell.

Opposite - arising in opposite pairs on the stem (usually refers to leaves).

Oval - leaf shape.

Ovary - in a plant, the structure containing the ovules, or immature seeds.

Ovate - roughly oval in outline.

Ovipositor - egg-laying structure found at the tail end of some female insects.

Ovoid - egg-shaped.

Palmate - leaf with finger-like lobes arising from the same point.

Palps - sensory appendages found around the mouth in insects and crustaceans.

Parasite - organism that lives in or on another organism, relying on it entirely for its nutrition.

Passage migrant - bird species seen mostly on migration, and that does not necessarily breed in Britain and Ireland.

Pedicel - stalk of an individual flower.

Pelagic - favouring the open sea.

Perennial - plant that lives for more than two years.

Perianth - collective name for a flower's petals and sepals.

Petals - inner segments of a flower; often colourful.

Petiole - leaf stalk.

Pinnate - leaf division with opposite pairs of leaflets and a terminal one.

Pod - elongated fruit, often almost cylindrical, seen in pea family members.

Pollen - tiny grains that contain male sex cells, produced by a flower's anthers.

Predator - animal that hunts and kills live prey.

Prehensile - capable of gripping; used in the context, for example, of a Harvest Mouse's tail.

Primaries - outermost flight feathers in birds.

Prostrate - referring to a plant that grows in a manner pressed tightly to the ground.

Pronotum - hardened dorsal plate covering the thorax of an insect.

Pubescent - with soft, downy hairs.

Pupa - stage in an insect's life cycle between the larva and adult; also called a chrysalis.

Ray florets - small flowers found on the outer fringe of the inflorescence in flowers of the daisy family.

Reflexed - bent back at an angle of more than 90 degrees.

Rhizome - underground stem.

Rosette - radiating arrangement of leaves.

Runner - creeping stem that occurs above ground and may root at nodes or tip.

Saprophyte - plant that lacks chlorophyll and derives its nutrition from decaying matter.

Secondaries - middle flight feathers in birds.

Sepal - outer row of structures surrounding the reproductive part of a flower.

Sessile - not stalked.

Shrub - branched, woody plant.

Sole - underside of the foot in molluscs.

Spadix - upright spike of flowers, found in arums.

Spathe - large bract surrounding the spadix in arums.

Species - unit of classification defining animals or plants that are able to breed with one another and produce viable offspring.

Species aggregate - A scientific term used (mainly in botanical classification) to describe a group of superficially very similar species whose separation and specific identification is best left to experts.

Speculum - species-specific patch of colour seen on ducks' wings.

Spike - simple, branched inflorescence.

Spikelet - inflorescence arrangement in grasses, sedges and rushes.

Spore - tiny reproductive body that disperses and gives rise to a new organism.

Stamen - male part of the flower, comprising the anther and filament.

Stigma - receptive surface of the female part of a flower, to which pollen adheres.

Stipe - supporting structure (akin to a stem) that supports the frond of lower plants such as seaweeds and ferns. Sometimes used to describe the 'stem' of toadstools.

Stipule - leaf-like or scale-like structure at base of a leaf stalk.

Style - element of the female part of the flower, sitting on the ovary and supporting the stigma.

Submoustachial stripe - line of feathers (typically contrasting pale) between the malar and moustachial stripes.

Subspecies - subdivision of a species, the members of which are able to breed with other subspecies but seldom do so because of geographical isolation.

Succulent - referring to a plant that is swollen and fleshy.

Supercilium - typically pale stripe of feathers that runs much of the length of the head above the eye in birds.

Tarsus - what most people refer to as a bird's 'leg', although strictly speaking it is anatomically part of the foot.

Tendril - slender, modified leaf or stem structure that assists climbing in some plants.

Tepals - both sepals and petals, when the two are indistinguishable.

Terrestrial - living on land.

Tertials - innermost flight feathers in birds.

Thallus - unspecialised vegetative body of a lower plant.

Thorax - middle section of an insect's body.

Tomentose - covered in cottony hairs.

Tragus - pointed inner-ear outgrowth found in some bat species.

Trifoliate - leaf divided into three sections.

Umbel - umbrella-like arrangement of flowers.

Vagrant - bird that appears accidentally outside the species' typical range, be it breeding, non-breeding or migratory.

Ventral - lower surface.

Vertebrate - animal with a backbone.

Whorl - several leaves or branches arising from the same point on a stem.

Wing coverts - feathers that cloak the leading half (as seen in flight) of both surfaces of the wings in birds.

Wingbar - striking bar on the wings of birds (typically either white or dark), formed by pale or dark margins to the wing-covert feathers.

Wingspan - distance from one wingtip to the other in birds.

Index

Photographic acknowledgements

All images used in this book were taken by **Paul Sterry** with the exception of those listed below; these can be identified using a combination of page number and subject.

From the files of Nature Photographers Ltd:
Janus Andersen: 238 Bean Geese flying; 249 Rough-legged Buzzard; 253 Rough-legged Buzzard; 253 Goshawk.
T. Andrewartha: 265 Temminck's Stint juvenile.
S.C. Bisserot: 131 Stinking Chamomile; 212 Greater Horseshoe Bat; 212 Greater Horseshoe Bat; 212 Lesser Horseshoe Bat; 212 Lesser Horseshoe Bat; 212 Daubenton's Bat; 212 Daubenton's Bat; 213 Natterer's Bat; 213 Whiskered Bat; 213 Whiskered Bat; 213 Bechstein's Bat; 213 Bechstein's Bat; 213 Brandt's Bat; 213 Brandt's Bat; 213 Lesser Horseshoe Bat; 214 Serotine; 214 Noctule Bat; 215 Leisler's Bat; 215 Leisler's Bat; 215 Barbastelle; 215 Barbastelle; 216 Greater Horseshoe Bat; 216 Lesser Horseshoe Bat; 216 Natterer's Bat; 216 Leisler's Bat; 216 Bechstein's Bat; 216 Daubenton's Bat; 217 Noctule Bat; 217 Brandt's Bat; 217 Barbastelle; 217 Whiskered Bat; 217 Serotine; 220 Toggenberg; 220 South Down sheep; 330 Great Crested Newt larva; 330 Great Crested Newt female; 381 Southern Hawker emergence; 393 *Tabanus bromius*; 396 *Bombus lapidarius*; 396 *Bombus terrestris*; 399 *Rhyssa persuasoria*; 452 Lesser Sand Eel; 455 Sea Hare; 457 Octopus; 459 Common Shrimp. **Frank Blackburn:** 222 Rabbit tracks; 287 Lesser Spotted Woodpecker. **K. Blamire:** 396 *Bombus pratorum*. **Mark Bolton:** 234 British Storm-petrel flying; 251 Peregrine (right); 279 Roseate Tern; 298 Ring Ouzel male; 322 Balearic Shearwater. **Derek Bonsall:** 375 *Cloeon dipterum*; 69 Harsh Downy-rose; 154 Dark Red Helleborine; 398 Artichoke Gall; 450 Ballan Wrasse. **Nicholas Phelps Brown:** 335 Human Louse; 392 *Tipula maxima*; 392 Midge larva; 393 Mosquito larva; 393 *Chironomus*; 393 *Chironomus* larva; 393 *Haematopa crassicornis*; 394 Common House Fly; 411 Jumping Spider; 434 Hydra; 434 *Polycelis nigra*; 435 ciliated protozoan; 435 *Spirogyra*; 435 Diatoms; 435 *Amoeba*; 435 Rotifers; 438 *Ephemerella* nymph; 438 Mosquito larvae; 438 Midge larva; 439 *Daphnia*; 439 *Cyclops*; 440 Hydra; 440 *Polycelis nigra*. **Brinsley Burbidge:** 49 River Water-crowfoot; 91 Marsh Violet; 101 Sea-milkwort; 107 Spring Gentian; 136 Oxford Ragwort; 170 Lesser Pond Sedge; 179 *Grimmia pulvinata*. **Robin Bush:** 65 Roseroot; 66 White Stonecrop; 75 Late Spider-orchid; 83 Knotted Clover; 92 Hoary Rock-rose; 105 Bearberry; 131 Corn Chamomile; 146 Sand Leek; 150 Man Orchid; 154 Red Helleborine; 155 Narrow-lipped Helleborine; 155 Irish Lady's Tresses; 175 Hart's-tongue Fern; 176 Holly Fern; 336 Swallowtail; 344 Wall Brown underside; 348 Chalkhill Blue male. **Andy Callow:** 335 Flea; 384 Azure Damselfly; 387 *Rhinocoris iracundus*; 388 *Deraecoris ruber*; 388 *Campyloneura virgula*; 388 *Orthops campestris*; 388 Rose Aphids; 393 Empid fly; 395 Snipe Fly; 396 *Osmia rufa*; 397 Field Digger Wasp; 398 Red Ant; 398 Black Ant; 402 *Rhagonycha fulva*; 404 2-spot Ladybird; 405 Tortoise Beetle; 407 Spiderlings; 409 *Araneus curcurbitina*; 410 Wolf Spider with young; 411 Red Spider Mite. **Kevin Carlson:** 302 Grasshopper Warbler; 327 Basking Adders. **Colin Carver:** 205 Mountain Hare; 206 Fox; 206 Stoat; 206 Weasel; 207 American Mink; 218 Sika Deer stag; 219 Roe Deer buck; 282 Stock Dove; 283 Reed Warbler and Cuckoo; 291 Shorelark; 320 Reed Bunting male; 321 Lapland Bunting male; 329 Grass Snake; 331 Edible Frog. **Bob Chapman:** 130 Water Lobelia; 280 Razorbill winter. **Hugh Clark:** 206 Hedgehog; 202 Water Vole swimming; 206 Wildcat; 207 Badger; 214 Pipistrelle flying; 215 Brown Long-eared Bat flying; 216 Long-eared Bat; 217 Pipistrelle flying; 227 Common Seal; 286 Kingfisher flying; 287 Great Spotted Woodpecker flying; 287 Green Woodpecker flying; 287 Lesser Spotted Woodpecker flying; 290 Swallow flying; 293 Grey Wagtail flying; 295 Robin flying; 307 Tawny Owl flying; 310 Treecreeper flying; 316 Chaffinch flying; 338 Small Tortoiseshells hibernating; 374 Springtails; 393 *Haematopa pluvialis*. **Andrew Cleave:** 9 Beech roots; 18 Sitka Spruce; 21 White Poplar; 26 Wild Pear; 36 Japanese Knotweed; 64 Great Sundew; 67 Lady's-Mantle; 68 Cloudberry; 74 New Forest Pony; 104 Cranberry; 105 Arctic Bearberry; 127 Hoary Plantain; 127 Hoary Plantain flower; 141 Lesser Hawkbit; 143 Autumn Squill; 150 Lady Orchid; 159 Limestone pavement; 160 Great

Wood Rush; 163 Deergrass; 164 Bristle Bent; 164 Black Bent; 167 Red Fescue; 169 Mat Grass; 169 Blue Moor-grass; 171 Floating Sweet-grass; 173 Hard Fern; 175 Lady Fern; 175 Polypody; 176 Black Spleenwort; 176 Green Spleenwort; 176 Stagshorn Clubmoss; 177 Beech Fern; 177 Limestone pavement; 179 Field Horsetail; 179 Wood Horsetail; 179 Water Horsetail; 180 *Racomitrum*; 199 Badger; 221 Dartmoor Pony; 221 Connemara Pony; 222 Otter holt; 226 Killer Whale; 227 Grey Seal pup; 233 Black-necked Grebe winter; 277 Great Black-backed Gull main; 277 Glaucous Gull main; 281 Little Auks flying; 429 Basking Shark; 442 Seashore background; 443 Common Blenny; 444 Peppery Furrow Shell; 445 Zonation; 452 Common Blenny; 452 Common Goby; 454 Common Periwinkle; 456 Peppery Furrow Shell; 456 Piddocks; 458 Masked Crab; 463 By-the-wind-sailor; 463 Beadlet Anemones; 463 Plumose Anemone. **Graeme Cresswell:** 225 Pilot Whale; 225 Pilot Whale; 226 Killer Whale; 226 Risso's Dolphin; 226 Risso's Dolphin; 226 Common Dolphin; 226 Common Dolphin; 227 Atlantic White-sided Dolphin; 227 Harbour Porpoise. **Ron Croucher:** 219 Roe Deer fawn; 306 Blue Tits in nestbox. **Geoff Du Feu:** 153 Fen Orchid; 200 Mole; 207 Otter; 207 Otter; 208 Small Tortoiseshell; 319 Crossbill female; 333 Horsefly; 393 Mosquito; 394 Greenbottle; 394 Lesser House Fly; 394 Bluebottle; 394 Common House Fly; 396 Leaf Cutter Bee; 399 *Pteromalus puparum*; 406 Spider silk and dew; 422 Frogbit. **R. Fisher:** 262 Dotterel juvenile. **Michael Foord:** 229 Striped Dolphin. **Phil Green:** 218 Sika Deer stag; 218 Sika Deer hind and calf; 311 Starling juvenile; 382 Scarce Chaser female; 410 *Pisaura mirabilis*. **Jean Hall:** 127 Sea Arrowgrass; 177 Field Horsetail; 220 Hampshire Down sheep; 331 Toad spawn. **Michael Hammett:** 375 *Perla bipunctata* nymph; 380 Common Hawker; 384 Variable Damselfly; 391 *Limnephilus elegans*; 391 *Limnephilus marmoratus*; 428 Lesser Spotted Dogfish; 430 Char; 450 Pollack; 450 Worm Pipefish; 450 Goldsinny Wrasse; 450 Corkwing Wrasse; 453 Plaice; 455 Sea Lemon; 457 Little Cuttlefish; 463 Dahlia Anemone; 463 Peacock Worm; 464 Brittlestar. **Barry Hughes:** 184 Woolly Milkcap; 188 Field Blewit; 248 Marsh Harrier female. **Ernie Janes:** 97 Giant Hogweed; 192 Meadow; 192 Oyster Mushroom; 203 Barn Owl; 204 Black Rat; 204 Brown Rat; 207 Polecat; 208 London Park; 215 Brown Long-eared Bat; 218 Fallow Deer buck; 219 Chinese Water Deer; 219 Muntjac; 220 Suffolk Ewe; 220 Friesian; 220 Guernsey; 221 Berkshire pig; 221 Middle White; 221 Wild Boar; 221 Przewalski's Horse; 221 Exmoor Pony; 246 Goshawk; 246 Sparrowhawk male; 256 Red Grouse female; 258 Grey Partridge female; 260 Curlew; 260 Knot; 266 Sanderling flying; 273 Gulls and plough; 285 Long-eared Owl main; 285 Short-eared Owl main; 306 Great Tits in nestbox; 306 Blue Tit fledgling; 319 Lesser Redpoll; 322 Curlew migrating; 333 House Spider; 348 Common Blue female; 375 *Dinocras cephalotes*; 407 Garden Spider web; 408 House Spider; 444 Razorshells. **Len Jessup:** 221 Saddleback; 423 Arrowhead. **Paul Knight:** 220 Dorset Horn sheep. **Keith Lugg:** 250 Merlin female. **Hugh Miles:** 256 Capercaillie female; 277 Long-tailed Skua. **Lee Morgan:** 68 Stone Bramble; 207 Otter top; 210 Greater Horseshoe Bat skull; 213 Natterer's Bat; 214 Serotine; 214 Common Pipistrelle; 214 Soprano Pipistrelle; 215 Grey Long-eared Bat; 216 Bat wing; 217 Bat handler; 464 Spiny Starfish. **Owen Newman:** 199 Long-eared Bat; 200 Hedgehog family; 201 Edible Dormouse; 201 Edible Dormouse hibernating; 203 Yellow-necked Mouse; 203 Wood Mouse family; 215 Brown Long-eared Bat; 250 Kestrel male. **Philip Newman:** 237 Barnacle Geese flying; 248 Hen Harrier female; 248 Hen Harrier male; 248 Hen Harrier female; 252 Hen Harrier male; 253 Kestrel male; 253 Kestrel male; 256 Red Grouse male; 256 Ptarmigan winter; 256 Black Grouse female; 256 Black Grouse male; 298 Ring Ouzel female; 299 Mistle Thrush wing stretching; 319 Crossbill male. **David Osborn:** 36 Curled Dock; 127 Ribwort Plantain; 153 Irish Marsh-orchid; 191 Chanterelle; 198 Grey Seal; 218 Fallow Deer dark buck; 219 Reindeer stag; 227 Grey Seal; 228 Common Seal; 234 Gannet flying; 234 Gannet main; 236 Bewick's Swans flying; 245 Goosander female; 246 Osprey; 246 Osprey; 250 Merlin male; 266 Sanderling summer; 269 Turnstone summer; 397 Bee Killer Wasp. **W.S. Paton:** 218 Red Deer

calf; 219 Reindeer hind; 220 Wild Goat; 258 Grey Partridge male; 284 Snowy Owl male. **Daniel Petrescu:** 253 White-tailed Eagle adult. **Richard Revels:** 204 Red Squirrel; 205 Hares boxing; 230 Puffin flying; 247 White-tailed Eagle flying; 280 Guillemot flying; 280 Puffin flying; 280 Razorbill flying; 298 Fieldfare flying; 333 Elephant Hawkmoth; 336 Green-veined White; 336 Small White; 337 Orange Tip; 337 Pale Clouded Yellow; 338 Small Tortoiseshell; 341 High Brown Fritillary upper; 341 High Brown Fritillary lower; 341 Silver-washed Fritillary upper; 341 Silver-washed Fritillary lower; 343 Large Blue; 344 Speckled Wood; 345 Ringlet upper; 345 Meadow Brown upper; 345 Gatekeeper upper; 347 Green Hairstreak; 347 Small Copper upper; 348 Common Blue male; 348 Chalkhill Blue female; 348 Chalkhill Blues under; 348 Adonis Blue female; 349 Holly Blue upper; 349 Small Blue upper; 349 Small Blue lower; 349 Brown Argus lower; 349 Northern Brown Argus; 350 Essex Skipper; 350 Small Skipper; 350 Large Skipper upper; 351 Grizzled Skipper; 351 Dingy Skipper; 355 Broad-bordered Bee Hawk; 355 Hummingbird Hawkmoth; 361 Cream-spot Tiger; 372 Lobster Moth larva; 372 Alder Moth larva; 373 Marbled White pupa; 373 Comma emerging; 373 Comma; 380 Migrant Hawker flying; 382 Scarce Chaser male; 382 Black-tailed Skimmer pair; 383 Common Darter pair; 383 Black Darter; 385 Blue-tailed Damselfly; 385 Large Red Damselfly; 385 White-legged Damselfly; 385 Common Blue Damselflies; 395 *Helophilus pendulus*; 396 *Bombus lucorum*; 396 Tawny Mining Bee; 397 German Wasp; 397 Common Wasp; 397 Median Wasp; 397 Sand Digger Wasp; 403 Cardinal Beetle flying; 405 Lily Beetles mating; 410 *Xysticus cristatus*; 439 White-clawed Crayfish. **Peter Roberts:** 275 Sabine's Gull adult flying; 277 Pomarine Skua. **Jim Russell:** 299 Mistle Thrush juvenile. **Tony Schilling:** 166 Wavy Hair-grass. **Don Smith:** 284 Tawny Owl flying; 284 Barn Owl flying; 459 Lobster; 462 Dead Man's Fingers; 463 Common Jellyfish. **R.T. Smith:** 287 Green Woodpecker male. **James Sutherland:** 458 Hermit Crab. **E. Thompson:** 284 Tawny Owl perched; 382 4-spot Chaser. **Roger Tidman:** 198 Fox; 218 Fallow Deer calf; 227 Common Seal pup; 229 Sperm Whale; 232 Red-throated Diver flying; 234 Leach's Petrel flying; 235 Bittern flying; 238 Greylag Goose flying; 239 Pink-footed Geese flying; 240 Gadwall flying; 241 Garganey flying; 242 Tufted Duck flying; 242 Pochard flying; 245 Goldeneye flying; 245 Smew flying; 246 Sparrowhawk female; 247 Golden Eagle juvenile; 247 White-tailed Eagle; 248 Marsh Harrier male; 248 Montagu's Harrier; 250 Kestrel flying; 252 Montagu's Harrier male; 252 Montagu's Harrier female; 253 Golden Eagle immature; 253 Golden Eagle adult; 253 Merlin; 253 Peregrine; 254 Mountain Hare; 255 Ptarmigan; 258 Red-legged Partridge; 260 Avocet; 263 Golden Plover winter; 264 Kentish Plover female; 264 Little Ringed Plover juvenile; 270 Bar-tailed Godwits flying; 270 Bar-tailed Godwit summer; 271 Woodcock; 277 Iceland Gull main; 277 Arctic Skua flying; 279 Sandwich Tern; 279 Sandwich Tern flying; 280 Guillemot winter; 282 White Dove; 284 Snowy Owl female; 290 Sand Martin flying; 290 House Martin flying; 291 Skylark flying; 293 Pied Wagtail adult; 296 Redstart female; 299 Song Thrush juvenile; 305 Melodious Warbler; 309 Bearded Tit male and female; 313 Raven main; 317 Goldfinch juvenile; 318 Hawfinch male and female; 320 Cirl Bunting female; 321 Ortolan Bunting; 323 Sooty Shearwater; 323 Cory's Shearwater; 323 Great Shearwater; 381 Emperor Dragonfly. **Derek Washington:** 279 Common Tern main. **Andrew Weston:** 26 Bird Cherry; 49 Common Water-Crowfoot; 147 Angular Solomon's-seal. **Patrick Whalley:** 309 Crested Tit. **A. Wharton:** 410 *Misumena vatia*; 154 Dune Helleborine. **Neil Wilmore:** 439 Ostracods.

From other sources:
Laurie Campbell: 198 Mountain Hare; 204 Red Squirrel leaping; 206 Wildcat lower; 206 Weasel top; 207 Pine Marten; 232 Black-throated Diver summer; 247 Golden Eagle adult; 431 Salmon. **Frederic Desmette (Windrush):** 319 Scottish Crossbill. **Göran Ekström:** 244 Velvet Scoters flying; 256 Ptarmigan flying; 305 Icterine Warbler. **Mike Lane (Natural Image):** 302 Savi's Warbler. **David Tipling (Windrush):** 256 Red Grouse flying; 257 Lady Amherst's Pheasant.